Emergency

Emergency
Crisis on the Flight Deck

Stanley Stewart

Airlife
England

'One thorn of experience is worth a whole wilderness of warning.'
James Russel Lowell

Copyright © Stanley Stewart, 1989, 1992

First published in the UK in 1989 (hardback) and 1992 (paperback) by Airlife Publishing Ltd.

British Library Cataloguing in Publication Data

A catalogue record for this book is available from the British Library

ISBN 1 85310 348 9

Printed in England by Livesey Ltd., Shrewsbury.

Airlife Publishing Ltd.

101 Longden Road, Shrewsbury, England.

Contents

Acknowledgements

Introduction

1   Pacific Search   1

2   The Bermuda Tangle   26

3   To Take-off or Not to Take-off . . .   51

4   The Windsor Incident   78

5   Don't be Fuelish   109

6   The Blackest Day   134

7   Ice Cool   166

8   Roll Out the Barrel   185

9   Strange Encounter   225

    Epilogue   259

    Abbreviations and Glossary   261

    Bibliography   264

Acknowledgements

Much help, assistance and advice was received from many people during the writing of this book and the author is deeply indebted to all those who so kindly contributed. Without the generous support of those involved in the incidents, and others in the aviation industry, this book would not have been possible. To all those who so kindly helped, the author would like to express his heartfelt thanks. **Pacific Search**: Captain Gordon Vette and Captain Jay Prochnow; **Bermuda Tangle**: Captain Steve Thompson, with assistance from Senior Engineer Officer Dave Hoyle (Tristar Flight Engineer Instructor); **To Take-off or Not to Take-off . . .**: Captain Wayne Sagar, with assistance from Senior Engineer Officer Simon Robinson (747 Flight Engineer); **The Windsor Incident**: Captain Bryce McCormick; **Don't be Fuelish**: Captain Bob Pearson and First Officer Maurice Quintal; **The Blackest Day**: Captain Pat Levix; **Ice Cool**: Captain Tom Hart; **Roll Out the Barrel**: Captain Harvey 'Hoot' Gibson and Mr Harold F. Marthinsen, Director ALPA Accident Investigation Department; **Strange Encounter**: Captain Eric Moody, Senior First Officer Roger Greaves, Senior Engineer Officer Barry Townley-Freeman, with assistance from Senior First Officer Frank Avery (747 Assistant Instructor).

I would also like to thank Captain Harry Orlady of NASA's San Francisco office of the Air Safety Reporting System for his help in tracing individuals and, last but not least, two ladies for their excellent work: Andrea Metcalfe for her careful reading and correcting of the manuscript and Sara Phillips for her conscientious work in transferring the manuscript to the word processor.

Any errors remaining are, of course, entirely my own.

Introduction

Flying is one of the safest forms of modern transport and, as a means of travelling quickly over long distances, its role is unsurpassed. The passenger of today is transported with amazing ease from one side of the globe to the other, with the problems of the earth being left far below. For present day flight crew, however, it is another world, with the view from the flight deck offering a different sight. Aircrew are only too well aware of the hostile nature of their working environment and, armed with such knowledge, are both ready and able to overcome the difficulties.

Aircraft frequently cross great empty oceans, vast featureless deserts, immense ice wastelands and enormous desolate regions in complete safety in spite of the adverse conditions of the terrain below. Aircrew are trained for all contingencies, and survival equipment for sea, ice and desert is carried aboard. On a journey half way round the world many areas of conflict may be crossed without a single sound of the calamity below being heard, the loudest report in the cabin being the pop of a champagne cork. Such danger zones can be traversed or circumvented in safety when approached with vigilance and care.

Natural disasters, civil strife and famine also prove of little effect for the high-flying traveller, although for staff and crew at transit stations the problems may be enormous. In the skies of starving Africa, passengers sitting miles above the horror indulge their tastes in international cuisine. Cosseted by eager airlines, the modern traveller is borne, with diligence and care, in genuine security over the trouble spots of the earth. Journeys over half the world are now so commonplace and scheduled arrivals so frequent that delays of an hour or so can annoy passengers. To arrive, say, a few hours late in Auckland after a 12,000 mile trip from London can be quite unacceptable to some. Yet if the same passengers stopped only for one moment to think of such a journey, they could not fail to be impressed by the accomplishment. The world may be shrinking, but it is not quite as small as we are led to believe. It is still a hazardous place for the unwary and even short flights are rarely as simple as they seem.

In spite of advanced technology and the magic of computers, the movement of something as big as a Boeing 747 from one side of the world to the other is an operation of complex proportions. Anyone who has contemplated driving their automobile in a foreign country will immediately recognise the problems; the difficulties in operating a big jet worldwide are immense; overflying rights, landing permission, insurance arrangements, fuel payments, cargo and passenger quotas are but a few of the challenges, all of which have to be negotiated between governments and often between countries barely talking to each other. The capital equipment required to service a large international airline operation is enormous: aircraft, offices, sales shops, hangars, terminal buildings and a plethora of expensive vehicles from mobile steps to push-back trucks. With air fares over long distances still comparatively low in respect of present-day incomes, it is a wonder that airlines make any profit at all. That they do and, in some cases, manage quite handsome returns in the face of fierce competition, is a great credit to the managers who run these large and costly outfits.

The fact that airlines function successfully throughout the globe is due to the dedicated and hardworking people within the industry who make the system work, in spite of the problems. Managers, office staff, sales persons, accountants, engineers, maintenance personnel, traffic supervisors, the backroom people of the world's airlines, all make a significant contribution, as well as the aircrews at the sharp end of the operation. For passengers, the smooth, comfortable and effortless transition from one place to another is not accomplished without a great deal of exertion from all concerned. Much of the effort is, of course, unobserved by the travelling public, not least the skills of the flight crew in flying the aircraft from departure to destination. It is an esoteric world where few of even the most well travelled of passengers have been permitted to enter.

The airline pilot's job today is essentially one of operations director and systems manager, but even on the most sophisticated of electronic flight decks the human contribution is significant. In spite of the advances in computers and electronics, machines can do no reasoning for themselves and creative thinking is still a necessary facet of the job. Aircraft computers can do only so much and modern electronic capabilities are not quite as fantastic as the public is led to believe. Malfunctions do occur and there are many traps for the unwary. Flight crews, of course, are alert to the problems, but are sometimes too eager to assure passengers by telling them how easy it has all become. Basic airmanship (i.e. the collective practical application of training, skill, experience and professional judgement) is still required to be exercised by all flight crews at all times.

All aircraft computers are required to be programmed before flight for each journey and the autopilot, when engaged, has to be instructed on every move. Automatic guidance of a 747, for example, down a narrow radio beam to accomplish an automatic landing in almost blind, foggy conditions with 400 people on board is not a task to be taken lightly. The automatics have to be very carefully monitored for malfunctions and the autopilot has to be told what to do at each stage of the approach. In the fog, of course, the wind is calm and the air still, and in such circumstances automatic landings can be effective. When the wind is blustery and conditions bumpy, however, especially with a strong cross-wind on landing, the autopilot cannot cope and the pilot has to take over and land the aircraft.

On most flights, crew operation is routine with standard procedures being followed but, of course, circumstances do change, even when flying repeatedly on the same route. Take-off and landing delays, work at airports, equipment malfunctions, re-routeings, adverse weather and so on, all present difficulties. Flight crews, however, not only have to perform the routine well, a highly skilled procedure in itself, but have to cope with any emergency which may arise. When severe weather strikes, systems malfunction, engines fail or aircraft fires erupt, it is sometimes only the skill of the flight crew which lies between safety and disaster. Flight crews have at all times to be alert to every situation. When a major emergency occurs decisions are made, sometimes in a split second, which can affect the safety of the aircraft and perhaps many hundreds of lives. Here the captain comes into his own and the training of the flight crew is put to the test.

Airline crews are well trained, highly motivated and dedicated professionals. Although mistakes are sometimes made and accidents occasionally happen, the high level of safety evident in the airline industry is a testament to the excellent standard maintained by all concerned. Too much publicity these days is given to the rare demise of an aircraft, with little being known of the incidents which, owing to aircrew professionalism, end safely and well. Many of the events told within this book are known in aviation circles, with only a few outside, and most of those involved remain unsung heroes. *Emergency* outlines a number of dramatic incidents and reveals crew procedures during the difficulties, inviting the reader into the exclusive environment of the flight deck to observe the operations.

The author, Stan Stewart, has flown for over twenty years and has operated heavy jets in British Airways for eighteen years. A graduate engineer, 747 pilot and aviation writer, he is uniquely qualified to write on aviation matters. *Emergency* is a celebration of the skills and abilities of pilots and flight engineers throughout the world, and the following chapters can be left to speak for themselves.

Chapter 1
Pacific Search

The vast Pacific Ocean is bounded by five continents and covers an area of seventy million square miles. Its deep waters stretch from the shores of Asia and Australia to the coastlines of North and South America, and to Antarctica in the south. Along the line of the equator from Indonesia to Equador the distance across the waves is over eight thousand nautical miles (nm).

The name, Pacific, belies the temperamental nature of the ocean, and on occasions storms of great ferocity lash the region. Even large modern airliners have to treat the area with caution, for all but very long range aircraft must island-hop to overcome the immense distances. Fortunately, the surface of the Pacific is dotted with thousands of islands, although they are mostly very small and widely scattered. The only regular flights to close the enormous Pacific gap are the 747SP (Special Performance) non-stop services of United Airlines and Qantas from Sydney to Los Angeles, a journey of 7,475 miles which takes thirteen-and-a-half hours to complete.

Aboard modern passenger jets, electronic navigation equipment, adequate fuel reserves and skilled crews have transformed flights between far-off islands to safe, routine and everyday events. For the light aircraft pilot, however, the Pacific still presents an awesome barrier. The journeys from island to distant island are long and tedious, with flight at slow speed. Navigation over the featureless stretches of sea is difficult, and any error or malfunction in locating the remote specks of land can prove disastrous. For one pilot venturing alone in a light aircraft across the vast Pacific in 1978, a simple instrument failure led to the aviator becoming hopelessly lost.

In 1978 Jay E. Prochnow was thirty-six years old and a very experienced pilot who had flown in the US Navy. He had also completed a tour of duty in Vietnam. At the time he worked as a delivery pilot for Trans Air of Oakland Airport, California. One assignment he had been given was to fly a Cessna 188 — a light single-engine single-seat aircraft used for crop spraying — from California across the Pacific to a customer in Australia. Such long distance flights, although hazardous for a small machine, are cheaper than inter-continental shipping costs. Extra fuel tanks were fitted to the aircraft for the ordeal. The only sources of navigation, however, apart from chart and compass, were the rather antiquated radio beacons known as non-directional beacons (NDB) scattered amongst the islands of the Pacific.

Although these beacons are relatively inaccurate for the distant plotting of position on a chart, some of the more powerful transmitters have signal ranges in excess of 300 nm over the sea by day and 700 nm by night. These radio beacons transmit a signal, not unlike a broadcast station, and Prochnow's aircraft was fitted with an automatic direction finder (ADF) which could be tuned to a particular NDB and the transmitter identified by its morse code emission. The ADF could then detect the incoming signal and point a needle in the direction of the beacon. The needle direction and compass reading could be utilised to establish the bearing of the aircraft from a distant beacon for plotting a position line on a chart, and three or more bearings could be taken from different beacons to fix the position roughly. At long ranges the needles of the ADF were known to fluctuate wildly, and precise track-keeping was very difficult using such a system. It was sufficiently accurate, however, to guide the aircraft to within range of the destination NDB and the pilot could then home in on the signal by simply following the direction of the needle.

A few days before Christmas, 1978, Prochnow arrived in Pago Pago in the American Samoa Islands, accompanied by a colleague in another Cessna 188 who was flying the same journey. By then they had successfully completed well over half the trip and if all went according to plan they would arrive in Australia in time to return home for Christmas. Two days later the two Cessnas took-off together at 03:30 local time with Prochnow in the lead. Unfortunately the second aircraft's fuel pump shaft sheared at lift-off and Prochnow watched with horror as his colleague was forced to ditch in the sea. Fortunately the downed pilot escaped unhurt. Prochnow returned to land and after a day's rest left on his own with full tanks in the middle of the following night for a now lonely and gruelling trip to tiny Norfolk Island, 1,475 nm away. Cruising at speeds of around 110 knots, the journey would take a minimum of fourteen hours to complete in still wind conditions, but would be nearer fifteen hours flying against the forecast light westerly winds. The flight time would allow a daylight landing in Norfolk in mid-afternoon at about 16:00 local time, giving sufficient leeway for adverse winds. The full tank fuel load gave a total endurance of twenty-two hours at normal cruise speeds, which seemed an adequate reserve for any contingency. If necessary fuel could be conserved by decreasing power a little and by slightly leaning the mixture, i.e. reducing the fuel-to-air ratio, thereby increasing the endurance.

Norfolk Island is situated at 29°S 168°E, about 800 nm off the east coast of Australia. The island lies about 600 nm NNW of Auckland, New Zealand, and 430 nm south of Noumea, New Caledonia. It would not be inaccurate to describe Norfolk Island as a tiny speck of rock lying in the middle of nowhere in a corner of the vast Pacific. There is no land around for over 400 nm in any direction. Although Norfolk Island is Australian territory, for convenience it is enclosed within the Auckland Oceanic Flight Information Region, and all flights in and out of the island airport communicate with the Auckland Air Traffic Control Centre (ATCC).

Prochnow took-off from Samoa in the darkness at about 03:00 local time on 21 December and climbed slowly under the weight of fuel to his cruising altitude of 8,000 ft (2,438 m). He turned due south-west to pick up the direct track of 220° magnetic from Pago Pago to Norfolk Island and settled down for the long flight ahead. With no autopilot to ease the monotony of hand flying, it was going to be a tiring and tedious journey. There would be no chance of a quick nap. He would have to navigate with a map on his knee and plot position while flying the aircraft, and would have to grab snacks from his pack meals whenever he could. The initial routeing lay over the Tonga, or Friendly Islands, then continued on southwest bound about 200 nm south of Fiji. For most of the first part of the route sufficient NDBs were available for reasonably accurate fixing of position and progress was good.

A few hours later the strain of flying through the blackness of the night was eased when dawn broke behind the tail of the Cessna at about 05:15 local, and bathed the ocean in light. Approximately 600 nm from Pago Pago lay the island of Ono-I-Lau, almost directly on route. Prochnow navigated toward the island using the various NDB beacons along the way to plot his track. In due course Ono-I-Lau was sighted and his position was confirmed. Beyond the tiny island stretched a vast area of ocean with the destination still over 850 nm away. Prochnow would now have to traverse several hundred miles of empty sea, without any source of navigation whatsoever except the basic magnetic compass, before picking up the signal of the powerful NDB at Norfolk Island. He could then home in on the beacon over the last two or three hundred miles and be guided to the local airport.

At almost the halfway point of the journey the little Cessna crossed the international date line at 180° E/W longitude, the line at which time is both twelve hours behind GMT and twelve hours ahead of GMT, i.e. at the same time on each side of the line, but on different days. As Prochnow traversed the date line he lost twenty-four hours and jumped one day ahead. The local date was now 22 December 1978 and the local time around 08:00. Approximately eight hours of flying remained.

A further 120 nm along track the small machine crossed the Tropic of Capricorn, the line on a chart marking 23½° South, the most southerly declination of the sun. This day the sun was at its highest point in the southern sky, for 21 December marks the Winter Solstice, the shortest day in the Northern Hemisphere's winter, and the longest day of the summer in the south. Prochnow's trip was well planned, for the maximum amount of daylight would be available for the long flight to Norfolk Island.

After several more hours of flying over the featureless sea the Cessna came within range of the destination radio beacon and the signal was identified by its 'NF' morse coding. The arrows of the ADF pointed directly ahead. Prochnow was now flying within the jurisdiction of the Auckland Oceanic ATCC and had established contact on high frequency (HF) long range radio. His estimated time of arrival (ETA) at Norfolk Island was given as 04:30 GMT, 16:00 local time.

The little Cessna droned on at 8,000 ft (2,438 m) as Prochnow followed the direction of the ADF needle. With the approach of the estimated time of arrival he peered ahead for a glimpse of the tiny island, but the bright sun made forward vision difficult. The ETA came and went with no sign of Norfolk, but the ADF needle still pointed steadfastly ahead. The winds could very well have been stronger then forecast which would easily affect the flight

Captain Gordon Vette.

time, although arrivals were normally within fifteen minutes of ETAs. It was just as well Prochnow had loaded plenty of fuel. He called Auckland and informed them of the situation but as yet felt no cause for alarm. Thirty minutes later the needle still indicated the island lying ahead. The bright sun began to fall in the western sky, further reducing forward visibility as Prochnow's eyes searched eagerly for a sight of the land. He tuned in two distant powerful radio beacons, one on Lord Howe Island and the other at Kaitaia in northern New Zealand, in an attempt to plot his position, but the resultant fix was right off his chart. Something was seriously wrong. On retuning the Norfolk beacon he saw to his horror that the needle now pointed in a completely different direction. In spite of the Norfolk Island NDB being correctly tuned and identified the functioning of the ADF appeared to be seriously amiss and the instrument seemed simply to point at random in any direction. Prochnow rapidly gathered his senses and assessed the situation. He had about seven hours of fuel remaining, perhaps more if he conserved his fuel. What should he do now? Which way should he turn for Norfolk Island? For some time he had been flying in the direction of the ADF needle which pointed where it chose. At this stage he could be sixty miles (ninety-six kilometres) or more adrift from his track to destination. He had no doubt that by now his situation was desperate. Prochnow radioed Auckland declaring an emergency and gave details of his predicament. He was hopelessly lost. Alone in a small aircraft in the middle of the empty Pacific, and with sunset only a few hours away, his chances of being found were very slim indeed. Immediately he began a square search of the area. If, while looking for Norfolk Island, his tanks ran dry he would be faced with ditching at night in the lonely Pacific, a prospect which was not appealing. If he was forced into such a situation his chances of survival would be nil.

At about 17:15 local Fiji time, Flight TE 103, an Air New Zealand DC-10, registration ZK-NZS (Zula Sierra), took off from Nadi airport for Auckland, some three hours' flight time away. The captain, Gordon Vette, a senior pilot with the company, turned the jet due south for New Zealand and climbed the aircraft to its cruising altitude of 33,000 ft (10,060 m). His co-pilot on the flight was First Officer Arthur Dovey and his flight engineer, Gordon Brooks. About twenty minutes later, as the big jet settled at cruise altitude for what was a short hop for the long-range aircraft, news of Prochnow's predicament came through from Nadi on very high frequency (VHF) short-range radio. Further information could be obtained on HF from Auckland ATCC who were handling the Cessna's flight.

By the time contact was established with the Centre, the news from Auckland was bleak, and it was obvious to the crew that the situation was very serious. Prochnow was completely lost and his reserves of fuel were being steadily consumed. Even if help were available immediately it could be a tight-run race for safety. In the busy air and shipping routes of the western world, maritime services and air-sea rescue units would be quickly mobilised to help, but in the remote Pacific it is not such a simple matter. An Orion aircraft of the RNZAF had been placed on standby at Whenuapai Air Base, twenty miles (thirty-two kilometres) north of Auckland, but it would be some hours before it could reach the search area once it became airborne. The Air New Zealand DC-10 would pass about 400 nm due east of Norfolk Island on its route from Fiji to Auckland, but if it proceeded directly to the search area it could be there in about one and a half hours. Since there was very little other aircraft activity in the region, would the crew of the New Zealand jet be able to help?

Captain Vette and his colleagues did not need to be asked twice, and soon the big jet was speeding towards Norfolk on its rescue attempt. By good fortune the DC-10 carried a lot of extra fuel and could remain airborne for some considerable time. In New Zealand aviation fuel is expensive and the jet was 'tankering' fuel to save costs. The flight was also scheduled to proceed to Wellington and the extra fuel load would save refuelling in the transit. But there was more in Prochnow's favour than he might have hoped, for Vette, although a senior captain, was an enthusiastic navigator and still kept his flight navigator's licence current. The DC-10, of course, had sophisticated electronic navigation units on board consisting of three completely separate area inertial navigation systems (AINS), and did not require a navigator, but the DC-8 aircraft in Air New Zealand's fleet at the time were not so electronically equipped. All co-pilots on such fleets were also trained as navigators and performed navigation duties as and when required. Vette used to navigate occasionally on the DC-8 to keep his licence up to date.

The AINS of the DC-10 could navigate with pin-point accuracy and displayed a continuous read-out of the aircraft position at all times. There was no equipment aboard, however, for homing in on the Cessna, and trying to find the light aircraft would be like looking for a needle in a haystack. The radar on the DC-10 flight deck was used only in scanning for weather and could pick out large storm clouds, but would never receive an echo from such a small machine. In this situation the specialist knowledge of the navigator could be put to good use. While the crew continued to co-ordinate the rescue plan with Auckland, Vette turned his attention to the passengers. All on board were expecting a short flight, and since the search could be a lengthy procedure the arrival in Auckland could be quite late. Fortunately there were only eighty-eight passengers. With a life at stake there would be none who could object to the crew's action.

'Ladies and gentlemen,' spoke Captain Vette from the flight deck, 'we've just received news from Auckland of a light aircraft lost in the region of Norfolk Island and since we're the only suitable aircraft in the vicinity we've altered course for the search area to offer assistance. The pilot is in serious danger of ditching and if he lands on the water his chances of survival will be remote. The search could be lengthy, but fortunately we have plenty of fuel. It could mean a rather late arrival in Auckland, however, but with a life at stake I'm sure you understand that all help should be given. If anyone objects, of course,' added Captain Vette jokingly 'we could always leave him to die!'

Meanwhile the DC-10 called Prochnow on a range of HF frequencies and finally managed to establish contact. The Cessna pilot was told of their intentions and that they would do all they could to assist. It was a relieved man who received news that help was on the way for Prochnow was by now three hours overdue on his original arrival estimate. The DC-10 crew were informed that his ADF appeared to be malfunctioning and that only four hours of the Cessna's fuel remained. In spite of his predicament Prochnow spoke on the radio with calmness, and his composed demeanour impressed the listeners. He was a very cool customer indeed.

Amongst the passengers on board the DC-10 was an Air New Zealand first officer, Malcolm Forsyth, like Vette a qualified and still licensed navigator. After the public address (PA) announcement to the passengers he came forward to the flight deck to offer his help. It was amazing that two current navigators should be on board and the good news was transmitted to Prochnow. The two navigators could co-ordinate the search while First Officer (F/O) Dovey and Flight Engineer (F/E) Brooks could fly the DC-10 and monitor the aircraft systems. All crew members would be required at times to operate the radios for a lot of co-ordination work would be required.

As the DC-10 flew southwards Vette questioned Prochnow on all the details of his flight — altitude, airspeed, estimated fuel consumption, estimated fuel remaining, estimated position, if he had any idea — anything they could grasp which might improve the chances of finding him. All communications were being conducted on HF long-range radio, whch is subject to static and background noise, and communications proved difficult. The level of noise in the Cessna cockpit was high and did not help the situation. As the flight continued, Vette still had the needs of his passengers in mind and they were kept fully informed and up to date on the progress of the search.

'When we reach the search area,' added Captain Vette 'you will all be able to help. The more pairs of eyes searching the skies the better the chances of finding the Cessna.'

Vette encouraged his passengers to feel part of the search team and little groups were invited in turn to the flight deck to witness the proceedings.

The next problem facing the DC-10 crew was to try and establish the position of the Cessna. A specialised navigation plotting chart was needed but none was carried on the DC-10. Once again luck was on Prochnow's side, for Vette searched his briefcase and found a spare navigation chart of the region. The Cessna pilot's fortunes were beginning to change. Captain Vette instructed Prochnow to take a series of bearings from what the experienced navigator knew to be powerful NDBs on the north island of New Zealand: Kaitaia on the most northern tip, Tauranga on the coast just north of Rotoru, and Gisborne on the east coast. Although Prochnow's ADF was suspected of being in error, a positive test of the integrity of the equipment was essential. Vette carefully plotted each reading on the chart but the result was nonsense, for it showed the Cessna to be somewhere south of Auckland! There was no doubt now that Prochnow's ADF was malfunctioning and would be useless in the search.

The Cessna's equipment was able to tune and identify radio beacons but unknown to anyone at the time the problem was caused by a most elementary fault. The needle of the ADF had simply become loose on the spindle and pointed totally at random in whichever direction it settled. Also unknown was the fact that the defect had taken the Cessna 200 miles (320 kilometres) to the south-east of

An Air New Zealand DC-10. (John Stroud)

Norfolk Island and that only by some miracle could Prochnow now be saved. That, however, did not take into account the determination of Captain Vette and his crew.

The next attempt at establishing position was made using long range HF direction finding (DF) equipment at Brisbane. The DF system is normally employed using VHF radio. An operator at an airport receives from a flight an incoming radio message from which equipment can detect the direction of the aircraft. This magnetic bearing information is then passed to the pilot for plotting on his chart. It is a useful and accurate method of establishing position. Using HF, DF is a totally different proposition, with radio messages being transmitted for perhaps thousands of miles. Several attempts at obtaining a bearing from the HF station at Brisbane while receiving Prochnow's radio messages gave a rough indication of direction, but a more accurate method of establishing position was required if he was to be found.

The DC-10 and the Cessna were still communicating on the rather difficult HF, but it was expected that shortly the two aircraft would be within VHF range. Prochnow was asked to call repeatedly on the VHF emergency frequency of 121.5 MHz to establish the exact point of contact. Since the maximum range of VHF transmission is of the order of 200 nm, at that juncture the distance of the Cessna from the DC-10 would be known. It would still place Prochnow at the centre of a circle of 200 nm radius, covering an area in excess of 125,000 square miles (323,750 square kilometres), but at least it would be a start. Since the position of the Cessna was unknown, it was considered essential to attempt to establish the relative positions of both aircraft. It could be that Flight TE 103, instead of speeding towards a rescue at Norfolk Island, was flying away from the light aircraft. Once again Vette's expertise was put to the test. The sun was setting low in the sky with the approach of dusk and its direction might just prove useful as a last resort.

'I have an idea,' called Vette over the radio above the noise of the HF static. 'Turn due west and face into the sun. When you are pointing directly at it read me back your compass heading.'

After a few moments came Prochnow's response: 'Heading two seven four degrees,' he radioed.

The large DC-10 also aimed straight for the sun and settled on a heading of 270°. That placed the Cessna, as expected, to the left, i.e. to the south of DC-10 'Zulu Sierra.' But was it east or west of the jet's position? A sextant on each aircraft could have measured the precise angular altitude of the sun above the horizon and the difference in altitudes could have been used to clarify the situation, but without such apparatus it was difficult to judge. Vette, however, recalled that a clenched fist held vertical at arm's length represented about ten degrees above the horizon and a finger a little more than a degree and three-quarters.

'Hold your arm at full length,' Vette instructed Prochnow, 'and measure the distance between the horizon and the centre of the sun using your fist and fingers.'

'That's impossible,' came the quick reply from the Cessna. 'The cockpit is small and with the windshield so close I can't even stretch my arm half-way out!'

It was a short moment of amusement that swiftly passed. As a compromise each agreed to hold his hand about one foot (thirty centimetres) before his eyes and to count the number of fingers.

'I make it almost four fingers,' radioed back Prochnow.

Vette's reading was almost two fingers thick, which placed the Cessna nearer the sun, i.e. to the west of the DC-10. The difference in measurement of about two fingers placed the altitude readings at just over three degrees apart. When a celestial body is observed from two separate points it can be shown mathematically that the difference in measured angular altitudes expressed in units of minutes of arc, is equal to the distance in nautical miles separating the observers' positions. Three degrees is equivalent to 180 minutes of arc which, with the reading a little more than that, placed Prochnow's position approximately 200 nm away. It was now established that the Cessna was to the south and west of the DC-10. If the calculations were correct the two aircraft should be in VHF contact very shortly. A few minutes later F/O Dovey picked up Prochnow's voice on the VHF emergency frequency of 121.5 MHz.

'I've got him,' shouted Dovey. 'He's coming over clearly now on VHF.'

The efforts of the crew were beginning to bear fruit and the co-ordinates of the first point of radio contact were accurately plotted on the navigation chart. At this busy moment in the flight the chief purser, Paul James, was asked to liaise with the passengers and to pass on the good news over the PA. His announcement was greeted with cheers. Vette now instructed Prochnow to turn eastwards towards them with his tail to the sun while Zulu Sierra flew towards the sun. They should now be flying directly at each other and closing the gap more quickly. With the DC-10 flying at a ground speed of 560 knots and the Cessna at 110 knots they should cover the 200 nm separation with ease.

'We'll be directly over Prochnow in eighteen minutes,' declared F/O Forsyth, the DC-8 co-pilot and navigator. Vette concurred. The next problem for the two aircraft was to spot each other, not an easy prospect with the sun low in the sky. Captain Vette had already encouraged his eighty-eight passengers to keep guard at the windows and he now instructed them in proper look-out techniques.

'Move your eyes only one or two inches along the horizon, then stop. Move another inch or so and then stop again, always moving in short bursts. Move down a space then repeat the sequence back across your vision, and so on.'

The Cessna was much lower than the DC-10, and using this method a full square search of the sky in view to each passenger could be made. It might just enable one of them to spot the tiny aircraft.

The DC-10 crew now turned their attention to making their own presence more conspicuous to Prochnow. Dovey turned 'Zulu Sierra' to see if they were forming a vapour trail for the Cessna to spot but nothing was in sight. The Met office in Auckland was contacted and confirmed that in the present conditions the formation of a contrail was unlikely, but it was worth a try. The first officer climbed, then descended the aircraft to different altitudes in an attempt to induce a trail, but to no avail. A few trips previously F/E Brooks recalled witnessing an aircraft dump fuel and how impressed he had been with the sight. Perhaps, suggested

Brooks, if the conditions were right, dumping fuel would leave a bright trail which would help Prochnow spot them more easily. At each wingtip, nozzles pointed rearwards and valves could be opened to allow pumping of fuel overboard at the rate of 2.5 tonnes per minute. Prochnow, in fact, had used the same trick in the Navy as a tanker pilot to help receiver aircraft spot the tanker.

At the rendezvous time Brooks threw open the fuel jettison switches and fuel poured from the nozzles. Two minutes later he shut the valves; five tonnes of fuel had gone!

'Can you see us?' called Vette. 'You should be able to see our fuel dump trail.''

Prochnow replied, disappointedly, that nothing was in sight. Perhaps they could try again. A second time Brooks opened the valves and on this occasion dumped fuel for three minutes. Over seven tonnes of fuel spilled into the sky leaving a distinct trail about thirty miles (forty-eight kilometres) long. That made a total of more than twelve tonnes of fuel dumped. With bated breath the DC-10 crew waited for Prochnow's call of a sighting but nothing was heard. The minutes passed and then came Prochnow's voice over the radio.

'It's hopeless,' he said, 'I can't see anything.'

The light was now fading and it was a bitter blow to the searchers that their efforts had been thwarted.

With darkness approaching Prochnow resigned himself to ditching in the Pacific for it now seemed that any attempt at finding him was going to fail. Carefully he prepared the inflatable life raft and emergency food and water supplies for survival at sea, and made sure all sharp objects were removed from his pockets. He had been in the air for nineteen hours and the strain was beginning to tell. He still had about three hours of fuel remaining, perhaps more, since he had been flying at economy cruise for some time with reduced power and leaned mixture, but at almost five hours overdue on his original arrival estimate he could by now be hundreds of miles from Norfolk. Ditching seemed almost a welcome relief to Prochnow as he was so tired and he had to fight to keep himself going.

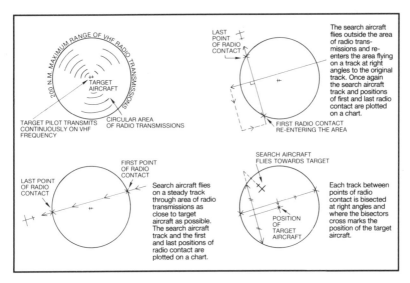

Aural boxing technique.

Vette and his crew were not prepared to give up and a further attempt at establishing position was begun using a technique known as 'aural boxing'. Prochnow would be required to make repeated radio calls, as he did when initially establishing VHF contact, and the points of lost communication and renewed radio contact would be recorded and plotted on the chart. The idea was to fly in a straight line, in this case at first roughly westbound, through the area of the circle of VHF radio transmissions from the Cessna formed by the sweep of the 200 nm radius of the maximum range of Prochnow's radio. The longest line through the circle would, of course, be the 400 nm diameter, taking over forty minutes to traverse, and since the DC-10 was likely to be fairly close to the Cessna, i.e. the centre of the circle, it could be a time-consuming process. The track towards the light aircraft had already been plotted, however, and the point of original radio contact had been marked on the chart, so time could be saved by re-establishing the original track and continuing the traverse. This meant, of course, flying away from Prochnow's estimated position, not an easy decision to make, but since until now all attempts at visual contact had failed it might just be worth a try.

When the traverse was completed and the point of lost VHF contact noted, the plan was then to sweep south-east bound outside the circle of VHF range and then to turn roughly due north to repenetrate the circumference flying at right angles to the original track. Once again the points of radio contact and lost communications would be recorded. The position of the Cessna would then be established by bisecting the lines of traverse and by noting the point at which the bisectors crossed. The Cessna's position was believed, with some justification, to be somewhere southeast of Norfolk Island, so while the 'aural box' method was being applied Prochnow was encouraged to fly in a north-westerly direction to help close the gap. Pin-pointing the Cessna's position would be more difficult with a moving target, but with time and fuel running out it was better than having Prochnow fly round in circles.

Something more immediate was required to try and establish Prochnow's range from Norfolk and once again the navigators conferred as to the best solution. The sun was almost setting over the Cessna and so Prochnow was asked to record the precise GMT time of sunset, i.e. the time at which the upper limb of the sun's circumference dipped below the horizon. Norfolk Island personnel were also asked to record their own GMT time of sunset, which was about 19:00 local, and the two times were compared. A small correction had to be applied to Prochnow's observation, for at 8,000 ft (2,438 m) sunset would be seen a little later than at the same position at sea level. When the calculations were complete the separation in sunset times was found to be twenty-two-and-a-half minutes.

The rising and setting of the sun, of course, results from the spinning earth, with the globe rotating one complete revolution per day, i.e. 360° in twenty-four hours, or 15° per hour, or 1° every four minutes. The earth is also divided into 360° of longitude, 180° to the east and 180° to the west. At a position, therefore, time — which is basically measured by the sun — is related to the longitude of the position, with every 15° of longitude representing one hour. A point 45°

east of the Greenwich Meridian, for example, is also three hours ahead of the time at Greenwich. Prochnow's time was ahead of Norfolk Island by twenty-two-and-a-half minutes which, with an earth rotation of 1° per four minutes, placed him 5.6° east of Norfolk. At about 30°S, 5.6° of longitude represented a range from Norfolk of about 290 nm. At a cruising speed of 110 knots, he could complete the journey in under three hours. Shortly before sunset his fuel remaining was three hours, perhaps more, so he might just be able to make it if only he could be found and pointed in the right direction.

With the sun below the horizon the lone pilot was once again enveloped in darkness. The Cessna droned on through the evening, heading roughly north-west in a desperate attempt to struggle closer to its destination. What Prochnow was feeling inside at that time can only be imagined for outwardly he remained cool. All his radio communications were calm and professional. It is a brave man indeed who can remain so composed when placed in such danger, for death was surely staring him in the face.

At the Auckland Search and Rescue Co-ordination Centre the co-ordinator, Bruce Millar, and his team were becoming increasingly concerned, so the RNZAF Orion on standby at Whenauapai was scrambled to join the search. The Orion would take about two-and-a-half hours to reach the Norfolk Island search area. Meanwhile, Prochnow continued with his series of radio calls on 121.5 MHz as the DC-10 sped away from his estimated position and out of range. As soon as communications were lost the position was plotted and 'Zulu Sierra' manoeuvred to sweep through the audio circle once again at right angles to the original traverse. It was almost an hour before the procedure was completed and the 'aural box' plotted on the chart. Allowing for movement of the Cessna, Prochnow's position was calculated at approximately 30°S 171°E.

'We estimate we'll be overhead your position in about five minutes,' radioed Vette to the Cessna. 'Turn in a circle and look out for our strobe lights.'

The strobes on 'Zulu Sierra' were powerful white flashing anti-collision lights which were so strong they could be seen clearly in daylight. If the Cessna was anywhere nearby surely the lights would be seen. Prochnow peered into the darkness, his hopes rising that at last he was to be found. The estimated time of rendezvous came, but once again the DC-10 was nowhere to be seen. With mounting despair amongst those involved, it had to be admitted that again an attempt at pin-pointing the Cessna's position had failed. Prochnow's spirits must have reached rock bottom. The DC-10 crew seemed to have exhausted all their options and now there was little they could do except sweep the area of the estimated position in the rather weak hope that Prochnow may spot their bright lights.

The crew and passengers of the DC-10 peered desperately into the darkness but there was little chance of one of them spotting the dim lights of the Cessna in the blackness of the night. In a little over an hour the RNZAF Orion would be on

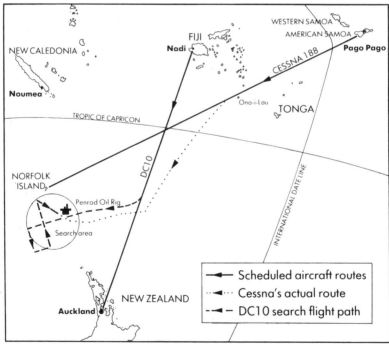

A map showing the relative positions of Auckland, Norfolk Island, Nadi and Pago Pago. The flight paths of the Cessna and DC10 are clearly marked.

station, and with its more sophisticated equipment it was hoped that the Cessna would be found quickly. Prochnow continued to fly north-west, hopefully reducing the gap between himself and Norfolk Island. If the Orion pinpointed his position swiftly and he was given a precise heading to steer for the airport, he might just make it. Without help of some kind he was doomed, for he would never find the island on his own in such a vast stretch of sea. Prochnow and the DC-10 crew had functioned well as a team and they had all worked hard in the search. Having given of their best it seemed a gross injustice that the Cessna remained unfound. Prochnow had been in the air for twenty-and-a-half hours and he was exhausted with the flying and the concentration. The strain on him was enormous. What was needed now, more than anything, was luck.

'A light!' shouted Prochnow over the radio. 'I see a bright light on the surface. I don't care what it is, I'm going for it.' Any ship nearby would increase his chances of survival in a ditching.

'Aim straight for it and read me back your heading,' instructed Vette, 'but make sure it's not a star low on the horizon.'

Prochnow reported his heading as 310°. As the Cessna flew closer to the source of the light he was able to give a description of the vessel.

'It appears to be a tall structure riding on a platform,' informed Prochnow. 'There are two or three lights on it.'

'Sounds like an oil rig,' said Dovey to the others on the DC-10 flight deck. Prochnow then reported seeing two tugs towing the vessel and all agreed it had to be an oil rig.

'Circle overhead and flash your lights to attract attention,' suggested Vette. 'We'll contact Auckland and obtain whatever information we can.'

The co-ordination centre was called on HF with a request for the marine division, part of the search organisation, to radio details of any oil rig platform being towed in the area. A few agonising minutes passed in waiting as the information was obtained, then back came the facts requested. It was identified as the *Penrod*, an oil rig being towed from

New Zealand to Singapore, and its position was given as 31°S,179°21'E. That established Prochnow's position as over 600 nm due east of Norfolk. He hadn't a chance. It also meant that the Cessna's position was hundreds of miles from where it had been placed by the navigator's estimates. Their methods had been rough and ready and only approximate fixes of position had been possible, but they were surprised that their calculations had been so far out. The Cessna was radioed with the bad news.

'You are too far east to make Norfolk on your available fuel,' informed Vette. 'It looks like you are going to have to ditch, but don't do anything till we get there. We can switch on our landing lights and brighten the area to help in your rescue when you go in.'

Dovey fed the co-ordinates of the *Penrod*'s position into the AINS and the big jet banked and flew towards the rig. Prochnow had already come to the same conclusion that ditching was inevitable, and in his mind was rehearsing procedures for the hundredth time. The rig had also spotted the Cessna circling overhead, flashing its lights, and the tugs had stopped their engines and hoved to. It was obvious that the light aircraft was in trouble and a boat had been launched in anticipation of ditching. The *Penrod* was now fully illuminated and the rig was ablaze with lights.

'It looks like a giant Christmas tree,' reported Prochnow.

In preparation for the ditching he lowered the Cessna's nose to inspect the sea.

'I'm going down now to take a look at the surface,' he called.

Prochnow gently descended the light aircraft towards the ocean but what he found when he skimmed over the sea made his heart sink. There was a huge swell running, with waves fifteen to twenty ft feet (five to six m) high. In the wake of the current flowing past the rig's platform the surface was a little flatter, but hardly suitable for ditching. If the Cessna struck the top of a wave it could easily flip over and trap Prochnow in the sinking machine. If the little aircraft slammed into the side of a giant wave it would almost certainly break up on impact. Neither prospect was very appealing, yet Prochnow's best chance now was to put the Cessna down on the sea while he still had some fuel left. He could then use the engine power to control his rate of descent. If he landed as close to the *Penrod* as possible there was just a chance that the aircraft would remain upright and in one piece and that the rig's boat could rescue him.

'The surface looks bad,' reported Prochnow. 'I don't like it at all.' There was a touch of anxiety in his voice now but his approach was still very professional. Under the circumstances he remained remarkably cool.

The DC-10 sped on towards the rig's reported position but it soon became obvious that something was amiss. Instead of Prochnow's radio calls becoming clearer his transmissions were beginning to fade. 'Zulu Sierra' was flying away from the Cessna!'

'There's something wrong with the rig's position,' radioed Vette to the search and rescue centre. 'Can you pass us a frequency so we can speak to him directly?'

Soon Vette was dialling the *Penrod*'s frequency of 119.1 MHz and contact was quickly established. Yes, they confirmed, a light aircraft was circling in the vicinity and they had launched a boat to rescue the pilot if he ditched.

'Can you confirm your position,' asked Vette. 'We have it as 31°S,179°21'E.' Moments later the source of the problem was revealed; the centre in Auckland had mistakenly read a nine for a zero and one digit had been incorrectly transmitted. It was a simple mistake that anyone could have made, but one which altered the entire aspect of the scene. The correct position of the *Penrod* was 31°S,170°21'E, and it didn't require the navigators to plot it on a chart to realise that the Cessna was much closer to Norfolk than the rig's incorrect position had first indicated. The new position of the *Penrod* was punched into the AINS, and as 'Zulu Sierra' turned back towards the target the Cessna was called with the good news.

'You may not have to ditch,' announced Vette. 'You are nearer Norfolk than we thought.'

The previous estimates of Prochnow's positions calculated by the navigators had, in fact, been remarkably accurate under the circumstances, and it had only been by the greatest of misfortunes that the two aircraft had failed to make visual contact and that Prochnow had been unable to spot the DC-10's powerful strobe lights. On retracing the steps on the chart it appeared that the fuel dumping had taken place directly over the top of the Cessna. It was possible that the fading light conditions had obscured the fuel trail from view. It was also very close to the aural box position adjusted for progress and the position line earlier indicated from Brisbane on HF.

A measurement by Forsyth from the chart revealed Prochnow to be only 150 nm from his destination, about one hour and twenty minutes' flying time away at the Cessna's cruising speed. Prochnow had now been in the air for about twenty-one-and-a-half hours, and only half an hour of fuel remained of his original endurance estimate. The Cessna had been flying at economy cruise for some time so as much as an hour's fuel could have been saved. That would give Prochnow the one-and-a-half hours' endurance he needed for Norfolk. Whether to make a go for it or to stay put was Prochnow's decision alone, and not an easy choice to make. If the Cessna ran out of fuel on the way Prochnow would be in a worse situation that he was now. He would have to glide alone in the dark into the ocean swell with very little chance of survival.

The DC-10 flew at speed towards the *Penrod* with growing anxiety amongst the crew. It was now a desperate race against time.

'What's your estimate of remaining fuel?' enquired Vette.

'I've got about one quarter of one tank left,' suggested Prochnow. Maybe about ten gallons (forty-five litres).' The Cessna's fuel consumption was eight gallons (thirty-six litres)) an hour, giving a remaining endurance of about one-and-a-quarter hours. It would be touch and go but he might just be able to make land.

The DC-10 had by now descended to 10,000 ft (3,050 m) to improve the chances of sighting the Cessna and already the bright lights of the *Penrod* rig could be seen ahead.

'We reckon you might just be able to make it,' announced Vette. 'What do you think?'

Prochnow took another look at the black and heaving sea

and didn't need to be asked twice. 'I'll go for it,' he said. 'Give me a heading and I'm off.'

'Steer 294° magnetic,' instructed Vette. 'We'll be with you shortly and we'll accompany you on the way.'

With a mixture of fear and hope in his heart Prochnow turned the little Cessna away from the friendly lights of the rig and back into the black night. F/O Dovey slowed the DC-10 to its minimum cruise speed of about 200 knots and turned slightly right to leave the *Penrod* on their left. Just then they spotted the little aircraft.

'The Cessna's been found!' announced Chief Purser James to the passengers. A great cheer rose from the cabin. While approaching the rig, the tiny Cessna could be seen from the flight deck, ahead and to the left, climbing away on its assigned heading. Prochnow was wisely gaining height to increase his gliding range in the event of the engine cutting with fuel starvation. The Cessna was difficult to see, even with its landing lights on, but this was a moment Captain Vette wished to share with his passengers.

'Ladies and gentlemen, if you look out of the left windows, about seventy degrees to the left and below, you'll see something very interesting. It's difficult to spot, but if you look carefully you'll see the dim lights of the Cessna.' Another cheer rose from the cabin. Eighty-eight pairs of eyes, with hands cupped at temples to shade the view, peered into the night for a sight of their quarry. It was a most satisfying moment for all.

The DC-10 sped past the tiny aircraft and Prochnow could clearly see the lights of the big jet above him and to his right.

'We'll set a course for Norfolk,' suggested Vette. 'Follow our strobe lights, and if you lose sight let us know. We'll then double back if needs be and pick you up again.'

The DC-10 remained just sufficiently clear to prevent its jet wash striking the Cessna, and Prochnow tucked himself in behind and followed the airliner's path. Vette was also able to inform the light aircraft pilot that news had just been received of the RNZAF Orion's imminent arrival, and it would shepherd him home for the last miles to Norfolk. The drama was not over for Prochnow, however, for there was still a chance that the episode could end in disaster. With the passing of each mile the chances of reaching his destination improved, but with the consumption of each unit of gasoline so did the risk of running out of fuel. It was a most anxious period of the flight. Prochnow had now been in the air for twenty-two hours and had reached the extent of his original estimated fuel endurance. He was now living on borrowed time and saved fuel, and at 100 nm from Norfolk he still had one hour to go. Prochnow held on as best he could and steadfastly followed the bright strobe lights of the DC-10 which could still be seen ahead.

'You are going to make it,' encouraged Vette. 'Just be sure you use all of your available fuel. Check every tank in turn and run each one dry.'

The RNZAF had caught up with the search by now, and as the Cessna pilot looked to his left he saw the Orion formate off his left wingtip. It was still going to be a close-run race, but at least help was at hand.

At fifty nm to Norfolk, 'Zulu Sierra's' strobe lights remained in Prochnow's sight as the DC-10 turned over the island. The Cessna hugged the side of the Orion. If a last minute ditching resulted, the Air Force crew were ready to aid his recovery. There was little more now that Vette could do and, with the DC-10 crew's tasks completed, the big jet turned for Auckland while the Orion oversaw the home-coming of their charge. Anxiously they listened on the radio for news. At last the Cessna, accompanied by the Orion, came into view at Norfolk, and after twenty-three hours in the air Prochnow approached the airport and turned onto short finals. At least now he was over dry land. Five minutes later, with virtually dry tanks, he touched down on the runway at the local Norfolk time of midnight, a relieved and happy man to be back on solid ground. The Cessna was eight hours late on arrival and twenty-two hours of fuel had been stretched to twenty-three hours and five minutes.

'The Cessna has landed,' announced Captain Vette to his passengers. 'He is safe at last.' The apology which followed for the delay was drowned by cheering and clapping from the cabin.

Vette and his crew, however, still had to complete the work on their own flight so, with the search satisfactorily concluded, they settled down for the one-hour journey remaining to Auckland. As the airliner sped on course, Prochnow called 'Zulu Sierra' on the radio while taxying in at Norfolk and expressed his sincere thanks for the DC-10 team's effort and for their work in helping to save his life.

'I think it's time to celebrate,' said Vette to his Chief Purser. 'Break open the champagne.' In the cabin, eighty-eight glasses were raised in salute to the gallant rescuers.

The entire search and rescue exercise had been an outstanding episode. The two navigators, Captain Vette and F/O Forsyth, had worked ceaselessly at their task for over three hours, while F/O Dovey and F/E Brooks had operated the aircraft over the period at a continuously high workload. It was an admirable achievement. Captain Vette remarked of his team, 'The fact that they were such exceptional airmen helped a great deal.' Prochnow's efforts were not to be forgotten either, for his calmness and professionalism throughout were a credit to the piloting fraternity.

At 01:29 local New Zealand time the following morning, Flight TE 103 arrived at the gate at Auckland international airport, three hours and fifty-four minutes behind schedule. For once, on a delayed service, no-one complained.

The Guild of Air Pilots and Air Navigators awarded Gordon Vette and Malcolm Forsyth the Johnstone Memorial Trophy in 1980 for outstanding air navigation. McDonnell Douglas, the DC-10's manufacturers, also awarded the two navigators, as well as Arthur Dovey and Gordon Brooks*, a certificate of commendation for displaying 'the highest standards of compassion, judgement and airmanship'.

* Gordon Brooks was killed in the crash at Mount Erebus, Antarctica, in 1979. Gordon Vette went on to write a book of the accident entitled *Impact Erebus*, published in 1983.

Chapter 2
The Bermuda Tangle

The island of Bermuda sits at the apex of a triangle of infamous dimensions, with its southerly points loosely touching the tip of Florida to the west and the island of Puerto Rico to the east. Along the line of its base lie the other islands of the Greater Antilles chain: Cuba, Jamaica, Haiti and the Dominican Republic. To the north of Cuba lie the Bahama Islands, with the most westerly islands hardly a stone's throw from Miami on the mainland coast of the United States. The capital of the Bahamas, Nassau, is situated on the northern shore of New Providence, one of the smallest islands of the group.

The area bounded by what is now known as the Bermuda Triangle achieved a certain notoriety when Charles Berlitz published a book on the subject in 1974. The accounts of ships and aircraft lost in strange circumstances are numerous. However these incidents, or the explanations of the events, may be received, there is no doubt that some of the losses are alarming. Reports dating from the mid-nineteenth century tell of some fifty ships and twenty aircraft mysteriously disappearing. Some vessels have been found abandoned, while other craft at sea and in the air have vanished without any distress signal being transmitted. In many cases no wreckage of aircraft or ship has ever been found.

One of the strangest aviation mysteries concerns the incident of US Navy Flight 19 which consisted of five Grumman TBM-3 Avenger torpedo bombers. The Avengers took-off from their base at Fort Lauderdale, just north of Miami on the Florida coast, on 5 December 1954. The flight was under the command of Lieutenant Charles Taylor and was scheduled to fly a long-range navigation exercise, ending back at base. By late afternoon the aircraft were reported missing and were never seen again. A US Navy patrol aircraft, a twin-engined Martin Mariner flying boat, took off from Banana River Air Station with thirteen crew members on board and began a thorough search of the region, but its efforts were also to end in catastrophe. The rescuers' aircraft vanished without trace. In neither case was a distress signal heard, nor at the time, was any wreckage found. In April 1987, the five Avenger bombers were eventually found lying close together at the bottom of the Atlantic.

Accounts of these incidents tend to play on the mysterious side of the disappearances, although logical explanations are available for their occurrence. The lead aircraft flown by Lieutenant Taylor was known to have compass problems and he was also unfamiliar with the area. It is believed he thought his position to be over the Gulf of Mexico when in fact he was over the Atlantic. If he had turned east, away from the setting sun, he would eventually have run out of fuel and would have been forced to ditch. The search aircraft was believed to have exploded in mid-air as a flash in the sky was reported by a ship's crew.

The Bermuda Triangle may now be tarnished with a mystical hue, but the number of incidents recorded within its vicinity are not excessive considering the very heavy traffic of the region. The figures, in fact, are similar to those of other busy air and sea routes. Each year many thousands of ships and aircraft navigate the area without incident and of those who experience difficulty, most successfully overcome their problems. A very few have come closer to danger than their crews would have liked, but with skill, experience and perhaps some luck, they have defied the allegedly mystical powers of the triangle. One such relatively recent incident occurred to a Lockheed L1011 Tristar in 1983.

Miami International Airport is Eastern Air Lines' main base and a large number of flights depart from there to cities all over the US, as well as to destinations throughout the Caribbean. Each day many services criss-cross the southern region of the Bermuda Triangle without incident and the airline has a very good safety record.

On the evening of 4 May 1983, an EAL Tristar, registration

N334EA, returned home to roost at Miami and was parked overnight at the company's terminal building. Normal line maintenance procedures were conducted and scheduled work card tasks were completed. Any outstanding maintenance log book items were also corrected. Apart from routine maintenance, however, there was also a requirement to conduct a master chip detector check. As early as September 1981, Rolls-Royce, the manufacturers of the three RB211 jets on the Tristar, became concerned over problems being experienced with the engine. Some in-flight failures had occurred necessitating shut-down. The master chip detector is positioned on the left side of the low pressure compressors and consists of a small magnetic probe which protrudes into one of the oil lines of the engine oil system. If any engine components are subject to wear, small metal particles are flushed through the oil lines and are attracted to the magnetic probe of the detector. The original inspection interval of 250 flying hours was reduced to twenty-five flying hours in an attempt to detect at an early stage any adverse engine trends.

An Eastern Airlines Lockheed Tristar L1011. (John Stroud)

The mechanic servicing the Tristar simply removed the chip detector from each engine and placed it in a plastic bag. Fresh detectors were then fitted in place. The amount of metal particles adhering to each detector was coded and recorded and found satisfactory. If larger amounts of metal particles had been discovered the detectors would have been sent to a laboratory for analysis of the contamination.

Early on the morning of 5 May, *N334EA* was prepared to operate Flight 855, a regular Eastern Air Lines scheduled service from Miami to Nassau. The flight time for the 156 nautical mile journey was only thirty-seven minutes. At 08:56 Eastern Daylight Time, Flight 855 departed from runway 27 right at a take-off weight of almost 149 tons, carrying a fuel load of twenty-one tons. On board were 162 passengers, most of whom were vacationers on their way to the Bahamas. In the left-hand seat of the Tristar was Captain Richard Boddy and with him on the flight deck was his co-pilot, Steve Thompson, also a captain, and his flight engineer, Second Officer Dudley Barnes. Captain Boddy was a likeable man who was known to be cool and calm under pressure. He had only been on the Tristar since March and Captain Thompson, a supervisory pilot, was conducting Boddy's second check flight on converting on to the Tristar. Since Boddy was not yet qualified on the type, technically he was only acting as captain and Thompson was in command. In the cabin were seven attendants, supervised by Senior Flight Attendant Shirley Alexiou, bringing the total on board to 172.

The Miami weather on departure was fine with a broken cloud layer at 2,300 ft (700 m) and a visibility of seven miles (11.25 kilometres). The forecast weather for Nassau was good although some scattered rain showers were expected: quite normal for the time of year. As the Tristar began its initial climb, however, a frontal system was passing through Nassau's International Airport and the sky was overcast. The cloud had reduced visibility to four miles (6.5 kilometres) and light rain was falling. If the passage of the front was delayed, weather conditions at their arrival time could deteriorate.

Captain Boddy climbed the aircraft on the westerly

runway direction at 170-180 knots following the Miami 5 departure routeing and at about 500 ft (150 m) turned onto a heading of 250° magnetic. Thompson radioed departure control and N334EA was instructed to turn left and to proceed all the way round to the south of the airport on to an easterly heading. At 1,000 ft (305 m) the captain retarded the engines to the climb power setting of 1.40 engine power ratio (EPR) and Flight Engineer Barnes routinely scanned his instruments. Passing 1,500 ft (460 m) Barnes radioed Eastern Air Lines on VHF radio box three with the departure time, then completed the climb check list. Once more the engineer scanned his panel. Boddy accelerated the aircraft, retracting the flaps in sequence, and by 5,000 ft (1,525 m) the Tristar was climbing normally at 250 knots. Out of 10,000 ft (3,050 m), the speed was further increased to 300 knots for the en route climb to the cleared flight level of 230 (23,000 ft or 7,010 m). At 09:01, Thompson called Miami Centre on 127.0 MHz and Flight 855 was cleared on a direct routeing to Nassau. Boddy turned the Tristar slightly right to track 110° and the lightly laden aircraft continued its rapid climb. Passing 18,000 ft (5,485 m) the altimeters were selected to the standard pressure setting of 29.92 inches of mercury.

At 09:08, the autopilot captured level 230 and Boddy adjusted the thrust levers to the cruise power setting of 1.30 EPR to maintain 300 knots. The take-off until level flight had taken only twelve minutes and with New Providence Island fast approaching the aircraft would not remain in the cruise for long. It was a short, busy little flight which kept the crew on their toes.

Permission for descent was requested from Miami and N334EA was cleared to 9,000 ft (2,740 m). At 09:10, after cruising for only two minutes, the thrust levers were retarded to flight idle. The nose dipped in descent at eighty miles (130 kilometres) from Nassau and the aircraft decelerated to 280 knots. Thompson changed frequency to Miami Oceanic Control Area (OCA) which would retain radio contact until Flight 855 descended below 6,000 ft (1,830 m). They would then be cleared to Nassau Approach on 121.0 MHz. Meanwhile, Flight Engineer Barnes contacted Nassau

Approach on the other radio to inform them of their imminent arrival.

'Nassau Approach, good morning, Eastern 855 is flight level two zero zero, descending, DME*, seven zero.'

* Distance Measuring Equipment.

'Good morning Eastern 855, I have the zero nine hundred weather for you: cloud ceiling one thousand with a visibility of four miles. To the east and south the cloud base is down to five hundred to eight hundred feet and there are thunderstorms to the north. Wind, three zero zero at three knots, temperature seventy five, altimeter two nine eight nine, and recently there has been some light rain. You're number two for the approach following a light twin which is thirty miles ahead.'

At Nassau Airport the runway in use was 32, 11,000 ft (3,350 m) long. Although, with the light wind, the Tristar could land almost straight in on the opposite direction runway 14, a large thundercloud to the north-west precluded its use. Neither runway approach, however, had an instrument landing system and guidance for final approach to land would have to be a non-precision approach using the VOR radio beacon. As a result the minimum weather conditions for landing at Nassau were stated to be: cloud base no lower than 400 ft (122 m) and visibility no less than one mile (1.6 kilometres). A cold front lay across the island and the large storm cell could be seen ahead on the radar. Because of the weather and other landing traffic Flight 855 was cleared for a DME arc approach to runway 32. The procedure was lengthy and involved flying an arc of ten miles (16 kilometres) radius from the airport all the way round to the south-east before turning onto final approach. There was no approach radar available and, with the weather close to limits, and deteriorating, it would not be an easy approach for Boddy to fly.

Flight Engineer Barnes continued to scan his control panel when able but his duties kept him busy. He quickly completed the landing data card then read through the descent and in-range check lists. He then radioed the EAL company office at Nassau with the estimated arrival time and with some routine passenger details. Suddenly, as the Tristar descended through 15,000 ft (4,570 m), Captain Boddy noticed the oil low pressure warning light for number two engine illuminate on the pilot's centre panel. He called out to the flight engineer. Barnes had not scanned his instruments for about five minutes as he completed his tasks, but the oil quantity and pressure gauges were positioned in the centre of his panel at just below eye level so were easy to check. Throughout the flight the oil quantity gauges had remained steady at eighteen quarts (seventeen litres), which was a normal reading, although the full level is twenty-one quarts (twenty litres). Numbers one and three oil pressure gauges showed in the green, or normal, range and the oil quantity indicators measured fifteen quarts (fourteen litres). Fluctuations of up to three quarts (2.8 litres) were not unusual and did not give cause for concern, but the number two readings were a different matter. Only eight quarts (7.5 litres) were indicated as remaining in the number two tail engine and the oil pressure was oscillating between fifteen and twenty-five lb/sq in (psi) (1.05-1.4 kg/cm²). The minimum acceptable normal operating pressure was thirty psi (2.1 kg/cm²).

Flight Engineer Barnes had never before had oil pressure trouble on the Tristar, but there was no doubt that he had a problem now. If the engine continued to run in this low oil condition it would not be long before it seized and Captain Boddy had little choice but to call for it to be shut down. The flight engineer initiated a precautionary engine shut-down by closing number two thrust lever and by selecting the fuel ignition switch to cut off to disconnect the fuel supply. As the engine ran down the captain called for the auxiliary power unit (APU) in the tail to be started. This was normal practice on the Tristar with one engine failed and the APU would help supply the systems normally powered by number two engine. N334EA was light and the flight could be continued on two engines without difficulty.

The weather at Nassau, however, was posing a problem. Rain was beginning to fall again and at times the visibility was dropping to as low as a quarter of a mile (400 metres), well below the minimum one mile (1.6 kilometres) permitted. Flying an approach on two engines in poor weather without radar is not recommended and it could be some time before conditions improved. Even if the weather cleared quickly, the light twin ahead would still delay their arrival. The time was now 09:15 and the range to Nassau fifty miles (eighty kilometres), with Miami just over 100 miles (160 kilometres) away. With the low visibility at destination it might be quicker and more prudent to return home. Weather conditions there were fine and there was also good engineering cover. If a major problem existed with number two engine it would be easier to repair it at base. The crew quickly conferred over their predicament and the captain made the sensible decision to return to Miami. They would be back on the ground there in, at the most, about thirty minutes.

Captain Boddy advanced numbers one and three thrust levers to 1.35 EPR and held the aircraft level at 12,000 ft (3,660 m) while Thompson called Miami OCA.

'Miami, Eastern 855, we have lost number two engine and we've stopped the descent at twelve thousand. We'd like reclearance to return to Miami.'

'Roger, Eastern 855, you're cleared to make a one eighty on to a heading of two nine zero and maintain twelve thousand. For further clearance contact Miami Centre on one two seven point zero.'

At 09:17, Flight 855 commenced a left turn at 300 knots, level at 12,000 feet, back towards Miami. The flight engineer called the company at Nassau on VHF 3 to say they were returning and Captain Boddy informed Senior Flight Attendant Alexiou of the bad news. Whilst turning through north, Thompson re-established contact with Miami Centre. The controller was ready to receive Flight 855 but did not yet have the Tristar back on his radar scope.

'Eastern 855, just continue on the to heading,' replied Miami. 'We'll have you back on radar shortly. Did you want any higher altitude?'

'Affirmative. We'd like to go up, ah . . . say, ah, oh, nineteen to twenty-one thousand.'

'Okay, Eastern 855. Ah what are you in, a left turn now?'

'Yes, sir.'

'Eastern 855, climb to flight level two zero zero for now.'

'Twenty thousand, roger.'

Once again the flight engineer was kept busy with his tasks as he completed the engine shut-down check list and the secondary items required to tidy his panel.

'Okay, 855, I have you back in radar here, leaving about one three six . . . Proceed on two seven zero heading . . . radar vectors towards Biscayne.'

'Okay, two seven zero, heading back to, ah, Biscayne, roger.'

'And any problem that we should know about?'

'We've, ah, we've had precautionary shutdown on number two engine.'

'Okay, you need any special handling?'

'Negative, sir.'

'Okay.'

As *N334EA* climbed through 15,000 ft (4,570 m), Captain Boddy faced a further problem, and the flight began to go seriously wrong. The low oil pressure light for number three engine also suddenly illuminated. With number two engine already shut down and number three engine now giving trouble, the crew were less than happy with the circumstances. The routine flight was turning out to be anything but routine. Boddy eased the power back slightly on number three engine to see if that would help, and reduced the rate of ascent. Almost immediately, before the crew could assess the situation they saw to their horror the number one engine low oil pressure light illuminate as well. Barnes quickly scanned his instruments; the oil pressures on both operating engines were low and dropping and all the oil quantity gauges indicated zero. The busy little flight of only ten minutes ago was turning into a major emergency. But this couldn't be right; the oil indicators had to be mis-reading. It was surely impossible for all three engines to lose all their oil in a similar fashion. No; it had to be an electrical fault causing incorrect indications. There was little point in going higher for the time being until everything was checked out, so Boddy levelled off at 16,000 ft (4,880 m), still flying at 300 knots, while Thompson radioed Miami with details of their problem.

'And Miami Centre, Eastern, ah 855, we have, ah, some rather serious indications of all three oil pressures on all three engines, ah, ah, down to zero. We believe it to be faulty indications, ah, since the chances of all three engines having zero oil pressure and zero quantity is almost nil. However, that is our indication in the cockpit at the present time.'

'Okay, fine. Why don't you turn right about fifteen degrees. We'll give you direct to Miami. Maintain flight level twenty thousand, or whatever altitude you wanna maintain, and we're just gonna have the emergency equipment standing by anyway for ya.'

'Proceed direct to Miami Airport. Roger, sir.'

While Thompson spoke on the number two radio box, Barnes switched VHF 3 to the Miami Technical Centre frequency and explained the situation. Could a common electrical source affect the engine instruments, he asked? Captain Boddy butted in.

'The low oil pressure and zero oil quantity for the three

Captain Steve Thompson.

engines seems a faulty indication since the likelihood of simultaneous oil exhaustion in all three engines is one in a million, I would think.'

The maintenance personnel at Miami referred to the appropriate manuals and soon came up with a possible cause for the malfunction. The number two AC electrical busbar was the common source for the oil quantity gauges and the problem might lie in that department. The flight engineer quickly checked the relevant circuit breakers but could find nothing amiss. The bizarre instrument displays remained a mystery. In spite of the indications, however, the flight seemed to be proceeding satisfactorily and Boddy maintained 16,000 ft while letting the speed drop back to 230 knots. In the meantime, Miami Centre called back requesting the numbers on board and Thompson mistakenly replied that there were 168.

'One hundred and sixty eight, okay. Does that include your crew?' asked Miami.

'Negative, sir, ah, stand by on the crew.'

With one engine shut down and the other two displaying alarming indications, the number of crew on board was the last thing on their minds. The Miami controller, of course, was thinking ahead and was concerned that the situation might deteriorate. If the rescue services were required for Flight 855 it would be useful to know the numbers involved. With no reply from Thompson on the number of crew, Miami repeated the request.

'Well, we've got three in the cockpit,' said Thompson. 'We'll have to count in the back and we'll give that to you in just a minute, sir.'

'Okay.'

'We have seven in the back.'

'Seven in the back, okay. Eastern 855, you can descend and maintain twelve, one-two, thousand at your discretion.'

Thompson had no time to reply to the descent clearance to 12,000 ft, for in an instant the course of events took a turn for the worse. At 09:33, without further warning, number three engine failed with a loud bang causing the aircraft to judder. It flamed out and combustion ceased. They were now flying on one engine.

'Ah, Eastern . . . number . . . Eastern 855, we've just lost our number two engine, sir.'

In the excitement of the moment Thompson had called that number two engine instead of number three, had failed.

'Okay, losing number two. You still got two turning?'

'Negative. We only have one now and, ah, we're gonna restart our number two engine.'

'Okay, fine. We're listening. Okay, cleared direct. Miami's altimeter two nine eight nine and you can descend at your discretion to any altitude you need. You're clear of traffic.'

'Okay. We need a bearing, ah, we'd like a heading to go to, ah, two seven left.'

'Okay, your position right now is seventy miles south-east of Miami. You're about fourteen minutes out, heading two eight five, two eighty-five, for Miami.'

'Two eighty-five, okay.'

'And you can plan a straight-in two seven left. We're telling Approach about it right now. Equipment will be standing by.'

'Roger.'

As number three engine failed, Flight 855 was about seventy nautical miles from Miami, and Captain Boddy began a gentle descent while letting the speed drop back slowly. To save weight with only one engine operating, Barnes started dumping some of the seventeen tons of fuel on board. It would take approximately five to ten minutes to dump about ten tons of fuel which would just leave sufficient for landing. At 15,000 ft (4,570 m) the number one engine power was increased to 1.40 EPR and, with the speed stabilised at 225 knots, the Tristar continued on a gradual descent at 600 ft (180 m) per minute. The aircraft was becoming lighter and could keep flying, albeit with difficulty, on one engine. The crew completed the single engine operating check list, which basically rearranged the power sources of the back-up systems. All secondary systems were appropriately distributed between the one remaining engine and the APU which was still running satisfactorily. The island of Bimini passed beneath Flight 855, but it remained out of sight under a low cloud cover. A 4,000 ft (1,220 m) runway was available at Bimini Airport for light aircraft.

The weather was good and, with Miami Airport nearby, the crew felt they could make it. There was, however, no doubt now that the engine instruments were reading correctly and that all the power plants had been starved of oil. How long would number one engine survive? The flight crew were not about to wait to find out and they began the drill to start number two engine which had been shut down earlier in the flight. The in-flight start sequence was not a simple exercise and each step was carefully undertaken from the check list. The number two engine instruments were checked to indicate the engine was turning and not seized, the thrust lever was checked closed and the tank pumps were checked on. The engine in-flight start table was entered with the aircraft altitude of about 14,000 ft (4,270 m) and a speed of 225 knots and the conditions were confirmed as lying in the middle of the in-flight start envelope for engine start using the windmilling effect of the airflow. Unfortunately the table was entered in the wrong way, with speed on the altitude scale and vice versa, and the indication that a windmilling start was possible was not correct. For such a procedure a speed of 285 knots was required and in the present circumstances a start could only be achieved using the engine starter motor. Having incorrectly checked the details the windmilling air start switch was selected to on to supply ignition. Nothing happened; the engine refused to start. The situation was becoming desperate and the captain gave the order to prepare for ditching.

The flight engineer now turned his attention to the ditching check list: the flight crew donned their life jackets, the aircraft systems were tidied up by appropriate switching, the seat belt and no smoking signs were switched on and in the cockpit the crew checked their seats and harnesses were secure. Barnes now called Senior Flight Attendant Alexiou to the flight deck and instructed her to initiate ditching procedures. Alexiou informed the other flight attendants of the situation and they returned to their stations and quickly made preparations to brief the passengers. The senior flight attendant announced to the passengers the captain's intention to land on water, repeating the instructions for the use of life vests which had been given, as required by law, before departure. On the ground few people listen to emergency briefings and it was deemed prudent to repeat the exercise. Many passengers were already aware that problems were being encountered and some now became hysterical. A handful were close to panic.

On the Tristar there are eight exits for evacuation, but on Flight 855 only seven attendants, so helpers were requested from the able-bodied passengers. It would have been a relatively simple exercise to instruct the volunteers how to open a door on coming to a stop on the water and that the slide/raft would then inflate automatically, but unfortunately no-one came forward. Some flight attendants also had to assist with the donning of life vests and had to give individual instruction to a few people. At least eight passengers inflated their life vests, in spite of instructions to the contrary, so if an evacuation became necessary their movements would be restricted. The passengers were also instructed to remove sharp objects, spectacles, high heels, etc, and to place their seat backs upright. They were shown the brace position for impact and told, when announced, to lean forward and grasp their ankles. They were instructed to remain seated until the aircraft had come to a halt and which exits to use for evacuation. The flight attendants then had to secure the cabin and to stow all loose objects. They were kept very busy at their tasks and, being unsure of how much time remained, were under a lot of pressure.

As the activity on the Tristar continued, Miami Centre called Flight 855 asking, politely, if they were not busy could they pass the quantity of fuel on board. The information might also prove useful to the rescuers. The flight crew were occupied with the emergency and Thompson missed the call.

'Say again, sir.'

'If you're not busy,' repeated Miami, 'we need fuel on board.'

'Okay, we have, ah, thirty-six thou.'*

* 36,000 lb : 16,400 kg

'Roger, Eastern 855, you're cleared to six, or any altitude you need. Advise.'

'We'll advise.'

'And Eastern 855, if you can, can you give me a status report how you're doing?'

'We're doing okay.'

'Okay, Eastern 855 your position now fifty-nine miles south-east of Miami Airport. Turn left about five degrees. You're about thirteen minutes out at your speed.'

'Roger.'

At 09:33, while approaching 12,000 ft (3,660 m) over the ocean, the inevitable happened. As abruptly as number one engine failed, number three engine also flamed out and began to run down.

'We're losing another engine. We've lost our third engine right now!'

'Okay, have you got the other one started?' asked Miami.

'Not yet.'

'Do you have any of them turning?'

'Negative.'

The big Tristar jet was now fifty-five miles (eighty-eight kilometres) from Miami and was gliding without engine power. The rate of descent increased to 1,600 ft (488m) per minute with the speed constant at 225 knots. Fortunately, the APU continued to function, pressurising the hydraulic systems via pneumatic pumps and supplying electrical power, so all the controls remained in operation and the aircraft could glide satisfactorily in this condition.

The engines windmilled in silence in the airflow, and it was obvious now that oil starvation had been the cause of failure. But why had three good engines similarly run out of oil? Unknown to the crew, the fault lay in the maintenance procedures which had been undertaken during the previous night. The master chip detectors on each engine had been removed for analysis of the deposits and placed in plastic bags. Replacement chip detectors were normally acquired from a cabinet in the foreman's office, but in the early hours of 5 May none were available. These chip detectors were usually fitted with two oil sealant 'O' rings (see diagram). Three chip detectors were drawn from the stock room, but none had the oil rings fitted, a fact which escaped the mechanic's attention. A 'serviceable parts' tag attached to the transparent plastic package containing each fresh chip

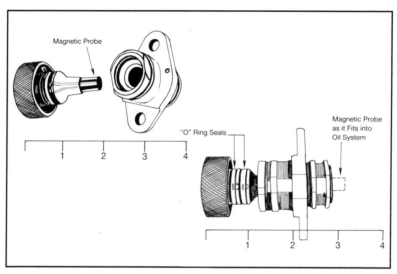

Master chip detector magnetic probe and position of the 'O' ring seals.
(NTSB)

detector served to support the illusion that 'O' ring seals were fitted. Working by feel in dark conditions when fitting the replacement chip detectors reduced the chance of the mechanics discovering the error. To check for leaks all three engines were turned for about ten seconds and no drips were noticed. It was only established later than a minimum of thirty seconds of engine motoring was required to induce leaking if oil seals were omitted. The work was duly signed off and the aircraft was released for service with the chip detector 'O' rings missing.

In the morning, when N334EA's engines were started and the oil systems pressurised, oil had begun to leak rapidly from the chip detector installations. After only a short period of flight, the engines were almost out of oil, and number two engine's remaining oil had been saved only by shutting it down. The power increase on numbers one and three engines on the climb back to Miami had probably increased the rate of oil loss in these engines and serious damage had resulted. The bearings had become molten and the movement of assemblies had resulted in the drive to the fuel pump being disconnected, stopping the engines. The engines had flamed out due to the lack of fuel but had that not happened they would have eventually seized anyway.

As the ditching drill was completed on the flight deck, Flight Engineer Barnes announced on the PA that ditching was imminent and the passengers assumed the brace position. A second attempt to restart number two engine was initiated. Meanwhile the rescue services were already in the air and on their way. As early as 09:28, the Coast Guard had been notified of the emergency by Miami and immediately a Falconjet and a helicopter had taken off from Opa-Locka Coast Guard station.

'The Coast Guard is coming out towards you now,' informed Miami.

'Roger.'

'Keep me advised of your intentions and we'll keep you on here as long as we can. Approach is also watching you at the same time. You're about twenty miles west of Bimini right now. And it looks like you're gonna have to ditch; just keep us advised. We should be able to hear you at least down to twenty-five hundred feet.'

'Yeah.'

Had the short runway at Bimini been in view as they had flown overhead they might have tried to make it back for an emergency landing, but with the island obscured by low cloud they didn't stand a chance.

At 09:35, Flight 855 approached 10,000 ft (3,050 m) with about six minutes' gliding time remaining. The fuel dumping had now been completed and the aircraft weighed about 135 tons. The second attempt to start number two engine proved in vain and ditching seemed inevitable, but at least the conditions were fine: good daylight visibility, fair weather and a calm sea. A gentle breeze was blowing and the water temperature near the coast was a comfortable 79°F. The crew had been well trained in ditching procedures and all those in the cabin were prepared for the worst. Captain Boddy would touch down on the sea with the landing gear retracted, landing flap set and with the aircraft nose pitched up at 12°. It had been calculated that on impact in that configuration the wing engines, flaps and horizontal stabiliser would probably be torn off. Even sustaining such damage, the

Tristar was still expected to remain afloat for at least twenty to twenty-five minutes. Although the event would be risky and an unpleasant experience for the passengers and crew, there was every likelihood that the ditching would be successful. Danger could be averted only if number two engine started, but so far it had resolutely refused to function. There was little hope of engines number one or three re-starting, but anything was worth a try. With time fast running out an in-flight start was attempted on each wing engine.

As *N334EA* continued inexorably on its descent towards the ocean, Miami called with the latest radar position report. 'Eastern 855 you're forty-nine miles south-east of Miami right now, twelve minutes out at your speed.'

'Okay, we don't believe we can make land, ah . . .'

'Okay, we've got all the help we can coming out as fast as we can.'

The Coast Guard aircraft was now close to the scene and the rescue pilot was listening out on the frequency. He interrupted the exchange.

'We're a Falconjet coming off the beach at, ah, flying out of Opa-Locka,' called the Coast Guard to Flight 855. 'Ah, what's your position, sir?'

Miami Centre had both aircraft on radar and required only a positive identification of the Falconjet to co-ordinate efforts.

'Okay, I'll tell you what,' suggested the Miami controller. 'Coast Guard, just give me an ident. I'll give you a position in relation to yourself. What code are you squawking?'

'Squawking one two seven seven.'

'Okay, squawk ident. Okay, I see you out there. He is one o'clock. Your position twenty miles. I'll vector you to him. I have you radar contact.'

'Roger, sir. Okay, we'll be turning to about a heading of one one zero.'

'Okay, one one zero. Let's make it a one two zero. We'll intercept closer.'

'Roger, that, one two zero. And, ah, what's the altitude?'

'Right now seventy-nine hundred and he's descending slowly.'

'Roger, were climbing through twenty-five hundred.'

'Roger. Try heading one three five Coast Guard, one three five.'

'One three five.'

'He's twelve o'clock, ten miles Coast Guard.'

'Roger that.'

'Coast Guard's gonna be right next to you here pretty soon, Eastern 855,' reassured the Miami controller.

'Okay, ah. Ah, say his altitude again, sir,' asked the Falconjet pilot.

'Eastern is sixty-nine hundred descending.'

'Roger, we're thirty-two hundred climbing.'

'Ah, roger.'

Another two aircraft already airborne from US Coast Guard Air Station Clearwater, a helicopter and a C-130, were now also diverted to assist in the operation. Three more Coast Guard choppers and another Falconjet were also scrambled and at sea a Coast Guard cutter and five other patrol vessels were standing by. A USAF helicopter and a C-130 were preparing to take off from their base at Homestead. It was an impressive rescue team that was being assembled, but it was just as well, for there was every indication that their services would be required.

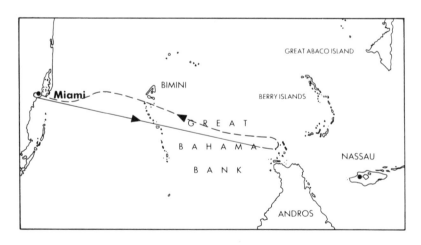

A map showing the relative positions of Miami and Nassau with the flight path of the Lockheed L1011 Tristar indicated.

At 09:37 the Tristar passed 7,000 ft (2,130 m), still flying at 225 knots, with about only four minutes' gliding time remaining. The attempts to start numbers one and three engines had failed and ditching seemed imminent. Boddy asked for guidance to land on the water as near to the shore as possible.

'Heading direct to closest land point, please,' radioed Thompson.

'Okay,' replied Miami, 'just as straight in is about the best you can do right now, Eastern 855, the way it looks here. Are you over land now, Coast Guard?'

'Ah, negative, sir. We're off shore about, ah, about fifteen miles.'

'Okay, Eastern is a ten eleven ah, turn further right another ten degrees. He's out there about twelve o'clock, be crossing left to right about seven miles.'

'Roger, just for your information there's a small scud, ah, thin layer, at about, ah, oh, four thousand feet.'

'Okay, suggest you stay down. He's fifty-two hundred descending, 855, suggest a heading of about another twenty degrees to the right.'

On *N334EA*'s flight deck the crew tried a third and final attempt to start number two engine. The unsuccessful windmilling start technique was abandoned and a full engine start procedure using the starter motor was implemented. It was a last effort before committing themselves to ditching. At the same moment the Coast Guard aircraft caught sight of the Tristar.

'We have him,' called the Falconjet pilot.

As Flight Engineer Barnes persevered with the in-flight drills, Miami continued to guide the rescue aircraft to its quarry.

'Eastern's forty-seven hundred descending, Coast Guard. He's about eleven-thirty, four or five miles.'

'Ah we have him at twelve-thirty, sir.'

'Okay, twelve-thirty. Keep him in sight. I'm gonna stop talking. Advise Approach what you have to do.'

'Ah, negative, sir. He just went into the clouds.'

'Okay, he's forty-four hundred feet. No traffic that I can see underneath you.'

'Roger that.'

'Okay, I'll tell you what,' continued Miami. 'If you want to turn to a heading of one eight zero now we'll bring you in behind heading one eight, make it one nine zero.'

'Roger, one nine zero.'

'He's forty-one hundred descending. Turn right heading two hundred degrees, Coast Guard.'

'Two zero zero.'

At 09:39, the last item of the in-flight start check was completed with the selection of the start switch to on. Shortly Boddy would ask for the full 33° flap setting and in just over two minutes they would hit the sea. Suddenly, as they had almost given up hope, number two engine ignited and began to run up. Slowly its speed increased to idle and the power remained steady. It had started! The Tristar was just passing 4,000 ft (1,220 m) and was still twenty-two miles (thirty-five kilometres) from Miami, but now they were in with a chance.

'We have an engine going now and we believe we can make the airport.'

'Okay, you say you have one of them turning, Eastern?'

'That's affirmative.'

'Okay, the Coast Guard's gonna come in behind ya. He's behind ya right now. Coast Guard 277, turn right heading two five zero.'

'Two five zero, turning.'

Captain Boddy opened up number two thrust lever to about 1.30 EPR for thirty seconds and the power held. He then increased the power to 1.50 EPR and at 3,000 ft (914 m) he managed to arrest the rate of descent. The immediate problems were now over but the flight crew still had to complete a one-engine approach and landing, a far from simple task on a big jet. If the approach was improperly handled they would be back in trouble. And for how long would number two engine continue to function? It had been shut down by crew action earlier in the flight so still had some oil left, but the captain was only too well aware that the oil would not last for long. If the engine flamed out on the approach over land they would be in a worse position. Runway 30 offered the fastest time to touchdown but no descent path signal was available from the instrument landing system (ILS) and its absence would further complicate the approach. The flight path to runway 30 also tracked a long distance over land. Runway 27L, on the other hand, had both centre line and approach path ILS signals functioning and the approach, mostly over water, gave a better chance of ditching if the remaining engine failed.

The Tristar crew completed the check list items of the one engine approach, switching half the fuel pumps off to reduce the electrical load, depressurising the aircraft and supplying all the pneumatic functions from the APU. Such an arrangement permitted maximum power availability from the one remaining engine. Meanwhile the Miami controller continued his dialogue.

'What are your flight conditions now, Eastern?'

'In the clear.'

'Okay, Coast Guard, you should have him in sight out there. You make your right turn from twelve o'clock a mile and a half.'

'Roger, sir, we have him.'

'Okay, I'm gonna stop talking, Eastern. You're cleared a straight-n, contact Approach if you can. Don't talk if you don't have to.'

'Cleared for a straight-in. We want 27 left. Ah, are we about pretty good for that?'

'Ah, just go on in. We're waiting for ya.'

'And this is the Coast Guard. Ah, we've lost him in the scud again.'

'Okay, you're coming in behind him, about a mile behind him. He's at three thousand now. He's gonna be 27 left.'

'Roger.'

'Okay, Coast Guard, you have him in sight?'

'Ah, negative, sir.'

'Okay, turn right, heading, ah, three zero zero degrees. He's looking, he's picking away from you, he's about three thousand.'

'We don't see him.'

'Okay. And he's holding two three zero knots. He's looking in pretty good shape.'

'Okay, we're at two four zero knots now and we'll follow

him all the way into the airport, if that's okay with you.'

'Okay, we'll tell Approach you're gonna be right behind him. Change to one two six eight five, that's one twenty-six eighty-five. That's Miami Approach, and tell them you're behind the Eastern.'

'Roger, we have a visual now.'

'Okay, very good.'

At 09:43, Flight 855 intercepted the instrument landing system for runway 27 left at a height of 3,000 ft (914 m) and at a distance of just under ten miles (sixteen kilometres) from the threshold. Boddy eased back on the power on number two engine and let the speed fall off gradually as he descended on the glide path. He called for the flaps to be set to 10° and the landing gear to be selected down.

'Well, we believe we've got it made,' radioed Thompson.

'Fantastic,' shouted the Miami controller.

'Of course, we're cleared to land?' added the captain, tongue in cheek.

At 1,000 ft (300 m) Boddy was satisfied with the situation and committed the aircraft to land. The passengers were still crouching in the brace position so Captain Thompson quickly informed them on the PA that they were about to land at Miami. It was a very relieved group of passengers and flight attendants who heard the announcement.

'Eastern 855, you need not acknowledge if you're still on the frequency,' radioed Miami. 'On landing call point nine.'

'Okay, sir, sure thank you for your help.'

'Certainly.'

At 09:46, the Tristar touched down safely on 27 left at Miami and the tension of the emergency drained from all on board. The landing was only just in time, for the oil quantity left in number two engine was less than had remained in engines number one and three when they had failed and it was very close to stopping. As the Tristar rolled down the runway it seemed the emergency was over, but there were more problems to come. The controller radioed to say that number one engine was smoking and there appeared to be a serious risk of fire. The fuel ignition switches had been left on in the forlorn hope of achieving a start-up and on the roll-out the windmilling engine was still pumping fuel. On slowing, the fuel residue began to smoke in the hot engine. Quickly the number one engine fire bottles were discharged to dampen any flames. On stopping the aircraft, Boddy received a further warning from ground personnel that number three engine was also smoking badly. The number three engine fire bottles were likewise discharged and the fire services, which were standing by, also doused the wing engines with water and foam. Eventually the smoke dispersed and the risk of fire was averted. Captain Boddy called for a tug to tow the aircraft to the terminal but when informed there would be a fifteen minute delay he decided to taxy using number two engine. Boddy opened up the number two thrust lever but there was insufficient power to move the aircraft, even with the lever pushed right to the stop. Within moments the power began to drop and number two engine failed. It had been very close indeed!

The round trip had only taken fifty minutes but had seemed to the passengers like hours. The crew had done a magnificent job in extremely difficult circumstances and their skill and composure had saved the day. They were lucky too, of course, for if number two engine had not re-started they would have been faced with a ditching. And if it had seized on the approach over land the situation would have been much worse. But there was more. The decision to return to Miami had also been opportune for if the flight had continued to Nassau, on the lengthy approach the engine would probably have failed before landing. If the short runway at Bimini had been visible the crew might have attempted a powerless, or deadstick, landing with possibly disastrous results. And, more significantly, if the in-flight start table had been entered correctly and the number two engine had been started earlier it would undoubtedly have failed before touchdown at Miami.

Flight 855 had defied the Bermuda Triangle and at the last moment had slipped from its grasp. Had all three engines failed simultaneously in a more remote part of the triangle, however, it might have been a different story. The Tristar could have been forced to ditch with little or no warning and if the landing had been on a stormy sea the aircraft could have sunk quickly in deep waters. The incident could well have been another Bermuda mystery. Fortunately it ended safely and Captain Boddy also passed his flight check!

The flight crew were each presented with an Award for Outstanding Airmanship from the American Airline Pilots' Association.

Chapter 3
To Take-off or Not to Take-off . . .

Pre-flight checks and procedures before the departure of a big jet are numerous and complex and about forty minutes are required to prepare something as large as a Boeing 747 for flight. The calculation of speeds for take-off form an integral part of the preparations and all particulars of aircraft weight, weather and runway details are carefully checked. The speeds important to pilots on take-off are assigned the letter 'V' for velocity and are designated V1, VR and V2. V1 is the go or no-go decision speed. In the event of an emergency occurring before V1, sufficient runway is available for stopping, but after V1 the aircraft is committed to take-off.

The second speed of interest to pilots is VR, or V rotate, which is the speed at which the pilot raises, or rotates, the aircraft nose to a predetermined pitch angle. As a result, lift from the wings is increased and the aircraft climbs into the air.

The third important speed is V2 which, after take-off, is the minimum safe climb-out speed required in the event of an engine failing at V1, the worst possible moment. The normal climb-out speed is V2 plus ten knots.

A performance manual is carried on all flight decks and contains the relevant runway pages and technical details required for take-off calculations. Each runway page lists maximum take-off weights for various conditions and is constructed with runway length, gradient and airport elevation in mind. The list is entered with wind component and temperature and a maximum permitted take-off weight for the prevailing conditions is obtained. The maximum structural take-off weight of the Boeing 747 100 series, using Pratt and Whitney JT9D-3 engines (known as the dash three), was about 323 tons, but short runways, high airports or hot days can all take their toll. On occasions, even using full power, the weight may have to be reduced by carrying a lower payload or less fuel. Even at sea level, a shorter runway or higher temperature can reduce the permitted maximum take-off weight considerably. Where the take-off is close to the maximum permitted in these conditions, even a slight change of wind or temperature can be critical. A one degree centigrade rise, for example, can result in a reduction of two tons from the permitted maximum take-off weight. In such situations the take-off calculations have to be conducted with diligence and care.

On 30 July 1971, a Pan American Airlines (PAA) Boeing 747, appropriately registered N747PA, landed at San Francisco International Airport, California, at 13:58 Pacific Daylight Time (PDT: GMT −7). The aircraft was the flagship of the 747 fleet and was named 'Clipper America.' The big jet was operating Flight PA 845, a regular, scheduled service originating in Los Angeles, California, and transitting San Francisco en route to Tokyo in Japan. Flight 845 took off from Los Angeles at 13:11 PDT for the short forty-five minute hop to San Francisco, and after a one hour stop-over was scheduled to depart for the ten hours forty minutes sector to Tokyo at 15:00 PDT. A fresh crew was to take over for the long Pacific flight and an extra co-pilot and flight engineer were to be carried to accommodate the extended duty day.

At just after 14:00, Captain Calvin Dyer and his flight crew of First Officer Paul Oakes, Second Officer Wayne Sager, first Flight Engineer Winfred Horne and second Flight Engineer Roderic Proctor, assembled in the operations room for the flight briefing. The crew were very experienced with the captain and both flight engineers having each accrued flying times in excess of 20,000 hours, but none of the five cockpit personnel had operated the Boeing 747 for longer than eighteen months. In fact, the inaugural commercial Boeing 747 service from New York to London had been flown by Pan Am only at the beginning of the previous year on 22 January 1970.

The duty dispatcher presented Captain Dyer and his crew with the paperwork for the flight: destination, diversion and en route weather, fuel requirements, air traffic control flight plan, flight logs, notices to airmen (NOTAMS) and take-off information. All the relevant details were outlined by the dispatcher and using a wind of 300° at fifteen knots, a temperature of 66°F (19°C), and a pressure setting of 29.99 inches of mercury, the take-off speeds were calculated for runway 28L, the longest runway at the airport. The actual take-off weight was 321.8 tons and was only just inside the maximum permitted take-off weight on 28L in the present conditions of 323 tons. The preferential noise abatement departure runway was OIR, but the dispatcher had opted for the longer 28L because of the aircraft's heavy weight. San Francisco Airport is situated on the western shore of San Francisco Bay and flights taking-off from the northerly runways were conveniently routed out over the water, while those landing on the westerly runways also approached over the bay.

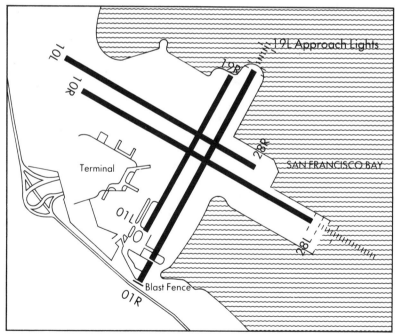

A map of San Francisco airport showing the location of the runway 19L approach lights.

The fuel requirement for the trip was 133.5 tons, close to the maximum possible of about 140 tons. Captain Dyer, in fact, had requested that the tanks be topped up for the long flight if the load dropped, but otherwise no more could be carried. Using standard 10° flap setting for take-off, the speeds of V1, 156 knots; VR, 164 knots; and V2, 171 knots were obtained. Normal operations were not possible without the use of 10° flap and the extra lift available more than compensated for any drag on the take-off run. In more critical conditions, flap 20° could be used on certain occasions further to improve lift. Runway 28L was 10,600 ft long (3,230 m) but on departure most of the runway would be used up before lift-off and the take-off run time would be of the order of fifty seconds.

Captain Dyer and his colleagues boarded N747PA at about 14:30 and began their pre-flight checks. As part of the procedures the take-off speeds were marked on the airspeed indicators using small, movable white plastic bugs on the rims of the two instruments. One hundred and ninety-nine passengers were travelling to Tokyo on Flight 845 and with five in the cockpit and fourteen stewardesses in the cabin, the total on the aircraft amounted to 218 people.

When boarding was completed, the engines were started in sequence, and almost precisely on time Flight 845 pushed back from the San Francisco gate at 15:01. During the process of push back, the first officer monitored the automatic terminal information service (ATIS) on 113.7 MHz, which broadcast weather and airport advisory material, and he soon became aware of a problem with runway 28L. The current ATIS, information 'X-ray,' had been valid from 14:02 and stated: 'Weather, clear; visibility 15 miles; temperature 66°F; wind 300° at 15 knots; altimeter 29.99. Landing runway 28R, departure runways one. Notice to airmen: runway 28L is closed; the first 1,000 feet of runway OIR is closed.' Unknown to the crew, the runway had been closed for repair since 08:30 because of break-up of the surface at the intersection with taxyway 'golf.' By misfortune, the dispatcher who had briefed the crew had been absent earlier to attend a company medical check, and on his return after lunch had not been apprised of the circumstances. He had been left unaware of the closure of the runway. The situation had been compounded by the omission of this information from the previous ATIS, 'whiskey,' which had been valid from 12:30.

In four consecutive ATIS broadcasts since 08:36 the non-availability of 28L had been announced, and it was unfortunate that by an oversight it had been omitted at the moment of the dispatcher's return. The exclusion was corrected in ATIS 'X-ray' which had been transmitted from 14:02, after the briefing of Flight 845's crew had begun. In addition, the information that the first 1,000 ft (300 m) of runway OIR was closed was also included in 'X-ray,' the first ATIS of the day to disclose this detail. The dispatcher had checked the feasibility of take-off from the shorter, 9,500 ft (2,895 m) OIR preferential runway and had found it possible, although more critical, and had opted for the longer 28L. The wind was also more favourable on this runway. Some delays could occur using the non-assigned runway but that would be acceptable under the circumstances. Now that the crew knew 28L was closed, they opted for take-off from the arrival runway 28R. Further delays would be inevitable using this runway because of landing traffic.

At 15:11, First Officer Oakes radioed the tower for taxy clearance using Pan Am's call sign, Clipper, and was instructed to proceed to the threshold of OIR.

'Negative,' replied Oakes, 'we require 28R.'

'OK, taxy to 28R.'

'Clipper 845, taxy 28R, and just confirm that 28L is closed.'

'That's affirmative.'

Oakes now contacted Pan American Operations (Panops) on the company frequency, stating that they had established 28L was closed and were planning 28R. Could the dispatcher check along with them the take-off details for the new runway? This message was received by the assistant dispatcher who simply informed the duty flight dispatcher responsible for PA 845 that a request had been received for a runway change. The duty dispatcher assumed radio control and, unaware of the specific choice of 28R, began discussing the feasibility of using OIR which had been checked earlier and found satisfactory.

'Take-off from OIR is possible using a zero wind component.'

'OK,' replied Oakes, 'but according to the ATIS the first 1,000 feet of OIR is closed.'

'It's not restricted as far as I can tell but, standby, I'll check it out.'

The dispatcher now listened for the first time to information 'X-ray' and afterwards contacted the tower controller by telephone to obtain confirmation of the details. The controller verified the closure, but added that the first 1,000 ft portion of runway OIR which was now shut to all traffic was not available to Boeing 747 aircraft at any time because of 'blast over-run'. He made reference to a statement issued by the airport authority on 11 February 1971. A highway ran to the south of the OIR threshold and a blast fence which had been constructed along the perimeter was too low to afford adequate protection from such a large jet. The jet efflux from the powerful engines of the Boeing 747 simply passed over the top. Even when a big fanjet engine is running at idle, a man can be blown over by jet blast within 150 ft (forty-five m) of the jet pipe, and at full power any vehicles proceeding along the freeway would be at risk.

This restriction to departing big jets from OIR had been issued as a Notice to Airmen (NOTAM) as early as 10 February 1970 by the airport authority and applied to all high thrust-line type aircraft. 'Take-off power was not at any time to be used until reaching the displaced threshold marker due to jet blast on the freeway.' Oakland Federal Aviation Administration Flight Safety Station (FAA FSS), who were responsible for dissemination of this information, did not consider the detail to meet the criteria of a NOTAM and downgraded the notice to an Airmen Advisory (AIRAD). AIRADs were accorded a more restricted and local circulation.

Early in 1970, the San Francisco Airport ATIS had originally broadcast a reminder of the take-off power restriction to Boeing 747 and other aircraft but had not done so for some time. To confuse the situation, the Boeing 747 restriction was rescinded on 16 April 1970 and reinstated on 11 February 1971. It was this last notice to which the tower controller had referred in his discussion with the dispatcher, but once again the detail had been issued by the Oakland FAA as an AIRAD and not as a NOTAM. Pan Am, in keeping with other aviation organisations, did not possess the teletype circuit required to receive AIRADs.

The airport authority became aware of the problem at an early date and made available a telephone number for airlines to call to obtain current information. Airline personnel, however, tended to use this facility only in unusual circumstances and normally relied on NOTAM and ATIS information. As a result, the dispatchers were not aware of the Boeing 747 take-off restriction on OIR, and never had been. They only became aware of the facts when confirming with the tower the details broadcast on the ATIS regarding the closure of the first 1,000 ft of the runway up to the displaced threshold. A trailer with flags had been placed on the centre-line of OIR about 500 ft (150 m) from the end and barricades had been erected across the taxyway to the original threshold to indicate closure. The first access open to the runway was taxyway 'M,' situated about 200 ft (sixty m) north of the displaced threshold which was marked by painted lines on the runway surface. All aircraft, therefore, irrespective of size, were compelled to depart from this point.

At 15:17 the dispatcher radioed PA845 with the information.

'Talked to tower: the 1,000 feet they were talking about that's closed is actually overrun. You couldn't start from that point in any event because of thrust damage. Start at the painted threshold and you still have 9,500 feet plus clearway ahead of you and under these circumstances the page using 3A power shows no take-off limitation at your gross, over.'

'We don't have those charts in our particular manual,' replied Oakes, 'we only have the dash 3.'

Although the aircraft was powered by upgraded 3A engines, only dash 3 engine details were available to the cockpit crew. The 3A charts were held only in the dispatch office. Because of the critical nature of the departure maximum available power would be required. It would also be necessary to use 'wet' take-off thrust, whereby a system would be activated to inject water into the engine to increase power. A flap 20° setting would be needed as well.

The 'clearway' the dispatcher referred to in his communication with the crew was, in fact, the flat expanse of San Francisco Bay which stretched to the north of OIR. The lack of obstructions on the departure path could be used in the calculations to improve the maximum permitted take-off weight from the runway. The only structure beyond the end of OIR was the approach light system for landing on the opposite direction 19L runway. The structure on which the lights were placed was constructed of 2 x 2 in (fifty mm) angle iron, and a maintenance walkway with a safety rail ran along the entire length of the system. The approach lights and walkway stood proud of the water surface by sixteen ft (4.9m) at the highest point, only five ft (1.5 m) above the runway which was eleven ft (3.35 m) elevation at the northern end. To add to the problem, however, the inclusion of the clearway computation in the take-off calculations also required special charts which were only available in operations and were not held on board Flight 845. Departure from OIR without a weight restriction, therefore, required the use of flap 20°, 3A 'wet' engine power and clearway computations, none of which could be calculated on the flight deck. The crew had to rely on the dispatchers to complete the work satisfactorily and to pass the results over the radio. After a final check of the details, the take-off speeds for runway OIR using flap 20° were radioed to Flight 845: V1, 149 knots; VR, 157 knots; and V2, 162 knots. First Officer Oakes copied down the figures and selected the flaps to 20°.

Although the flight dispatchers appeared to have overlooked the original runway change request to 28R, the crew were still contemplating its use. In 1971 the length of 28R was promulgated as 9,700 ft (2,956 m), but it was known that there was uncertainty regarding this figure. The actual length available was, in fact, 9,500 ft (2,896 m), the same as the noise abatement preferential northerly runway, and a flap 20° take-off was also required for departure. Runway 28R, however, was favoured by the wind which by that time had backed to a westerly direction at twenty knots. This shift in the wind had placed OIR outside the FAA recommended crosswind limits for a noise abatement preferential runway although this factor alone did not preclude its use.

The assistant tower controller on duty at the time was aware of the problem, but also knew of the complaints of noise from the public that would be received by the airport authority if a switch was made to 28R. Also, landing on the northerly runways was prohibited because of the proximity of the built-up area. With 28L closed, only one runway would be available for departure and arrival for the entire airport if a change to 28R for take-offs was implemented. Serious delays would result, and under the circumstances it was felt prudent to retain OIR for departures.

If Captain Dyer still preferred take-off from 28R, the chance to exercise his choice was fast receding. Flight 845 had been taxying and holding for over fifteen minutes while trying to sort out the confusion and precious fuel reserves were being consumed. If Dyer insisted on the use of 28R, further, inevitable and possibly lengthy delays could result. In their present position on the airport they were also blocking the progress of other aircraft and a decision had to be made quickly. The flight dispatcher had already confirmed the suitability of OIR for take-off with no delays expected on that runway. The calculations had been completed and the speeds passed to the aircraft, so to avoid further inconvenience the captain opted for the northerly departure. At 15:19, taxy clearance to the displaced threshold of OIR was received and Flight 845 proceeded on its way.

Captain Dyer and his crew continued with the before take-off checks, aware of some of the problems using OIR, but not all. The temperature remained the same at 66°F (19°C), but the wind had changed to a less favourable quarter with the speed increasing in strength. Initial calculations had been based on a wind component of zero, but the crew were now presented with a slight tail wind. Even under the most relaxed circumstances the maximum permitted tail wind of the Boeing 747 was only ten knots. The take-off limitations of OIR had been improved by use of the clearway — the area of water to the north of the runway — but unknown to the crew and the dispatchers the clearway did not meet FAA criteria in spite of the relevant charts being available in the manuals.

The height of the opposite direction 19L approach light system, standing proud of the runway by only five ft (1.5 m) at its highest point, *did* constitute an obstruction and negated the use of a clearway in the calculations. Barges were also known to cross the take-off zone. But there was more. A further, more serious and insidious error lurked within the manuals. The page for OIR clearly stated the length of the runway to be 9,500 ft, but this was highly misleading. The total distance of the runway *was* 9,500 ft, but the displaced threshold lay 1,100 ft (335 m) from the beginning of the runway and only 8,400 ft (2,560 m) was available for take-off for Boeing 747 aircraft. It appeared from the chart that 9,500 ft was available from the displaced threshold. The dispatchers had only just been made aware by the tower controller of the Boeing 747 take-off restriction precluding use of the runway surface south of the displaced threshold, but there was no indication that this had not been taken into account in the charts. The airport authority had properly disseminated the details and it was not unreasonable to assume that any runway restrictions appropriate to the Boeing 747 had been considered in the manuals for the aircraft type. The airline, however, had never received the information. Captain Dyer and his crew had been assured by the dispatchers of the adequacy of OIR, but unknown to anyone involved in Flight 845's departure, the big jet was about to depart from a runway which had a take-off run 1,100 ft less than stated.

As *N747PA* taxied towards the OIR threshold, the crew continued with the before take-off checks. The 20° flap setting and the stabiliser trim were checked, the flight instruments were confirmed operating satisfactorily, the flying controls were tested and the fuel and hydraulic systems were arranged for departure. The take-off speeds pertaining to the flap 10° take-off on 28L had been set on the airspeed indicators with the white plastic bugs during the engine start check which had been completed before push back. It was now necessary to set the lower take-off speeds relating to the flap 20° requirements, but in the confusion the crew omitted to reset the bugs. Confirmation of the bugged take-off speeds did not re-appear on the check list after the initial setting during the engine start check and no reminder was available to prompt the crew. VR remained at the original speed of 164 knots instead of the correct 157 knots, and more runway would be needed to gain the higher speed. It now appeared the trap was well and truly set.

Although, by any standards, a departure on OIR under the present circumstances was unacceptable, there *was* sufficient runway to complete the take-off. A rejected take-off at V1 could not be accommodated on the length available for there was insufficient stopping distance, but there was just enough runway for a take-off run. It would be close, very close indeed, but it was possible. In fact, the distance required for Flight 845 to accelerate to lift off by rotating at the flap 20° speed of 157 knots was only 7,400 ft (2,255 m), well inside the 8,400 ft (2,560 m) available. Rotating at the higher, incorrect speed of 164 knots, also with flap 20° set, required 8,400 feet, right on the limit; an extremely critical but not impossible situation. The position of the aircraft at the start of the run, the rate of application of power, the rotation technique and any slight change of wind would all play a part and any one might tip the balance. It would be touch and go.

Flight 845 now approached runway OIR unaware of the insidious danger of the situation. Captain Dyer stopped by the side of the runway and checked the departure routeing which had been issued earlier as a San Francisco 3. After take-off from runway OIR the big jet was to turn on to a heading of 030° magnetic for radar vectors to the assigned route. A few moments later the tower controller radioed the instruction to continue.

'Cleared into position on runway OIR to hold.'

Entering the runway from taxyway 'M' required a back-track to the displaced threshold. Dyer taxied the aircraft beyond the take-off point and turned to face north with the nosewheel sitting on the painted line. Every foot would count on the take-off run and this start position was in the 747's favour. The last of the before take-off checks was completed and Flight 845 was ready to go. The tower controller frequently called out the wind velocity to departing aircraft as a check and at 15:26 the information was relayed to Dyer.

'Clipper 845, the wind's two seven zero at twenty two.'

Two minutes later, at 15:28, the instruction to depart was issued by the tower. 'Clipper 845, cleared for take-off. The wind two six zero at twenty.'

The wind had backed a further 10° to 260° and resulted in a tail wind in excess of three knots. It seemed an insignificant amount blowing the jet down the runway, but under the critical conditions it could just tip the balance. It would be desperately close.

Captain Dyer opened up the thrust levers with his right hand to 1.1 engine pressure ratio (EPR) while his left hand rested on the nose wheel steering tiller.

'Power stable,' called F/E Horn.

Captain Dyer quickly released the brakes and steadily advanced the thrust levers close to maximum power. The big jet leapt forward and began to accelerate rapidly.

'Set full power,' called Dyer.

F/E Horne finely adjusted the levers to the maximum setting while Captain Dyer's hand rested gently on top. As the airspeed indicators became effective the co-pilot checked both instruments and the needles were seen to move. At eighty knots the rudder became fully effective and at the call of 'eighty knots' from F/O Oakes, the captain transferred his left hand from the nose wheel tiller to the control column and continued to steer with rudder. The co-pilot released his hold on the wheel.

'I have control,' confirmed Dyer.

A good degree of right rudder was required against the effect of the cross-wind as the captain steered the 747 down the runway with delicate movements of the rudder pedals. The captain's right hand still rested on the thrust levers ready to close them in the event of a rejected take-off while his left hand grasped the control column and held down the windward wing. The control felt uncomfortably crossed with right rudder and left aileron selected. F/O Oakes and F/E Horne scanned the instruments and monitored progress; the aircraft seemed to be accelerating normally. At the bugged speed of 156 knots the first officer called out 'Vee one.' The aircraft was now committed to take-off and the captain moved his right hand from the thrust levers to the control column. Both hands now carefully held the wheel and the flight engineer assumed control of the throttles. The aircraft continued to accelerate and reached 157 knots, the correct VR speed for flap 20° at which rotation should have occurred. A further four seconds and another 1,000 ft (300 m) of runway would be needed to achieve the bugged VR speed of 164 knots. By now, however, the increase in speed seemed to be agonisingly slow . . . 158 knots . . . 159 . . . 160 . . . and the end of the runway was rapidly looming. The crew were seriously alarmed by the situation and the co-pilot began to sense that they were not going to make it. Five men held their breaths as the runway end rushed towards them.

'Rotate,' Oakes yelled to his captain.

The shout was uttered not because the bugged rotation speed of 164 knots had been reached but because the end of the runway was 'coming up at a very rapid speed'. At 15:29 and a speed of 161 knots Dyer pulled back steadily on the control column. The aircraft was now above the flap 20° speed required for take-off of 157 knots and would fly satisfactorily, but the moments to lift off seemed an eternity. At fifty-two seconds on the take-off run, as the speed edged through 165 knots, the end of the runway flashed beneath the nose. The main wheels still touched the asphalt surface. A fraction of a second later the landing gear left the paved runway and for a moment the 747 seemed to be airborne safely. Climb was neglible but the aircraft did not sink. Dyer continued to rotate to the required nose-up attitude. Eagerly the machine clawed for height and seemed to hang in the air.

Just as all seemed well the flight crew were suddenly alerted by a slight jolt which belied the danger of the situation. With the upwards rotation of the nose the rear section of the 747 had lowered and, unknown to the crew, the landing gear had smashed into the approach lights of 19L. Serious damage was being sustained. At just after rotation, the speed rose to 175 knots but with the impact it dropped to 160 knots. Desperately Dyer fought to save the big jet's life.

The Boeing 747's landing gear comprises sixteen wheels set in four bogies of four wheels each. Two bogies extend from the belly of the fuselage, known as the body gears, while the other two extend from the wing roots and are known as the wing gears. The body gears are positioned inboard and aft of the wing gears. On impact with the approach light structure the right body gear struck the first platform of the system and was forced backwards and upwards through the fuselage and into the cargo compartment. The right rear wing spar was completely severed and seriously affected the load bearing capacity of the wing. A large area of the cabin floor above was raised about one to two ft (thirty-sixty cm) and six passenger seats in the middle were dislodged by the floor bulge. Fortunately the seats were unoccupied. The left body gear also hit the first light platform and piling and went progressively deeper into the pier striking the second and third platforms on its way. It was half severed with the sequence of impacts and dangled uselessly from the aircraft with two wheels ripped from its bogie. The bulkhead to which the left body landing gear was mounted broke loose and the number one hydraulic system line fractured. Fluid spilled from the break and all control surfaces supplied by the number one hydraulic system failed.

At a point 300 ft (ninety-one m) along the lighting system, just past the third platform, the underside of the rear fuselage came into contact with the structure. The tail of the big jet dragged through the lighting system and large pieces of the two-by-two angle iron and parts of the handrail on the walkway pierced the fuselage skin. Angle iron sections and handrail lengths lanced the aircraft like giant spears. Seven pieces of steel, mostly handrail sections, shot deeply into the rear of the aircraft as debris scattered in all directions. All the metal spears entered the fuselage on the right, except one length which penetrated the aircraft skin just below door five left, shot through the tail section and exited at the tail cone. On the right side, another piece, fifteen ft (4.5 m) in length, also entered the tail section, piercing the fuselage just forward to door five right and becoming lodged in the tail cone below the auxiliary power unit. Two long sections, sixteen-and-a-half and seventeen ft (five and 5.2 m) long

Runway 19L approach light structure damage. Note the twisted angle iron sections.

became jammed in the fuselage side at the base of the centre cargo door and trailed bent and twisted in the airflow.

Of the three remaining metal lengths, all penetrated the passenger area causing extensive damage. One handrail section, like a lethal missile, streaked right through the rear section of the cabin. It punctured the skin just aft of the centre cargo door, passed through the cargo compartment and entered the cabin under seat 54F. It then shot through the aft cabin partition, through the three rear right-hand toilets, pierced the rear pressure bulkhead (an umbrella-shaped dome which plugs the end of the pressurised cabin), and exited the fuselage close to the tail fin, just below the lower rudder. The two other metal pieces, one a thirteen ft (four m) section of angle iron, pierced the fuselage just forward of the centre cargo door and tore a large hole in the skin. The projectiles then ripped through the cargo hold and punctured the cabin floor. The thirteen ft angle iron piece impaled the empty 'F' seats at rows 45, 46, 47 and 48, and remained embedded in the seats. Fortunately this cabin penetration, like the incident described above, caused no injuries to aircraft occupants, but the trajectory of the third, smaller seven ft (2.1 m) length was not so favourable. It struck the cabin floor below seat 'G' at row 46 with great force and continued through the same seat of rows 47 and 48. A passenger seated in 47G received a bad leg injury below the knee and another passenger in 48G had his left arm cut and crushed. The handrail length then smashed through the overhead hat racks and the ceiling above row 52, knocking sections of the ceiling panelling on to the seats. It then punctured the fuselage roof in the middle and became embedded in the centre of the tail fin. It was by the greatest of good fortune that so many seats were unoccupied in the damaged areas and that no one was killed.

The tail of the Boeing 747 continued to plough through the structure. As handrail and angle sections lanced the fuselage, wood and metal debris flew at random with great force causing serious damage to other aircraft sections. The inboard trailing edge flaps, the left and right horizontal stabilisers and the right elevator were all struck. Flying fragments caused severe structural damage to the stabiliser, and the hydraulic lines of numbers three and four systems were severed. One missile penetrated the right stabiliser and passed into the right elevator, only narrowly missing the line for the number two hydraulic system. Within a matter of seconds, with the number one system already fractured, the big jet had lost three of its four hydraulic systems and had come within four in. (10 cm) of all the aircraft flying controls being completely disabled.

Now with only number two hydraulic system functioning the Boeing 747 was left with a minimum of flying controls: the left outboard and right inboard ailerons, the right inboard elevator and the lower rudder. It would fly in this condition but handling would be quite different and control response would be slow and sluggish. It would not be easy to control. The fuselage tail continued to trail in the approach light pier and with only one hydraulic system operating, Flight 845 was close to disaster. With great skill Captain Dyer wrestled to save the stricken machine. At a point about twenty ft (six m) beyond the sixth light platform the big jet at last broke free from the tangled metal and eased slowly into the air. Oakes quickly checked the altimeters and was relieved finally to call 'positive climb'. Dyer climbed away at the achieved speed of 160 knots. The take-off had been a close run event and the aircraft had been badly damaged but now they were back from the brink of disaster and were climbing safely.

The flight engineer quickly checked the extent of the damage. From his panel he could see that the hydraulics were in a bad way and the hydraulic pressures of systems one, three and four were dropping rapidly. The hydraulic fluid quantity indications were also decreasing.

'We're losing hydraulic systems one, three and four,' called Horne.

Immediately he selected the air-driven pumps to off and depressurised the engine-driven pumps. With the aircraft settled in the climb the landing gear would normally have been selected up but with serious hydraulic problems and possible gear damage it was left extended. At this stage the crew were still unaware of the destruction caused by the impact.

Once more, Flight Engineer Horne checked the hydraulic systems and found one, three and four still leaking, so he switched the engine pumps to off.

'I've had to shut down one, three and four hydraulic systems, captain, so we're down to number two hydraulic system only.'

Dyer could feel the aircraft was handling in a sluggish manner but he climbed at a steady 160 knots and held the aircraft close to the runway heading as it flew out over the bay. At 1,000 ft (300 m) he gently eased forward on the control column to increase speed to 165 knots and continued striving for a safe height. Suddenly the cabin crew in-flight director appeared on the flight deck and informed the captain of the serious damage that had been sustained. Long pieces of metal were sticking through the cabin and two passengers had been hurt. Captain Dyer sent back the second officer and number two flight engineer to assess the damage and to report on the condition of the injured cabin occupants. Dyer maintained control of the stricken aircraft and continued to check out the effectiveness of the flying controls. Flight Engineer Horne returned to the check list to confirm which services would be available and which secondary systems would have to be employed.

The immediate problem was the lack of hydraulic power from numbers one and four systems to raise the trailing edge flaps. The leading edge flaps operated using pneumatic power and would retract and extend as scheduled with movement of the trailing edge flaps. The normally hydraulically powered trailing edge flaps, however, would have to be raised using the secondary electrical system. It would take time to bring in the flaps in this manner, but it was essential to retract them to accelerate to a safe flying speed. Dyer would also be unable to raise the landing gear, but since, under the circumstances, it was sensible to leave it extended that difficulty was eliminated. Normal and secondary braking systems were also lost but that was not a problem for the moment. Fortunately the reserve brakes were still available from number two system and the source could be opened up just before landing. Flight Engineer Horne quickly checked through the major systems that had been lost and read them out to his captain.

'Okay,' replied Dyer. 'Let's have the alternate trailing edge flap check list.'

Horne opened the book at the required page and stood by for the drill. At 15:32 Captain Dyer decreased the ascent and began accelerating. The flight engineer called out the appropriate drill while First Officer Oakes selected the trailing edge flap alternate switch to arm. He then placed the alternate trailing edge flap switches to up and, fortunately, the flaps began to run in. After a few minutes the drill was completed.

The Boeing 747 was now flying with clean wings but with the damaged landing gear still extended. Captain Dyer levelled off and flew the aircraft between 2,500 and 3,000 ft (760-910m) at about 280 knots while the crew continued to assess the situation. With the aircraft safely airborne the next concern of the crew was to get it back down again without incident. The most sensible course of action was to land back at San Francisco as soon as possible and plans were made accordingly.

Meanwhile, in the cabin, passengers were moved forward

from the damage area and an announcement was made calling for medical assistance. Two doctors came forward and, with the aid of one of the stewardesses who was a nurse, administered to the injured. There was little they could do except stem the bleeding and make the passengers comfortable.

N747PA was much too heavy for an immediate landing and arrangements were made to dump fuel to reduce the weight to a structurally safe level. The 747 still weighed about 320 tons and the weight would have to be decreased to below the maximum permitted landing weight of 265 tons before commencing an approach. Oakes contacted radar control with the request to dump fuel and Flight 845 was vectored on to a westerly heading to steer the aircraft away from the built-up area and out over the Pacific Ocean. Captain Dyer discussed the fuel situation with his flight engineer and decided to discharge eighty tons of fuel. That would bring the aircraft down to a manageable 240 tons and would still leave fifty tons of fuel for the return and landing back at San Francisco.

As the Boeing 747 flew westbound over the Pacific, the two extra flight crew who had been sent aft to check the damage, returned with their report. It appeared from the metal lengths in the cabin that the aircraft had struck the 19L approach light system and there was little doubt that the landing gear was badly damaged in the process. The injured passengers were being attended to by qualified personnel and they were being made as comfortable as possible.

'I think we better get an observer in another aircraft to check out the landing gear,' suggested Dyer.

Oakes called control with the request and immediately the Coast Guard were alerted and Pan Am engineers summoned. The controller also informed the airport authorities who confirmed that the lighting pier at runway 19L was damaged and that a landing gear bogie was lying in the water. Within a short time a US Coast Guard aircraft took off from its base with airline personnel on board and flew in pursuit of the Boeing 747. Radar control vectored the Coast Guard flight on to the tail of *N747PA* and it flew below the big jet to allow the engineers to inspect the damage.

'The right body gear is missing,' radioed the pilot from the inspection aircraft, 'and the left body gear is hanging down with two wheels missing.'

Some minor damage to the flaps, tailplane and elevator was also reported, but the engineers were able to inform the 747 crew that the nose gear and wing landing gears appeared intact and looked able safely to sustain a landing.

As the Coast Guard aircraft flew off and returned to base, Flight Engineer Horne completed the drill in preparation for fuel dumping. An altitude of 6,000 ft (1,830 m) was the normal minimum height for jettisoning fuel to allow the kerosene to disperse, but dumping at 3,000 ft (915 m) over the ocean would cause little problem if particles rained on the sea. On each wingtip a large nozzle protruded rearwards at the trailing edge from which fuel could be discharged, and powerful pumps could push the fuel overboard at an average of 2.3 tons per minute.

'We're now ready for dumping,' radioed Oakes, 'and we'll need just over thirty minutes to lose the fuel.'

With permission received, Horne opened the appropriate valves and switched on the pumps in sequence. Fuel poured from the wingtips in two powerful streams.

As the kerosene discharged from the aircraft, Dyer spoke to the passengers on the PA system. He explained the structural damage which had been caused and assured the cabin occupants that the aircraft was flying safely and that a return to San Francisco would be commenced as soon as possible. They were dumping fuel to bring the weight down for landing and that would take about thirty minutes. Once the drill was completed they would be ready to commence an approach and they would be back on the ground in about one hour. Dyer also instructed his cabin crew director to prepare the cabin for an emergency landing. The passengers were briefed on crash landing drills and also on ditching procedures in case the aircraft landed in the bay. The use of life jackets was demonstrated and they were donned by all on board. Shoes, pens, combs and other sharp objects were removed, blankets and pillows were issued for protection, and the 'brace' position was demonstrated. The passengers were informed that the aircraft would be evacuated via the slides and groups of cabin occupants were assigned specific exits. With the landing gear so seriously damaged there was a likehood of it collapsing on touch-down and it would be as well to get people out of the aircraft as quickly as possible. There would also be a serious risk of fire, even if the gear held on landing. Amongst the passengers was a positioning crew, 'dead-heading' to Tokyo, and they were commandeered into service to help the operating crew with the preparations.

Meanwhile, on the flight deck, Dyer and his crew rechecked the control availability on the number two hydraulic system and referred to the flying manual for landing with the body gear up.

The fuel dumping took longer than expected and about forty-five minutes was required to reduce the aircraft weight to 240 tons. At the appropriate moment, Flight Engineer Horne switched off the pumps and shut the discharge nozzles to stop the jettisoning while Oakes radioed control that dumping was completed. Flight 845 was radar vectored for landing in a westerly direction and in the clear, daylight conditions Dyer would be able to make a safe visual approach over the bay. The approach and landing checks were completed in good time and as the aircraft slowed the flaps were successfully lowered to the 20° setting using the alternate electrical system.

Approaching the airport from the east at about six miles (9.5 kilometres) and level at 1,900 ft (580 m), the previously closed runway 28L, being the longest, was opened up to receive the stricken aircraft. At 17:07, just over one-and-a-half hours after take-off, N747PA established on the 28L instrument landing system and Dyer stabilised the speed at 145 knots. The flaps were selected to the 30° landing position and the crew were relieved to see them properly set. If the flaps had jammed at one of the earlier stages the touch-down speed would, of necessity, have been fast and the damaged landing gear would not have helped the situation. The threshold reference landing speed was 123 knots, but with the strong twenty-knot headwind the captain would increase this speed by ten knots to touch down at 133 knots. The aircraft seemed to handle satisfactorily, but the crew were greatly concerned about the damaged landing gear. They were more than aware that the gear might not support the aircraft on the ground and if it collapsed under the strain on landing the machine would career along the ground on its belly. The chance of fuel spillage and a major fire developing would be high and the occupants would be in great peril.

Flight 845 was cleared to land and approaching 200 ft (sixty m) Oakes called 'brace, brace' over the PA. The passengers and cabin crew took up their crash landing postures. At the same moment Dyer stretched forward with his right hand and opened up the reserve brakes. Slowly he eased back on the power to reduce to 133 knots, but at the lower speed the effectiveness of the single elevator section was reduced. Also, unknown to the crew, the stabiliser leading edge forward of the inboard elevator was badly damaged, exposing the tailplane spar, and further affected the operation of the remaining elevator. He pulled back on the control column to reduce the rate of descent but the response was extremely sluggish. The captain increased the power for more speed to arrest the sink rate, but his efforts were in vain. The aircraft's momentum carried it downwards and at 17:11, one hour and forty-two minutes after departure, the 747 contacted the right side of the runway with great force about 2,000 ft (610 m) forward of the 28L threshold. N747PA bounced back into the air. Moments later it landed heavily again a short distance from the first point of contact. By great fortune the landing gear held and the aircraft rolled on down the runway.

Captain Dyer immediately pulled the thrust levers to the reverse thrust selection but only number four engine indicated it had gone into reverse condition. Since asymmetric reverse thrust is not permitted the captain did not apply reverse power, which was just as well. As he pushed on the toe brake pedals the brakes locked on. The anti-skid system was defective and six of the eight tyres remaining on the wing landing gears blew out. Flat spots were milled on the wheel rims as the bogies scraped along the ground and a fire erupted on the left wing gear. The aircraft started to veer to the right and had reverse thrust been applied on number four engine only, the 747 would have swung violently.

Dyer tried desperately to hold the aircraft straight using rudder but once again his efforts were thwarted. *N747PA* ran off the right side of the paved surface just before cross taxyway 'L', about 4,000 ft (1,220 m) from the threshold, and ploughed on through the grass. The 747 then ran across OIR, the runway from which it had departed, and came to a halt, 5,300 ft (1,615 m) from the approach end of 28L, on a grass square in the middle of the airport at the intersection of all four runways. Fortunately the fire at the left wing landing gear was extinguished by the dirt and dust as the shredded tyres dug through the earth.

On stopping, Dyer gave the order to evacuate and as Horne shut down the engines Oakes made the call for the passengers to depart. Present day aircraft have emergency evacuation alarms but such systems were not available on earlier models. Immediately the two extra flight crew members left the flight deck and ran down the stairs to help. They arrived in the cabin before the evacuation had commenced and each opened up a door at the front. On opening the doors, the escape slides, positioned in packs on the inside of the doors, automatically deployed. In moments the chutes were inflated but the strong breeze blowing could be felt by the open doors and soon the slides were being caught by the gusts. The aircraft had stopped at an angle to the runway and the left side of the 747 was being exposed to the twenty-knot wind.

As door one left slide deployed the wind lifted the chute and twisted it backwards. It was unavailable for use. The same fate befell the slide at door two left. It flew backwards and flapped across the wing parallel to the fuselage. Evacuation from doors three and four left was not possible owing to further problems. Door three left is an over-wing exit and the escape route is deployed in two portions; one over the wing and another slide section over the trailing edge to the ground. The slide portion failed to deploy because the gas bottle trigger mechanism for inflation situated in the left body gear wheel was damaged on take-off. At door four left the entire slide pack had been dislodged from the door on impact with the lighting pier and was completely useless. At that stage door five left had not been

The Pan Am 747 shortly after landing. The wing gear only is down with the tyres on the left wing gear blown. Only number four engine is in reverse. (Captain Wayne Sager)

opened and in the first forty seconds after stopping no slides on the left, exposed side of the aircraft were usable.

At forty-three seconds after the aircraft came to a halt the first passenger slid down door one right escape route. Door two right chute inflated satisfactorily but the wind blowing below the 747's fuselage lifted it horizontal. One passenger was persuaded to venture on to the slide and his weight brought it to the ground. That exit was then used successfully for escape. Someone on the ground ran round to the door one left escape slide to pull it straight and at fifty-six seconds after stopping passengers began using this exit.

At door three right over-wing exit the escape route deployed satisfactorily, but the door four right slide failed to inflate. Its gas bottle had shifted on impact and the trigger mechanism had misaligned. Door five right was also late in being opened and as yet neither of the two aft slides had been deployed. As a result, only the left most forward door and the three forward doors on the right were being used as escape slides. Since most passengers were seated in the economy section, only a few people departed using these doors. At about one minute after the aircraft stopped, both doors five left and right were opened and normal slide inflation was achieved. Passengers began to evacuate by these doors and cabin occupants near the failed exits in the centre of the aircraft started moving rearwards towards the usable escape slides.

In a matter of seconds the unexpected happened. Devoid of any landing gear support at the aft section of the aircraft, the 747 slowly began to tilt rearwards. No-one knew at the time, not even Boeing, that with body gear support lacking the aircraft could tip backwards. The movement was so slow that few people realised what was happening but eventually it settled on its tail. The nosewheel lifted clear of the ground and N747PA's nose stuck in the air. As the tail dropped, door five left's slide jack-knifed and was jammed below the fuselage. The two most forward chutes slanted almost vertically and a few passengers received back injuries while using the slides in this condition before the exits could be blocked. Escape from the four exits forward of the wings now became impossible. As the front slides became unusable, all five escape routes on the left side were once again rendered inaccessible, although the sill of door five left was only five feet from the ground and a few people simply jumped on to the grass.

On the right side only two doors, the over-wing and aft exits were available for evacuation. The door five right slide

The arrival of the rescue services. Note the door two left slide blown backwards over the wing. (Captain Wayne Sager)

After the event. Note the steepness of the slides at the forward doors and the damage to the wind blown upper deck slide. (Captain Wayne Sager)

was lying at a very shallow angle and people were able to run straight along it on to the ground. In spite of the major problems being encountered, however, the cabin crew accomplished their tasks well and the evacuation of the remaining passengers was completed in good time using the only two doors available. In all, twenty-seven people were hurt departing the aircraft, but fortunately no further outbreaks of fire occurred and all on board escaped to safety.

The incident began as a result of a sequence of errors and omissions by a number of people, including the flight crew, leading to impact with the lighting pier, but with skill a potential disaster had been averted. The accident had been relegated to a minor incident and only two people had been hurt on impact. Captain Dyer and his colleagues may have been partly responsible for their predicament but their misdemeanour should not detract from the professional way in which they and the entire crew handled the emergency.

Firemen stand by. Note the damage to the right leading edge stabiliser just forward of the right inboard elevator, the only elevator control remaining.
(Captain Wayne Sager)

As always, however, they were lucky too. A piece of metal debris had passed only four in. (10 cm) from number two hydraulic system and had it severed the line all flying controls would have been lost. And had the wing folded when the right rear spar separated on initial impact the result would have been inevitable.

Chapter 4
The Windsor Incident

'Shall we give it a try?' asked Captain McCormick.

'OK, I'll switch off the hydraulics.'

The instructor turned to the flight engineer's panel and selected the hydraulic pumps to off.

'You're on your own now.'

With hydraulic pressure reduced to zero the flying controls became ineffective and McCormick was left with engine power only to guide the machine. Both men watched the performance with interest and were pleased with the result. The DC-10 'flew' quite well and could be controlled with reasonable success on engine power alone. A bit of practice was all that was required.

Captain Bryce McCormick was a very experienced pilot with 24,000 flying hours to his credit. In the spring of 1972 he was fifty-two years of age and had been with American Airlines since before the end of the war. In his career he had flown a number of aircraft: Convair 240, DC-3, DC-4, DC-6, DC-7 and Boeing 707. In early 1972 McCormick had converted to the DC-10 and by the end of March had completed his ground school, simulator and flying training. In April he recommended line flying duties on his new aircraft type.

McCormick liked the DC-10 and thought it a fine machine, but his admiration was not entirely without misgivings. The DC-10, like the Lockheed Tristar and Boeing 747, was one of the new generation of 'wide bodied' aircraft which were introduced into service in the early 1970s. These aircraft were not only bigger with more powerful engines and larger passenger capacity, but also incorporated new design concepts. Conservative pilots like McCormick, steeped in traditional values, felt uneasy about accepting some of the

Captain Bryce McCormick.

more radical changes. All jet transport aircraft, for example, have hydraulically operated flying controls, but the three 'wide bodies' had no manual back-up facilities to operate the controls in the event of complete hydraulic failure, unlike the earlier generation Boeing 707 which McCormick had flown.

Back-up systems normally consisted of cables running from the control column which were connected directly to the elevators, ailerons and rudders. With total hydraulic power loss the pilot could, albeit with some difficulty and a lot of muscle, successfully retain control of the aircraft. The 'wide bodies' also employed control cables, but these only fed demands to hydraulic power control units which in turn deflected the control surfaces. Any pilot of one of the newly introduced 'big jets', faced with the unlikely event of total hydraulic power loss, would have no flying controls available to fly the aircraft. What would happen in these circumstances? Could the aircraft be turned, climbed and descended by simply varying the power of the appropriate engines? It was a question that McCormick had pondered on a number of occasions during his DC-10 conversion course and one that he was determined to find out. He now sat in an American Airlines DC-10 flight simulator at their training school at Fort Worth, Texas, and, with the aid of an instructor friend, was experimenting in guiding the aircraft on engines alone.

'It handles pretty well,' commented McCormick.

Both men were pleased with the outcome but not entirely surprised by the success of the exercise. The DC-10, in particular amongst the three 'big jets', has its engines exceptionally well placed for steering by engine power only: one on each side slung in pods below the wings about one third of the way out from the wing roots with the lines of thrust sufficiently far apart for directional control and also directed below the fuselage, and one placed high on the lower half of the tail fin with its line of thrust directed above the fuselage. The Tristar is similar, but the DC-10's highly placed tail engine makes pitch control easier. By using asymmetric power, i.e. by increasing power on one wing engine and decreasing the other, the aircraft could be turned, not unlike the differential steering of a tank by varying the speeds of the tracks. Likewise, the aircraft could be climbed by pitching the nose up using increased power from the low slung wing engines and decreased power of the highly placed tail engine, and vice versa for descent. McCormick spent thirty to forty-five minutes practising the technique and became quite adept at 'flying' the DC-10 simulator in such a manner. By the end of the exercise he could control the simulator from initial climb to approach phase using only the thrust levers without touching the flying control column. McCormick was satisfied with the outcome of the experiment and his success in handling the stricken machine helped allay some of his concern. He was not to realise at that time, however, that in only a few short months his expertise would be put to the test in earnest.

On the morning of 12 June 1972, American Airlines Flight 96 took off from Los Angeles, California, on a scheduled service to New York's La Guardia Airport, routeing via

Detroit Metro, Michigan, and Buffalo in the western tip of New York State. Flight 96, a DC-10-10, registration *N103AA*, was commanded by Captain McCormick, with First Officer R. Paige Whitney as his co-pilot, and Flight Engineer Clayton Burke. In the cabin eight stewardesses were led by Chief Flight Attendant Cydya Smith. The DC-10 departed forty-six minutes late from the American West Coast because of passenger handling difficulties and air traffic control delays. After an uneventful four-hour flight, *N103AA* arrived in Detroit at 18:36 Eastern Standard Time. Of the few passengers aboard, thirty-eight disembarked, while sixteen remained, to be joined by forty others for the next leg. Captain McCormick hoped for a good turn-round time to help make up some of the delay. With only fifty-six passengers travelling on the next sector and little fuel required for the short forty-five minute hop, there was every likelihood of a swift departure.

Soon ground preparations were completed and the passenger doors closed. The two main cargo doors on the right side, situated in the mid-and-forward positions, were checked closed, but the smaller aft bulk cargo hold door on the left side was causing difficulty. It was not the first time that

An American Airlines DC-10. (John Stroud)

problems had occurred in the locking of this door and all four airlines operating the DC-10 at that time (American, Continental, National and United) had experienced trouble. The door was shut by an electric motor and actuator and was then secured in position by latches being wound on to spools (see diagram). The door handle operated a locking mechanism and when placed flush with the door slid locking pins into place to prevent movement of the latches. Closing of the handle also shut a small vent door and extinguished the 'door open' light on the flight deck. In such a condition the door was safe and would not burst open in flight under the force of the air in the pressurised cabin.

The problem being encountered with the aft cargo door was that the actuator being powered by the electric motor was not driving the latches fully home and it was necessary on a number of occasions to use a hand crank to complete closure. McDonnell-Douglas were already aware of the circumstances and had issued service bulletins recommending rewiring of the electric motors with heavier gauge wire. Captain McCormick's aircraft, unfortunately, had not yet

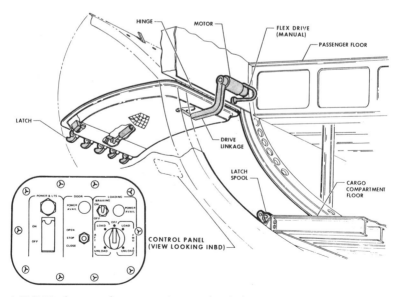

A DC-10 aft cargo door operating mechanism. (NTSB)

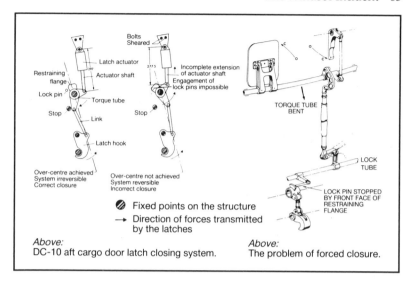

Above:
DC-10 aft cargo door latch closing system.

Above:
The problem of forced closure.

The aft cargo door latch closing system — the problem of forced closure. (NTSB)

been modified. On the transit in Detroit the ramp service agent closed the door electrically and listened for the motors to stop running. Cut-out of the electric motor should have indicated the correct positioning of the latching mechanism, but once again problems occurred. The latches were not driven fully home and the latch linkages were not moved to the over-centre position (see diagram). Flanges obstructed movement of the locking pins and prevented the door handle from being placed flush with the surface. Using a certain amount of pressure the ramp agent could feel movement of the handle and simply assumed that the mechanism was stiff. Eventually he managed to force the handle into position using his knee and with the lever properly stowed assumed the locking procedure to be completed. The latches, however, were still improperly seated and the door remained unlocked. With the locking pins jammed against the restraining flanges the forced closure of the door handle had buckled the locking rods. The linkage had moved sufficiently, however, to close the vent door, although in a slightly cocked position, and to extinguish the 'door open' light on the flight deck. Not surprisingly the ramp agent was unhappy with the circumstances, and he called over a mechanic to examine the vent door which could be viewed through the vent aperture. The mechanic could see the vent door slightly out of position but all the indications appeared normal. The state of the door, however, was anything but normal and was in a most unsafe position. In the cockpit the flight engineer checked the 'door open' light extinguished and could only assume that the door was closed and locked. Captain McCormick and his crew were unaware of any danger.

On Flight 96's departure from Detroit, compressed air would be tapped from the engine compressors to pressurise the cabin for relatively normal breathing for those on board as the aircraft climbed into the thinner air. The aircraft hull and doors were designed to contain the pressurised air in the cabin from bursting outwards into the rarefied atmosphere. To help ease the exertion on the structure, however, cabin pressure is reduced as the aircraft climbs (for example, to an equivalent cabin altitude of 6,000 ft (1,830 m) when cruising at 35,000 ft (10,670 m), but the outwards force is still quite considerable. Would the partially locked aft cargo hold door be able to withstand the strain of several tons of pressure as the cabin pressurised after take-off? If not, and the latches released under the pressure, the door would blow off. The rapid venting of the pressurised air to the atmosphere would cause untold damage and could place the aircraft in great jeopardy.

Meanwhile, unaware of the insidious danger of the situation, the DC-10 crew continued with their departure preparations. As the mechanic inspected the aft cargo door, Whitney called on the radio for departure instructions. The time was just after 19:03.

'Metro clearance delivery from American 96, information "delta", and airways to Buffalo.'

'Delta' was the current weather details transmitted by the automatic terminal information service and indicated the cloud base to be 4,500 ft (1,370 m), the visibility reduced to 1-1½ miles (1.6-2.4 kilometres) in smoke and fog, a southwest wind at six knots, a temperature of 61°F and an altimeter pressure setting of 29.85 inches of mercury. The weather, although not good, would give no problems on departure. Clearance delivery replied instructing American 96 to turn right after take-off on to a heading of 060° for radar vectors to airway Jet 554. The flight was to maintain 4,000 ft (1,220 m) and could expect to be cleared to flight level 250 (25,000 ft, 7,620 m) ten minutes after departure. The departure frequency was given as 118.95 MHz and the squawk, or radar identification code as 0200. Whitney repeated the clearance as read, but the controller corrected himself and re-issued a departure frequency of 118.4 MHz.

In the cockpit the crew completed their pre-start checks. A few minutes' delay ensued as the aft cargo door inspection continued, but finally the door condition was accepted and the all-clear was received on the intercom from the ground engineer. The engine start sequence was commenced and Whitney contacted ground control on 121.9 MHz.

'Ground, American 96, push from gate ten.'

Clearance to push was given, and as the push-back truck revved its engines the DC-10 moved slowly backwards from the gate. The time was 19:11 Eastern Standard Time and the turn-round had been an excellent thirty-five minutes. McCormick was pleased to be making up some time. A few minutes later, with all engines running, Whitney requested permission to taxy and ground quickly replied.

'American 96, taxy to runway three right.'

On the taxy out the take-off performance on runway 03R was checked and the 'before take-off' check list commenced. Approaching the threshold Whitney called the tower on 121.1 MHz.

'American 96,' replied the tower controller, 'into position and hold.'

The DC-10 lined up on the runway and with the anti-skid selected the flight engineer pronounced the checks completed. First Office Whitney was to fly the aircraft on this sector and he stood-by for the take-off. The Captain took over operation of the radio.

'American 96, maintain runway heading and contact departure. Cleared for take-off.'

Whitney opened up the power to the required setting and

the lightly laden big jet rapidly accelerated down the runway. At 19:20 the wheels lifted from the surface and N103AA climbed easily into the air. McCormick established contact with the departure controller.

'American 96, Detroit departure, radar contact. Climb and maintain six thousand.'

'OK,' replied McCormick, 'you want us to climb and maintain six thousand, American 96.'

'Roger, American 96, now turn right heading zero six zero.'

Whitney increased the DC-10's speed and the flaps were selected from the take-off position to up.

'American 96, turn right heading zero nine zero.'

Flight 96 was being carefully radar vectored to pick up the easterly airway Jet 554 en route to Buffalo.

'American 96, turn right heading one one zero. Join Jet five five four when you intercept.'

The DC-10 now entered cloud, and as the aircraft climbed rapidly towards its cleared altitude of 6,000 ft (1,830 m) McCormick replied asking if they were to maintain this height.

'OK, I'll have something higher for ya in just a moment,' replied departure. 'Climb and maintain, ah, flight level two one zero.'

Clearance for further climb arrived just in time for the '1,000 to go' alert tone sounded in the cockpit.

'OK, ah, climb two one zero, American 96,' repeated McCormick. 'We are out of, ah, fifty five hundred now.'

'American 96, call Cleveland Centre now on frequency one two six point four. Good day.'

The US/Canadian border lies lengthwise along the centre of Lake Erie and most of Flight 96's journey would be spent in Canadian airspace. Since the DC-10 was operating between two American airports, however, US controllers continued to direct the flight. N103AA was already over Canadian territory and was now close to the town of Windsor in Ontario.

In the cockpit McCormick quickly selected the Carleton radio beacon on both navigation receivers and set the course indicators to 083°, the Jet 554 airway centre line, to permit interception of the airway. As the DC-10 settled on the heading of 110° Whitney engaged the autopilot, but out of habit kept his hands on the control column. The Cleveland frequency was dialled on the radio box and McCormick called the Centre.

'Good evening, Cleveland Centre, American 96 is out of, ah, seven thousand now for two one zero.'

'American 96, squawk code, ah, one one zero zero and ident. Maintain flight level two three zero, report reaching.'

The DC-10 continued towards Jet 554 and, still in cloud, climbed at the maximum permitted speed below 10,000 ft (3,050 m) of 250 knots to its newly assigned flight level. At 19:24, only four minutes after lift-off, the lightly laden aircraft passed through 10,000 ft. Whitney lowered the nose of the DC-10 using the autopilot vertical speed control to accelerate the machine to the normal climb speed of 340 knots. The aircraft's rate of climb reduced to the 1,000 ft (305 m) per minute selected and the speed began to increase. As the flight approached the top of the weather the cloud thinned and above could be glimpsed the first signs of the sun. A few moments later, passing 11,500 ft (3,500 m), the DC-10 broke from the cloud layer into a bright mid-summer's evening. 'What a beautiful day,' thought Whitney to himself.

N103AA passed over the town of Windsor, situated by the shore of Lake St Clair lying over two miles (3.2 kilometres) below, and continued towards Jet 554 as it headed east-south-east between the Great Lakes of Huron to the north and Erie to the south. At only four-and-a-half minutes after take-off the DC-10 climbed through 11,750 ft (3,580 m) and the increasing speed edged past 260 knots. Captain McCormick looked out into the clear sky and could see far above a giant Boeing 747 flying majestically through the air.

'There goes a big one up there.'

Whitney leaned forward to catch a better view of the other aircraft, still resting his hand gently on the control column. It was a pleasant moment in the trip with the weather and the busy departure behind and a clear flight ahead, but it was a calm which did not last long. Suddenly an enormous 'thud' was heard from the rear of the aircraft and in an instant their

The flight crew: Captain Bryce McCormick (left), First Officer R. Paige Whitney (centre) and Flight Engineer Clayton Burke (right).
(Captain Bryce McCormick)

peace was shattered. The air 'fogged' and a great rush of air swept past the flight crew. Dirt and dust flew in their faces and stung their eyes and skin as if a firecracker had gone off below their noses. For a moment they were blinded.

'Oh . . .!' someone shouted.

The rudder pedals exploded and smacked at great speed to the full left rudder position. The captain was resting his feet on the pedals and his right leg was thrown back against the seat with extreme force. His right knee was driven to his chest and his headset was knocked from his ears. First Officer Whitney, still leaning forward, was thrown violently rearwards and hit his head as he crashed against his seat. At the same moment the three thrust levers snapped to flight idle with the number two (tail) engine throttle hitting the stop with a loud crack. The aircraft was felt to 'jerk' momentarily and the autopilot disengaged with the red disconnect light flashing. McCormick feared a mid-air collision had occurred and suspected the windshield had shattered. When his eyes cleared he could see it was still in place but, disbelievingly, he stretched out his hand to touch the window.

'What the hell was it, I wonder?' he called.

One of the crew replied with a long whistle. Captain McCormick now noticed a red failure flag on his airspeed indicator and speculated that the radar dome might have blown off. With the aircraft nose cover missing erratic speed indications could be expected. Whitney still maintained his hands on the column but the DC-10 yawed and banked slowly to the right with the nose dropping sharply out of control.

'Let me have it,' yelled the captain as he grabbed the control wheel. Quickly he tried to 'feel out' the controls.

'I think it's going to fly,' reassured Whitney.

At three seconds after the initial explosion an engine fire warning sounded, together with the *beep, beep, beep* of the cabin altitude warning horn. The DC-10 appeared to have suffered an explosive decompression and the intermittent horn indicated that the cabin air pressure had reduced to an equivalent altitude in excess of 10,000 ft (3,050 m). McCormick had no doubt that he had a 'pretty sick airplane' on his hands and the first priority was to keep it in the air. 'Fly the airplane, fly the airplane,' he kept saying to himself. The engine and pressurisation problems could wait.

'We'll pass the fire warning,' he called.

Capain McCormick was an advocate of the theory that he who hesitates, survives. When faced with a sudden emergency he firmly believed that a pilot's first action should be to do nothing — think the problem out, then act. In normal circumstances, with an intermittent horn warning of reduced cabin pressure, the crew would commence an emergency descent, but the captain was reluctant to force the aircraft into a steep dive until the damage could be assessed. If he did drop down quickly he might not be able to recover. Flying at around 12,000 ft (3,660 m) few breathing problems would be experienced, although strictly speaking flight crew should don oxygen masks above 10,000 ft. The engine fire, too, was a serious problem, but a more pressing matter was the need to keep the aircraft flying. There would be little point in extinguishing a fire if the DC-10 tumbled out of control. To the crew their actions felt strange, like moving in slow motion.

'We've hit something,' said the flight engineer.

The co-pilot had been looking out at the time and had not seen anything of another aircraft so thought it more likely to be disintegration of number two engine. That would explain the fire warning and the rudder problem.

'We've lost . . . lost an engine here,' he said.

'Ah, which one is it?' asked the captain.

Flight Engineer Burke could see the problem from the engine instruments.

'Two. Number one is still good . . . and, ah, Captain . . . we'll have to . . . to check this out.'

McCormick could see the aircraft descending towards the cloud tops which lay below at 10,000 ft and he wanted to avoid penetrating the weather until he was sure he had control of the stricken machine. Steadily he heaved back on the control column but the response was very sluggish. The speed had dropped to 220 knots and simultaneously he pushed the three thrust levers forward. The numbers one and three (wing) engines responded but the number two (tail) engine remained in flight idle. The number two throttle could be moved backwards and forwards quite easily and it obviously wasn't attached to anything. Immediately the DC-10 responded to the power increase from the wing engines and the nose pitched up. McCormick only just managed to stay in the clear.

'OK,' said Whitney, 'we apparently . . . master warning . . . this board's got an engine fire over here. Yeah, we got two engines, one and three. Do we have, ah, hydraulics?'

'No,' replied McCormick, 'I've got full rudder here.'

The rudder was still jammed with the left pedal fully forward and the captain assumed the condition to be the result of hydraulic failure. The flight engineer checked out the systems, however, and found them satisfactory.

'Hydraulic pressure is OK,' he said.

In the cabin there was just as much confusion as there was in the cockpit, and some of the passengers were hysterical. Chief Flight Attendant Smith rushed forward to the flight deck to check with the captain.

'Is everything all right up here?'

'No,' called McCormick. The co-pilot turned and shock his head. 'You go back to the cabin.'

The DC-10's speed established at 250 knots and Captain McCormick managed, with difficulty, to hold the aircraft clear of cloud at 12,000 ft. While the other two continued to check out the systems he retrieved his headset from the back of the seat and quickly radioed Cleveland Centre declaring an emergency. There was little information he could give as they didn't know what had happened except that they had a serious problem.

'Ah, Centre, this is American Airlines Flight 96, we got an emergency.'

'American 96, roger,' replied Cleveland. 'Returning back to Metro?'

'Ah, negative, I want to get in to an airport that's in the open. Where's one open?'

Having just left Detroit Metro with a visibility of one to one-and-a-half miles in smoke and fog and having climbed through a 6,000 ft thick cloud layer, McCormick was reluctant to return there until he was sure of the integrity of the flight controls. It would be more sensible to avoid flying on instruments if there were control problems and to head somewhere clear of weather. He would make a decision once he had checked out the DC-10's controllability. The aircraft was continuing to fly, albeit with difficulty, but there was no doubt that the controls were severely impaired. The rudder pedals were solidly jammed, with the left pedal fully forward, but with the rudder itself trailing slightly to the right. The effect yawed the aircraft nose to the right and the aircraft flew in an askew condition. The elevators controlling climb and descent were extremely stiff to move and produced a very sluggish response.

McCormick tried to trim out the elevator control column forces by operating the electric stabiliser trim switches, but to no avail. Trimming consists of varying the angle of the tailplane to the airflow to stabilise an aircraft in flight, and as such the variable tailplane is known as the stabiliser. On McCormick's aircraft the stabiliser position indicator remained fixed and the stabiliser did not function. The captain then tried to operate the manual trim control but as he did so the handle came away in his hand. So much for that! There was little McCormick could do now to ease the elevator strain but he felt he could overcome some of the difficulty in pitch control by varying the power of the wing engines. The off-set rudder kept rolling the aircraft to the right, but, with the control wheel turned 45° to the left, it was possible to use the ailerons, which are normally employed to bank the aircraft for a turn, to hold the wings level. In this condition, turning of the aircraft, especially to the right, was a delicate operation and only slow 15° banked turns could be achieved.

In a few moments, by experimenting with the controls, McCormick quickly brought the stricken airliner under control. He sensed now that he could keep the aircraft in the air, but could he land it in such a condition. That was going to be a different problem. Flying the DC-10 on engines was one thing, but guiding a large jet in such a manner on to a 200 ft (sixty m) wide runway was a different matter. Unfortunately there was little choice but to try.

McCormick's thoughts now turned to his passengers whom he suspected would be in some distress. Quickly he picked up the public address (PA) handset to offer reassurance. He had little to tell them but he felt any words of encouragement would be better than none. Trying to hide his own apprehension he spoke as calmly as possible saying that they had a mechanical problem and that they would be returning to land.

'American 96,' called Cleveland Centre, 'start right turn. Heading'll be one seven zero, maintain ten thousand.'

As the stricken aircraft banked slowly to the right, the flight crew pondered the problems which lay ahead. As yet no one had any idea of what had happened. Back in the cabin Chief Flight Attendant Cydya Smith took stock of the situation and began assessing the damage. She knew that the captain would expect a full report as soon as possible. As she proceeded rearwards, comforting and assuring passengers, she was amazed at the destruction she could see towards the rear. At the time of the 'explosion' she had been standing in the service centre preparing coffee and drinks for the passengers. The door of the lift to the lower galley suddenly burst open, narrowly missing her head, and out billowed what appeared to be a 'smoky substance'. She was thrown off balance but managed to steady herself by grabbing a handrail. She felt a sensation of weightlessness which was followed by complete silence. Instantly she thought the aircraft had suffered a depressurisation and quickly checked to see if oxygen masks had dropped in the cabin. They remained in position and she had no difficulty in breathing. (At cabin altitudes above 10,000 ft [3,050 m] the warning horn on the flight deck sounds but it is only at cabin altitudes above 14,000 ft [4,270 m] that passenger oxygen masks drop automatically.) It was then that she rushed forward to the flight deck to check on the situation and on the way had been puzzled to see the captain's hat lying on the floor of the front cabin. In the cockpit she had seen from the expressionless look on the co-pilot's face as he spoke to her that something was seriously wrong so she had returned immediately to the cabin. She then spoke on the PA, instructing the travellers to stay in their seats, not to smoke and to remain calm.

Suddenly Stewardess Carol Stevens grabbed her arm and told her that one of the flight attendants at the back was trapped by the edge of a large hole and that she urgently needed help. Stevens had been sitting with her seat belt still fastened when she heard a 'varoom' type of noise and the cabin had fogged. Air had rushed from front to rear of the aircraft. When the mist cleared she had seen the devastation in the rear section of the cabin. A small lounge bar was situated in this area for the use of economy passengers, but since few people were on board and the sector was short it was considered not worth offering the facility. No one was seated in the area, which was just as well, for the floor had caved in and ceiling panels were hanging down. The bar unit had collapsed and had fallen into the hole on the left side of the aircraft. On the floor and halfway into the cavity lay Stewardess Bea Copeland, prostrate on her side but facing forwards with her head towards the centre. She was trapped by the wreckage and was shouting for help. Stevens had tried to call Cydya Smith to tell her of the girl's predicament but the cabin interphone failed to work. She was hesitant to leave her seat since she assumed the cabin had decompressed and she was expecting an emergency descent. Eventually she noticed some of the other attendants out of their seats so she had run forward to raise the alarm.

On take-off Bea Copeland had been positioned by the aft left exit and had remained in her seat with her seat belt fastened. There were few passengers on board and, with time to spare, she had sat chatting to Stewardess Sandra McConnell who sat by the aft right exit. The bar area was situated immediately forward of the girls' positions. Copeland had only just released her seat belt when suddenly she had been lifted from her seat and thrown to the floor. A ceiling panel had detached and had fallen on her head and debris had trapped her foot. From where she lay she could see down into the cargo hold. The large bar unit had crashed behind her, partially blocking the hole and preventing her from falling further. She had tried to shift the panel covering her head but was unable to move it so she begun calling for help. On the right side Stewardess McConnell had been thrown from her seat with the 'explosion' and had struck the divider position behind the bar. She had landed on the damaged floor and had felt the area around her begin to crumble. From where she lay she could see daylight through the side of the aircraft and she could feel herself slipping slowly into the hole.

Chief Flight Attendant Smith, with Carol Stevens and a male passenger who offered to help, rushed back to assist the trapped stewardesses. One of the other attendants, unaware of the rescue in progress, called the flight deck on the interphone to ask if someone could lend a hand. In the cockpit the flight crew had their own problems, of course, but they seemed to be overcoming the initial difficulties. As Cleveland Radar vectored N103AA southwards for a return to land, the crew quickly rearranged their duties. Until the moment of the incident the co-pilot had been flying the aircraft. The captain had taken control but had continued to operate the radio while the other two inspected the systems. They had checked out the number two engine fire warning and found it to be false. McCormick continued to fly the aircraft while Whitney now took over communications.

'We've got one seven zero heading, sir,' called the co-pilot to Cleveland Control, 'and, ah, maintaining twelve thousand.'

'96, roger, type of emergency?'

Whitney misheard and simply replied, 'Yeah, yes sir.'

'We have a control problem,' McCormick added. 'We have no rudder, got a full jam. We've had something happen, I don't know what it is.'

'American 96, understand . . . cleared to maintain, ah, niner thousand, altimeter two nine eight seven. Be, ah, radar vector back towards the ILS course runway three. You want the equipment standing by?'

Cleveland was suggesting the DC-10 return to Detroit Metro, landing in the same direction as take-off with radar guidance onto the instrument landing system (ILS) for either of the parallel 03 runways.

'OK, sir,' replied Whitney, 'ah, say again the heading and we'll let down slowly to nine thousand.'

'Heading'll be two zero zero . . . two zero zero for American 96. And do you wish the equipment to be standing by?'

'Affirmative.'

'Understand. Full equipment standing by at Metro.'

'Yes, sir.'

The interphone now chimed in the cockpit and Flight Engineer Burke answered. It was the call requesting help for the trapped stewardess. The floor had collapsed, explained the caller, and she was lying by the edge and in need of assistance. Clayton Burke discussed the situation with the captain and since the immediate flying problems seemed to be resolved the flight engineer left his seat to help. Aware of the need to be properly dressed in the cabin he searched for his hat which was missing from the clips on the flight deck door. He looked in the crew wardrobe but it was nowhere to be seen. Unknown to Burke the initial rush of air had blown the crews' hats from the back of the door as it flew open and they lay scattered on the cabin floor. Stewardess Carol McGhee had been sitting strapped in by the front entrance at the time of the incident and had seen the flight deck door burst open and the escape hatch from the downstairs galley shoot up and strike a passenger on the head. A dusty rush of air had gushed from the flight deck together with the flight crews' hats which had flown out at head height as if worn by ghosts bolting from the cockpit. Hence the captain's hat seen by Cydya Smith lying on the floor and the flight engineer's inability to find his own.

At the back of the aircraft Bea Copeland still struggled on her own to free herself from the wreckage. Several times she tried to move the panel covering her head and eventually she managed to push it away. Her shoe was stuck in the debris and she was unable to move it, but with some effort she managed to pull her foot out of the shoe and set herself free. She then began to climb out of the area and as she struggled over the bulkhead was met by Smith and the others. Her rescuers helped her onto the rear lounge seats and then grabbed her hands and pulled her into the economy section. They now looked for McConnell who had been sitting by the aft right exit but they could see no sign of her and feared she had fallen into the hold. Smith called out her name repeatedly, but to no avail.

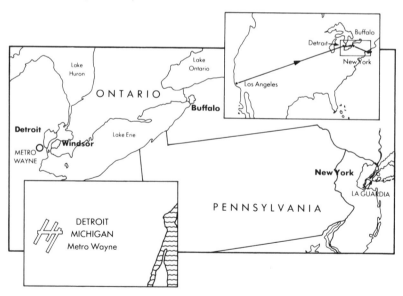

Map showing the relative locations of Detroit, Windsor, Buffalo and New York. Inset shows the DC10 flight path from Los Angeles.

'Sandy, Sandy, are you OK?'

Suddenly as if from nowhere, she appeared in view. She had managed to pull herself out of the hole and, somewhat dazed, had clambered into a rear toilet. The floor appeared very weak and she had felt that it would have collapsed if she had walked over it. Fortunately those on the scene were now able to help and with some difficulty manoeuvred her over the rear seats and into the cabin area. She was badly shaken, as was Copeland who had been trapped on the other side, but otherwise both girls were unhurt. It had only been a few seconds before that the call for help had been put through to the flight deck so one of the girls now rang to say that the two flight attendants had been rescued. Flight Engineer Burke had only just started to look for his hat when the good news was received, so he simply resumed his seat.

At the back of the aircraft Chief Flight Attendant Smith instructed the stewardesses to move away from the damaged area and to shift all the passengers nearby to forward sections. She then returned to the flight deck to report to the captain. Smith informed McCormick of the destruction at the rear of the cabin and that 'there was a hole in the fuselage in the very aft of the left-hand side'. The situation was worse than any of them had expected and they had done well to keep the aircraft flying. The reason for the DC-10's plight was still a mystery, for the problem with the locking of the left rear cargo door was unknown to the crew.

As the fuselage interior had pressurised on the climb out from Detroit the partially locked latches had been subjected to an increasing force. About five minutes after take-off, as Flight 96 passed through 11,700 ft (3,570 m), the pressure on the door was in excess of five tons. The door's electric actuator bolts had suddenly sheared under the stress and the door latches had sprung. The door had blown open, causing explosive decompression of the aircraft. The door had been torn off by the airstream, damaging the left tailplane in the process. The pressurised cargo air had immediately exhausted to the atmosphere via the gaping hole and the cabin air pressure had placed an undue load on the floor. With insufficient venting in the cabin floor area the floor had simply collapsed, tumbling the bar unit into the gap.

Unfortunately the structural devastation was not the only damage sustained. Through the beams of the cabin floor ran a number of vital control cables, hydraulic pipes, fuel lines and wiring to the empennage control surfaces and the number two engine in the tail fin. Many had either been severed or jammed, or had their operation severely curtailed. Ironically, McCormick's original fears of total hydraulic failure remained unfounded in this circumstance for the three hydraulic system lines, like the fuel line to number two engine, remained intact. They did not run through the collapsed central section of the floor and, with the lines coiled at various points to allow for stretching, had survived the impact of the decompression.

The flying controls at the tail, however, functioned via power control units which, like the number two engine controls, were operated by cables which ran from the flight deck through the central floor section along the entire length of the aircraft. As the pilots manipulated the control columns, or engine thrust levers and fuel controls, the taut and finely adjusted control cables moved in response. When the rear cabin floor collapsed the left rudder cable was broken, allowing the right rudder cable to slacken. The intact right cable was then forced downwards by the fallen floor which had deflected the rudder to the right, jamming it in that position. Three of the elevator control cables were severed and only the right outboard elevator panel remained functioning. The downwards load of the collapsed floor also resulted in the elevator being extremely difficult to operate. The tail engine thrust lever cable and the fuel shut-off valve cable were also broken. Severed wiring had resulted in the spurious number two engine fire warning.

McCormick informed his chief flight attendant that they expected to be landing in about eight to ten minutes and he instructed her to prepare for an emergency landing. Smith enquired if they'd be evacuating via the chutes but since the captain was uncertain of what would happen when they touched down he was unable to say. As the discussion ensued Cleveland Centre called again.

'American 96, you say you believe you hit something?'

'Ah, I don't know, sir,' replied Whitney. 'Just standby

one. Standby one until we assess the situation. We've got here, ah, definite problems.'

McCormick interrupted the exchange.

'OK, now we have got, ah, problems. I got a hole in the cabin, I think we've lost number two engine, we got a jammed full left rudder and we need to, ah, get down and make an approach. I guess Detroit Metro would be best and, ah, can you vector us around?'

McCormick felt the DC-10 to be flying satisfactorily, so Control's earlier suggestion of landing back at Detroit seemed a reasonable course of action.

'American 96, roger,' replied Cleveland. 'Turn further right now, heading'll be two zero zero.'

'OK, two zero zero, American 96.'

'Roger. And that the left, ah, rudder is jammed?'

'Er, it's possible,' continued McCormick. 'We don't know

The collapsed floor at the rear of the cabin. (Captain Bryce McCormick)

what the problem is. We've got a hole in the side of the airplane and we've got a full left rudder here. But we're under control and we're heading, ah, two three zero at the present time, letting down slowly to nine thousand.'

'Roger, two three zero, and continue around the, ah, turn to a heading of two seven five.

'Continue around to two seven five, sir.'

'Can you make a standard rate of descent?' asked Cleveland.

'Ah, negative, we gotta go a little slower.'

'Understand, slow descent. And what about turns?'

'Ah, turn we can give you, ah, is close to fifteen degrees maximum.'

'OK, that'll be fine.'

'I have no rudder control whatsoever, so our turns are gonna have to be very slow and cautious.'

'Understand.'

The crew discussed their predicament further and agreed that with judicious use of engine power the captain could more readily control the descent and turns. It was an amazing and fortunate coincidence that this incident had happened to a pilot who had previously practised such procedures. McCormick was thankful for his earlier experience in the simulator and was glad that he could now put the lessons learned then to good use.

'Thank goodness it's one and three we've got,' said Whitney to his captain.

'American 96,' continued Cleveland, 'descend to five thousand. Say altitude now.'

'Twelve to five,' replied Whitney.

'American 96, if you are able to, squawk code zero two zero zero on your transponder.'

'OK.'

'American 96, altitude now?'

'Ah, eleven thousand two hundred.'

'OK. About two hundred feet per minute, then?'

'Yes, sir.'

American 96, you wanna make a left turn now to a heading of, ah, two four zero.'

'Two four zero on the heading?' questioned Whitney.

'Yes sir. Two four zero, so it can give you, ah, a little more room for descent.'

'Yeah. Give me plenty room to start a long approach.'

As McCormick gingerly descended the stricken aircraft he was able to experiment with the use of engine power to aid control. The attempt seemed to be effective. Number two engine appeared to be inactive but they were uncertain if it had stopped at the time of the decompression or if it was still operating in idle. Whitney and Burke executed the appropriate drill and shut the engine down. What effect it had on the engine was unknown but it seemed to ease the heaviness of the elevators. In the cabin the flight attendants were equally busy with procedures as they prepared the passengers for an emergency landing. Chief Flight Attendant Smith gathered the stewardesses together on her return from reporting to the captain and gave them details of the

Detached ceiling panels expose the roof wiring. (Captain Bryce McCormick)

situation. She then spoke on the PA while the other flight attendants gave a demonstration of the brace position. She instructed the passengers to lean forward when told with their seat belts tightly fastened and to place their heads on cushions on their laps with their arms folded across their heads for protection. She then pointed out the six exit locations they would use, disregarding the two rearmost doors near the damaged area, and instructed them in the use of the emergency escape slides. The passengers were then requested to remove spectacles, pens, combs and other sharp objects which could cause injury and the attendants collected them in plastic bags. Shoes were also taken off to avoid ankle and foot injuries. With the cabin secure and the briefings delivered, Smith returned to the flight deck to inform the captain that her preparations were completed. She also asked McCormick if he would give the 'brace' command on the PA if required.

The return to Detroit was taking a little longer than anticipated and, with just over ten minutes to touchdown, the approach procedures were commenced.

'OK, give me about fifteen degrees on the flaps now,' called McCormick to his co-pilot. 'Watch it carefully.'

'We'll be landing about two ninety two thousand pounds,' added Burke.

It was possible to lower the weight further by dumping fuel overboard but since there was little in the tanks and the aft damage was unknown the thought was discarded. The apprehension on the flight deck could be felt and the air was very tense. At that moment one of the stewardesses popped her head through the cockpit door.

'Do you guys have a problem up here?' she asked.

In the strained atmosphere, with the captain struggling to maintain control, the question seemed quite ridiculous and the crew laughed.

'Yes, we have a problem,' they called.

She asked if the escape slides would be used but the captain said he didn't know and that if necessary he would activate the evacuation signal.

Captain McCormick spoke once more to the passengers on the PA as calmly as possible, assuring them that the aircraft was under control and that they were returning to Detroit. He apologised for the inconvenience and said that American Airlines would do all they could to provide transport to their destinations. The composed and routine tone of his voice had a comforting effect and helped ease the tension in the cabin.

'American 96, Cleveland, call me out of ten thousand.'

'We're out of eight seven for five thousand, right?' replied Whitney.

'American 96, roger. Turn back right now, heading'll be two eight zero.'

'Two eighty. I'm guessing that we're gonna have to have a long turn on the final for the ILS. I have no control on rudder and steering directions.'

'Roger, I'm planning about twenty miles. It'll be enough?'

'Ah, I hope so, thank you. I'll keep you advised.'

'American 96, now cleared to maintain three thousand.'

'Three thousand.'

'American 96, contact Metro Approach now, one two five one five. Good night.'

'Thank you. One two five one five.'

'We've got a nice rate of descent,' commented the co-pilot to McCormick. 'Even if we have to touch down this way we're doing well.'

'American 96, Detroit,' called Metro Approach.

'Loud and clear, sir. And we're through fifty five hundred for three thousand.'

'American 96, turn right heading three six zero, descend and maintain three thousand. Vector for the ILS for three left final approach course. Altimeter two nine eight five. Visibility one and one half. Braking, clear for all types of aircraft.'

'We'll probably have no brakes, ah, so we'll try reverse . . . we're heading north and we're outta forty eight hundred for three thousand.'

'American 96, turn right heading zero two zero. Your position is thirteen miles from the marker. You're cleared for ILS three left approach.'

'Zero two zero, and cleared for ILS three left approach, American 96.'

The left tailplane damage caused by the ejected cargo door.
(Captain Bryce McCormick)

The DC-10 joined the ILS localiser radio signal indicating the extended centre line of runway three left at about eighteen miles (twenty-nine kilometres) from touchdown, thirteen miles (twenty-one kilometres) from the marker beacon positioned at the five miles (eight kilometres) to go point. The aircraft speed was 150 knots with a rate of descent, or sink rate, of 6-700 ft (180-210 m) per minute.

'Well, gimme the gear,' called McCormick.

Whitney leaned forward and selected the landing gear lever to down.

'OK,' said McCormick, 'here we're coming into the ILS. I'm gonna start slowing her down. Give me twenty two on the flaps.'

The aircraft then intercepted the ILS glide path radio signal marking the descent profile at about ten miles (sixteen kilometres) from the threshold, and continued its approach to touchdown.

'All right, we got the green lights,' commented the captain, confirming the landing gear was down and locked.

'American 96,' radioed approach 'you're two and a half miles from the marker, contact tower on one two one point one. Good night.'

'Good night, sir.'

Whitney switched to the tower frequency as the DC-10 neared the marker beacon.

'American 96 is approaching the outer marker inbound.'

'American 96, roger,' replied the tower, 'continue your approach.'

The DC-10 was now only three minutes from touchdown and the atmosphere on the flight deck was very tense.

'American 96, the wind is one two zero at five, cleared to land.'

'American 96,' repeated the co-pilot, 'cleared to land.'

'Give me thirty five on the flaps,' called McCormick.

With full landing flaps set and the landing gear down the sink rate doubled to 1,500 ft (460 m) per minute, the captain had no choice but to increase power to maintain an acceptable descent of 800 ft (240 m) per minute. The speed rose to 165 knots, thirty-five knots faster than the required threshold speed for the aircraft's 130-ton weight. The touchdown would be fast, but if they didn't land too far down the runway the DC-10 should be able to stop in the 10,500 ft (3,200 m) length available. It was necessary to fly the entire approach with the nose yawed five to ten degrees to the right to keep in line with the runway, so what would happen when the wheels touched was anyone's guess. The captain tried to reduce speed a little by pulling back on the power but the sink rate rose dramatically. Whitney was calling out the rates of descent and was alarmed to see it momentarily increase to 1,800 ft (550 m) per minute. McCormick restored the power to regain the approach path and settled for the high landing speed.

'I have no rudder to straighten it out when it hits,' McCormick reminded the others.

The DC-10 crossed the runway threshold at 100 ft (thirty m) and the captain began to pull back slowly on the control column. The movement was so stiff he had to ask his co-pilot to give him a hand. McCormick squeezed on a little more power to lift the nose for a gentle touchdown. The aircraft was flying flat and fast and floated some way down the runway, but eventually the wheels contacted smoothly with the surface at 19:44, twenty-four minutes after take-off. The aircraft landed 1,900 ft (580 m) deep into the runway at a speed of 160 knots, with the nose pointing slightly to the right. McCormick was only just thinking to himself that it had been a good landing when 'all hell broke loose'.

Almost immediately Flight 96 ran towards the right side of the runway and for the first time since the incident the captain felt he had completely lost control. The DC-10 veered off the runway and ploughed through the grass. The spoilers on the wings deployed automatically to impair the lift and help the aircraft settle on the ground while the captain pulled on full reverse power on the wing engines. McCormick shouted at his co-pilot to take control of the thrust levers and the captain grasped the flying controls with both hands,

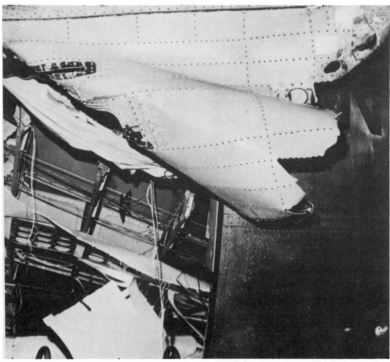

View from the left side of the fuselage showing the collapsed floor inside and the skin torn by the ejected door. (Captain Bryce McCormick)

holding the wheel left in an effort to keep the wings level. Whitney pulled the left wing engine throttle to maximum reverse power and cancelled the right wing engine reverse in an attempt to guide the machine. The asymmetric reverse power had the desired effect and countered the influence of the trailing right rudder. The swing to the right stopped and the DC-10 paralleled the right side of the runway.

The captain desperately tried to brake using the toe pedals on the rudder bar but with the bar askew to the extreme left it was very difficult to apply. The DC-10 raced on at speed between the runway and the parallel taxyway, digging up the turf in its path. Each time the aircraft met a cross taxyway McCormick tried to ease the strain on the nose wheel by pulling back on the control column, but the landing gear smashed into the hard surfaces with great force. The ground was extremely rough and with the bumpy ride the crew feared that the gear couldn't take much more punishment. If it collapsed now the flight could yet end in disaster.

Gradually the aircraft began to decelerate and the effect of the rudder displacement decreased. Under the influence of the asymmetric reverse thrust, aided by the full left deflection of the rudder bar which steered the nose wheel about ten degrees to the left, the DC-10 slowly eased to the left and back onto the paved surface. Eventually the aircraft came to a halt, severely shaken but still intact apart from the hole at the rear, about 1,700 ft (520 m) from the end of the runway. The left main landing gear and the nose wheel rested on the runway with the right gear still on the grass. McCormick, alarmed by the rough landing and fearful of ruptured fuel lines and the risk of fire, ordered an emergency evacuation. He activated the evacuation signal while the other flight crew carried out the appropriate drills. Quickly Whitney and Burke completed the check list.

'OK,' said the co-pilot, 'now the engines at your discretion.'

'SHUT 'EM DOWN!' called back the captain.

In the cabin the flight attendants rushed to their tasks and using the chutes at the forward six exits evacuated the aircraft in thirty seconds. The last of the travellers were off before the crew had finished their duties in the cockpit and the captain could see his passengers on the left looking up at

him from the far side of the runway. Around the aircraft the
fire trucks were standing by. At the last moment Chief Flight
Attendant Smith called into the flight deck.

'All passengers are off, captain. Goodbye. See you later.'

McCormick quickly left the cockpit and marched down the
aisle of the aircraft to check that all the passengers had
departed. As he passed the first door a fireman yelled at him
to jump but he called back saying he still had things to do.
When he reached door three left he stepped on to the wing
to check for fire but could see that the aircraft was safe. He
thought that number two engine would be damaged but
when he looked up he could see it was intact. Then he
noticed the aft left cargo hold door was missing and the
damage it had caused the tailplane. He stepped back into the
cabin and, joined by his co-pilot and flight engineer walking
down the opposite aisle, the three made their way to the rear
of the aircraft. Only now could the flight crew realise the full
extent of what had happened. They could see the amount of
damage the aircraft had suffered and now appreciated how
well they had done to effect a safe landing. McCormick's
totally professional approach to his job had won the day.
They had been lucky, too, of course, but the entire crew had
done a magnificent job. All passengers had escaped unhurt,
except for a few minor injuries sustained during the
evacuation.

With the immediate danger over the three flight crew took
their time in leaving, but when they returned to the cockpit
to collect their belongings they were unable to find their
hats. Unknown to them, one of the stewardesses had found
the hats lying on the cabin floor and had thrown them in
to a forward toilet. Eventually the hats were discovered
and, with uniforms complete, Captain McCormick and his
colleagues disembarked.

Captain McCormick and the crew received distinguished
service awards from American Airlines and the captain also
received a card for he and his wife to travel first class with
the airline with reserved seats. Only the president of the
company can displace them. McCormick's fellow pilots also
voted him president of the Grey Eagles, an organisation of
senior pilots.

Chapter 5
Don't be Fuelish

Weather, followed by fuel, are the two most important factors of any flight, the forecast of the former often deciding the quantity of the latter. In the 'good old days' it was simply a matter of filling the tanks, a procedure which became somewhat modified with the introduction of jets. Even then it was standard practice to carry more than requirements, adding extra for 'mum and the kids' and a 'bit for gran' as well. Fuel was plentiful and cheap and no-one thought much about it. The first oil crisis which began in the 1970s ended that attitude, with fuel prices trebling in three years. Costs to industry and the private individual soared. The United States was hit especially hard, the outcome of the crisis being smaller, more fuel-efficient cars and lower speed limits on the highways. An advertising campaign was introduced across America encouraging people to save fuel, with billboard signs saying, 'Don't be fuelish', being pasted throughout the nation.

Governments of the world encouraged their public to be fuel-conscious and industry searched for any means of saving fuel. Airlines found their fuel bills soared to thirty per cent of operating costs and companies were forced to examine their fuel policies closely. The world's airlines had no alternative source of energy to which they could turn and the only option open to them was to carry less fuel. The minimum legal fuel load requirements for a journey are: sufficient fuel for the flight from A to B, extra fuel for a diversion to C (in case of problems such as bad weather at B), some reserve fuel to cover contingencies, and a little for taxying. A typical London-New York flight for a Boeing 747 would require: eighty tons of fuel for the journey from Heathrow to Kennedy, thirteen tons in case of diversion, say to Boston, four tons of reserve fuel for contingencies, and one ton for taxying. The total fuel requirement would be ninety-eight tons but a flight would normally use only eighty-one tons — trip plus taxying fuel — and would land with seventeen tons still in the tanks.

Any fuel carried over and above the minimum fuel load is referred to as excess fuel. The B747 is a very long-range aircraft and the maximum fuel load which can be carried in the tanks is about 140 tons; a weight equal to a fully laden Boeing 707. The equivalent volume is 39,000 gallons (177,290 litres). On a typical transatlantic journey, therefore, the average flight has forty-two tons spare fuel capacity. The greater the aircraft weight, however, the higher the fuel consumption and the problem for airline accountants is that any excess fuel carried above the minimum requirements has a significant portion (i.e. three per cent per hour) of that quantity used up just to carry the excess. If ten tons of excess fuel are carried on a ten-hour flight, for example, then three tons of that fuel will actually be used carrying itself. Only seven tens will be available for use over the destination. In fact, it has been calculated that to carry regularly the weight

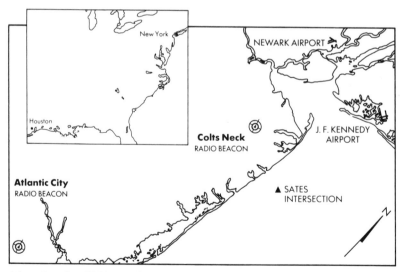

Map showing JF Kennedy airport, the Atlantic City and Colts Neck radio beacons, and the Sates intersection.

of just one small sachet of sugar increases fuel consumption by one gallon per year.

The problem for airline captains, on the other hand, is that one of the most useless things in aviation is fuel left in the refuelling bowser. Fuel on the ground may be expensive, but fuel in the air is priceless. Although, once airborne, aircraft are able to dump fuel if required — on a 747 at the rate of two tons a minute from nozzles in the wingtips — mid-air refuelling of civilian aircraft is not possible and it is essential that sufficient fuel is carried. Captains may be under pressure to carry the lowest amount of fuel commensurate with safety and the law but, with unpredictable bad weather in certain areas and frequent unexpected delays due to traffic increases in others, the dividing line between carrying sufficient fuel and running out in flight can be very thin indeed. Not surprisingly, crews still tend to err on the side of safety and, of course, if poor weather or landing delays are likely the captain has no choice but to carry excess fuel. On occasions, however, even the most conscientious of commanders can be caught out.

On 20 October 1979, a Pan Am 747 approached New York's John F. Kennedy Airport with the weather reported as poor. Low cloud and bad visibility were causing landing difficulties. The flight had departed about three hours earlier from Houston Intercontinental Airport, Texas, for the 1,200 nm journey, and now faced landing delays at its destination because of worsening conditions. The 747 had pushed back from the gate at Houston with 42.3 tons of fuel in the tanks but owing to congestion at the airport had used 1.9 tons of fuel in taxying. The flight had eventually taken-off at the weight of 225.7 tons, with 40.4 tons of fuel remaining: 28.9 tons for the trip, 8.8 tons in case of diversion and 2.7 tons in reserve. The 747 had passed over the VOR radio beacon at Gordonsville, Virginia, cruising at 37,000 ft (11,280 m) and had been cleared to proceed direct to Waterloo VOR, situated in Delaware on Delaware Bay, 128 nm south-west of Kennedy. The captain was now further cleared to descend to 21,000 ft (6,400 m) and was instructed to join the stack at Waterloo with turns to the left. Because of the poor visibility causing landing delays at New York he would have to hold at that point before proceeding. The stacks or holds are flown in a precise racetrack pattern, normally over a radio beacon, with aircraft being stacked one above the other at 1,000 ft (300 m) intervals. Normally the holding procedures are conducted near airports at lower levels, so the instruction to Pan Am to hold at such a height so far out clearly indicated the backlog of flights. It looked as though it was going to be a long delay.

The captain spoke to the passengers on the public address system and told them the bad news. He had enough fuel to hold for about forty-five minutes before commencing an approach to New York, so there was no undue worry. In the meantime they would just have to go round and round until their turn came. On the descent to 21,000 ft, the crew established contact with Pan Am ground personnel on the company radio frequency and a series of messages passed between the two in an attempt to establish a landing time. It was also agreed that if Kennedy closed they would proceed to Newark, New Jersey, where the weather was known to be better. The 747 joined the holding pattern over the Waterloo beacon, still descending, and after one or two left circuits the Captain was instructed to change to right-hand turns. The Pan Am aircraft changed its turns to the right and levelled at 21,000 ft.

While established in the hold awaiting onward clearance, the crew began to work out the exact fuel required for a diversion to Newark. There was every intention of landing at New York, of course, but it was as well to be prepared. At the beginning of the hold the tanks still contained fifteen tons of fuel and using a 'flight conduct summary' table for fuel management, a diversion fuel figure of 6.5 tons was computed. The calculations were based on an estimated landing weight at Newark of 195 tons, and using a stated distance of 25 nm between the two airports. This distance, however, was as the crow flies and, since aircraft don't normally fly in straight lines, the use of the summary tables resulted in a somewhat pared diversion fuel figure. It would be unusual, of course, for a flight to be right down to this fuel figure before commencing a diversion, so there would normally be something in hand.

After holding for fourteen minutes, the Pan Am aircraft was cleared to Sates intersection, thirty nm south of New York, via a 'Kennedy Two' arrival route, which proceeded over the radio beacon at Atlantic City. There would be a further delay at Sates after which the flight would be radar vectored to runway 22L for landing. At Sates the 747 joined the other holding aircraft. As flights were cleared from the bottom of the stack to commence their approach, Pan Am was instructed to descend lower in 1,000 ft steps. Eventually, after thirty-three minutes of holding at Sates, the 747 departed the stack on a radar heading, level at 7,000 ft (2,130 m). 8.2 tons of fuel remained in the tanks and, with aircraft ahead landing safely, the captain was satisfied with their progress. The crew kept a close ear on the weather information which indicated the visibility along the runway, or runway visual range (RVR), to be varying from 550-850 m (1,800-2,790 ft) at the landing end, to as low as 200 m (650 ft) at the rollout end. With both autopilots engaged the 747 could land automatically with an RVR as low as 400 m (1,300 ft), so there was some room to spare.

In case a go-around was required with the low fuel state, however, the flight engineer suggested preparing the aircraft for the 'minimum fuel go-around' procedures. There was a danger, if a missed approach procedure was initiated, of the fuel pumps being uncovered in the tanks by the low fuel levels. An engine could be starved of fuel and could run down. All boost pump switches were turned on and all cross feed valves were opened. It was also recommended on the go-around not to accelerate rapidly and to reduce the nose-up angle to eight degrees.

The Pan Am flight was instructed to call Approach Control, and the first officer established contact.

'Descend to 3,000 feet and reduce to 180 knots,' radioed Kennedy Approach.

The captain commenced descent and called for 5° flap to be set as he slowed the aircraft. Shortly they would be turned on to the final approach course, but before the controller could issue a heading the weather took a sudden sharp turn for the worse. The landing runway visual range dropped to 600 m (1,970 ft) and then to 300 m (985 ft). The visibility was now below the aircraft's landing limit and there was no way they could land. Nor could anyone else, for that matter, and they could hear on the radio the aircraft ahead requesting a diversion to Newark.

'Remain in line,' called back Approach, 'and continue to fly the runway course for traffic sequencing.'

Kennedy Approach then advised all aircraft that flights on the approach to land would have to be vectored along the 22L runway centre line for spacing before proceeding. The Pan Am 747 flew down the runway direction at 3,000 ft (910 m) and, with the weather still below limits, the first officer informed the controller that they would also like to go to Newark. Only 6.7 tons of fuel remained according to the total fuel weight indicator, or fuel totaliser gauge, with a calculated minimum fuel figure requirement of 6.5 tons. It was going to be close.

'Turn on to a heading of 240°,' called back Approach, 'vectors for Colts Neck, and contact Departure.'

Colts Neck VOR radio beacon lay to the south west, about twenty-five miles (forty kilometres) in front of them, and was situated at about the same distance to the south of Newark. They appeared to be going the long way round. The first officer changed frequencies and called the next controller.

'Proceed direct to Colts Neck,' replied Departure Control, 'maintain 3,000 feet and increase to 250 knots. Leave Colts Neck heading 335° for radar vectors to runway 22L at Newark.'

Over the Colts Neck radio beacon the captain turned the aircraft to the right on to the assigned heading, but as he did so he was instructed to roll out on 320°. The aircraft proceeded on the north-westerly heading maintaining 3,000 feet and 250 knots. The 747 still had to route all the way round to the north of the airport before turning back onto a south-westerly direction for landing. The total distance covered would be nearer eighty-five nm than twenty-five nm. Passing several miles to the west of Newark the flight engineer scanned the indicators once more. The totaliser gauge now showed 2.7 tons of fuel remaining, about fifteen to twenty minutes' flying time to dry tanks, which would be just enough to get there with a fraction to spare. The fuel quantity gauges confirmed the amount remaining in the tanks but the fuel left seemed uncomfortably low. Again he checked the calculations. The figures tallied but the amount, even if they had gone on a direct route, hardly seemed enough. What the flight engineer did not know at the time, nor the captain or anyone else for that matter, was that the 'flight conduct memory' table on board the 747 which had been used for the fuel management computations, was in error. The correct calculated fuel figure for diversion was 7.9 tons, 1.4 tons more than assumed. The Pan Am aircraft was seriously short of fuel.

Suddenly, to the flight engineer's horror, he saw the low pressure lights of the forward and aft boost pumps illuminate for number one tank. The fuel was so low the pumps were being starved of supply. The number one tank fuel gauge indicated 800 lb (360 kg), just over one third of a ton, but not much more fuel could be pumped from the bottom. The other tanks held only a little more, and he surmised that their pumps would also run dry when the fuel dropped to that level. By a quick reckoning he estimated that only 1.3 tons of the 2.7 tons indicated remaining was usable. The other 1.4 tons couldn't be taken into account because it had to be assumed it wasn't there. The pilots, already under pressure to land quickly, were shocked by the news. There was little more than five minutes' flying time remaining and a potential disaster was only moments away. The captain thought of turning quickly to the right and landing in an easterly direction on runway 11, but there were no instrument approach charts available for that runway and the aircraft was still in cloud. By now the first officer was in contact with Newark Approach and he requested their time to touchdown.

'Five minutes,' replied the controller.

There was barely enough time and the captain declared an emergency. The strain on the crew was now enormous and they were under tremendous pressure. Newark Approach radar vectored the 747 directly on to the 22L runway course and within minutes they were established on the instrument landing system. Just before the outer marker the aircraft broke cloud and the runway could be seen ahead. About two minutes remained to touchdown. Newark Tower was contacted and the controller requested the nature of the emergency.

'Fuel,' replied the co-pilot briefly, his heart in his mouth.

The captain maintained the flap set at 20° and waited till he was certain of getting in before selecting the full 30° landing flap setting. After passing the outer marker the landing gear was lowered but the captain flew the aircraft high on the approach path and kept the speed about thirty knots fast. If the engines suddenly stopped through fuel starvation he hoped to be able to glide to a landing.

'What fuel do you have remaining?' asked the Tower controller. With virtually nothing left there wasn't much that could be said.

'The gauges are unreliable so we don't know,' replied the first officer. 'One tank is feeding all engines.'

Approaching 200 ft (sixty m) the flight engineer informed the captain that all the fuel low pressure lights on all the boost pumps were illuminated. The 747 was moments from running out of fuel. Passing 200 ft the captain felt that a landing was certain and he called for the selection of full flap. The engines were now running on thin air. If they cut out at this moment he felt he could reach the threshold, but it would be a close run event. The three cockpit crew held their breaths in the tense atmosphere. Fifteen seconds later the wheels gently touched the runway surface and a mightily relieved crew felt the aircraft settle on terra firma. All four thrust levers were pulled into reverse and the engines responded with a roar, but it was a noise that did not last long. The aircraft slowed and the captain cancelled reverse thrust, but as he did so engine numbers one and four ran down from fuel starvation. The 747 taxied from the runway and proceeded along a parallel taxiway where, after about a mile, number two engine also ran down from lack of fuel. The captain shut down number three engine and waited for a tow truck to pull them to the terminal. In the meantime, the flight engineer checked his fuel panel and noted that the fuel totaliser indicated 1.4 tons remaining. According to the fuel quantity gauges about 1.5 tons of fuel remained in the tanks so the figures cross-checked. The four main fuel tank gauges read, from left to right: 800 lb (360 kg), 400 lb (180 kg), 1,100 lb (500 kg) and 1,100 lb again. The small amount of fuel which was indicated remaining should have been usable but was obviously not available. This discrepancy in the gauges and the error in the fuel management chart had placed the 747 great jeopardy. If the engines had stopped only a few minutes earlier while the aircraft was still in cloud the 747 would have landed short of the runway and a disaster would have been inevitable. By good fortune and a quick thinking crew a catastrophe was only narrowly averted.

The instances of airliners running out of fuel in flight are extremely rare, which is just as well, for a big jet trying to land without engine power, especially with flaps and landing gear set, has the flying properties of something close to a brick. The light aircraft pilot may be used to practising forced landings from height, usually simulating an engine failure, but such a prospect for the airline pilot is a different matter. If the incident occurs at night or in cloud the chances of success are about nil.

A big jet can, in fact, glide reasonably well and, indeed, when descent is initiated from cruise level the thrust levers are closed to flight idle and the aircraft literally becomes a giant glider. The descent rate is normally about 2,500 ft (760 m) per minute for a 747, so the rate of descent without idling engine power would be something greater. If all engines suddenly failed on a 747 at cruise height, therefore, the flight could probably remain airborne in the glide for about twenty minutes or so — as seen in Chapter 9.

At the lower levels, however, it would be necessary to select the landing gear down for an emergency powerless landing and that would be another story. The big jet would drop like a stone. A touchdown without power is referred to as a deadstick landing and such a procedure on a big jet aircraft can be considered extremely hazardous. Great skill would be required to judge height and rate of descent and to execute a safe deadstick landing. There would only be one chance and more than a little luck would be needed as well. It would be hard to find quickly somewhere suitable to land for, unlike the light aircraft pilot, a grass field would not suffice. If a big jet suddenly lost all power in flight, attempting a forced landing and accomplishing the task safely would be very difficult indeed. Not, perhaps, entirely impossible.

On a warm afternoon of 23 July 1983, Captain Robert Pearson and his co-pilot, Maurice Quintal, boarded Air Canada's Boeing 767 wide-body jet in Montreal, Quebec. The aircraft, registered C-GUAN, had only just arrived from Edmonton and Ottawa and, with a fresh crew, was about to turn round and fly the same journey back.

The Boeing 767 is an ultra-modern, two-crew, twin-engined machine with advanced electronics, and the aircraft was brand new to Air Canada. The flight deck has an uncluttered appearance with most details being presented on cathode ray tube type displays. The 767 is a pleasure to fly and is liked by pilots. On arrival in the cockpit on the afternoon of 23 July, however, Captain Pearson found they had a problem. The fuel indicating system on C-GUAN was faulty and there was no display of fuel load on the flight deck. The 767 fuel gauges are situated in the middle of the pilot's overhead panel and display fuel quantities in the two main wing tanks and the central auxiliary tank, as well as total fuel load. The minimum equipment list carried on board each aircraft stated that departure with this defect was not permitted. Only one main wing tank fuel gauge was allowed to be inoperative, and then the refueller had to dip the tank with a stick to confirm the level. The procedure, of course, was designed to safeguard against insufficient fuel being loaded.

In spite of the total fuel gauge failure, however, the 767 had received special dispensation to operate in that condition from Maintenance Control. Aircraft are often dispatched with systems unserviceable which are considered more of an inconvenience than a safety factor, but the minimum equipment list was designed to prevent departure in an unsafe condition. There are occasions, however, when dispensations are granted as long as proper precautions are taken.

The problem with the fuel quantity indicating system on

Air Canada's 767, *C-GUAN*, had begun at the start of the month when a fault on the equipment was found to be intermittent. In fact, inconsistencies had been discovered months earlier on 767s operated by other airlines and Boeing had issued instructions for the systems to be checked regularly. A United Airlines 767 on a delivery flight had shown that discrepancies could exist between the quantity of fuel displayed on the gauges and the actual amount in the tanks. A situation could arise where the pilots thought they had more fuel than was actually available.

On 5 July, in Edmonton, an Air Canada certified aircraft technician, Conrad Yaremko, conducted a routine check on *C-GUAN*'s fuel gauge system. The aircraft had arrived from Toronto with a fuel processor channel unserviceable and he found that during the check all fuel quantity gauges went blank. Two fuel quantity processors channels on the aircraft compute fuel quantities which are then displayed on the gauges. Each processor computer channel sums the quantities independently and the two systems then cross-check each other. If one fails the other can operate on its own.

Yaremko discovered, however, that there was a fault in the number two processor channel and that this was

An Air Canada Boeing 767. (Air Canada)

affecting the entire system and causing all fuel indicators to go blank. He found that if he isolated the number two processor channel computer by pulling its electrical circuit breaker he could get the system to work. All indicators would display as normal, but since there was only one processor channel operating there would be no computer cross-check of the calculated amount. It was necessary, therefore, as in the case of a main wing tank gauge failure, for a mechanic to dipstick the tanks to confirm the displayed quantity. The aircraft duly returned to Toronto and the number two processor channel was checked. Surprisingly, it was shown to be within tolerance. The circuit breaker was reset and the processor appeared to be working normally.

On 14 July, en route from Toronto to San Francisco, the number two fuel processor channel computer failed again and all fuel gauges went blank. In San Francisco the equipment in the electronics bay was re-racked and the system operated normally. On 22 July, again in Edmonton, in the evening before the day in question, *C-GUAN* arrived on another flight from Toronto. The system was operating normally but, during the routine check, the fuel gauges blanked once more. Yaremko checked the equipment and found the same number two processor channel failed again. He was unable to clear the fault but no spare fuel processor unit was available. He ordered one from the stores to be available the following evening when the flight returned once more to Edmonton. Having encountered the trouble before he simply isolated the number two processor channel and the fuel indications returned. In this condition the flight could depart in the morning.

Early the next day, 23 July, the 767 was satisfactorily dispatched but, before the flight left, Yaremko spoke to the departing commander, Captain John Weir. The two men agreed that the condition satisfied the minimum equipment list but that a fuel tank dipstick check would also be required. During the conversation Yaremko mentioned that on 5 July the aircraft had arrived from Toronto with the same problem and he had cleared the fault in a similar manner. In a misunderstanding, however, Weir thought he referred to the previous evening when the same aircraft had also arrived from Toronto, and the captain was left with the impression that the fault had been running from the day before.

The 767 left Edmonton on time and the flight operated via Ottawa to Montreal, arriving in the afternoon. On going off duty in Montreal, Captain Weir met Captain Pearson and First Officer Quintal on their way to work and passed on the news of the problem. He mentioned that it would be necessary to dip the tanks and added, incorrectly, that the fault had been like that from the previous afternoon. He suggested that they fuelled in Montreal for both return sectors because by doing so they would save themselves some bother during the transit in Ottawa. The conversation touched on previous problems with the fuel indicating system in general and Captain Pearson, in a further misunderstanding, mistakenly believed that the fuel indicating equipment was completely unserviceable. What's more he believed he was being told that it had been like that for the last day and a half.

When Captain Pearson and his co-pilot entered C-GUAN's flight deck, they saw before them what they expected to see; the fuel quantity indicators were blank. But the flight had arrived with the system working and only the number two fuel processor channel computer unserviceable. This was the failure that Captain Weir had been referring to in his brief conversation with Pearson and that, of course, was why the tanks had to be dipped. In Edmonton the previous evening, Yaremko had tripped the number two processor channel electrical circuit breaker and after doing so had attached to it the relevant inoperative label. Why had the fuel indicators now gone blank? Unknown to Captain Pearson, an unfortunate sequence of events was beginning to unfold.

Before the new crew arrived at the 767, a Mr Ouellet had been sent to the aircraft by his foreman to dipstick the tanks after fuelling was completed. While waiting to perform his task he sat on the flight deck and noticed the inoperative label on the number two fuel processor channel circuit breaker. Although Ouellet was not qualified to check the system he thought he might be able to help and he reset the breaker. Immediately the fuel indicators went blank. On checking the equipment he found it deficient but, like

Yaremko, he was unable to obtain a spare. He was told that one was being positioned to Edmonton for the flight's return. At about this time he was called to dipstick the tanks, and, unfortunately he forgot to re-trip the number two fuel processor channel circuit breaker. The breaker remained pushed in and its position was masked by the inoperative tag which was still attached. To the eye it appeared that the breaker was tripped and all the screens remained blank.

Captain Pearson checked the minimum equipment list (MEL) which confirmed that he was not permitted to depart with the fuel indicating system inoperative. At that time, however, the MEL on the newly introduced 767 was incomplete with many items blank and alterations taking place to it constantly. In the few months since the 767's introduction, over fifty changes had been made. The list was not considered reliable and it was Maintenance Control's practice then to authorise flights contrary to the MEL.

Captain Bob Pearson.

Pearson checked the paper work and the fuel flight plan simply stated: FUEL QTY PROC # 2 INOP. DIP REQD. The maintenance log was also inspected and two entries were apparent. One by Yaremko in Edmonton read: SERVICE CHK — FOUND FUEL QTY IND BLANK — FUEL QTY # 2 C/B PULLED & TAGGED — FUEL DIP REQD PRIOR TO DEP. SEE MEL. Yaremko, of course, referred to the indicators going blank during the check but that was not clear from the text. The other entry by Ouellet stated: FUEL QTY IND U/S. SUSPECT PROCESSOR UNIT AT FAULT. NIL STOCK. The maintenance log had then been signed out as satisfactory.

The captain also discussed the situation with the mechanics in Montreal and he was informed by them that a special dispensation had been received to operate in this manner. The case was supported by the chat with Captain Weir, the blank fuel gauges, the maintenance log and the belief that the machine had flown in this fashion from the day before. If others had flown it this way then he would have to do so as well and Captain Pearson accepted the aircraft. A fuel processor was to be ready in Edmonton for the flight's return and that provided the justification for the dispensation. The idea that the blank fuel indicating system was acceptable, however, was *not* correct and the aircraft, in spite of Maintenance Control's approval, was not permitted to depart in that condition.

Captain Pearson and F/O Quintal both believed they would be operating in accordance with the applicable rules and regulations or they would not have considered proceeding with the flight. By an unfortunate sequence of events the captain had been deceived into accepting an aircraft which was unsatisfactory. There was still the fuel dip check to be performed and that would cross-check the fuel quantity on board against the volume pumped into the tanks from the bowser plus the load remaining on arrival. If the calculations were performed carefully, what could possibly go wrong? What, indeed!

Captain Pearson made the decision to carry sufficient fuel out of Montreal for the two sector flight via Ottawa to Edmonton. The required fuel load amounted to 22.3 tons, or more correctly 22.3 tonnes — one tonne being equivalent to one metric ton or 1,000 kilogrammes — for the calculations were being conducted in kilogrammes. The system of units used in aviation throughout the world is not standardised and the present situation can only be described as messy.

The International Civil Aviation Organisation (ICAO) units are mostly metric, exceptions being distance which is stated in nautical miles and speed which is given in knots. Very few nations comply completely with the ICAO basic standard and most signatories abide by what is known as the Blue Table, with height, for example, being measured in feet. The Republic of China and the Soviet Union use metric units throughout with height expressed in metres and speed in kilometres per hour. Wind speeds are given in metres per second. Flying in a metric country with aircraft which comply with the Blue Table is not easy.

The USA uses its own modified 'Imperial' system where many of the same problems apply. Almost all foreign aircraft have to convert just about everything for calculations: degrees Fahrenheit to degrees Celsius, pounds to kilogrammes, inches of mercury to millibars and US gallons to litres. The modern world trend, however, is towards metrication in spite of America steadfastly adhering to its own version of 'avoir dupois'. The names of units are also moving to a standardised form and are now being referred to mostly by the names of people prominent in their particular field: centigrade has become Celsius, after its Swedish inventor, a cycle per second has become Hertz, after the famous German physicist; and more recently, a millibar has become a hectoPascal, after the French scientist Blaise Pascal.

In the 1970s, nations as far apart as the UK and Australia began the painful task of converting to the metric system, and in the early 1980s Canada also commenced the transition. For most countries the changeover was smooth and gradual, if not always without problems. The aviation industry welcomes any move towards a standard, if not least because the profusion of units within its own ranks was, and is, still confusing. Fuel, for example, is pumped aboard an aircraft by unit volume, and at a busy international airport like London's Heathrow, airlines ask for their fuel in litres, US gallons or Imperial gallons. The fuel quantity is then converted to pounds or kilogrammes for trim purposes and if a dipstick check is required the tank depth is measured in centimetres or inches. The general attitude is that it is a wonder that more mistakes are not made.

Air Canada's aircraft had been refuelled for some time using litres which, in a mixture of units, were than converted into pounds. In line with the nation's transition to metric, however, the airline was in the process of changing, and the 767 fleet was the first Air Canada type to have its gauges calibrated in kilogrammes. On Captain Pearson's 767, C-GUAN, the refuellers discussing the fuel load with him were aware the amount he required was a kilogramme measurement, but when the maintenance personnel calculated the quantity in litres to be pumped aboard they used the incorrect conversion factor. The conversion employed was for changing litres into pounds (1.77 instead of 0.8 for kilogrammes), and only sufficient fuel was loaded to give a combined litre volume in the tanks equivalent to 22,300 lb (10,115 kg). The tank quantities were checked using a dipstick which measured the depth of fuel in centimetres which the ground personnel, using a table for conversion, then changed into litres. This volume compared with that

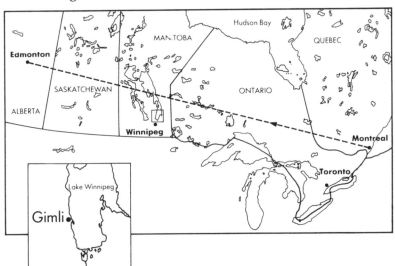

Flight 143's routeing and the position of Gimli airfield.

loaded by the refueller, plus the amount in the tanks on arrival. When the captain conducted a cross-check, however, he was also given the incorrect conversion factor, and when he checked the calculations he confirmed the mathematics were correct. The refueller had also used the same figure before converting to litres and both were in agreement. Having both used the wrong factor, however, the volume of litres in the tanks was only equivalent to 22,300 lb instead of 22,300 kilogrammes. One pound is less than half a kilogramme so the fuel load on board was less than half required. With no fuel quantity indicators on the flight deck, the crew had no warning of the situation.

Before departure, the flight management computer (FMC) was loaded with the fuel quantity. A page in the Boeing operating manual at the time gave instructions which showed how fuel quantity from dipsticking the tanks could be entered in the computer. After engine start, fuel consumption would be recorded, permitting a total fuel readout to be constantly displayed throughout the flight, and *en route* fuel checks could be conducted normally. The FMC, however, was calibrated in kilogrammes and when the fuel load of 22,300 was entered, it was accepted by the computer as kilogrammes.

A radio call was received from the maintenance section before push back clearing the flight for the trip.

Captain Pearson took off from Montreal on the afternoon of Saturday 23 July operating Flight 143's first short hop to Ottawa. The aircraft arrived safely at the transit station and the turn-round was conducted expeditiously. No extra fuel was added but, as a precaution, the fuel tanks were dipped again to check the quantity before departure and to update the FMC. By an amazing stroke of misfortune, the same incorrect conversion factor was used by the ground personnel and the error was not discovered. Maintenance staff there also approved dispatch of the aircraft in the unsatisfactory condition. Flight 143 departed from Ottawa in the evening for the longer westbound sector to Edmonton. Only sixty-one passengers were on board, being looked after by six flight attendants led by Robert Desjardins, which gave a total, including the two cockpit crew, of sixty-nine. The lightly-laden 767 climbed rapidly to 41,000 ft (12,500 m) and the aircraft quickly settled in the cruise. It was a fine evening, the visibility was good, and the pleasant circumtances offered a feeling of well-being to the passengers and crew. In the tanks, however, was less than half the fuel needed for the journey and Flight 143 had no chance of reaching its destination.

The first sector had been flown by Captain Pearson and now F/O Quintal handled the aircraft on the way to Edmonton. As Flight 143 approached the border of Ontario with Manitoba, the 767 was about halfway on its journey. At 20:00 Central Daylight Time (CDT), Quintal spoke to the passengers.

'Good evening ladies and gentlemen, this is your first officer. We're presently coming up over Red Lake and are cruising at 41,000 feet. In Edmonton it's a beautiful day.'

Below, the countryside stretched before the cockpit and Red Lake could be clearly seen.

'I'm going to sit here and watch the trout swimming in the lake,' joked Pearson.

Ten minutes later, as the 767 crossed the border into Manitoba, the jesting stopped. If the fuel monitoring system had been operating properly, the first sign of trouble would have been a visual and aural low fuel state warning as the fuel quantity dropped below two tons. This would have been just sufficient fuel for an immediate descent and landing. With the equipment unserviceable, however, the first indication of a problem was when the instruments indicated a low fuel pressure in the left forward pump. Beep, beep, beep, sounded the warning.

'Holy . . .!' shouted Captain Pearson.

The first officer checked out the systems.

'Something's wrong with the fuel pump.'

'Left-forward fuel pump', confirmed Pearson. 'OK, what have we got here? I hope it's just the . . . pump failing, I'll tell you that.'

Moments later a second warning light illuminated indicating low fuel pressure in another pump. The captain's first reaction was that he had a computer problem but it would be safer to have it checked before going any further. It was too much of a coincidence that two fuel pumps should fail on a brand new aircraft, so Pearson decided to divert.

'Let's go to Winnipeg.'

Winnipeg was the nearest major airport and lay 120 nm away to the south-west. Clearance was received for the aircraft to proceed direct and to commence descent. Captain Pearson resumed control, closed the thrust levers and began the descent from 41,000 ft towards Winnipeg. Seconds later, warning lights indicating loss of fuel pressure in both right-hand pumps also illuminated. At that point the crew realised that so many pumps failing simultaneously on a brand new aircraft was highly unlikely and that they must have a fuel problem. The chief flight attendant, Desjardins, was called to the flight deck and told to prepare the passengers for an emergency landing. A few minutes later the number one engine was starved of fuel and ran down.

'We've just lost an engine,' radioed Quintal to Winnipeg Control. 'Request the fire trucks standing by for our arrival.'

Three minutes later the number two engine ran out of fuel and flamed out. The aircraft was now totally without engine power and with no fuel left in the tanks it was not possible to restart the engines. The 767 could only glide to a landing. At the moment of total engine failure the 767 was descending through 25,000 ft (7,620m) and was sixty-five nm from Winnipeg. A small air turbine dropped from the belly of the aircraft to supply power to operate the flying controls while emergency instruments in the cockpit were fed from the batteries. The cathode ray tube screens went blank and all that was left was a standby magnetic compass, an artificial horizon, an airspeed indicator and an altimeter. The captain expected to see the compass display of the radio magnetic indicator still functioning, but it did not work. Although it was mid-evening it was still bright daylight outside and the visibility was excellent. In the cockpit, however, 'it became the blackest place in the world,' Captain Pearson was to comment later. A Mayday call was transmitted.

'Centre, 143, this is a Mayday, and we require a vector on to the closest available runway. We are 22,000 feet on . . . with both engines failed due to, looks like fuel starvation, and we are on emergency instruments . . . now please give us a vector to the nearest runway.'

In the cabin the attendants secured galley equipment and briefed the passengers on the emergency landing and evacuation procedures. The cabin occupants were informed only that the aircraft had fuel trouble, but with both engines stopped it was obvious there was a major problem. It was very quiet. Two points were in the passengers' favour, however: Captain Pearson was a very experienced glider pilot and F/O Quintal knew the area well. He had been stationed near Winnipeg while serving in the Canadian Air Force. Their knowledge and skills would be greatly needed this day. The captain slowed the aircraft to about 220 knots which he guessed would be the best speed for a long glide. Steering the aircraft according to the controller's instructions proved to be another matter. The standby compass was situated in the centre of the flight deck at the top of the glare shield and was difficult to read. It was swinging wildly and became impossible to use for flying headings.

'So I steered by the clouds underneath us,' explained the captain after the event. 'I would ask Winnipeg Centre for a heading, they would say "left twenty degrees" and I would turn left about that much, judging by the clouds, and then I'd ask Winnipeg how my heading was. Using the clouds, I kept eyeballing it.'

Judging the rate of descent and planning a descent profile proved to be the most difficult task of all. No vertical speed indicator was functioning. In an attempt to calculate a descent path, F/O Quintal repeatedly asked the controller for distances to go to Winnipeg and compared those with the altitudes indicated. At the same time he also had to complete the emergency check lists, so he was kept very busy. As the 767 descended through 14,500 ft (4,420 m) its distance to Winnipeg was forty-five nm and the aircraft was dropping much faster than either of the pilots expected. By 9,500 ft (2,900 m) the Winnipeg radar controller informed them that they still had thirty-five nm to go. Flight 143 had dropped 5,000 ft (1,520 m) in ten miles and at the most they could only glide for about another twenty nm.

'We'll never make it,' said Quintal.

A hasty replanning of the situation was required. 'What about Gimli?' suggested the co-pilot.

Gimli Air Force base was about forty nm north of Winnipeg, situated on the west coast of Lake Winnipeg, and Quintal had spent some time there during his military service. The 767 was only just near the south of the lake and they could see its waters on the right hand side. Gimli couldn't be far away but some cloud obscured the view.

'Gimli's about twelve miles away from your position,' reported the radar controller.

F/O Quintal informed his captain that the base had two parallel runways and both were long enough to use. Pearson made an immediate decision to aim for Gimli, which at that time was in their four o'clock position, and they were instructed by radar to turn right 120 degrees. Descending below the cloud, with the sun about ten degrees above the western horizon, they proceeded almost northbound under radar assistance for several miles. Suddenly Quintal spotted Gimli Airfield. The time was 20:32 CDT and in the cabin all preparations were complete. Now it was up to the captain. He had only one chance at a landing. He would also be unable to select flap without power so the landing would be fast, but before touchdown they would have to lower the landing gear. It could be dropped by the force of gravity but since it would increase their rate of descent enormously it would have to be left to the last moment. Captain Pearson then asked the controller for more details of Gimli. The Air Force base was disbanded, he was told, but the right runway was still used by light aircraft and was 6,800 ft (2,070 m) long. It would be best to go for that one.

'There will be nobody on the runway when we get there, eh . . . nothing?' questioned Pearson.

'I don't know,' replied the controller, 'I can't tell you for sure.'

Quintal could see the runway clearly ahead now and he called out to his captain.

'We're going to make Gimli OK,' radioed Pearson to Winnipeg.

'Great! We show you about six miles to touchdown.'

The only problem now was that the 767 was too high

and fast. Pearson could make a quick orbit but if he did so he would lose sight of the runway and he might lose too much height. There was only one thing to do and that was to side-slip. The technique is used effectively by light aircraft pilots to lose height quickly but it is not recommended on a big jet. The captain pushed on right rudder and at the same time banked the wings in the opposite direction to the left. The crossed controls had the desired effect, and as the aircraft crabbed sideways through the air the drag plummetted the machine earthwards. In the cabin, the passengers felt the 767 plunge close to the ground and most people assumed the captain had lost control. On the contrary, however, the speed and height were both dropping rapidly and their situation was improving.

'Five miles to touchdown,' called the controller.

'Roger,' acknowledged Quintal, 'we have the field in sight.'

With the speed back to 180 knots and sufficient height lost, the captain straightened the aircraft on the final approach. He was satisfied now that he could commit himself to a landing and he called for the landing gear to be lowered. Quintal selected the landing gear lever down but nothing happened. His heart sank. Quickly he checked the handbook but he could find no reference to free-falling the wheels. The secondary procedure for landing gear extension was, in fact, located at the back of the section on hydraulics, but by an oversight the detail had been omitted from the index. He was unable to find the relevant information so, of his own volition, he selected the alternate gear extension switch to the down position. At last he heard the gear fall down into place but when he checked the indicators he was shocked to see the nose wheel had not locked down. Frantically he searched the pages of the handbook once more in a desperate attempt to find the required drill, but before doing so the time ran out.

The captain had judged the descent beautifully and he crossed the threshold at 180 knots, about fifty knots faster than normal, at just the right height. It was then that the mistake was noticed. The right runway which had been recommended for landing was darker in texture than the other and had remained unseen as it blended with the countryside. The runway on which the 767 was about to land was the lighter, left runway. Its colour had been easier to pick out from the air but, although it was longer at 7,200 ft (2,195 m), it was disused. The problem with landing on the left side was that the abandoned runway was employed over weekends as a racing car circuit and the last race of that Saturday evening had only just finished.

Captain Pearson touched down perfectly within 800 ft (245 m) of the threshold at about 175 knots but as he did so the two pilots saw to their horror that people and vehicles milled about at the far end of the runway. Children were playing and cycling in the area. Beyond the activity there were tents and caravans in which the racing drivers and their families were staying for the weekend. The 767 sped towards the gathering with no reverse power or ground spoilers available to help slow the machine. In one camper vehicle parked near the runway a racer's wife, Jo Ann Barry, was washing dishes after their evening meal. Suddenly she heard a boy shout that a jet was landing.

'I opened the camper door and there was this huge plane coming at us.'

Pearson hit the brakes hard and the aircraft reduced speed, but as it did so the unlocked nose wheel collapsed. The nose dropped to the ground and the nose wheel was forced back into the housing. Showers of sparks were thrown into the air as the nose section scraped along the ground. As it turned out, the fallen nose gear was a blessing in disguise for the friction slowed the aircraft rapidly and the 767 shuddered to a halt well short of the race meeting.

As the aircraft stopped, the forward section filled with smoke and the captain ordered an emergency evacuation. The flight attendants operated the escape slides: six in total with two at the front, two at the overwing emergency exits and two at the rear. With the nose tilted down the forward chutes sloped to the ground at a gentle angle but the rear slides were very steep. In the event, however, there was no problem with so few passengers and all aboard evacuated quickly with only minor injuries being sustained. No flames could be seen but thick, oily smoke continued to billow from

The 767 at rest at Gimli. Apart from the damage in the nosewheel area, no other damage was sustained. (Captain Bob Pearson)

the front of the aircraft and about ten large fire extinguishers, borrowed from the Winnipeg Sports Car Club, were required to dampen the emission. The fuel tanks, when checked later, were found to be completely dry.

The incident proved a lucky escape for all concerned and but for the skill of the pilots the result could have been much worse. Captain Pearson had displayed his abilities to the full and had executed a brilliant forced landing.

At the summing up of the subsequent Federal Government Public Inquiry Mr Justice Lockwood said of the crew, 'Thanks to the professionalism and skill of the flight crew and of the flight attendants, the corporate and equipment deficiencies were overcome and a major disaster averted.'

Chapter 6
The Blackest Day

The day did not start too well. To begin with, Pan Am's Flight 93 was late leaving Brussels and was behind schedule on its arrival into Amsterdam. The journey had been routine but there had been no chance to make up time on the short hop. The ground staff would try their best to speed the transit in Amsterdam but the Boeing 747 was a new aircraft type for the airline and had not been in service for long. Preparing the 747 would take time. The next stop was New York, and at least 100 tons of fuel would be needed for the Atlantic crossing; just pumping the fuel quantity into the tanks would take over thirty minutes. The passenger load was light, with only 152 travelling on the Amsterdam-New York sector, so time could be saved on boarding, but the number was disappointingly low for the airline. Pan Am's brand new 747, named *Fortune*, could hold 360 people, double that of the Boeing 707, and less than half the available seating capacity would be filled. The profit margin on this flight would be thin.

The day was a cool, cloudy Sunday in northern Europe, 6 September 1970, and 49-year-old Captain John Priddy, Flight 93's commander, would be pleased to be on the way. His co-pilot was First Officer Pat Levix and his flight engineer was Julius Dzuiba, known to his colleagues as Zuby. In the cabin he had fourteen flight attendants, led by Flight Service Director John Ferruggio, bringing the total on board to 169. Among the passengers was France's new deputy delegate to the United Nations, M. Francois de la Gorce, a deadheading crew returning to New York as passengers and Captain Priddy's wife, Valerie.

At about 13:30 GMT, 15:30 local time at Amsterdam's Schiphol Airport, Flight 93 was finally ready for departure.

The big jet pushed back from the gate and taxied out to the runway. The air traffic control flight clearance was given and confirmed the routeing as across the North Sea to overhead London, then westbound over the UK to begin the Atlantic crossing. Approaching the threshold Captain Priddy was instructed to hold position to await landing traffic and was told take-off clearance would be given shortly. As the crew waited for the arriving aircraft to land, a radio message was received from the control tower with the controller using Pan Am's call sign, 'Clipper'.

'Clipper 93, change to Ground Control for a minute, they have a message for you.'

The first officer re-selected the ground frequency and established contact.

'We have just been informed,' radioed ground, 'that there are two passengers on your flight that were refused passage by El Al.'

'For what reason?' asked Priddy.

'We have no idea.'

It seemed just another little problem to add to those of the day for the information was too vague to be conclusive.

'If you can give us the names of the passengers we'll check them out,' suggested the captain.

'The names are Diop and Gueye.'

Captain Priddy called his flight service director to the cockpit and they discussed the situation. Ferruggio did not recognise the names and he knew nothing about them. The captain felt it advisable to have a talk with the two suspects so he and his service director left the flight deck to seek them out. Meanwhile F/O Levix contacted the Pan Am operations office, known as Panops, in Amsterdam and spoke to a representative on their company frequency. Any additional information they could be given would be appreciated.

Priddy and Ferruggio descended the stairs from the upper deck to the first class cabin and the service director paged the two men on the public address (PA) system. The captain and his chief flight attendant walked through first class, right down one aisle to the back of the aircraft, then crossed over to the opposite side and walked all the way back. No one came forward. A second PA announcement was made and, this time, as Captain Priddy returned to the first class cabin, the two men revealed themselves.

'Hey, that's us,' they shouted.

They were sitting in the middle of the last row of seats. Immediately it could be seen that they were young men of Middle East or Arab origin and that they were very well dressed. They seemed pleasant people, and the captain felt uncomfortable as he approached them. He didn't know what he was looking for and he felt he had no right of search. And, after all, they had paid first class fares. As tactfully as possible he explained that an alert had been received and that their names had been mentioned. The two suspects spoke very good English and they talked with the captain for a few minutes. They were polite, well mannered and helpful. Captain Priddy could see no cause for alarm.

'Well, there's been some sort of misunderstanding,' he assured them, 'but I'll either have to take you back or give you the option of being searched.'

'If you want to search us, go ahead,' the men replied obligingly.

The captain searched their persons and belongings but nothing was found. They each had only a small Samsonite briefcase containing a few papers and it was obvious they were clean. Captain Priddy was satisfied they did not pose a threat.

'I'm sorry,' apologised the captain, 'apparently there has been some mistake.'

On returning to the flight deck, Priddy found that no further information on the suspects was available and, since the warning seemed a false alarm, he decided to depart.

Flight 93 took-off from Schiphol Airport at just before 14:00 GMT (now used throughout) and climbed south-west across the North Sea, reaching level cruise approximately 25 minutes later. At about 14:30, the Pan Am aircraft flew overhead London and, with the aircraft settled comfortably in the cruise, the decision to continue seemed justified. Captain Priddy contacted Panops at Heathrow and discussed the situation with them. He explained that he had searched the two suspects but had found nothing. Suddenly the flight deck door burst open and the captain stopped mid-sentence.

Pan Am Boeing 747. (John Stroud)

Standing behind the flight crew at the back of the cockpit were Diop and Gueye, each brandishing a pistol in one hand and a grenade with the pin pulled out in the other. The nearer of the two men held the first class purser before him as a shield, with an arm around her neck. Pan Am Flight 93 had been hijacked!

The day for Captain Priddy and his colleagues could now not get much worse, but just how black it had become for aviation at large was not known to Pan Am's crew. The first blow of the day was struck at 11:15 when a TWA Boeing 707 was hijacked on departure from Frankfurt, West Germany. The flight had originated in Tel Aviv and was bound for New York. The aircraft was commanded by Captain C. D. Woods who was forced by a group of armed guerrillas to fly to an unknown destination in the Middle East. Only fifteen minutes later, at 11:30, the hijacking of another Boeing 707 began, this time of airline El Al, by two passengers, a man and a woman. Their names were Patrick Arguello and Leila Khaled. Like TWA's aircraft, El Al's Flight 219 had also originated in Tel Aviv and was bound for New York, but it had, instead, transitted Amsterdam.

About thirty minutes after departure from Schiphol Airport, Arguello had let out an 'animal type bellow' and, with a small .22 pistol in his hand, he had rushed towards the front of the aircraft. He was quickly followed by Khaled who pulled a couple of grenades from her bra and charged down the aisle with one in each hand. As Arguello approached the flight deck door he was intercepted by Steward Shlomo Vider and a scuffle ensued. Armed sky marshals raced into action but before they could do anything Vider was shot in the chest. In the first class cabin a guard lunged at Khaled but she managed to pull a pin from one of the grenades. He grabbed her elbows from behind but as he pushed her down the grenade tumbled to the floor. Miraculously it failed to explode.

Another security guard rushed forward from the rear, his pistol firing, and Arguello dropped, fatally wounded. Khaled was securely bound with string and a necktie and Flight 219's commander, Captain Uri Bar-Leb, informed London he required an emergency landing at Heathrow. The captain also radioed Tel Aviv with news of the attempted hijacking and received a request for an immediate return to Israel. The authorities there would have liked very much to get their hands on the hijacketer, but Captain Bar-Leb insisted on landing in London. The wounded steward was seriously ill and urgent medical attention was needed.

On Flight 219's arrival at Heathrow a brave crew member ran from the aircraft with the unexploded grenade and placed it some distance away on the tarmac. Vider was rushed to hospital, the body of Arguello was removed and Khaled was placed in custody. The 707 was thoroughly searched and after a short delay was permitted to proceed on its way. Steward Vider later made a full recovery from his injuries.

In the afternoon, three hours after the incident, Pan Am's Flight 93 had followed El Al's route out of Amsterdam and had suffered a similar hijack attempt. The 747, however, did not carry armed sky marshals and the Pan Am big jet had been commandeered without resistance. Captain Priddy, at this stage, had no knowledge of the previous events and it was only later that the authorities were to realise how closely connected were the El Al and Pan Am hijack incidents.

As if the morning's attempts at air piracy were not enough — with one successful and one failed — another hijack occurred in the early afternoon before Pan Am became involved. At 12:15, a Swissair DC-8 en route from Zurich to New York was also overcome by a group of armed guerrillas. The hijacking had taken place near Paris and the commander, Captain Fritz Schreiber, was also forced to fly to an unknown destination in the Middle East.

Captain Priddy's 747 was the fourth aircraft that Sunday to be boarded in as many hours by armed hijackers and 6 September 1970 was a black day indeed. It was the worst for piracy in the history of civil aviation. Had the Pan Am captain been made more fully aware of the earlier events and had the warning of the Schiphol Airport police regarding the two suspects been more specific, he would never have left the ground. The police certainly knew of the attack on El Al before the 747 departed. The polite and helpful young gentlemen Captain Priddy had interviewed earlier now stood on his flight deck, armed and nervous guerrillas, determined to carry out their task. With the pins removed from the grenades each held, any attempt at opposition would have been suicidal and there was no choice but to obey the hijackers' demands.

Soon after the 747's take over the guerrillas ordered Flight 93 to turn back to Amsterdam. 3100 was selected on the transponder, at that time the hijack squawk code, and London Airways requested the 747 to descend to 27,000 ft because of traffic. A few minutes later the hijackers changed their minds and decided on Beirut, the capital of the Lebanon, as their destination. Priddy received clearance to climb back to a more suitable level and continued the flight eastwards to Beirut.

'Are we going to land there?' asked Captain Priddy.

'I don't know,' said Gueye, 'I'll tell you when we get there. I have to talk to my people on the ground.'

The reply was an indication that perhaps not all had gone according to plan and further instructions were required. The 747 had plenty of fuel for the expected Atlantic crossing so the four-hour flight to Beirut in the opposite direction did not give the captain cause for concern on that account. He hoped, once on the ground in Lebanon, to be able to persuade the guerrillas to release the passengers. Unaware of the facts, he was not to realise the outcome of the events.

As the 747 flew towards Beirut the hijackers began to relax a little, but one or other of them remained on the flight deck for the entire journey. At all times the grenades were grasped with the pins removed. On one occasion one of the guerrillas played with the spare pin in his fingers and accidentally dropped it. For a while it couldn't be found and a shiver ran down F/E Dzuiba's spine at the thought of being unable to secure the grenade. The flight engineer rummaged on the floor with his flashlight and the commotion attracted Priddy's attention.

'What the hell's going on?' asked the captain.

'Never mind, keep flying,' called back Dzuiba. 'I'll find it.'

Eventually a relieved flight engineer recovered the pin and returned it to its owner.

Meanwhile in Amsterdam, the authorities had not been inactive and details of the hijackings began to be pieced together. Diop was identified as Mazn Abu Mehana, a Palestinian from Haifa, and Gueye as Samir Abdel Meguid, a Palestinian from Jerusalem. Both men, like their colleagues who had hijacked the other aircraft, were members of the Popular Front for the Liberation of Palestine (PFLP), an organisation dedicated to furthering the Palestinian cause by whatever means available. Aircraft of the enemy, Israel, or of their supporters, American or European, were considered legitimate targets.

Diop and Gueye, it transpired, had originally been booked on El Al's Flight 219 which had departed Amsterdam with Arguello and Khaled aboard a few hours ahead of the Pan Am 747. A team of four was considered the minimum required to tackle an El Al aircraft with armed sky marshals but, with the two men denied access to Flight 219 and no time to find replacements, the attempt had failed. El Al's suspicions regarding Diop and Gueye had arisen ten days earlier when they had tried to make a booking on the Amsterdam-New York flight. Security officers checked the men out and found the facts didn't tally. Both men carried

Senegalese passports and claimed to be students travelling to South America. The journey they had chosen via New York to Santiago in Chile, however, was not the cheapest route to fly and, what's more, they had bought one-way first class tickets paid for in cash. Students from Senegal did not normally travel in this fashion! Diop and Gueye were eventually told that the El Al flight out of Amsterdam to New York was full and they were unable to carry them. The airline simply marked the paid-for El Al tickets with an 'open endorsement' which allowed them to fly with any other airline.

Three days before Sunday the 6th, the two men booked on Pan Am's Flight 93. At check-in, the counter clerk noted the El Al 'open endorsement' tickets but received the two men for the flight. Pan Am representatives, however, became concerned that they might have unwittingly accepted rejects from another airline and they checked the names with El Al. At about the same time, news of the attempted hijacking of Flight 219 reached Amsterdam and El Al security agents became aware of the potential danger to the 747. Pan Am's Flight 93 was already taxiing out for take-off at this stage so the Schiphol Airport police were contacted immediately. Unfortunately the police did not take the threat as seriously as the airline, and the information, when passed via the ground controller to Captain Priddy, was vague and incomplete.

When the Pan Am captain descended the stairs to interview the suspects, unknown to him at the time the two men had been sitting in the third row on the right side of first class. After the PA announcement calling for them the guerrillas had slipped back to the last row of seats in the middle. Captain Priddy had searched the right men in the wrong place.

The original intention of the PFLP was to hijack three aircraft that September Sunday: the TWA 707, the El Al 707 and the Swissair DC-8. When Diop and Gueye were found to be spare they had been switched to Pan Am to cover any possible failure of the attack on Flight 219. The plan had been carefully organised and executed and success in three out of four attempts was quite remarkable. All four aircraft had been hijacked at the beginning of transatlantic flights and carried plenty of fuel. The one problem facing Pan Am's hijackers, however, was that they expected the aircraft to be a 707. The Boeing 747 had been only newly introduced into service and it had caught them by surprise.

As Flight 93 flew towards Beirut, Diop and Gueye re-arranged the seating of the passengers. With one on the flight deck at all times, the other, pistol in hand, herded all the cabin occupants into the rear of the economy section. Passports were collected and were thoroughly checked. Any potential trouble makers — military personnel, diplomatic staff, crew members — were then isolated from the main group and seated in first class. A special watch would be kept on them. All passengers were firmly ordered to stay in their seats and to remain still. If any rescue was attempted the perpetrators would face grave consequences.

After several hours the captain was finally permitted to talk to the passengers and he informed them on the PA that the aircraft was no longer under his command. He explained, in as calm and reassuring a manner as possible, that the flight was now not going to New York, but to Beirut Airport, 2,000 miles (3,707 km) to the east of their departure point.

As the Pan Am 747 approached Beirut Airport, Captain Priddy was instructed to remain at height and to circle overhead. He obtained permission to descend to 27,000 ft (8,230 m), below the airway structure, and at the speed of 250 knots he prepared to begin orbiting. Meanwhile, in Jordan, 100 miles (160 km) to the south, another drama was unfolding. The two aircraft which had been hijacked earlier — the TWA 707 and the Swissair DC-8 — had been forced to land at a disused desert airbase in northern Jordan, known as Dawson's Field, situated near the town of Zarqa, fifteen miles (twenty-four kilometres) to the north of Amman. The landings on the rough and uneven desert strip had been hazardous and the passengers had been badly shaken. One woman aboard the TWA aircraft had sustained a broken wrist. The parked aircraft were immediately surrounded by armed commandos of the PFLP.

Detachments of Israeli airborne helicopter troops crossed the border and landed near the towns of Irbid and Jerash in preparation for an attempt to release the two airliners, but their efforts were too late. Armed factions of the PFLP also feared that the Jordan Army, with which they had been in recent conflict, would attempt a rescue, but they too were helpless. King Hussein opposed the PFLP action in hijacking the aircraft and Jordan in fact, heavily infiltrated by Palestinian guerrillas, was on the brink of civil war. About 300 passengers, 155 aboard the DC-8 and 145 aboard the 707, were held hostage in the desert against their will.

Unaware of the circumstances, Captain Priddy began circling the 747 overhead Beirut Airport at 27,000 ft. The time was now about 18:30 GMT, 20:30 local Lebanese time, and darkness had fallen, but the lights of the town could be seen five miles (eight kilometres) below. Priddy established contact with Panops Beirut and was instructed by the hijackers to order an Arabic speaker to be brought to the mike. The guerrillas wanted to converse in their own language to conceal their discussions from the crew. An Arabic speaker was quickly acquired and he was told by one of the hijackers to summon a PFLP official to the airport to discuss the situation. A few minutes later the hijacker called back again.

'Brother, have you made contact with any of the responsible officials?'

'We are now in contact with a responsible official,' replied the Arabic speaker, 'and he shall communicate with you on this frequency when he comes to the office.'

As the arrival of the PFLP official was awaited a conflict of interest arose between Captain Priddy in the circling 747 and the Lebanese authorities at the airport. Priddy, unaware of the hostages held in the desert, was convinced the best course of action was to land in Beirut. He was confident that once on the ground he could persuade the hijackers to release the passengers, or at least the women and children. He had learned that Diop and Gueye were fairly knowledgeable about aircraft operations but their one area of weakness seemed to be in fuel management of the 747. He became aware that they had no idea of how long they could continue flying with the remaining fuel and he sensed he could force them into landing at Beirut by declaring a fuel shortage. On the ground, a Mr A. Bedran, the deputy manager of Beirut Airport, and a Colonel Salloum of the Lebanese Army, had other ideas. To them the 747 was trouble and they wanted nothing to do with it. They had their own plans for dissuading the aircraft from landing.

'For your information,' radioed Mr Bedran, 'the main runway at Beirut Airport is closed at the present time due to work in progress. The other runway cannot tolerate the weight of the Boeing 747. We suggest that the aircraft be directed to Damascus Airport, which airport is presently ready to receive this type of aircraft.'

'We will not land at Damascus Airport, but will land at Beirut Airport. Did you hear? Answer, answer! We will land at Beirut Airport and will not land at any other airport unless contact is established with a responsible official from the Popular Front. I want an answer.'

'We are in contact — attempting to contact — the Popular

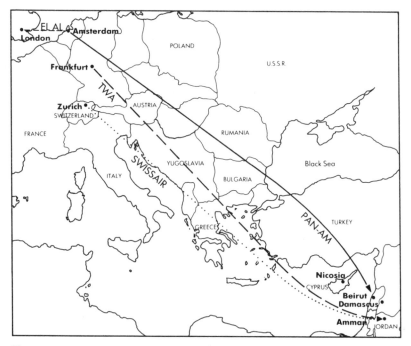

The routeings of the four aircraft hijacked on 6 September.

Front to send a representative to the airport,' reaffirmed the deputy airport manager. 'Therefore please fly overhead Beirut until contact is established and agreement reached to head to another airport as Beirut Airport is not technically ready to receive this aircraft.'

News was then received that representatives of the PFLP would arrive imminently and that they would be in the control tower within ten minutes. If the 747 crew also selected the tower frequency of 118.9 MHz on another radio box the hijackers could communicate with them shortly. It then transpired that the Palestinian officials were arriving from Bhamdoun, a mountain town some distance away, and it would take at least thirty minutes just to drive into Beirut.

An hour passed and still there was no sign of the representatives. The Pan Am 747 continued to fly in circles above the airport and by now it was obvious to all concerned that the big jet's arrival overhead Beirut was unexpected.

'It's taking a long time to get these people,' commented Priddy.

'Well, you know the traffic in Beirut on a Sunday night,' Gueye replied casually as if in everyday conversation.

Meanwhile Bedran, the deputy airport manager, tried again to stress that Beirut was not technically suitable to receive the 747. The main runway had been closed for over a month, the hijackers were informed, and the first 2,300 ft (700 m) of the second runway were unusable. It would be extremely difficult for the 747 to land. He did not reveal, however, that it *was* possible for the big jet to use the shortened runway for landing. Whether Bedran's remarks were bluff or fact was not known to anyone aboard the Pan Am aircraft but when the deputy airport manager offered the information that Damascus had been contacted and the airport could receive the big jet the hijackers became hysterical. Bedran had no authority to speak to Damascus, one of the guerrillas shouted over the radio, and on no account was the 747 going anywhere until a responsible Palestinian official was contacted. Captain Priddy, however, was concerned that there might be some truth in the statements and he, in conversation with Panops Beirut on the other radio box, enquired if Baghdad would be a suitable alternative. Unfortunately, neither the airport there nor Iraqi Airways were able to handle the big jet.

On the tower frequency, the hijacker, still in conversation with Bedran, asked about one of the other events of that Sunday.

'An El Al aircraft was hijacked today en route to London. Where did it land?'

'It looks like it has landed at London,' replied the deputy airport manager.

'Who is responsible for its hijacking?'

'Oh, by God, we don't know, excepting that it has landed at London at the present time.'

'Thanks.'

Diop and Gueye had, of course, more than a passing interest in the event and it appeared from what had been said that the attempt had failed. Unaware of the other successes they, too, felt that their day had not gone according to plan.

About one-and-a-half hours after the initial request for a PFLP representative, two responsible officials, Abou-Khaled and Abou-Ahmad, finally arrived in the control tower. Abou-Khaled spoke over the radio to Gueye, now identified in Amsterdam as Samir Maguid, but neither of the men knew each other personally and both were very cautious.

'Hello, hello.'

'Yes, go ahead,' said Gueye.

'Hello, hello, this is Abou-Khaled speaking, do you hear me?'

'Who?'

'Walid Kaddoura speaking, do you hear me? Hello, who is talking to me, brother? Who is talking?'

'Samir.'

'Samir?'

'Yes.'

'We have news from Amman that you could not land at the airport there,' said Abou-Khaled.

'Of course we cannot land there.'

'Who is talking, brother? What squad?'

'Well . . . do you want to disclose names?'

'No, the name of the squad, the name of the squad! You

originated from Amsterdam, is that right?'

'Yes, from Amsterdam.'

'What happened?' asked Abou-Khaled.

'What happened when we entered the Amsterdam Airport is that those responsible on the aircraft were suspicious and started to search us but could not find the weapons. When the aircraft took off, its destination was diverted in accordance with instructions we received.'

'Can you go to any other place? Can you go to any other place?'

'Where do you want us to take it?'

'Is there enough fuel?' questioned Abou-Khaled.

'Is there enough fuel? I will ask the captain . . . Hello, hello, he informed us that he has fuel enough for forty minutes.'

'Can you go to Amman? Is the fuel sufficient to reach the destined place?'

'Yes, it is, but landing is impossible. We can reach the destined place, but landing there is impossible.'

'Brother, if anything happens to the aircraft there, it does not matter. Is this possible?'

'It doesn't matter if anything happens to the aircraft there?' asked Gueye, surprised.

'It does not matter if anything happens to the aircraft. It did happen to the TWA's Amman landing to a certain extent, but it did not matter.'

'Does this mean that we can land there at the destined place? What I want you to understand is that this is a 747 aircraft and it weighs many times the weight of a 707 aircraft. It is impossible to land at the destined place under any circumstance whatsoever. The destined airport definitely cannot stand it, and if it lands there it will be a wreck.'

The radio conversations were in Arabic and were, of course, not understood by the Pan Am crew, but the language difference could not disguise the hijackers' concern. Not only did Captain Priddy not want to be hijacked, but the guerrillas had not wanted to commandeer a Boeing 747! If the aircraft had been a 707 it would by now be sitting at the 'destined place' — Dawson's Field — with the other two machines, but it was impossible to land such a big jet on the rough desert strip. The PFLP, therefore, were left with a rather large embarrassment circling five miles overhead Beirut: they didn't want the 747 nor did they know what to do with it.

Captain Priddy's estimate of forty minutes' fuel was on the low side and he had in fact much more than he was telling. Since Beirut Airport appeared unable to accept the 747, the hijackers suggested to the PFLP officials that they land at either Baghdad or Cairo for refuelling. With what they assumed was little fuel remaining, however, they required an answer quickly. Priddy had already established that Baghdad Airport could not handle a 747 but he did not understand the discussion in Arabic. When questioned he simply said he could remain overhead for about a further twenty minutes after which he would have to land at Beirut. If they wanted to go elsewhere they would have to decide soon.

'So I beg of you,' pleaded Gueye with Abou-Khaled, 'to give an answer in less than fifteen minutes.'

'Within five minutes we will give you an answer, brother.'

By now the Beirut Airport manager had joined those in the tower and Abou-Khaled confronted him face to face. He called his bluff and demanded to know the true situation at the airport otherwise he would order the destruction of the 747 in the air. A Colonel Hamdan of the Lebanese Army telephoned the Transport Minister, Pierre Gemayal, and relayed the threat. Could permission be given for the 747 to land? With little choice the Minister relented and allowed Beirut Airport to accept the Pan Am aircraft.

Meanwhile, in Flight 93's cockpit, the hijackers were becoming impatient.

'Hello, Abou-Ahmad, what shall we do? Have you received an answer, or not yet?'

'Abou-Ahmad talking to you. I will give you an answer in three minutes.'

'Okay, I am listening to you. Didn't you expect us here, or what? Is that why you cannot give us a quick answer? Four minutes are over, and we are still waiting.'

'They will allow you to land at the airport to refuel,' a relieved Abou-Ahmad at last reported.

'Land at what airport?'

'Instructions will be given to land at Beirut Airport to refuel, which will take forty-five minutes, after which we will give you instructions as to where to go.'

'Does this mean that the Lebanese authorities agreed to our landing at Beirut? They said technically the airport was not fit for landing.'

'Technically you can land, you can land,' affirmed Abou-Ahmad. 'The authorities have approved landing at the airport for refuelling.'

'Please inform them of the following message: If any attempt is made to approach the aircraft, we will destroy the aircraft completely.'

It was agreed that, on landing, the two PFLP represent-atives in the tower could board the aircraft for discussion, and descriptions of the men were radioed to the hijackers for identification. In the darkness it would be difficult to see who was approaching the 747. Gueye, however, was still unhappy about the fact that he was not known to those in the tower and he expressed his concern.

'I am astounded that you do not know who is talking to you as long as you know that I originated in Amsterdam.'

'Brother, it was not planned for you to land at Beirut, remonstrated Abou-Ahmad. Ashraf* and Abou-Hani* did not tell you to land at Beirut. This is an emergency. I am not supposed to know.'

'I know it is an emergency. We also did not expect it to be a 747.'

'Therefore, it *is* an emergency, and we do not know.'

Captain Priddy now prepared the aircraft for landing and the tower controller butted in to the radio conversation to pass on the relevant details.

'Okay, Clipper 93 — calling for landing — you have to be very smooth on landing and you have to choose either 03 which is landing from over the sea or runway 21 from over the sea. The full length will be available, I mean runway 03. On runway 21 the first 600 metres are to be avoided, and anyhow on runway 03 and 21 there will be a VASI* which will guide you in for landing.'

* Amman PFLP leaders.

'What is the field elevation?'

'The field elevation is eighty-seven feet, sir.'

'Eighty-seven feet, okay,' repeated Priddy. 'What is the wind direction and velocity?'

'Easterly at five knots, runway length available 3,180 metres.'

The Pan Am 747 had no airport charts on board for the arrival at Beirut and all relevant details had to be passed by radio, but Priddy was not over-concerned about the landing. He had already been in touch with a Pan Am unit in Frankfurt on HF long range radio and had received much information. He was suspicious the poor conditions at Beirut had been exaggerated by the authorities to prevent a landing and he did not expect any problems.

Suddenly the PFLP officials in the tower spotted Lebanese Army vehicles and troops massing on the manoeuvring areas and they became extremely agitated. Word had also spread amongst the Palestinian community in Beirut of the excitement at the airport and a number of heavily-armed PFLP commandos had positioned themselves in front of the terminal building. An ugly scene was brewing and there was a great deal of tension.

'Do not land pending further instructions!' ordered Abou-Ahmad from the tower. 'There are army vehicles at the airport. Do not land now!'

'Delay the aircraft landing? Why?' demanded Gueye, confused.

'In case an attempt is made to approach the aircraft, they are not going to be happy at all.'

'Who is attempting to approach the aircraft? Answer! Who is attempting to approach the aircraft? Answer!'

'Can you postpone landing for the time being, brother?'

'Why? Give me convincing reasons.'

'Brother, brother. Abou-Ahmad is talking to you. We now see on the ramp army vehicles moving here and there, therefore any attempt to approach the aircraft from the army will not be tolerated by us at all. So please postpone the landing a few minutes to enable us to agree with them to pull out all army vehicles.'

* Visual Approach Slope Indicator.

'Okay, will inform the captain of this. I want to tell you something which you must accept and understand quickly, in that, if any obstacles develop and we could not land at Beirut, in about ten or fifteen minutes we will not be able to go to any other airport as the fuel would not suffice.'

'Okay, I heard you, you will definitely land at Beirut Airport for refuelling, but we beg you to postpone landing for a few minutes in order for us to come to an agreement with the army to withdraw all his vehicles.'

'Fifteen minutes enough?'

'Fifteen minutes in order.'

'Okay, will postpone landing for fifteen minutes.'

'Clipper 93 calling Tower,' radioed Priddy.

'Go ahead.'

'Okay, this is the captain speaking. How about clearing that runway so that we can get down, huh? Do not cause us any more trouble. These folks are serious and are willing to destroy the airplane.'

'Okay, sir. We are going to give the military here orders to clear all the military around the runway so as to give way for landing safely.'

'Okay, as fast as you can, huh?'

'Okay, as fast as we can.'

'We just want to land at Beirut,' restated the captain.

'Do you prefer to land on runway 03 or runway 21?'

'We'd rather land on 03.'

'Roger. Stand by, please.'

Unknown to any of the communicators, other ears were listening to the conversation. The border with Israel lies less than sixty miles (ninety-six kilometres) to the south of Beirut and Israeli controllers were monitoring the Lebanese airport's frequencies. Suddenly a new voice was heard on the air.

'Hey, Clipper 93, this is Tel Aviv Control. Come on down! We've got two beautiful, long, lighted runways — we got emergency equipment standing by — we'll give you fighter escort. You just make a dash for it.'

The two hijackers went through the roof and Gueye became hysterical. He waved the grenade around and threatened to detonate it immediately. The Israeli controller kept trying to persuade the 747 to turn south for Tel Aviv, in spite of Captain Priddy's protestations, and Gueye was going to blow them all up on the spot. Eventually the captain managed to persuade the Israeli controller to keep quiet and he was able to pacify the hijackers. Alarmed now at how easily the situation could get out of hand Priddy stressed his concern once again.

'From Clipper 93. They say they know the airport and if there are any tricks they will blow up the airplane. So cut out the horsing around down there and get the runway cleared and get us down, huh?'

'Okay, that is what we are doing now.'

The tower controller repeatedly asked the captain for his endurance — how long he could stay in the air — but Priddy, not wanting to reveal that he still had plenty of fuel, simply kept saying that an immediate landing was required. He also radioed Panops Beirut on the other box, asking them to impress upon those concerned that the hijackers were serious in their intentions to blow up the aircraft.

'Understand,' replied Panops. 'For your information, I personally saw some groups of forces, policemen and army groups, pulling out of the airport in order to avoid a clash with the men on board. I would say that it is about to be cleared.'

Ten minutes later, after almost two hours of circling, descent clearance for landing at Beirut Airport was finally received. F/O Levix changed frequency to radar control and the Pan Am aircraft was vectored on to the approach for runway 03. After a short while the 747 at last lined up on the final approach to land and a large crowd, including many armed PFLP commandos, gathered to watch its arrival.

'Clipper 93, eight miles from touchdown,' informed the radar controller.

'Okay, we will stay with you for the moment . . . now we have the runway in sight.'

'Understand you have the runway in sight. Take over visually to runway 03 and call Tower, frequency 118.9.'

'Okay.'

Levix changed frequency but before he could establish contact the tower controller transmitted.

'Clipper 93, Beirut Tower, your range please.'

'Okay, we are about eight miles out and we have the runway in sight.'

'Roger, we have the VASI on now. Cleared to land on runway 03, wind east at five knots.'

'Okay, understand the wind is east at five knots and cleared to land, runway 03.'

'Okay, and I suggest you try to stop as soon as possible, please. We will take you to the parking.'

'Okay, I will try to run short if we can. I will stay on your frequency for taxy and instructions. Do you understand?'

'It is correct,' replied tower, 'remain on this frequency.'

'Okay.'

'East section cleared to parking. Use the first left inter-section and pick up the follow-me jeep.'

'Okay, we understand that.'

'Clipper 93, this is Beirut Tower.'

'Okay, go ahead.'

'After you stop, call the frequency 121.9, please, and let somebody of the people who are with you talk to us.'

'Okay.'

At 20:40 GMT, 22:40 local time, the Pan Am 747 touched down safely, if a little roughly, on the uneven surface and observers watching from the terminal clapped as it landed. A follow-me truck led the aircraft from the runway to the east parking area and as the big jet approached the terminal it turned to face away from the building. Captain Priddy set the parking brake and called for the engines to be shut down.

Abou-Khaled and Abou-Ahmad descended from the tower and mounted a jeep while another PFLP party member, Salah, maintained contact from the control tower on the ground frequency of 121.9 MHz. The hijackers could also communicate by radio with the officials in the jeep and as the vehicle approached they were ordered to stop and disembark. Anxiously the hijackers, on edge and suspicious of every move, watched from the flight deck. The jeep sped rapidly back towards the terminal and the two PFLP men approached the big jet on foot. Neither group could see clearly what was happening in the darkness but the men on the ground were accepted and the left forward door was opened. After conferring with the officials, instructions were given to the captain to order maintenance steps which were eventually placed by the door and the PFLP men boarded the aircraft. The door was closed and the steps were removed.

As the four men on the aircraft discussed the situation, Salah, in the tower, awaited instructions from the party members in Amman. A reply was expected in approximately half an hour regarding the fate of the Pan Am aircraft and its occupants. In due course orders were received and Salah transmitted the details to Abou-Khaled in the 747 cockpit.

'Take the following instructions,' commanded Salah. 'News received from Amman to the effect that you first refuel. This means that the first thing you do is refuel, and if they refuse refuelling here, we will blow up the aircraft at Beirut Airport, of course after you disembark. Also, of course, we have notified the airport manager of this, and he is making the necessary arrangements for you to refuel. After you refuel, the same guys who are commanding the aircraft now will head on it to Cairo. In Cairo they will precisely act in like manner as Abou-Doummar and Leila Khaled have acted. Clearly, they will act the way Abou-Doummar and Leila Khaled have acted, but they have to avoid all the mistakes that occurred in the operation of Leila Khaled and Abou-Doummar at Damascus Airport. You remember what happened at Damascus Airport? — the way Abou-Doummar and Leila Khaled acted — but they have to avoid the mistakes that occurred. This means that the operation must be accomplished in perfect manner, the front and the rear, that is, the nose and the tail.'

In August of the year before, Abou-Doummar and Khaled had hijacked a TWA 707 en route from Rome to Tel Aviv via Athens. The jet had been forced to land at Damascus where a bomb had been detonated only moments after the passengers and crew had been evacuated. Fortunately for the airline the damage was not irreparable and about one-and-a-half months later the aircraft had been returned to service. There was to be no mistake this time. The 747 was to be effectively wired with explosives at both the front and the rear of the fuselage and to be completely destroyed.

The destination, Cairo, had not been chosen without reason. The Egyptian president, Nasser, had permitted peace negotiations to be conducted with the Israelis in New York and the parties were moving towards agreement. The Palestinians, however, had been excluded from the talks and the blowing up of a 747 at the Egyptian capital's airport, would effectively demonstrate their displeasure.

Discussions between the PFLP men in the control tower and the 747 now centred on the refuelling of the aircraft. The hijackers were extremely nervous that the process might be used to launch a rescue attempt and instructions were given that only the minimum number of refuellers necessary would be permitted to approach the aircraft. They were also to be closely monitored. The Lebanese Army, although withdrawn, could be quickly recalled, and the large number of armed PFLP men in the airport area created a very tense atmosphere. The situation was very delicate and in the darkness any wrong move could turn the parking apron into a battle ground. The hijackers also had a problem in explaining to Salah that they had no explosives on board and these items — code named 'sandwiches' — would have to be loaded. Eventually the message was understood and arrangements were made to secure the explosives from downtown.

'During this period,' radioed Abou-Khalad from the 747, 'kindly secure the items required by the guys here for travel to Cairo. There is possibility that one of us would travel with the guys to Cairo. Did you hear me, Salah?'

'I heard you, Abou-Khaled,' replied Salah, 'and have arranged everything. All the items will be available quickly. Samir El-Saheb and Captain Ali are going to supervise five other elements to refuel the aircraft. And "sandwiches" will be sent to you so that the comrades may have dinner.'

The Lebanese authorities, only wanting to see the back of the Pan Am flight, offered Captain Priddy as much fuel as he wanted. Full tanks would have been ideal, of course, but the aircraft was restricted in the fuel load it could carry because of the short distance to Cairo. The flight from Beirut to Cairo was less than a hour and the 747 would have to land at its destination at or below the maximum landing weight. A figure was agreed and refuelling began. Meanwhile, Salah passed over the radio further instructions from Amman to one of the hijackers on the 747 flight deck.

'You will execute the plan you will now work out at your end. That is to say that you, Abou-Khaled, the comrade you have, and Abou-Ahmad will work out a plan as to how you should act there. With regards to the passengers, deplane them quietly and smoothly. After the passengers deplane, determine the distribution of your responsibilities as to who goes in the front and who in the rear. The quantity of the sandwiches which we will give to you — you will amuse yourselves with them on the way — arrange them properly in order for your plan to be a complete success. This means that you will work out all the details with Abou-Ahmad and Abou-Khaled. Tell them this for me. You have received the instructions. Make out the arrangements, work out a detailed plan. That is your mission. Consider all plans previously made as cancelled. All of you aboard the aircraft work out a complete plan as to how you should act there. Secure the passengers deplaning. Of course, it is important for us that the passengers deplane safely.'

As the preparations for departure progressed the tension on the aircraft eased and the passengers were able to relax a little. Drinks were served and they were even permitted to stretch their legs on board. A political statement was read over the PA by one of the guerrillas. While waiting, the hijackers then took it in turns to walk through the cabin to talk to the passengers. Where did they come from? Had they any questions? The Palestinians' manner was friendly and polite, but the pistol in one hand and grenade in the other were readily evident. Few cabin occupants wanted to converse with them.

The explosives arrived about halfway through the refuelling process and arrangements were made to bring them to the 747. The 'sandwiches' were packed in a large seaman's bag which weighed about eighty lb (thirty-six kilogrammes) and two men were needed to carry it. The maintenance steps were drawn up again to the left forward door and nine guerrillas, including one woman, brought the heavy explosives bag and a quantity of small-arms on to the aircraft. The Palestinians aboard welcomed their comrades warmly and they embraced each other by the door. When all were gathered in the 747 the door was closed and the steps were removed once more. A conference was held in first class and one of the joining group who had accompanied the refuellers to the aircraft, a Captain Ali, discussed the distribution of the 'sandwiches' on the aircraft. He was the explosives expert who was going to travel with the two hijackers on the trip to Cairo. Captain Priddy and his colleagues were waiting in first class when the group of guerrillas boarded and the meeting was held, but with all conversation in Arabic they understood nothing. Priddy asked if he could go back to talk to the passengers and he had a few words with each and every one. The captain was later followed into the cabin by several members of the guerrilla group who, like the hijackers, were friendly towards their hostages and encouraged questions. They also tried to justify their behaviour and to advertise the Palestinian cause.

By now the refuelling was nearing completion and Captain Priddy returned to the flight deck to supervise the details. He sat in the cockpit with the hijacker, Gueye, and tried to persuade him to release the passengers. Couldn't he fly to Cairo with just the crew as hostages?

'No, you don't understand,' admonished Gueye.

'What will happen when we get to Cairo?'

'Everyone will get out of the airplane, I promise you,' assured Gueye.

The fact that the aircraft was going to be blown up was left unmentioned, as were a number of other details, but Gueye had never lied to the crew and the captain felt encouraged. As Priddy checked the calculations, Salah in the tower spoke some last words to the hijacker, urging him to be extremely cautious and offering him encouragement.

'Now I am going to go down,' replied Gueye, ' and I will leave the captain for a little while in the cockpit, in order to make the arrangements for the Cairo trip. This means that I will be away from the cockpit for fifteen minutes in order for him to get ready and also to send the crew to him.'

By now Diop had returned to the flight deck and he was

left in charge as Gueye descended the stairs to supervise the final preparations. The two other flight crew were sent back to their posts and Captain Priddy and his team began to get the 747 ready for the flight. Departure was expected in forty-five minutes. About half an hour later, at 22:52 GMT, just before 01:00 local time on Monday morning, the refuelling was completed and the fuel bowser and ground personnel were withdrawn from the aircraft. The forward left cabin door was opened once more and a group of PFLP represent-atives on the ground spoke to those on the aircraft. All of the joining group were instructed to leave, except Captain Ali, the 21-year-old explosives expert, and the maintenance ladder was repositioned at the door for them to disembark.

All preparations were now completed and as the door was closed and the area around the aircraft cleared, the engines were started. A few minutes later all engines were running and at 23:30 the 747 taxied out to the runway. F/O Levix radioed for air traffic control instructions from the tower and the controller read back a flight clearance to Amman.

'We want clearance to Cairo,' corrected Levix. One of the hijackers also butted in saying that they were not going to Amman.

'Well, you're cleared for take-off,' sighed the tower controller, 'for wherever you want to go.'

Approaching the runway a clearance to Cairo was finally received and one of the PFLP representatives in the tower informed the hijackers that radio communications were to cease. The 747 was to take-off and proceed en route in radio silence. On no account were other stations to be informed of the aircraft's destination.

Captain Priddy started the take-off roll at 23:36 and one minute later, almost three hours after touch-down, the Pan Am aircraft lifted off for Cairo. The flight soon settled down to routine and all was as usual except for the lack of radio communications and the one hijacker who remained constantly on the flight deck. The passengers were served a meal and the children were permitted to run free. A pretence of normality returned to the situation and all on board remained calm. Only one young man, Captain Ali, busied himself on the aircraft, swiftly planting the explosives in closets, toilets and other parts of the structure. Between the charges he cut and laid the explosive fuze, a simple burning type which would have to be ignited by a naked flame to start the sequence. At the beginning he cut just sufficient for an eight-minute delay. He knew his business well and he was cool and efficient. His quiet manner masked his actions and many passengers were unaware of what he was doing. Ali concentrated on the front section of the 747 and as he worked in the first class cabin he was observed by those people segregated in that section. Amongst them was the Service Director, John Ferruggio, who later asked one of the hijackers what was going to happen.

'We will give you eight minutes,' he answered.

'Eight minutes for what?' questioned Ferruggio.

The hijacker refused to reply.

'What about this?' asked the service director, pointing to the explosives.

'What do you care about this imperialistic piece of equipment?' sneered the guerrilla as he looked about the 747.

'I think it's a beautiful piece of equipment and, imperialistic or not, I do care.'

The hijacker paused for a moment then added matter of factly.

'It's going to be blown up. You will have eight minutes to get out.'

At this stage the 747 was passing to the south of Cyprus, in the Nicosia flight information region, and about forty-five minutes flight time remained to Cairo. Ferruggio was concerned about the short period allowed for evacuation and he asked if he could brief the crew and passengers on procedures. He summoned all the available crew, including the four flight crew travelling passenger, to first class and told them what he knew. He suggested that the passengers be broken into groups and each section be instructed by flight attendants on evacuation procedures. The four off-duty flight crew were each asked to position by a door carrying a baby. When the chutes were deployed they were to jump first with the infant and the mother was to follow immediately. The parent was to take the child on the ground and the crew member could then help the evacuees at the bottom of the slide.

The task of dividing the passengers into groups and instructing them in evacuation procedures was carefully undertaken. An impression of urgency had to be imparted without telling everyone that from the moment the wheels touched the runway they would have only eight minutes to get out before being blown to pieces. Remarkably, much to the credit of the crew, the passengers remained calm and prepared themselves mentally for the immediate evacuation. All knew exactly what to do. They could take with them only the valuables they could wear and the ladies could carry nothing larger than a small, soft shoulder bag. All shoes were to be removed and thrown into blankets in piles for later return. When they got on the ground they were to run as fast as they could away from the aircraft.

As the 747 entered Cairo flight information region, the aircraft began descent in the darkness maintaining radio silence. The night was clear and approaching the city at 00:50 GMT, 02:50 local Egyptian time, Cairo Airport could be seen below. The tower controller tried many times to establish contact but the flight crew was forbidden to respond. By now Captain Ali had effectively completed his grim task and he had joined Gueye on the flight deck. Diop remained on guard in the cabin.

'Circle over the city,' ordered Gueye.

Ali and the hijacker spoke excitedly to each other in Arabic, pointing to sights they could see on the ground. The guerrillas were taking no chances and were checking carefully that Priddy had led them to their desired destination. As they flew over the Nile in the centre of the city Gueye confirmed their position.

'OK, it's Cairo, go ahead and land.'

Captain Priddy flew the 747 onto final approach at 01:20 and Gueye broke radio silence, telling the controller in Arabic that they were about to land and confirming that the aircraft would be blown up on the ground. Cairo had already been informed from intercepted radio communications that the aircraft would probably be destroyed and they were anxious not to have their entire runway blocked. The controller asked that the 747 be stopped at the far end of the runway, and the hijacker obligingly agreed. The captain, however, had a different thought. Priddy and his colleagues had been isolated in the cockpit when the plan to blow up the aircraft had first been revealed to Ferruggio, and they had no idea of the hijacker's intentions. Captain Priddy knew, however, that the handling facilities at Cairo for a 747 would be non-existent. He felt that the best course of action would be to halt at a suitable turning point on the runway, manoeuvre without the aid of the towing truck, and taxy back to the terminal building.

On the final approach to land Captain Ali left the flight deck and returned to the forward cabin. He spoke to First Class Purser Augusta Schneider and casually asked her for a match. There was little she could do except hand him a box. Ali simply struck a light and ignited the fuze. The aircraft at this stage was still airborne and the lit fuze began to burn away time. The evacuation would have to be very quick indeed.

Captain Priddy landed safely in the darkness at 01:24, and as the aircraft rolled down the runway he began braking to stop at a suitable point.

'Go all the way to the end of the runway,' instructed the tower controller. Gueye also shouted at him to go all the way but Priddy butted in on the radio.

'I want to stop short enough that I have room to turn round.'

Priddy kept braking and stopped the aircraft. Gueye screamed at him to keep moving and the controller shouted over the radio to continue to the end. In the cabin, Service Director Ferruggio felt the aircraft come to a halt and, using a megaphone, he immediately ordered the evacuation. All doors flew open and the chutes deployed satisfactorily. Instantly the dead-heading flight crew, each holding a baby, slid to the ground, followed closely by the mothers and the other cabin occupants. The flight attendants were cool and efficient and the evacuation was orderly. It was just as well, for half the fuze time had already gone and only four minutes remained to detonation.

In the cockpit, Captain Priddy was still being shouted at

by both Gueye and the controller to continue further down the runway and, unaware of the circumstances, he opened up the thrust levers. Service Director Ferruggio was horrified to hear the noise of the engines and to feel the 747 begin to move while people rushed down the slides. He mounted the stairs to the upper deck two at a time and from half way up bellowed through the megaphone.

'For God's sake, stop. We're evacuating.'

Priddy jammed on the brakes and called for the emergency evacuation check list. Quickly the flight crew completed the checks, shutting down the engines and systems. Ferruggio rushed back below, the trip up and down the stairs taking only about thirty seconds, to find all the cabin occupants, including Ali and Diop, had gone. They had been evacuated in less than a minute and a half. The service director grabbed two blankets laden with shoes and threw them down the slide, then followed the bundles himself. When he hit the ground he joined the passengers running clear. By now the security forces were closing in on Ali and Diop who were mingling with the passengers and the two men opened fire. Ferruggio could hear the crackle of shots behind him as he ran and he feared that Gueye had shot the flight crew still sitting in the cockpit. Less than two minutes' fuze time remained.

On the flight deck, Captain Priddy and his colleagues, still very much alive, had completed their tasks and were ready for departure. Gueye still stood behind them, one eye on his watch, and he told the flight engineer, Dzuiba, to go. The two pilots rose at the same time but Gueye shouted at them.

'No, you two guys stay.'

Dzuiba left, and Priddy and Levix had no choice but to remain seated. The seconds ticked by on Gueye's watch. Were they going to be made to die with the hijacker?

'OK, go,' Gueye cried at last, 'and good luck.'

The two men rushed down the stairs leaving Gueye on his own. They quickly ran along the entire length of the 747 and back, Priddy down the right aisle and Levix down the left, to check that all the passengers had gone, but when they completed their search they found Purser Schneider still picking up shoes. Priddy shouted at her to get out and she jumped down the nearest chute while the two men slid from the left forward door. By now Gueye was in hot pursuit and he followed the two pilots to the ground. The security forces at the airport were returning the guerrillas' fire and a lot of shooting could be heard. Tracer bullets arced through the night, striking the side of the big jet. In the darkness the situation was very confusing. Priddy and Levix took to their heels and ran for their lives. Any second now the 747 would go up and they would need to be well clear to avoid injury. As the men were only twenty-two yds (twenty m) from the fuselage, still inside the wingspan of the aircraft and fleeing past the outboard engine, the bombs planted at the front of the big jet exploded. The entire cockpit area was blown to pieces and the pilots could feel the blast in their backs, but fortunately they escaped injury. The night sky was lit up with the eruption. Moments later an enormous blast occurred at the rear, and soon after the entire top section lifted off. In seconds the 747 was a burning inferno.

Somehow everyone managed to scramble clear and only minor injuries were sustained. When the captain reached an airport bus nearby he found passengers lying on the floor,

The charred remains of the brand new 747 on the runway at Cairo the following morning. (Captain Pat Levix)

staying low to avoid the gunfire. The Egyptian driver was frozen with terror. From too close a range the occupants caught glimpses of the 747 disintegrating, but the driver refused to move. Priddy quickly took over the driver's seat and drove the bus to safety. Captain Priddy's action was typical of his conduct throughout the hijack, for his behaviour had been outstanding. His calm voice over the PA and his reassuring manner in the cabin, much of the time with a pistol at his head, had sustained the passengers throughout the event and his coolness had brought them through safely. It took over an hour to round up all the passengers and crew but eventually each and every one was accounted for. The brand new aircraft, unfortunately, was totally destroyed and only part of the tail was left standing. The three guerrillas were arrested by the security forces and seven passengers were hospitalised, while the remainder were accommodated in the airport hotel. Later that day the runway was cleared and a Pan Am 707 was dispatched to Cairo from London to collect those people fit to travel and to speed them on their way. The hijack had been a harrowing ordeal for the passengers and crew of the 747 but at least they were now free from danger. The entire crew had behaved in an exemplary fashion and their efforts had undoubtedly saved many lives.

At Dawson's Field in northern Jordan, the hostages aboard the hijacked TWA 707 and Swissair DC-8 aircraft were still being held and faced the first of many a hot and uncomfortable day in the baking desert. That afternoon, about 100 women, children and elderly passengers were freed. In return for the safe release of the approximately 200 remaining detainees and the two airliners, the PFLP demanded the release of seven Palestinian captives: three guerrillas serving twelve-year sentences in Switzerland for a machine-gun and grenade attack in February of the previous year on an El Al aircraft at Zurich Airport, three terrorists from a rival group, the Action Organisation for the Liberation of Palestine, being detained in Bavarian jails in West Germany for a bomb attack in the same month on a bus load of El Al passengers at Munich Airport, and the lovely Leila Khaled, being held by the police in the UK.

A few days later, on 9 September, in order to press the British for the release of Khaled, a BOAC VC-10 was hijacked. The aircraft was flying en route from Bombay to London, via Bahrain and Beirut, and was commandeered as it departed Bahrain with 105 passengers and ten crew on board. The captain, Cyril Goulbourn, was forced to refuel at Beirut and was made to fly at gunpoint to the desert strip. The PFLP then held three aircraft at Dawson's Field and over 300 hostages. The airliners were surrounded by heavily armed PFLP commandos with anti-tank and anti-aircraft guns. Beyond the Palestinian positions, the Jordanian Army, helpless to interfere, ringed the defensive circle at a discreet distance with tanks and other armoured vehicles.

Representatives of the five nations involved, Britain, Israel, Switzerland, the United States and West Germany, met to discuss joint action and, with little bargaining power, moved to succumb to the PFLP demands. Hostages continued to be released in batches to join the women and children freed earlier until by 13 September only 56 people, mostly Israeli or dual American-Israeli nationals, were still held. On that day, amid negotiations for their release and return of the aircraft, the three airliners were blown up and totally destroyed. In retaliation the Israelis arrested 450 prominent Arabs as an assurance against the safety of their nationals. Two weeks later, by 29 September, negotiations were finalised and only six hostages remained. In Amman, battles flared between Palestinian commandos and the Jordanian Army and open war erupted. The next day, President Nasser died of a heart attack. By the end of the month all the hostages had returned home safely and on 1 October the seven terrorists held by Britain, Switzerland and West Germany were freed. Hundreds of detainees and prisoners were released by the Israelis. So ended the blackest period in the history of civil aviation.

Chapter 7
Ice Cool

Alaska lies to the north-west of Canada and is America's most northern state. Its vast area stretches from 130° west at the border with British Columbia to 173° east at the far end of the Aleutian Island chain. At the Bering Strait, lying between the USSR and the USA, only fifty-six miles (ninety kilometres) of ice separates the two antagonists. Alaska, in fact, was purchased from the Soviet Union in 1867 for $7.2 m, and to this day its sale must upset the Kremlin. The land is rich in minerals, and oil is a major industry. Fishing is prominent too, and there is also a strong military presence. The Distant Early Warning (DEW) radar line stretches from Alaska to north-eastern Canada and defends the North American continent from attack by Soviet missiles.

Alaska is a cold and mountainous land and the early pioneers who first trekked north in search of gold, and then of oil, were as rugged as the landscape they conquered. Alaska, however, is also a land of beauty, and the State boasts America's highest mountain, Mount McKinley, standing at 20,650 ft (6,194 m). The scenery, as well as the ice cold air, takes your breath away.

Anchorage, the State capital, lies on the shores of Cook Inlet, so named after Captain Cook, who anchored in its sheltered waters on his third voyage in 1778. Hence, also, the name of the town. In the pioneering days Anchorage was a tough and wild place, but today it is a modern and sophisticated city of 200,000 inhabitants. Situated at 61°N, just south of the Arctic Circle at 66½°N, which marks the extent of perpetual daylight in the northern summer and perpetual darkness in the northern winter, the days are long in the brief summer and short in the many months of winter. The weather is harsh, being cold and snowy for most of the year, with only a three-month respite from June to August. Travel in the rough and extensive territory is mostly by air, with many of the excellent cross-State highways being impassable for months during the winter. Light aircraft in Alaska are used as much as automobiles elsewhere. Alaska Airlines connects the north-western State with 'outside' America while many small carriers ably operate within the region. One such local airline is Reeve Aleutian Airways (RAA) which, from its base in Anchorage, serves the 10,000 inhabitants of the Aleutian Island chain. The communities are remote and stretch from the southern shore of Alaska to Shemya Island, 1,500 nm away. RAA was started in 1932 by Bob Reeve at 'about the time that the aircraft was replacing the dog team'. Today the company is run by his son, Dick, who is owner and president of the airline. Reeve Aleutian operates nine aircraft, two Boeing 727-100s, four Electra L-188s and three Nihon YS-11s, and employs 240 staff. The 727 operations are limited to Cold Bay, which has a 10,500 ft (3,200 m) runway, constructed during World War Two, Adak, a US Navy base, and Shemya Island, a US Air Force Station at the end of the chain. The Electra and YS-11 turboprops fly the entire route network from King Salmon, near the mainland, through such places as Port Heiden and Dutch Harbor, to Shemya at the tip of the Aleutians.

Along the line of the islands, cold air from the frozen

A Reeve Aleutian Airways YS-11. (Reeve Aleutian)

Bering Sea to the north meets warm air from the Pacific currents to the south, and intense frontal activity results. Weather conditions are difficult and changeable with fog in the summer and storms in the winter, and operations are rarely planned for night: the environment in the daytime is enough to cope with. The Aleutian Island chain, however, hooks down as far south as 53°N and the temperature variation is not as bad as might be imagined. Winter temperature lows in the islands are of the order of −7°C and summer highs are around 13°C. In the moist atmosphere, however, the temperature range is ideal for icing and it is a major hazard to aircraft.

Most freight to the Aleutians is shipped by sea, but RAA carries a lot of mail and all the airline's aircraft are in a combined cargo/passenger arrangement with the cargo compartment at the front. At short notice the configuration can be changed, if required, to carry extra passengers. Load factors tend to be low but, surprisingly, Reeve Aleutian carries 50,000 passengers annually, with half that total being carried in the three months of summer. The Electra is the workhorse of the airline and seems to be ideally suited to the conditions and terrain, with the YS-11 giving useful support. Both can fly in weather and to airports unacceptable to the 727. The YS-11s are costly to run and maintain, and in the combi role have only 2.2 tons cargo capacity and eighteen seats available, but the airline owns them outright and can make a useful profit from their operation.

Weather conditions, especially in winter, are a major problem, and the long experience of the airline in flying in such harsh surroundings is of paramount importance. Stormy, violent and changeable weather, bad visibility, poor landing aids, gale force winds and icing in sub-zero temperatures all combine to form a working environment for Reeve Aleutian crews which is a tough one indeed. Fortunately the pilots are as robust as the machines they fly and as cool as the ice with which, in the long winter, they have to contend daily.

Early on the morning of 16 February 1982, a Nihon YS-11A, registered *N169RV*, was prepared to operate Reeve Aleutian Airways Flight 69 from Anchorage to Cold Bay, via

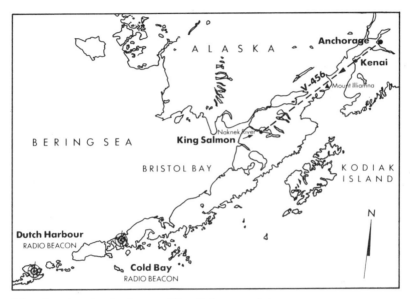

Map showing the Anchorage-King Salmon routeing.

King Salmon and Dutch Harbor. The aircraft had been in the hangar all the previous day and while there fuel had been drained from the fuel sumps to check for water. No contamination was found. At 06:00 local Alaskan time (now used throughout) on the day of departure the YS-11 was removed from the hangar and towed in the darkness to Gate 18 at the terminal. The sky was overcast at 1,200 ft (366 m) with five miles (eight kilometres) visibility in light snow, but fortunately the wind was light. The temperature indicated − 21°C. The wing fuel tanks were topped up with 900 gallons (4,090 litres) of Jet A fuel to give a total fuel load of 1,350 gallons (6,137 litres). Once again, following standard practice, the fuel sumps were drained for a further contamination check, but no water was detected. The examination was conducted with great care for any water remaining would quickly freeze in the very cold conditions and would block filters and disrupt the fuel flow.

N169RV was a sturdy, twin-engined turboprop powered by Rolls-Royce Dart 542 engines. The aircraft was the Japanese equivalent of the HS748 and had been manufactured in 1971. At just over ten years of age it had completed only 4,385 flying hours. The machine had been acquired by RAA two years previously and had given good and reliable service. Flight 69 was carrying little cargo and the seating area had been expanded to accommodate thirty-six passengers who were travelling on the route. By 07:45 all the cabin occupants were on board and the commander, Captain Thomas Hart, prepared for departure. Captain Hart was forty-five and had 13,500 flying hours to his credit, 4,200 of which were on the YS-11. He was a check captain with the company and was also qualified to fly the Electra, DC-3 and C-46 aircraft. His co-pilot was 34-year-old Roger Showers. First Officer (F/O) Showers had only just converted on to the YS-11 but he had flown 6,500 hours on other types. He was a qualified flight instructor, a certified mechanic, and was also rated to fly helicopters. In the cabin one flight attendant, Cecilia Allen, looked after the passengers. She was known to the crews as Ceecee and she had kept secret that she was pregnant and was about to leave the company.

As the door was closed the co-pilot copied the clearance. The take-off runway in use was 32 and Flight 69 was cleared for an Anchorage 4 departure: after take-off the YS-11 was to climb on runway heading till 400 ft (122 m) and then to turn left on to a track of 300° magnetic. On that course the aircraft would be issued headings on to the assigned low-level airway, Victor 456, which routed via Kenai, then southwest down the Aleutian Peninsula abeam Mount Iliamna and on to King Salmon on the Naknek River. The planned cruise altitude was 12,000 ft (3,660 m).

At 07:55, in the darkness with light snow failing, Flight 69 lifted off from runway 32 at an all-up-weight of 54,220 lb (24,594 kg); 8,800 lb (3,992 kg) of fuel was contained in the wings. About ten minutes after departure the YS-11 established on airway Victor 456, climbing to 12,000 ft towards Kenai. The outside air temperature began to drop rapidly and the crew had to keep an eye on the conditions. The potential problem in the cruise was not so much with airframe and propeller icing, for the air in the frozen north in winter is mainly free of moisture, but with the freezing of water droplets suspended in the fuel.

Checks of the fuel for free visible water had, of course, been conducted in the hangar, and after refuelling, but it was impossible, using such a method, to detect in the fuel dissolved water or water suspended as submicroscopic particles. Such dissolved or suspended water is usually present in jet fuel but its extremely small quantity does not affect normal combustion. When the fuel in the tanks is subjected to very cold temperatures at cruise altitude, however, it is a potential source of ice in the fuel system. Tiny ice crystals can form which can block fuel filters and the engines can stop with the disrupted fuel flow. To prevent fuel icing, therefore, fuel heating systems are installed on most aircraft.

On the YS-11 each engine is supplied with fuel from its respective wing tank. An engine-driven fuel pump first draws fuel from the tank via a fuel heater which is a fuel-air heat exchanger using hot compressed air bled straight from the engine compressor. The heated fuel is then fed through a filter and into a pump from where it is delivered at high pressure to the fuel control unit. The metered fuel is then directed to fuel nozzles in the combustion chambers. The

Captain Tom Hart.

fuel heater is controlled by a three-position switch on the flight deck, marked manual, off and auto. The normal setting for the switch is auto.

Any icing of the fuel filter results in a blockage which causes a difference in pressure between the inlet to the fuel heater and the outlet of the fuel filter. When the pressure differential reaches a certain limit a detector illuminates a warning light which is positioned by the fuel heat switch. With the switch selected to auto, compressor air is automatically bled from the engine and fed to the fuel heater. The fuel heat switch 'auto' position, therefore, is a safety setting designed to operate with inadvertent icing. With the switch in the manual position, compressor air is fed continuously to the fuel heater.

As the YS-11 climbed towards its cruising altitude a request was made to climb to 14,000 ft (4,270 m) and Anchorage control approved. Approaching 10,000 ft (3,050 m), however, passing over the Kenai VOR, the crew noted that the OAT was −25°C, lower than expected, and F/O Showers moved the fuel heater selector switches from auto to manual. Immediately a rise in the fuel temperatues was checked on the gauges as the fuel heaters began to operate. A few minutes later Flight 69 levelled in the cruise at 14,000 ft and, with the temperature now indicating −37°C, the fuel heaters remained on. A serious risk of fuel icing existed without their use. Some thin stratus cloud was encountered en route but no airframe icing was evident.

Flight 69's cruise towards King Salmon was routine, except for the very low temperatures, and by over halfway along the journey, abeam Mount Iliamna and the Iliamna Lakes, the OAT had dropped to −40°C. Later the temperature rose slightly to −38°C, but it was still well below normal for that altitude. F/O Showers established contact with King Salmon Approach at fifty miles (eighty kilometres) out and descent clearance was requested.

'Roger, Reeve 69,' radioed back approach, 'descend and maintain 7,000.'

After forty-five minutes in the cruise, at 08:50 and forty nm north-east of King Salmon, Captain Hart eased back the throttles to a reduced setting. The descent from 14,000 ft was commenced at a speed of 240 knots with constant engine power.

King Salmon Airport is used by both civilian and military traffic and the responsibilities are shared; the Federal Aviation Administration provides air traffic control personnel and the US Air Force supplies the fire and rescue services. There are two paved runways: 11/29 being the main runway at 8,500 ft (2,590 m) long and 18/36 the shorter at 5,000 ft (1,525 m). The airport lies almost at sea level and is bounded to the south and west by the meandering Naknek River which in winter freezes in parts to a depth of three feet (0.9 m). The weather in King Salmon was good and on the descent the crew copied the details from the automatic terminal information service (ATIS). The wind was 320° at ten knots and runway 29 was in use. Flight 69 was heading south-west and the aircraft would simply be able to position on a right base leg before turning right through 90° on to finals to land toward the north-west. The cloud cover was scattered at 5,000 ft and the visibility was a good fifty miles. The altimeter setting was 29.32 inches of mercury.

The crew set and checked their altimeters and approaching 7,000 ft (2,134 m) the captain advanced the throttles and held the aircraft level with the speed still steady at 240 knots. After one minute, Reeve Flight 69 was instructed to descend to 5,000 ft. Again Captain Hart eased back on the power and at the same speed slowly descended the YS-11 to 5,000 ft. The captain called for the descent and approach check lists and the drills were completed. Flight Attendant Allen reported that all the passengers were strapped in and that the cabin was ready for landing. Captain Hart set the power to 12,500 rpm and 850 lb (385 kg) per hour fuel flow, and began to decrease speed to 210 knots. The controller was slow in clearing the aircraft for further descent and the YS-11 was becoming a little high on the approach. The time was now just before 09:00 and the half-light of the northern winter day had dawned, but the visibility was very clear in the unpolluted atmosphere. As Flight 69 neared 5,000 ft the aircraft was positioned ten to fifteen miles (sixteen to twenty-four kilometres) to the north-east of King Salmon and the airport could be seen clearly ahead and to the right.

'Approach,' radioed Showers. We're just levelling 5,000 feet and we have the airport in sight.'

'Roger, Flight 69, you're cleared for right base for a visual approach to runway 29.'

The approach controller's instruction automatically cancelled the YS-11's flight plan and Captain Hart was now free to control his own progress. He could judge his own heights and speeds and could position the aircraft as he desired for approach and landing by visually sighting the airport. The YS-11 continued descent at 210 knots and at just after 09:00, as the aircraft passed 2,600 ft (792 m), five miles (eight kilometres) from touchdown, instructions were given to contact the tower.

'King Salmon Tower, Reeve 69,' called Showers, 'five miles right base for 29, been cleared for a visual.'

'69, King Salmon Tower, roger. Report turning final, runway 29.'

'69.'

'Three zero at one zero,' called back the tower, giving the wind velocity, but making a slip.

'Three zero at one zero?' questioned Captain Hart.

'Say again the last part of transmission,' radioed Showers.

'Roger, the wind at this time is three two zero at one zero, you are in sight and cleared to land runway 29.'

'Three two zero at one zero,' said F/O Showers to his captain, 'that was in the ATIS.'

'Thought he was giving us a squawk,' admitted Hart, referring to the four-figure code normally issued by radar controllers for setting on the transponder. 'He's not very clear today,' he continued, 'sounds like he's talking through a wet sock.'

'No, it's not very clear,' confirmed Showers.

The YS-11 was now two to three minutes from landing and, with the aircraft a bit high and fast for the short distance remaining, Captain Hart levelled off at 1,800 ft (549 m) and let the speed drop. The power setting was left constant but the speed fell to below 200 knots and continued to decrease. Height and speed would also be lost on the 90° right turn onto finals, so the situation did not pose a problem. Meanwhile, the co-pilot began to perform the before-landing check. The list would not be completed till the landing gear was down and locked but some items could be initiated by the co-pilot of his own volition. Once the gear was set he would read the check list in a challenge and response manner whereby he would call the items and both pilots would confirm the tasks accomplished. The sequence of the before-landing check list was as follows and, for visual approaches, was to be performed upon entering the traffic pattern and completed before turning onto final approach, except for the final flap settings.

Item	Response
Landing gear	Down and three greens
High pressure (HP) fuel cocks	High stop withdrawal lock (HSWL)
Landing lights	On below 165 knots
Fuel trim	Set
Prop lights	3 on and 3 off
Fuel heaters	Off
Flaps	Off
Water/methanol	On
Spill valves	Manual

Most of the items such as the landing gear, landing lights and flaps, etc, would be set in the normal flow of events but at this stage the first officer could set the HP cock to HSWL and could switch off the fuel heaters. The 'high stop' referred to in the second item above is a mechanical propeller pitch stop which is a safety device engaged during take-off and cruise. The propellers are variable pitch and if, for example, an engine fails at take-off, the stop prevents the windmilling propeller blades from decreasing to a flat pitch, i.e. from turning flat to the airflow, and so creating enormous drag. If such an incident happened during take-off or in the cruise, control of the aircraft could be lost, and the stop is designed to prevent its occurrence. The stop is removed during the before-landing check to permit finer pitch angles for landing and is accomplished by moving the HP cock to the high stop withdrawal lock. Two blue lights then illuminate to confirm the high stops have been removed.

Another selection for the HP cock is the 'feather' position and is used with an engine failure to reduce the windmilling propeller drag to a minimum. The HP cock is first selected to 'feather' to cut off the fuel to the engine and this is followed by the pressing of a 'feather' button. The propeller blades are then hydraulically driven in line with the fore and aft axis of the aircraft to allow the blades to cut through the air like a knife, reducing drag to a minimum.

The selection of fuel heat to off as part of the before-landing check is also a precautionary measure. As mentioned previously, hot air is bled directly from the engine compressors to feed the fuel heaters, and extracting air in this manner causes a reduction in engine power. At the full power setting, for example, power is reduced by four per cent when compressor air is bled for fuel heat. For maximum power availability in the event of a go-around, therefore, the fuel heaters are turned off during the approach to land.

With the HP cock selected to the high stop withdrawal lock and the fuel heaters switched off, Showers routinely scanned the engine instruments. The aircraft was now two minutes from landing; the first officer had performed the drills properly and the two pilots had complied correctly with procedures. Unknown to the crew, however, the fuel in the tanks, in spite of stringent checks, had been contaminated with more than the usual minute quantities of water. The amount would probably have remained unnoticed in normal circumstances but, operating the YS-11 in such extreme temperatures, fuel freezing problems were likely, even though the water content was very small and the fuel heaters were switched off at a very late stage. During the flight, with temperatures down to −40°C, the fuel in the wing tanks would have been very cold because of the long exposure to such icy air. As long as fuel heat was supplied any ice crystals in the system were melted and easily passed through to the engine as liquid. The fuel heater on this occasion, however, only raised the temperature of the fuel entering the system to slightly above freezing. As soon as the fuel heaters were turned off, the microscopic water droplets suspended in the cold fuel began to freeze. Ice crystals began to form in the small orifices and screens in the fuel control units and pumps, and began to impregnate the fuel filters and to restrict the fuel flow.

'Barrier one,' called the tower controller to another aircraft, 'what's your position now . . . ah, I've got you in sight. Remain clear of runway one one and two nine, landing traffic a YS-11 on right base.'

After a few seconds 'tower' informed both flights that men were working near the runway.

'Barrier one and Reeve 69, there will be men and equipment on the departure end of the runway on the left side. They are clear and outside the lights.'

First Officer Showers ignored the radio call for, as the controller spoke, he could see that there was a problem with the right-hand, number two, engine. The torque pressure indicator was dropping, as was the fuel flow. Approaching 09:02, with the height just under 1,800 ft (550 m) and the speed dropping below 190 knots, Captain Hart felt the aircraft yaw as power dropped on number two engine and the nose swung to the right. He pushed on the left rudder to hold the machine straight.

'We've lost one!' shouted Showers.

The number two fuel flow was well below 500 lb (227 kg) per hour when it should have been 850 lb (385 kg) per hour.

'Fuel flow . . . ' he called, pointing to the gauge.

The number two torque pressure was also down, indicating forty lb per square inch (psi) (2.8 kg/cm^2) instead of the 100 psi (7 kg/cm^2) it should have read with the throttle position at 12,500 rpm.

'Torque's very down.'

The captain advanced the number two throttle but there was no response. It was obvious the engine had failed. He turned on both relight switches and pushed the throttle forward once more, but to no avail. With the little extra height and speed in hand the YS-11 was, fortunately, in a good position for a single-engine approach and landing. The captain felt it was better to eliminate the drag from the right engine, thereby easing the control difficulties, rather than continue attempts to restart it, and he decided to shut the engine down.

'Okay, feather the . . . ' called Hart.

Showers selected the number two HP cock to 'feather' and pressed the 'feather' button. Fuel was cut to the engine and as it ran to a stop the propeller blades moved to the feather position.

'Feathered?' questioned the captain.

'It's feathered.'

Neither of the fuel filter differential pressure warning lights, which would have indicated a filter blockage, illuminated, nor did the low fuel pressure warning lights. Since the crew had complied properly with procedures the two pilots did not at first associate the engine failure with icing.

Feathering of the number two engine was completed as the speed dropped below 170 knots and under the circumstances the YS-11 seemed safe. The aircraft was still on base leg at just below 1,800 ft and as the speed dropped to 165 knots the captain continued the descent.

'Tell 'em we've got one shut down,' said Hart, 'and would like the fire trucks out.'

'And, ah, Anchorage King Salmon, Reeve 69,' radioed Showers, 'ah, get the fire trucks out. We've lost one engine on . . . we're turning final at this time.'

At this stage the flaps were selected to the approach flap setting and, with the YS-11 still a little high, Hart called for the landing gear to be lowered. As the wheels dropped he advanced the number one throttle to increase power on the left engine to stabilise the approach. Suddenly popping sounds could be heard from the number one engine and a smell of smoke could be detected on the flight deck.

' . . . . it!' exclaimed the captain, 'we're losing the other one.'

Fuel icing was also affecting the left engine but in a different manner. As the throttle was moved forward, ice crystals in the fuel control servo system caused large fluctuation in the fuel flow. Blocked orifices or screens in the system were preventing the fuel pump output from matching the throttle position and too much fuel was flowing into the engine. Excessive fuel was being admitted to the combustion chambers and an intense rise in heat was occurring without a corresponding increase in the propeller rpm to drive air through the engine for cooling. The insufficient cooling airflow was causing a severe over-temperature condition in the turbine and the engine was literally burning itself out.

The number one engine power began to drop as the popping sounds continued and the smell of smoke on the flight deck increased. The engine fire was close to the air conditioning duct for the cabin and fumes were being drawn into the aircraft. No engine fire warning was given. Immediately Captain Hart began the right turn to line up on the runway. At 1,200 ft (366 m) his height and position seemed good for a safe landing but the rate of descent increased rapidly and as a precaution the captain brought the flaps up.

'Say your fuel aboard and persons,' requested tower.

The controller was obviously trying to obtain the information for the emergency services but Showers was too busy to respond. As the wheels fell into place and were checked locked the co-pilot confirmed their condition.

'Gear down and three greens,' he called to his captain.

The number one engine now began to surge and the power dropped and then increased at random. With the number two engine shut down and the landing gear lowered the rate of descent increased dramatically.

'Okay,' said the captain, 'tell the girl in the back we've got a problem.'

Showers pressed the call button to the intercom in the cabin.

'Hello,' replied Flight Attendant Allen.

'Yeah,' said Showers, 'we've got a little problem here. We've shut down one engine and we're losing the other one. Prepare the passengers for an emergency landing and evacuation.'

Cecilia Allen went about her duties and as she did so the tower controller radioed with permission to land.

'Reeve 69, cleared to land runway two niner. The ah, emergency equipment has been advised.'

'69, roger.'

'Get the other one going,' said the captain to his co-pilot.

The number one engine continued to pop and surge and the YS-11 was losing height rapidly. Large fluctuations were noticed in the left engine fuel flow, approximately centered around 1,000 lb (454 kg) per hour, and at drops in the power the captain could feel the aircraft yaw to the left. The number one engine power now began to fade and there was a danger they might not reach the runway.

'Get the other engine going,' urged the captain. 'Get it going.'

The first officer had already moved the number two engine HP cock lever forward to the start position and he was in the process of pushing the 'feather' button to unfeather the engine. They were quickly running out of time but the number two engine resolutely refused to start. With both props now unfeathered and the landing gear lowered the drag on the aircraft was enormous. Rapidly the YS-11 dropped towards the ground.

'We'e not going to make it,' stated the captain simply, 'we're not going to make it.'

Keeping a cool head and thinking quickly, Captain Hart turned the aircraft 90° to the left and headed south-west towards the Naknek River. There was no time now to do anything except land on the frozen surface. That, however, was going to be easier said than done, for not all the river was frozen. At the sharp bend in the river to the south of the airport the stream slowed sufficiently for the river to freeze to a depth of about three ft (0.9 m) in the extreme temperatures. About one mile (1.6 kilometres) up and down stream from this point, however, the water was still unfrozen and only one week earlier, in warmer conditions, a water channel had been open along the length of the river. Captain Hart aimed for the two-mile (3.2 kilometres) stretch of ice.

'69 has lost both engines,' radioed F/O Showers, 'on final here to the river.'

'69, roger,' replied the controller, as he watched the stricken aircraft dive to the south of the airport. Immediately he alerted the fire and rescue services and directed them toward the Naknek River.

Captain Hart aligned the YS-11 with the banks of the ice-covered river and, with the speed about 120 knots, prepared for a rough landing.

'You want the gear up?' asked the co-pilot.

If a wheel caught in a rut on the ice it could tear off the undercarriage, but the smooth belly of the aircraft would glide easily over the frozen surface. There was also another consideration: the ice in the centre of the river was clear and appeared green in texture as the water could be seen flowing below the surface. It did not look to either pilot as if the ice would hold but there was nowhere else to go. If the surface gave way it would be better to land with the gear retracted rather than have an undercarriage leg smash through the ice.

'Yeah, put it up,' said Hart.

The first officer selected the landing gear lever to the up position but the retraction was slow because the hydraulic pumps were not operating fully with the engines windmilling. At the same time the captain selected both HP cocks to 'feather.' The right prop feathered immediately but the left side was slow to respond and Hart had a great deal of difficulty in controlling the machine. Both hands were needed to fly the aircraft. As the YS-11 rapidly approached the icy surface, the ground proximity warning system sounded its alarm.

'Pull up, pull up, pull up . . .'

Captain Hart wrestled to level the wings but he was unable to raise the left wingtip. With great skill he managed to pull the powerless machine out of the dive but just before landing the left aileron struck the ice. Fortunately little damage was caused. A moment later, with the gear still retracting, Hart touched down on the frozen river as gently as possible. The belly of the aircraft gradually settled onto the surface as the wheels retracted completely and the aircraft began to slide across the ice. Both props were now in the feathered position but with the right side stopped and the left one still turning. Immediately the prop blades on both sides dug into the ice and slashed the surface. The aircraft bumped across the ice, which was remarkably smooth, and banging and scraping sounds could be heard in the cabin. The nose gear doors were crushed, the fuselage belly skin was buckled in a number of places, but fortunately did not rip from the airframe, and the rotating safety beacon and several antennas were broken and bent. The YS-11 continued to slide and rumble over the ice. Captain Hart, still concerned that the ice might give away, applied right rudder and steered the aircraft towards a sand bank on the north side. Slowly the machine curved in an arc to the right. Suddenly, a red warning light illuminated and a bell rang indicating a fire in number one engine.

'We got a fire,' called the captain.

The required drill was to place the HP cock to the 'feather' position to cut off fuel at the control unit and to pull the fuel shut-off 'T' handle to cut off fuel at the engine, as well as to isolate the services. The fuel cocks were already in the 'feather' position so it was simply a matter of pulling the 'T' handle and pressing the button.

'Pull 'em both?' asked Showers.

The captain nodded and the first officer pulled the fuel shut off handles.

'Gang bar, fire bottles,' said Hart.

Showers pushed up the electrical gang bar and shut off the aircraft electrics.

'Fire bottles,' repeated the captain as the aircraft continued to bump over the ice.

The first officer pressed both fire buttons and discharged the extinguishant from the bottles into the engines.

As the YS-11 skidded over the frozen river it continued in a gentle curve to the right. After sliding for about half a mile (800 metres) it eventually came to an abrupt stop facing towards the west with the left wingtip resting on the surface. The aircraft still weighed in excess of twenty tons and, in spite of the three feet of frozen water covering the river, the ice began to crack as it settled. Immediately, Captain Hart ordered an emergency evacuation. The number one engine was still burning in spite of their efforts and, with the possibility of ruptured fuel tanks, there was a serious risk of fire. Showers quickly left his seat while the captain remained to secure the flight deck. The co-pilot ran to the back of the cabin and assisted Flight Attendant Allen in evacuating passengers via the aft cabin door on the right side.

In the cockpit Captain Hart completed his tasks then quickly radioed King Salmon tower with news of their predicament.

'Hey, this is Reeve 69,' called Hart. 'We're down on the ice, nobody's hurt, we're still on fire over here, though. And, ah, we had a fire in the air and lost power on the engines and couldn't get to the end of the runway. We had to make a quick left turn here but we're still on fire.'

Unknown to Captain Hart, help was already on the way and the fire trucks were speeding to the scene. Hart rapidly left the flight deck and went back to the cabin where he found some passengers struggling to open the right over-wing exit. Flames could be seen on the left side through the passenger windows and smoke was entering the cabin. It was imperative to get the people out quickly. The captain swiftly released the catch of the over-wing exit, pulled the handle, and threw the door out of the opening. He then assisted the passengers in evacuating by that escape route. In a matter of moments all thirty-six passengers and three crew were out of the aircraft and standing safely on the ice. Only one passenger received a minor knee injury while fleeing from the aircraft. Almost immediately the fire trucks also appeared on the scene and the fire in the number one engine was quickly extinguished. Fortunately the frozen surface, in spite of the cracking, held the additional weight. In due course transport arrived to take the passengers and crew to the terminal building, but not before some people suffered frostbite in the severe temperature.

A rescue operation was mounted to save the YS-11 and stakes were driven into the frozen river to secure the aircraft to prevent its movement on the shifting ice. Air bags were

The YS-11 back on its feet. Note the smooth surface of the ice.
(Reeve Aleutian)

The YS-11 being towed up the river bank on the specially constructed ice ramp. (Reeve Aleutian)

then placed below the wings and nose and inflated in turn to raise the machine to a sufficient height for the lowering of the landing gear. The YS-11 was then towed three miles (4.8 km) across the ice and was hauled from the frozen surface up a specially constructed ice ramp on to the river bank. From there it was moved to the airport for repair. The entire rescue operation took five days and was only just completed in time, for the next day the ice river melted. Two weeks later, YS-11 *N169RV* flew out of King Salmon Airport; an engineering feat almost as remarkable as the landing on the ice. The cause of the water contamination in the fuel was never discovered.

The incident had developed very quickly and a normal approach had swiftly developed into a potentially disastrous situation. Keeping cool heads, Captain Hart and First Officer Showers had reacted calmly and efficiently to the emergency and the captain's quick thinking and great skill had saved his passengers and aircraft from an almost certain catastrophe. Both pilots received Superior Airmanship Awards from the American Airline Pilots' Association.

Chapter 8
Roll Out the Barrel

Aircraft manufacture is a highly refined engineering art and modern jets are built to be extremely robust. The strength of contemporary airframes lies in their monocoque structure, whereby the skin takes much of the load although supported by a reinforcing frame. The central section of the aircraft is the strongest part with the form of the machine being more or less built around it; at this location the wings are attached and the landing gear fitted so the region is subjected to all flight and ground loads.

During normal operational flying aircraft are exposed to a variety of stresses: extremes of weather, large temperature changes — from +40°C on the ground to −70°C in the air — cabin pressurisation, wing flexing, landing and take-off loads and engine vibration. The aluminium-copper alloys mostly used in aircraft construction are light and flexible yet strong enough to cope with such conditions. Heavy landings, violent weather or severe turbulence can also be handled by modern jets with safety, although airframe checks may subsequently be required. Normal maximum manoeuvring loads in flight for modern jets are +2.5g to −1.0G, where 'g' is the force of gravity and negative 'g' the impression of weightlessness.

On completion of a new aircraft design, a full test programme is implemented, with an airframe being tested to destruction in a test rig. The wings are bent upwards by large jacks to test strength, and to witness the wingtips almost touching before failure occurs would dispel any fear of flying. On occasions, admittedly inadvertently, aircraft are put to the test in earnest during flight. Fortunately, in the few incidents recorded, the airliners involved have survived and the events have borne witness to the impressive strength of modern machines. One such incident occurred to a Boeing 747 SP (special performance) on 18 February 1985.

China Airlines Flight 006 was en route from Taipei, the capital of Taiwan, to Los Angeles, California, with the aircraft under the command of Captain Ming Yuan Ho. The B747 SP was shorter than the standard 747 model and was specifically constructed for such long-range flights, being designed to fly at high altitudes to conserve fuel. At just after 10:00 local time, Flight 006 was cruising at 41,000 ft (12,500 m) over the Pacific, with the aircraft lying 300 nm north-west of San Francisco. One and a half hours flying time remained to Los Angeles. The journey so far had been routine and the 254 passengers had just finished breakfast. Extra crew members were carried for the long duty day, and the five flight deck members and fifteen flight attendants brought the total on board to 274.

In the rarefied atmosphere at 41,000 ft and at a weight of 245 tons, the 747 SP's margin between the maximum and minimum flying speeds was narrow, and could have been as small as thirty knots. At too fast speeds, high speed buffetting caused by shock waves on the wings occurs and the aircraft can be damaged. At too slow speeds, low speed buffetting caused by turbulent airflow over the wings occurs and lift is lost from the wings. If the speed is allowed to

China Airlines 747. (John Stroud)

decay further the wings can stall and the aircraft can drop from the sky.

The China Airlines 747's autopilot was engaged and the autothrottle of the Performance Management System (PMS) was selected to control the speed at Mach 0.85 (i.e. 85 per cent of the speed of sound at that level). Flying conditions in the daylight were clear, with cloud 10,000 ft (3,050 m) below covering the ocean, but the changeable wind was causing turbulence and was upsetting the delicate balance of the flight. The speed fluctuated from Mach 0.84 to Mach 0.88 and the autothrottle moved in response in an attempt to stabilise the speed. Suddenly, as the thrust levers retarded automatically to flight idle to reduce excessive speed, number four engine suffered a problem. On achievement of the desired speed numbers one, two and three engines increased to the required setting, but number four engine

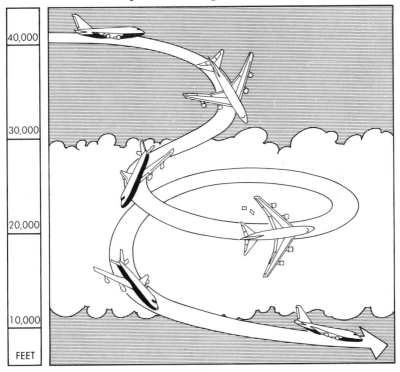

Flight 006's spiral dive.

failed to respond. It remained in what is known as a 'hung condition'.

Before flight, Captain Ming had noted from the technical log that number four engine had a history of ceasing combustion, or flaming out, when decelerating. In the turbulent conditions, with the autothrottle advancing and retarding the thrust levers in an attempt to hold speed, the number four engine had required special attention. If engine flame-out occurred, there would be insufficient power from the remaining engines to maintain speed in the rarefied atmosphere and the delicate balance of the flight would be impaired. Captain Ming would have no choice but to descend into the more dense air at a lower level.

The flight engineer attempted to restore power to number four engine but his efforts failed and ninety seconds later the engine flamed out. The speed began to decay and the aircraft yawed and rolled to the right. With no automatic rudder, the autopilot applied full left aileron to compensate for asymmetric thrust. Captain Ming disengaged the PMS and the height lock and eased the nose down in descent using the autopilot control wheel. The actions of the captain, however, were insufficient to contain the situation and when Ming disengaged the autopilot to revert to manual control he, and the other 273 occupants aboard Flight 006, received the shock of their lives. Instantly the control wheel snapped to the central position and the aircraft bank increased abruptly 63° to the right, with the speed dropping to Mach 0.75. The big jet began a steep dive towards the ground. Flight 006's nose pitched down to 67° and the aircraft rolled on its back to a bank angle of 160°. The 747 plummetted earthwards at 15,000 ft (4,570 m) per minute and entered the cloud layer which lay below. The crew were caught completely by surprise and in seconds were totally disorientated.

In the cabin all hell broke loose. 'Dishes crashed against the walls and floors,' recounted William Peacock, a Vietnam veteran and colonel in the Marine Corps Reserve. 'Baggage compartments opened up and window shades were forced down by vibration.' 'You could feel bits popping off,' commented another passenger. Those people not in their seats were thrown about the cabin and fifty passengers received minor injuries. Two flight attendants suffered serious back damage. The aircraft continued its spiralling plunge, screaming almost vertically towards the earth as it dropped at three miles per minute. As the 747 spun through the cloud the airframe and wings were subjected to five g, five times the force of gravity and twice the structure's design limit. The aircraft shook violently and the vibrations caused large portions of the horizontal stabiliser to break off. Both outboard elevator sections detached, the auxiliary power unit broke free and the right wingtip high frequency radio antenna separated. By rights the aircraft should have broken apart completely, but somehow the machine held together.

Captain Ming tried desperately to regain control but, with no visual reference and pinned to his seat with the 'g' forces, he faced a difficult task. There seemed little chance of recovery as the aircraft shuddered in a steep diving turn when, to complicate matters, the landing gear suddenly deployed. The powerful uplatches locking the undercarriage in the bays had released under the excessive 'g' forces and the landing gear had fallen down into place. The two body landing gear doors were ripped from their hinges and flew off into space. The lowered undercarriage, however, had an unexpectedly beneficial effect as the wheels biting into the airflow seemed to stabilise the aircraft. This gave Captain Ming the chance he needed and by an extraordinary effort he righted the falling machine. Passing 11,000 ft (3,350 m), with only forty seconds to go before the 747 plunged into the sea, he managed to pull the aircraft out of the dive. The three good engines were still functioning and the captain applied power. As Ming regained control, Flight 006 broke from the cloud and the aircraft was levelled off at 9,500 ft (2,895 m). Flying in the clear conditions, the crew were able to regain their composure and to take stock of the situation. Quickly Captain Ming spoke to the passengers and in a 'pretty shaky' voice instructed the cabin occupants to fasten their seat belts.

The captain made the decision to land as soon as possible to inspect the damage and Flight 006 was cleared to divert to San Francisco, the nearest suitable airport. Ming climbed the aircraft back to 27,000 ft (8,230 m) and one hour later, without further incident, the 747 commenced its final approach. Fortunately the crew were able to lock the landing gear down and to obtain a safe display on the indicators. Flight 006 landed safely and on touchdown Captain Ming and his crew received a spontaneous round of applause.

Mystery surrounded the incident for some months until a subsequent inquiry verified the facts. It was concluded that

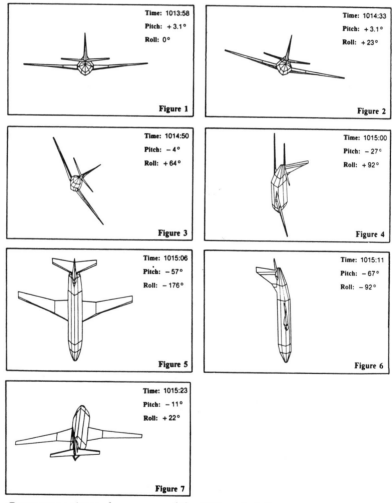

Computer-animated sequence of the 747's spiral dive. (NTSB)

the actions of the captain when dealing with the engine failure in the delicate circumstances of the high altitude cruise had been on the tardy side. As a result the autopilot inputs required to maintain control had been excessive. When Ming disconnected the autopilot he was caught completely off guard by the out-of-balance forces, and the machine had spun from his grasp. In defence of the captain, the International Federation of Airline Pilots (IFALPA) stated that too often on long distance flights aircraft commanders are compelled to fly at very high altitudes to lower fuel consumption in order to complete their journeys. Captain Ming had also done a fine job in regaining control of the stricken aircraft, albeit with some luck from the falling landing gear, and he deserved his applause on landing. The passengers, perhaps, should also have clapped the Boeing Aircraft Company, for the 747 was the real hero of the event. The aircraft was subjected to probably the greatest 'g' forces sustained by any wide-bodied civilian airliner, and survived. No clearer demonstration of the strength of modern aircraft could have been presented and the successful outcome was a fine testament to the skill of the aircraft designers.

Unfortunately the gross jet upset suffered by the China Airlines Boeing 747 was not the only such event recorded in aviation history in the last decade. In 1979, another Boeing machine, in this case a 727, was involved in a similar upset. The mystery surrounding that incident was never properly resolved and to this day the repercussions of the episode still affect those involved.

On 4 April 1979, TWA Flight 841, a Boeing 727-100, registered *N840TW*, prepared for departure from John F. Kennedy (JFK), New York, for a scheduled service to Minneapolis/St Paul International Airport in Minnesota. In command of the aircraft was Captain Harvey 'Hoot' Gibson. His co-pilot was First Officer Jess Kennedy and at the flight engineers' station was Second Officer Gary Banks. The crew was a very experienced trio with Captain Gibson having amassed a grand total of 15,710 flying hours at the relatively young age of forty-four. He had learned to fly at only thirteen years of age and after a period serving as an air traffic controller in Chicago had joined TWA in 1963. Gibson was not only qualified to fly the 727 but also had his licence endorsed to fly the DC-9, L1011 Tristar and the Boeing 747. He was one of aviation's 'all-rounders' and he flew helicopters and piloted balloons as well. Captain Gibson had not flown the B727 regularly for some time, however, for he had been operating as co-pilot on the B747 for just over one year. In December 1978 he had broken his ankle and the injury had kept him off work for three months. He had only just returned to his duties as captain on the B727 three weeks earlier and had completed ground school refresher, simulator and aircraft handling details. On 28 March he successfully completed a route flight check of over five hours' duration.

First Officer Kennedy was forty years of age and had joined TWA in 1967. He had operated as a flight engineer for eight years and in 1978 had transferred to the right-hand seat as co-pilot. He had a total of 10,336 flying hours, of which an impressive 8,348 hours had been acquired on the B727. Second Officer Banks was thirty-seven and had operated as a flight engineer for nine years. He was also a qualified commercial pilot. In the cabin were four flight attendants providing for the eighty-two passengers, bringing the total on board to eighty-nine.

The trip for the flight crew began in Los Angeles, California, the day before, on 3 April, and it was Captain Gibson's first time in command without supervision after his spell as first officer on the 747 and his sick leave. Duty had commenced at about 08:30 local time (11:30 Eastern Standard Time, now used throughout) and after a series of short sectors had terminated in Colombus, Ohio, at 22:00. After a night's rest the crew had recommenced duty the following day, Wednesday 4 April, at 13:45, and in the afternoon had operated Flight 841 via Philadelphia, Pennsylvania to JFK, New York. The B727 had arrived at Kennedy in the evening at 17:20. With only eighty-two passengers scheduled for the next sector to Minneapolis/St Paul, a comfortable turn-round should have been accomplished, but it was not to be. New York in the 1970s suffered from a surfeit of aircraft movements, and the mid-week traffic congestion was causing extensive delays.

Night fell as Flight 841 prepared for departure, and while

waiting for take-off a score of anti-collision lights could be seen twinkling ahead in the darkness. After forty-five minutes of taxying, three quarters of a ton of fuel had been consumed from the sixteen-ton load on board, even with one engine shut down to conserve fuel. Finally the B727 was given clearance for take-off, and at 20:25 Flight 841, weighing sixty-five tons, lifted off from JFK. The aircraft climbed steadily in the blackness to its planned cruising height of 35,000 ft (10,670 m) and twenty-nine minutes later, at 20:54, levelled in the cruise. The cockpit crew settled down to their usual routine and all was normal except for an exceptionally strong headwind.

The scheduled flight time for the 900 nm journey was just over two hours, but the very high winds being encountered slowed their progress. The ground speed was down to about 380 knots and at this rate the flight time would be nearer two-and-a-half hours. The B727 would be further delayed and additional fuel would be consumed from the load already eroded by the waiting time at Kennedy. Captain Gibson hoped the winds might slacken later in the flight and in the meantime the crew ate their evening meal.

Flight 841's routeing crossed the border into Canada over the northern end of Lake Erie, then passed over London, Ontario, before re-entering the States over Michigan. About one hour after departure the 727 entered Canadian airspace and First Officer Kennedy established contact with Toronto Control. Captain Gibson was still concerned about the very strong headwind which blew in excess of 100 knots and, with the aircraft now light enough to climb, he felt it might be prudent to change altitude. The decision, however, was not taken lightly. The 727-100 was not considered by pilots to be a comfortable aircraft to fly at such a high altitude, but it was felt the slacker winds expected at the upper level would make the climb worthwhile.

'Toronto Centre,' radioed Kennedy, 'have you any reports of winds from flights at other levels?'

'Negative,' replied Toronto, 'no reports.'

'We're encountering headwinds in excess of 100 knots and we'd like to climb to 390.'

'Roger, TWA 841, climb and maintain level 390.'

Captain Gibson disengaged the autopilot and initiated a manual cruise climb, gradually ascending the aircraft to the upper level. At 21:38, 39,000 ft (11,890 m) was achieved, and the autopilot and altitude hold were reselected. Second Officer Banks adjusted the thrust levers to maintain Mach 0.81. The winds at the new level were down to 85 knots, which helped improve the ground speed, and the fuel consumption was better. The outside air temperature indicated −57°C. The flight engineer estimated that about six-and-a-half tons of fuel had been used and at 21:40 he noted the aircraft weight of fifty-eight-and-a-half tons in the flight data log. The night was clear with cloud cover far below and a half moon shone about 50° above the horizon. The ride was smooth and the climb to 39,000 ft seemed justified. First Officer Kennedy set about recalculating the aircraft's ground speed and adjusting the leg times in the flight log while the captain monitored progress.

Flight 841 crossed the Canadian/US border into Michigan state and continued north-west between the city of Detroit to the south and Lake Huron to the north. The city of Saginaw, situated just south of Saginaw Bay, lay ahead. The sky was clear and bright, almost like daylight with the brilliant, near-full moon, and a thin but solid overcast lay directly below at 35,000 ft (10,670 m). Captain Gibson began sorting his charts for the onward route and he turned briefly to his left to extract fresh maps from his flight bag which lay by his left side on the cockpit floor. The time was 21:47. As his attention was momentarily distracted he suddenly felt a strange buzzing sensation. Within seconds the buzzing increased to a light buffet and he could feel the effect through the airframe.

Quickly Gibson looked at the instruments and checked his visual reference in the clear night. The aircraft wings were level but the autopilot had turned the control wheel 20°-30° to the left. The situation seemed very strange indeed. The other two on the flight deck were still busy with their calculations and at first had not noticed the development. Captain Gibson disconnected the autopilot and maintained the wings level by holding the control wheel to the left. Suddenly, the aircraft yawed sharply to the right, followed

TWA Boeing 727. (John Stroud)

by another severe right yaw. All on the flight deck were surprised by the manoeuvre. The second yaw continued with the aircraft rolling and skidding in a turn to the right. The nose dropped swiftly and a rapid roll to the right developed. The captain gradually pulled back on the control column to the limit with a slight backwards pressure and applied full left aileron. Neither input was sufficient to halt the manoeuvre. As the aircraft banked through 30°-45° to the right, Gibson applied full left rudder. For a moment the roll rate slowed, then the rapid right roll continued. The sky was crystal clear in the bright moonlight and, with the situation deteriorating so swiftly, all control inputs and decisions were made with reference to the visual horizon. Gibson managed a last glimpse of the artificial horizon as the machine banked through 45°-60° to the right and felt the aircraft was about to turn over. In a desperate effort to save the situation he took his right hand from the control column and quickly retarded the thrust levers to flight idle. His efforts, unfortunately, were to no avail. The aircraft continued to roll rapidly in a right diving turn, with the nose dropping sharply, and there seemed nothing the captain could do.

'We're going over,' he shouted.

The 727 flipped over on its back and, with the nose below the horizon, plunged earthwards close to the city of Saginaw. The engines hiccuped with the disturbed airflow and two quick, loud bangs of the compressors stalling could be heard in the cabin. For less than a second a feeling of absolute terror gripped Captain Gibson and then, strangely, a near calm returned almost as quickly. It was as if he knew that in spite of any recovery attempt his efforts would not be sufficient and that it was all over. Suddenly he felt relatively cool and his mind was clear. The rapid roll rate continued and by the end of the first roll, as the wings passed through level again for the first time, the nose had dropped 30°-40° below the horizon.

'Get them up,' shouted Gibson, referring to the speed brakes.

Kennedy did not understand the command so the captain pulled the speed brake lever himself. The operation seemed to have no effect so he recycled the speed brakes to check the function. The fast roll rate continued unabated and by only one-and-a-half rolls the nose was pointing near vertically downwards towards the cloud layer below. What had begun as a barrel roll had quickly developed into a vertical, screaming dive. The aircraft was now pulling 3.5g or more (727s are stressed to +2.5g or −1.0g) so the machine was already being strained beyond design limit. The 727 continued to race earthwards almost vertically with the wings rotating at about 50° per second. Gibson fought desperately to regain control. As the aircraft broke through the cloud layer at 35,000 ft (10,670 m) the captain could see, framed in his windscreen, the ground spinning directly below. It was not a panoramic sweep of the horizon as experienced in a flat spin in a light aircraft, but a vertical, rotating dive. Gibson could see a large dark patch, possibly a forest or lake, and one large city and three small cities spinning before his eyes.

The 727 plunged towards the ground at an average descent speed of 46,000 ft (14,020 m) per minute, with the rate of descent at moments reaching 76,000 ft (23,165 m) per minute. At such speed the aircraft broke the sound barrier and sonic booms could be heard clearly on the ground. On board, the noise level was painful to the ears. Still the 727 raced downwards in the darkness, rolling continuously, and the aircraft buffeted and shook with the fast speed. The artificial horizon which, in level flight, normally showed the top half sky blue and the bottom half earth black, displayed only a solid black. By 30,000 ft (9,145 m), the airspeed indicated about 450 knots on the instruments and the altimeters ran down so quickly the numbers were blurred and were difficult to read. The situation appeared hopeless and all Captain Gibson's efforts seemed of no effect. Still the machine dived and rolled inexorably towards its destruction. The spinning 727 still subjected the airframe to an increasing gravity force and in the cabin passengers were pinned to their seats. Those people standing at the beginning of the manoeuvre were forced to the floor. No one could move and fortunately only minor injuries were sustained. Many passengers found difficulty in breathing and some passed out for a few seconds under the strain.

On the flight deck Gibson watched the airspeed race past the limit and reach 470 knots in the dive. The captain estimated the aircraft to be descending through 20,000 ft (6,095 m), still out of control after about five to six rolls, and drastic action was needed. Quickly he called for extension of the landing gear. It was not permitted to raise or lower the undercarriage above 270 knots because of risk of damage to the gear doors, but in this situation that was hardly a consideration. Kennedy immediately selected the lever to the down position and the crew heard what sounded like an explosion as the landing gear dropped into the excessive airflow. The noise was unbelievable and for a moment Gibson thought the wings had separated. Both main landing gear doors and operating mechanisms were severely damaged and hydraulic line 'A' was ruptured. The right landing gear nearly came off, being blown past the over-centre position. It twisted rearwards into the trailing edge flap track and canoe fairings, jamming the flaps. The lowering of the undercarriage, however, appeared to have the desired effect and the speed at last began to reduce. The 'g' force decreased and the captain noticed some response return to the control column. At the completion of the last spin he managed to stop the aircraft rolling in just over one second, but as he fought to gain recovery the wings continued to rock about 20° to either side with little control. The nose, however, still pointed downwards almost vertically and with the height passing through 11,000 ft (3,350 m) and the aircraft still dropping rapidly, swift action was required.

Maintaining the wings level, he heaved steadily back on the controls and desperately tried to pull the aircraft from its dive. The 'g' forces quickly increased as the 727 began to respond and at about 8,000 ft (2,440 m) the captain appeared to have regained control. *N840TW*, however, was still descending and it was another 3,000 ft (914 m) before it bottomed out of the dive. At 5,000 ft (1,525 m) Gibson continued to pull back on the control column and he could feel his stomach drop to his knees as the aircraft was subjected to six g. The wings were tearing at their roots as they withstood six times the aircraft's normal weight. Miraculously they held. The Boeing jet had dropped 34,000 ft (10,360 m) in forty-four seconds, an average descent rate of 46,000 ft (14,020 m) per minute, or over one mile about every eight seconds.

Gibson had pulled the nose from near vertical to arrest the descent, but before the captain could react it passed through the level position and the aircraft shot skywards again. The nose pitched up to about 50° above the horizon and the speed dropped rapidly. Captain Gibson was so intent on not hitting the ground he nearly looped the plane. He had to work hard to prevent the machine from stalling out of the sky and plummeting earthwards once more. Climbing through 8,500 ft (2,590 m) the airspeed decreased to 280 knots and by 11,000 ft (3,350 m) it was down to as low as 160 knots. Using the moon as reference and with Kennedy and Banks both calling out pitch attitudes from the artificial horizon, Captain Gibson managed to regain his bearings and gradually he eased the nose forward. The 727 descended below 10,000 ft (3,050 m) again but, with the speed increasing to 180 knots, he was able to ease the stricken machine back into the climb. By 13,000 ft (3,960 m) he managed to level the aircraft with guidance from the other two and he was able, finally, to make a full recovery. It was an amazing feat of airmanship.

Under the circumstances there was no choice but to land as soon as possible and the captain informed the passengers that everything was now under control and that they would be making an emergency landing at Detroit. By now the flight engineer had taken action over the warning light indicating failure of hydraulic system 'A' and the flap extension drills were commenced. The trailing edge flaps were jammed by the twisted right landing gear and would not move and the leading edge flaps could only be extended by the alternate system.

On completion of the leading edge slats extension, Gibson once more experienced control difficulties. At speeds below 200 knots and above 220 knots the aircraft rolled uncontrollably, this time rapidly to the left, and it was necessary to fly at about 210 knots to maintain control. Attempts to retract the leading edge slats failed. Only inboard ailerons were available, compounding the problem, for the outboard ailerons, which normally come into play with the selection of trailing edge flaps, were still locked in with the flaps being jammed. Part of the rudder was also inoperative with 'A' system hydraulic failure.

The aircraft buffeted severely, making it impossible to read some of the instruments, and Gibson was unable to verify his height. In the distance the crew could see the lights of Chicago, where the captain had once been an air traffic controller, but it was feared too far for a landing. The aircraft might not hold together till then, so the nearer Detroit Airport had been chosen, although the weather there was much worse with a low cloud base. The damaged landing gear also displayed an unsafe condition with three red warning lights indicating all three gears unlocked. Nothing could be done for the two main landing gears, but a manual wind down facility was available on the flight deck for extending and locking the nose gear. Second Officer Banks began to crank the mechanism. As the three cockpit crew performed their emergency checklist procedures, the flight attendants in the cabin prepared the 82 passengers for an emergency landing.

After about thirty minutes, approaching Detroit, Banks succeeded in locking down the nose gear and stowing the doors. The nose gear light now glowed green and the stowed nose gear doors reduced the buffeting by about forty per cent, which at last enabled the captain to read all the flight instruments.

The aircraft was proving very difficult to handle and

Captain 'Hoot' Gibson at the controls of a Ford Trimotor.

Captain Gibson gingerly commenced an approach to Detroit's 03L runway, fighting the aircraft down through the cloud. He did not intend to land from this first approach but planned to execute a low level pass over the airport to allow an examination of the landing gear. Over 90° of aileron control wheel input were required, with the rudder held fully to the right, to maintain control. Even with maximum right trim selected, the strain on Gibson's right leg was enormous and he could feel his leg pulsating to the beat of his heart. The weather did not help either with freezing drizzle falling from a cloud base only a few hundred feet above the ground. The 727 broke from the cloud at about 400 feet and Gibson flew down the runway on a low altitude pass at about 50-57 ft (15-23 m) while crash rescue personnel shone searchlights on the landing gear. The gears were checked in place by the tower controllers but much damage could be observed and the right gear appeared misplaced.

On being so close to the runway, Captain Gibson had the greatest urge to jump from the aircraft and place his feet firmly on solid ground rather than go back up into the weather. The captain commenced a left circuit of the airport, staying low and trying to keep in visual contact with the runway. Turning to the left at the start of the downwind leg he concentrated hard on maintaining sight of the lights as he executed the uncomfortable manoeuvre. The aileron control was still held almost fully to the right, with hard right rudder, and his right leg still pulsated with the strain. Suddenly the captain lost control and he was unable to level the wings. The left turn continued and the 727 headed back towards the airport. The co-pilot also grabbed the control wheel, holding on full right aileron, but was unable to stop the turn. Gibson pushed the number one thrust lever to high power and retarded the centre and right engines to idle. Gradually the asymmetric power took effect and, just before crossing the runway, he managed to gain control and to bring the wings level.

The next approach, this time for a landing, was flown at about 205 knots and the 03L threshold was crossed at the same fast speed. Gibson flared the aircraft with full right rudder and aileron applied and at 22:31 executed a beautifully smooth landing. The 727 landed with the left gear touching down first at a speed of 197 knots. Fortunately it held. On closing the throttles, however, the machine rolled uncontrollably to the left. Reverse thrust was selected as normal and the nose wheel was lowered to the ground. As yet the right gear had not touched and the crew feared it had gone. As the 727 edged towards the left side of the runway the captain slowly lowered the right wing and at a considerable angle the right main landing gear finally touched. By good chance it also held and the aircraft continued its rollout down the runway with a pronounced right wing low attitude. Sparks showered from the dragging right landing gear and the tower controller radioed advising of a possible fire. Nose wheel steering was not available with the loss of hydraulic system 'A' and Gibson had to steer the aircraft off the runway along a high speed turn off using differential braking and asymmetric reverse power. On the taxiway, the 727 was met head-on by fire trucks racing towards the stricken aircraft and Gibson finally brought the machine to a halt. It was a fine demonstration of flying skills in very trying circumstances.

Foam was sprayed on the damaged areas and one of the firemen on the ground called the captain on intercom to advise that 'fuel was running all over' and that an emergency evacuation using escape slides might be prudent. Captain Gibson felt that his passengers had been through enough and that with fuel and fire trucks all over the place in the darkness it would be safer to walk the passengers off the aft exit. He replied accordingly. The plane was only about half full anyway and it would not take long to disembark the passengers from the rear. Quickly all on board evacuated by the aft steps and a relieved group of passengers and crew were finally taken to the terminal. They had faced death and survived.

An attempt was made later to tow the 727 from the high speed turn off but when the aircraft had been moved less than ten feet the right landing gear started to separate. The machine was jacked up to examine the damage but as the wing was raised the right landing gear broke from the wheel well and fell in three separate pieces.

A further inspection of the aircraft the next day revealed extensive damage. Wing skin panels were buckled, bolts had been sheared and large pieces of the machine had virtually been torn off. Wrinkling of fuselage panels was evident at the wing roots and clearly indicated the strain to which the wings had been subjected. Fuel leaked from around several structural fasteners in the left wing. A speed brake section had been wrenched from its hinges, an aileron bolt had been severed, a trailing edge flap carriage had been damaged and a flap transmission mechanism had been broken. All landing gear doors were extensively damaged and hydraulic system line 'A' was fractured. A flap track fairing was missing, as were some engine panels, and a leading edge lift device had sustained damage. To be precise, the number seven leading edge slat mechanism had suffered sheared bolts and twisted tracks and more significantly the slat itself was missing. Could this be the cause of the spiral dive of the 727? A search of the Saginaw area was conducted for the missing aircraft parts and the number seven slat, broken in two, the flap track fairing, and most of the speed brake section were found about six to seven miles (ten kilometres) north of the city. The stricken 727 lay at the side of the runway while the landing gear was made safe and was eventually towed to the hangar for repairs. Several days after the incident the aircraft was sufficiently repaired to permit a ferry flight to the TWA maintenance base at Kansas City, where major repair work was conducted.

Captain Gibson and his crew received just praise for their skills in saving the stricken 727 and the captain received a letter of commendation from the Federal Aviation Administration (FAA) via TWA's vice-president of operations. The cause of the incident was unknown but both the FAA and TWA exonerated the crew from any blame.

Immediately the National Transport Safety Board (NTSB) began an investigation of the accident, including the part played by the crew in the upset. As a formality, Captain Gibson was required by the FAA to be re-rated in the 727 simulator, as well as on the aircraft, and to attend a high altitude awareness course. He passed with flying colours and TWA received a further letter from the FAA complimenting the captain on his abilities. Ten days after the incident Gibson was promoted to Lockheed L-1011 captain.

The investigation of the upset proved to be a lengthy affair and it was to be more than two years before the NSTB reached a conclusion. Fortunately the B727 involved in the incident was not compromised by the delay and, although damaged extensively, *N840TW* was repaired within a matter of weeks. It was duly returned to service, none the worse for its ordeal, in late May 1979. Sadly, the same could not be said for the flight crew. As the inquiry continued, the three men found the situation changing for the worse. Adverse rumours began to circulate that more was involved than met the eye and Gibson and his colleagues were placed unfairly under suspicion, as much by the piloting fraternity as anyone else. The trio faced a harrowing ordeal and the mental anguish suffered made their return to duties and a normal life substantially more difficult.

The initial suspicion that the number seven slat was responsible for the gross upset prevailed, but how and why the slat had extended whilst cruising at 39,000 ft (11,890 m) remained a mystery. Lift devices on the leading and trailing edges of the wings of modern jet aircraft are essential for take-off and landing but are not designed for use at such high altitudes. On modern jets the wings are swept back at a large angle (on the B727, 34°) to allow the aircraft to fly high and fast by delaying the onset of shock waves as airflow over the wings approaches the speed of sound. At slow aircraft speeds, however, the lift producing qualities of the wings are poor. To improve lift, high lift producing devices in the form of slats (see diagram) are required at the wing leading edges*, and flaps are required at the wing trailing edges. When extended these devices increase the wing surface area and the camber of the wing shape. With slats and flaps fully extended the wing area is increased by twenty per cent and the lift by over eighty per cent. To improve lift at take-off, some flap and all slats are extended, any increase in drag being more than compensated for by increase in lift, and take-off without slats and flap is not possible at normal operating weights. On landing, slats and full flap are always selected in normal circumstances. Leading edge slats and trailing edge flaps are set together by the operation of a single lever on the flight deck. With selection of 2° of trailing edge flap the numbers 2, 3, 6 and 7 slats automatically

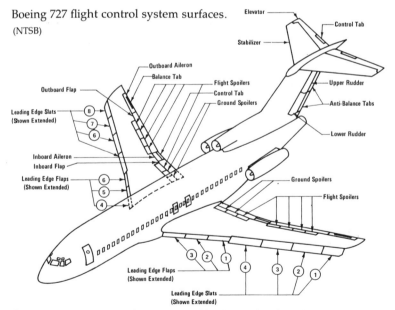

Boeing 727 flight control system surfaces. (NTSB)

* The Boeing 727 also has a type of leading edge flap, but these are not relevant to the story.

extend, and with selection of 5° of trailing edge flap the remaining slats, numbers 1, 4, 5 and 8 also automatically extend. With 5° of trailing edge flap set, all the leading edge devices are extended, and further operation of the flap lever selects only additional trailing edge flap extensions until full landing flap is set.

At cruise speeds and levels there is no requirement to select slats or flaps, and operation of these lift devices on the B727 is expressly forbidden above an indicated airspeed of 230 knots and an altitude of 25,000 ft (7,620 m). Rumour later had it, however, that flaps on the B727 were being extended by crews at higher altitudes than permitted to improve performance, and that the habit was not just confined to a few people but was fairly widespread. Since the use of leading edge devices was dangerous at the higher flight levels and speeds, 727 crews, it was alleged, were pulling the electrical circuit breaker which shuts off the hydraulic valve feeding the leading edge slats, thereby isolating the mechanisms in the retracted position, and then extending the trailing edge flap to 2° by use of the flap lever. The hydraulic valve cut-off circuit breaker was situated beside the crew wardrobe, on the right of the cockpit in a recess aft of the flight engineer's panel.

The effect of the small amount of trailing edge flap in the cruise, it was rumoured, tucked the nose down and improved speed and fuel consumption. The implications were, therefore, that the pilot of Flight 841 had deliberately selected the trailing edge flaps and that by some error the number seven slat had extended and become isolated in that position. In a landing configuration, an extended number seven slat would provide extra lift and would raise the right wing, rolling the aircraft to the left. In a fast, high altitude cruise with a low nose attitude, however, the device would create the opposite effect. When extended the number seven slat would produce negative lift on the right wing and the aircraft would roll sharply to the right.

Flight and simulator tests subsequently appeared to demonstrate that extension of the number seven slat had caused the incident and that it had detached near the end of the last vertical roll, at about the time of lowering the landing gear, and had allowed Captain Gibson to regain control and save the aircraft. Whether the extension was an unscheduled movement or, as whispered, an irresponsible crew action, however, remained unresolved. No proof of any crew involvement was forthcoming although the stigma remained. Rumours implicating the pilots in the event continued unabated and it appeared, on hearsay alone, that a crew who should have been treated as heroes were being seriously maligned. The crew protested that they had never even heard of such a malpractice, but it was to no avail. At the time of Gibson's incident few, if any, knew of the action of selecting flaps in the cruise, but afterwards rumour had it that the practice was widespread. Suddenly everyone seemed to know someone who had tried it although no one who had actually used the procedure emerged. In answer to the accusations of tampering with the flap controls, Captain Gibson, in a deposition to the inquiry in Los Angeles, California on 12 April 1979, one week after the incident, issued a sworn affidavit.

'At no time prior to the incident did I take any action within the cockpit either intentionally or inadvertently, that would have caused the extension of the leading edge slats or trailing edge flaps. Nor did I observe any other crew member take any action within the cockpit, either intentional or inadvertent, which would have caused the extension.'

The first and second officers also both issued sworn statements at the same place and time denying tampering with the flaps. The deposition was conducted in the full glare of publicity with TV cameras and a battery of twenty-six microphones.

These testimonies seemed sufficiently clear but they did not satisfy the NTSB. A statement, allegedly made by one of the investigators a few months later, clearly laid blame on the crew's shoulders. 'I think those guys were fooling around up there and I don't think we really know what they were doing yet,' a 'spokesman' revealed to *Aviation Consumer* magazine on 15 October, 1979. Such a statement, if accurately reported, was a disgraceful breach of ethics, for the investigation was still at a premature stage. Soon the national press echoed the comment and a countrywide accusation of guilt resulted. The crew seemed convicted before any hearing had begun. Meanwhile, Captain Gibson and his colleagues, desperately trying to pick up the pieces of their careers, were restrained from issuing public statements in their own defence.

A number of points surrounding the incident were also found questionable by members of the inquiry. To begin with, the cockpit voice recorder (CVR) tape on the 727 had been erased after Flight 841's emergency landing at Detroit. Cockpit voice recorders are installed on aircraft to record flight deck conversations which might be useful to investigators following an accident. CVRs are not permitted to be used by the FAA in any disciplinary action. The recorder tape operates on a continuous, thirty minute loop and would not have recorded details of the dive, but the NTSB were suspicious that conversations on the flight deck after the incident could have revealed details of the crews' actions and were deliberately erased. In 1979 the CVR had been only recently introduced and crews distrusted the equipment. Most pilots feared misuse of its information and regularly erased the tape after a trip. Under questioning by counsel, Captain Gibson admitted that he, too, normally erased the tape after a flight but that in this instance he could not remember doing so.

'Did you erase the recorder?'

Captain Gibson: 'Not to my knowledge.'

'Did anyone erase it?'

Gibson: 'Not to my knowledge. I didn't see anyone erase it.'

'Do you usually erase the recorder?'

Gibson: 'I usually do, yes. I don't recall erasing it.'

'Can you erase it in the air?'

Gibson: 'No.'

'What is required to erase the CVR?'

Gibson: 'The parking brakes have to be set.'

'How many minutes of recording are there on the CVR before previous contents are erased?'

Gibson: 'Thirty minutes. It was forty-five minutes to an hour [after the incident] before the aircraft was shut down and we got off.'

'So if the tape had not been bulk-erased at the time of shut-down would there have been anything meaningful on the tape?'

Gibson: 'No, the tape could only have made my other two crew members look good. They did a real good job. All that would have been on the tape would have been the other crew members complying with the check list.'

'Can you explain why it is your habit and routine to erase cockpit voice recorder data on landing?'

Gibson: 'It is an accepted practice, and as far as I am concerned at the time it was done by everyone. It is done by an awful lot of people. When they put the cockpit voice recorder on the airplane I would say 100 per cent of people always erase it on landing after they park their brakes.'

'Why do you do it?'

Gibson: 'Because I might say something unkind about some of the people in management, and they might take that tape out and send it someplace.'

Both Kennedy and Banks also swore that they had not erased the tape. It seemed reasonable that, in the highly charged atmosphere after such an event, crew members might by force of habit, take actions which they might later forget, but the investigators were not to be convinced.

'We believe the captain's erasure of the CVR is a factor we cannot ignore and cannot sanction. Although we recognise that habits can cause actions not desired or intended by the actor, we have difficulty accepting the fact that the captain's putative habit of routinely erasing the CVR after each flight was not restrainable after a flight in which disaster was only narrowly averted. Our scepticism persists, even though the CVR would not have contained any contemporaneous information about the events that immediately preceded the loss of control, because we believe it probable that the twenty-five minutes or more of recording which preceded the landing at Detroit could have provided clues about causal factors and might have served to refresh the flight crew's memories about the whole matter.'

The investigating team clearly felt that the erasure of the CVR was deliberate. If the inadvertent operation of the number seven leading edge slat was also to be proved part of a deliberate act by the crew, then any mechanical failure of components would have to be discounted. The NTSB instigated a thorough investigation of the possibilities of an unscheduled extension of the slat. If some kind of mechanical fault had resulted in accidental operation of the number seven slat, the Safety Board would do their utmost to find the cause. A total of 118 trials were conducted in a Boeing flight simulator to try to identify the condition that precipitated the aircraft's upset and to duplicate and evaluate its manoeuvre.

'The flight simulator traces showed that the simulated aircraft could be returned to wings-level flight with relatively little loss of altitude provided corrective action was begun before the roll and airspeed were allowed to increase excessively. In the simulations, the pilot could delay reaction for about sixteen seconds and regain control with an altitude loss of about 6,000 ft (1,830 m). However, when the pilot delayed corrective action for seventeen seconds or more, a manoeuvre was entered that approximated Flight 841's airspeed, altitude, and g-traces. In this manoeuvre, the aircraft continued throughout the descent to roll to the right, in spite of full left aileron and rudder, until the slat was retracted to simulate its loss from the aircraft.'

During none of these tests, however, was yaw induced at the beginning of the upset.

A thorough inspection of the damaged aircraft and the retrieved debris was conducted but no evidence of a malfunction which might have caused an inadvertent slat extension was discovered. An examination of the slat locking devices also proved fruitless. The slats are locked in the retracted position by hydraulic pressure from system 'A' and by a mechanical locking mechanism. Both systems would have had to fail before the slat could have extended and there was no evidence to suggest that this had occurred. Loss of hydraulic system 'A' and/or failure of a mechanical lock would have illuminated warning lights in the cockpit, and neither were reported by the flight crew.

A repair of the hydraulic system line 'A', which ruptured on extension of the landing gear, was implemented, and the slat and flap systems were found to function satisfactorily. Number seven leading edge slat, the offending device, could not be tested since it had separated from the wing, and its extending and retracting hydraulic mechanism, or actuating system (see diagram), could not be checked since it was broken. The investigating Board acknowledged, however, that failure of the actuating system, namely, fracture of the actuating piston or separation of the piston from the actuating rod, could have resulted in inadvertent extension, but Boeing convinced the NTSB that the safety margins, particularly regarding an actuator piston fracture, were massive and that in sixteen years of service history and over thirty-six million flight hours such a fracture had never occurred.

There is always a first time of course, but the event was deemed by the Board to be highly unlikely and was not considered. Later it was revealed that at the time of the inquiry no less than six cases of cracked actuator pistons had been reported to Boeing by Lufthansa. Separation of the actuator rod from the actuator piston seemed to the Board to be the only reasonable single failure which could have permitted an unscheduled slat operation, but this was also discounted. The Safety Board contended that the distortion of the remaining section of the broken actuator cylinder and the fact that the piston did not stay in the remnant of the cylinder indicated that the piston rod was attached to the piston when the slat extended. The problem facing the NTSB, however, was that the forward two-thirds of the actuator cylinder, the actuator piston and the actuator piston rod were missing and in their reports the investigators could only speculate as to the reason for the failure. The Safety

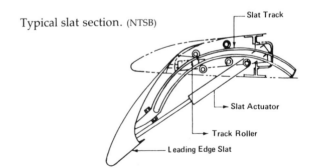

Typical slat section. (NTSB)

Slat Track

Slat Actuator

Track Roller

Leading Edge Slat

Board had to reach their conclusions only on the limited evidence available and it seemed a firm dismissal of any mechanical failures, especially the only single failure which was considered reasonable, when the suspect parts were never recovered.

Over the years, a large number of 'service difficulty reports' had been received by the Federal Aviation Administration (FAA) outlining slat problems, most of them minor, but some significant. Whether the 727 involved in the incident, N840TW, was the subject of such reports is not known, but it was an old aircraft and was TWA's first 727-100 in service. Up to 1973, there had been approximately fifteen reported cases of uncommanded single leading edge slat extensions at low level with the slat separating in flight. To resolve the situation, in 1973 Boeing issued an Airworthiness Directive requiring a stronger actuator piston rod end, and the problem was assumed to be solved. In the five years preceding Gibson's upset, only two instances of importance were reported. One notable event arose in 1976 when an unscheduled extension of a leading edge slat occurred. Significantly, the extension resulted from the failure of a slat actuator support fitting. The slat remained attached and the aircraft upset was contained.

In 1978, one B727 operator declared that in cruise at 25,000 ft (7,620 m) the numbers six and seven leading edge slats had been unintentionally extended by the crew. The captain stated that he believed the trailing edge flaps were partially extended and when he attempted to retract them the leading edge devices were accidentally operated. The captain immediately retracted the leading edge slats, but the numbers six and seven slats remained extended. The roll to the right was contained, and when the aircraft slowed the slats retracted. Whether rumours of flight crew tampering with the flap controls surrounded this upset was not revealed, but the NTSB appeared to be drawing comparisons between this event and Captain Gibson's. Since the investigators had discarded the likelihood of mechanical malfunction, the Safety Board were forced to consider that Flight 841's crew, in spite of their statements, had deliberately operated the flaps.

The NTSB advanced the theory, therefore, that the electrical circuit breaker for the leading edge lift devices had been pulled to prevent their extension and the trailing edge flaps had then been set to 2° to improve aircraft performance. The Board then tried to imagine a scenario in which the number seven slat could have been inadvertently extended and were aided in their quest by what appeared to be an unusual source: Captain Gibson himself. In an unofficial interview given shortly after the incident, an FAA representative had asked Gibson if all the crew members had been in their places when the upset occurred. The captain replied in the positive, adding that the flight engineer had just returned to his seat. The FAA representative assumed from the statement, incorrectly, that Banks had actually left the cockpit and, unknown to Gibson, the NTSB were informed accordingly. The suspicions of the Safety Board were corroborated by a female passenger in economy class who stated that she saw a male crew member carry trays back to the galley from the forward end of the aircraft. The unidentified man was wearing epaulets and was assumed by her to be one of the flight crew. Later it appeared that one of the stewards, who also wore epaulets at that time, had carried the cockpit crew's meal trays back from the flight deck. The Board, therefore, attempted in secret to build a scenario around an erroneous premise and surmised that the setting of the flaps might have occurred with Second Officer Banks absent from the flight deck. The relevant circuit breaker was to be found aft of the flight engineer's station and, on his return, it was suggested, he may have found it tripped. Unaware of the circumstances, Banks may have simply pushed it home, resulting in extension of the leading edge devices. The captain would then have selected the flap lever to up but, in something similar to the previous case, the number seven slat remained extended. An examination of the offending slat mechanism seemed to support the theory. A sheared bolt was shown by metallurgical tests to have been weakened by fatigue and had probably broken before the event. The failure would have resulted in sagging of the inboard end of the slat, which seemed to be confirmed by slat wear. The number seven slat would, in this condition, have extended misaligned and aerodynamic loads would have prevented its retraction.

A broken bolt was also found on the outboard aileron's mechanism which had been weakened similarly by fatigue. It was not possible to determine when the bolt had failed but, if it had sheared before the barrel rolls, the free play in the mechanism would have permitted the aileron to float upwards about one inch (twenty-five millimetres). This would have induced a roll to the right which would have been noticeable to the pilot and would not have helped in controlling the gross upset. All the events could, of course, have occurred simultaneously, but the NTSB investigators concluded that, even had the broken aileron bolt existed at the moment of the upset, it would not have contributed to the loss of control of the aircraft. There were those who disagreed. The US Air Line Pilots' Association (ALPA) were also participating in the investigation and they were becoming increasingly disturbed by the trend of the inquiry. ALPA instigated its own study and claimed that in the five years before the incident no less than 400 'service difficulty reports' of B727 slat problems had been submitted to the FAA.

'ALPA is concerned that this incident is not limited to a single aircraft on a single airline. Rather it appears to be symptomatic of the fatigue problem that has not been properly addressed by the aircraft industry. Recent emphasis on this aging problem associated with the older jets through-out the world underlines ALPA's concern.'

The Pilots' Association felt that the large number of slat problems which had been reported tended to support the mechanical failure theory. The Safety Board's suggestion of malpractice, therefore, appeared hasty and their rejection of the only single failure considered reasonable seemed open to argument, especially since the relevant parts were missing. The only previously reported incident of an unscheduled slat extension also related to problems with the actuator, although in that case it was failure of the actuator mount fitting.

'We believe,' reported ALPA 'that pre-existing fatigue, corrosion and component failures within the number seven slat and right outboard aileron mechanisms caused the slat to extend. We further believe that free play in the right outboard aileron played a significant part in the controllability of the aircraft and initiation of the manoeuvre.'

Flight tests were conducted by Boeing in October 1980 to examine the effects of slat extension in conditions as close as practicable to *N840TW's* incident, but these also resulted in discrepancies. The NTSB noted that the decrease in airspeed which accompanied the selection of numbers 2, 3, 6 and 7 slats on the test aircraft compared with the airspeed decrease indicated by Flight 841's flight data recorder in the moments preceding the upset. This, the investigator claimed, was evidence that the relevant leading edge devices had originally extended before the number seven slat had become isolated and was conclusive proof of their theory. Captain Gibson's control deflections of elevator, aileron and rudder in an attempt to maintain level flight, however, could, likewise, have caused a similar reduction in speed. There were also significant differences in other areas, namely the 'g' forces at the onset of buffet. The effect of selecting leading edge slats in the test in fact produced more than moderate buffeting, while Captain Gibson had testified that only light buffeting was experienced. The 'g' traces supported the captain's statement. Gibson's colleagues on the flight deck were not at first aware of the developing situation before the spiral dive, a fact which also seemed to be in keeping with the buffeting being light. If the buffeting had been moderate their attention would have been more readily drawn to the circumstances. No passengers aboard *TW 841* reported any shaking of the aircraft prior to the upset. Light buffeting, however, pointed to extension of only the number seven slat, but no test was conducted with the aircraft in that condition.

On 29 January 1980, almost ten months after the upset, the three flight crew were interviewed for the first time by NTSB investigators in Kansas City, Missouri. The line of questioning was directed at the single issue of the location of the second officer at the time of the incident and clearly indicated the train of thought of the investigators. Demands to be questioned on all aspects of the case by Gibson and the FAA were ignored. Captain Gibson and his crew once again gave testimony but, on this occcasion, only regarding the position of the flight engineer, and no other depositions were made.

The US ALPA, with increasing concern at the direction of the inquiry, issued a statement to the NTSB.

'It has been alleged during the investigation that the crew, through some unorthodox procedure, inadvertently extended the leading edge slats, recognised their mistake, and took action to retract them. This allegation further assumes that due to pre-existing damage to the number seven slat it did not retract, but went to the fully extended position. The crew members vehemently deny that this happened. TWA undertook a flight test to determine the effect of extending the leading edge slats at the same height and airspeed. According to the pilot of this flight, Captain George Andre, the aircraft experienced moderate buffet. This statement contradicts any possible extension of other than the number seven slat as the initial onset of this incident.'

At no time during the investigation, or at any other time, were the crew confronted with the evidence upon which the Board based its findings: the analysis of the flight data recorder, the results of the simulator exercises and the outcome of the flight tests. Captain Gibson and his colleagues had to await the publication of the official report before such details were revealed. It was to be June of the following year, 1981, and a further sixteen months of anxiety for the crew, before the NTSB's findings were released.

'The Safety Board determines that the probable cause of this accident was the isolation of the No 7 leading slat in the fully or partially extended position after an extension of the Nos 2, 3, 6 and 7 leading edge slats and the subsequent retraction of the Nos 2, 3 and 6 slats, and the captain's untimely flight control inputs to counter the roll resulting from the slat asymmetry. Contributing to the cause was a pre-existing misalignment of the No 7 slat which, when combined with the cruise condition airloads, precluded retraction of that slat. After eliminating all probable individual or combined mechanical failures or malfunctions which could lead to slat extension, the Safety Board determined that the extension of the slats was the result of the flight crew's manipulation of the flat/slat controls. Contributing to the captain's untimely use of the flight controls was distraction due probably to his efforts to rectify the source of the control problem.'

It was a tremendous blow to Flight 841's crew and one that was followed by vociferous denials from the three concerned, as well as hostile reaction from ALPA. But there was more. The report had been signed by three NTSB members, one of whom was Francis McAdams, the only pilot and aviation professional on the Board. McAdams was unhappy with some aspects of the case, especially the fact that neither he nor anyone else on the Board had seen or interviewed any of the accused, and he appended his own statement to the report.

'Although I voted to approve the Board's report which concluded that the extension of the leading edge slat was due to flight crew action, I do so reluctantly.

'The report as written, based on the available evidence . . . appears to support the Board's conclusion. However, I am troubled by the fact that the Board has categorically rejected the crew's sworn testimony without the crew having had the opportunity to be confronted with all of the evidence upon which the Board was basing its findings. At the time of the first deposition . . . no evidence was available to the crew or to the Board. Although the crew was deposed a second time, their testimony was limited to one issue, i.e. the physical location of the flight engineer at the time of the incident. I had recommended that since the Board was ordering a second deposition it be conducted *de novo* so that the crew would have been aware of all the evidence. The Board did not agree.

'Furthermore, I do not agree that a probable cause of this accident, as stated by the Board, was "the captain's untimely flight control inputs to counter the roll resulting from the slat asymmetry." In my opinion, the captain acted expeditiously and reasonably in attempting to correct for the severe right roll condition induced by the extended slat.'

The report, although firm in its conclusions, remained unconvincing, and the crew was left with a 'not proven' verdict. Under a cloud of uncertainty the three continued their duties with TWA, still proclaiming their innocence. The strain, however, proved too much for the second officer and Banks retired early to become a college lecturer.

As time went by the struggle to clear their names did not diminish and much information was gathered by ALPA in support of their cause.

At the beginning of the inquiry, suspicions of malpractice were initially aroused by what appeared to the investigators to be the deliberate erasure of the cockpit voice recorder (CVR) by Captain Gibson. The crew, the Board surmised, must have had something to hide. Erasure of the tape by Gibson, however, was clearly shown to be impossible. The CVR can be erased by the pilot only when the aircraft is safely on the ground with the parking brake set. Squat switches on the landing gears contact when the oleos compress with weight, closing circuits which confirm the machine is on the ground. The high speed lowering of the landing gear during the upset resulted in damage which tore the squat switches and circuitry from the structure and the 727 landed with both main gear lights glowing red. The CVR erasure circuit was incomplete and, even if the captain had pressed the button on the flight deck, nothing would have happened. Gibson could *not* have deliberately erased the tape and the Board's initial suspicion that he did so because he had something to hide was unjustified. What is more, this was known and understood by the investigators early in the inquiry. Also after erasure, the tape recycled and a further nine minutes of conversation, including statements relating to the incident, were recorded. If it had been Gibson's intention to erase the tape, and it had been physically possible to do so, it is most unlikely that any evidence would have remained. So how was the tape erased? *Popular Mechanics* magazine interviewed a CVR technical expert and revealed that on a slow transfer of electrical power from the 727 engine to the auxiliary power unit (APU), which Banks accomplished on this occasion, erasure of the CVR is possible. This is the only way in which erasure of the CVR could have occurred. The timing of the tape recycling and the moment of slow power transfer also corresponded, as did the ending of the recording, nine minutes later, and the shut down of the APU.

The question of whether the flight engineer had left the flight deck was also suitably resolved. Gibson and his colleagues denied that Banks had left the cockpit and it was established that shortly after top of climb the flight engineer had risen only momentarily from his position to place finished meal trays at the back of the flight deck for collection. The flight attendants, when interviewed, supported the flight crew's statements and fourteen passengers, including four to five first class travellers who were in a good position to view movement from the flight deck, were also questioned. All swore that Second Officer Banks did not leave the flight deck near the time of the incident. The NTSB refused to consider this evidence and the statements were ignored. During the investigation, Captain Gibson kept hearing that someone had told the Board of Banks' absence, although the informer's identity could not be revealed, and it was realised only much later that it was Gibson's own comment which had led to the misunderstanding. The FAA representative who had spoken unofficially to the captain immediately after the accident, in spite of claiming that 'he was not wired for sound and wouldn't take notes', did write down in pencil the next day his recollections of the interview. It was the FAA representative's mistaken belief that Banks had left the flight deck, unwittingly reported to the NTSB, which had started the Board's 'what if the flight engineer had left the cockpit' scenario. With Banks on the flight deck the NTSB had no case.

The Board's version of the slat extension also did not stand up to scrutiny and when examined closely seemed absurd. The NTSB imagined the 727 levelling off at 39,000 ft (11,890 m), the power being set and Banks leaving the flight deck for a brief spell. Since all on board, except one lady passenger in economy class, testified to not having noticed his departure, his absence must have been short indeed. One of the pilots was then imagined jumping from his seat and in the darkness moving aside the crew's coats and swiftly finding and pulling the relevant circuit breaker. Very quickly, and in secret, the trailing edge flaps are set to 2°. Almost immediately Banks is imagined returning and noticing in the darkness, *beside* the crew's coats, that the circuit breaker is pulled and, *without saying anything*, pushing it back in again. According to the NTSB, half the leading edge slats then fully extend causing, by their own admission, moderate buffeting, and the three crew members are imagined sitting doing nothing for six seconds as they watch the leading edge flap lights sequencing. Then, with the leading edge slats fully extended, someone realises the error of their ways and is imagined selecting the flap lever up, thereby retracting all the lift devices except the number seven slat which inadvertently remains fully extended. This account, unfortunately, fails to appreciate the fact that if the pilots had just set the flaps they would have been highly aware of the circumstances and at the first sign of buffet one of them would have selected the flap lever to up long before full leading edge slat extension had occurred. It also fails to appreciate that modern airline crews do not operate in this manner and that in any usual situation, circumstances are discussed fully before action is taken. And, if the practice of selecting two degrees of flap had been widespread, as alleged, Banks, had he noticed the popped circuit breaker, would have known exactly what was going on and that would have been the last thing he would have touched. The NTSB scenario did not fit the facts and, even if the crew had deliberately set 2° of flap, an action not entertained here, the extension of number seven leading edge slat was highly unlikely to have happened in this manner.

There were also a number of other factors relating to the investigation which were disturbing. One hundred and eighteen simulator tests were undertaken by Boeing and the NTSB in order to resolve the situation but all tests were conducted according to the flight data recorder (FDR) and not one trial was performed in the manner described by the flight crew. The scientific officers analysing the accident data accepted only the scientific details as revealed by the FDR, even though it was known the equipment could be unreliable after being subjected to such a violent manoeuvre. The recorder indicated, for example, a rapid heading change of about five degrees to the right in less than a second at the start of the upset which was identified as a sharp roll. The flight crew denied that the wings rolled, stating that the upset was preceded by a fierce yaw to the right. The flight attendants also stated that they noticed no rocking of the wings and the passengers interviewed said the same. Not one simulator exercise was conducted with yaw as described by Gibson. All tests were performed in a 727-200 series simulator whereas Captain Gibson's machine was a 727-100, a shorter version which was considered by pilots to be less stable at high altitude. Having taken the decision to fly at the unusually lofty level for a 727-100 of 39,000 ft (11,890 m), the crew were then highly unlikely to jeopardise further the aircraft's stability by deploying the trailing edge flaps. As Captain Gibson commented, 'It is very difficult to get three crew members to agree to take a 727-100 to 39,000 ft. I seriously doubt that the three most stupid pilots in the industry would ever consider experimenting with flaps at 39,000 ft in a 727-100.'

All NTSB trials were restricted to the suspected Board scenario and all fault analysis was confined to calculated problems which could have resulted in the extension of a slat. The most improbable cases were simply discarded. The U.S. Air Line Pilots' Association (ALPA) demonstrated that 'a fracture of the actuator piston circumferentially through the lock key hole', no matter how unlikely, could have resulted in unscheduled extension of only the number seven slat. Although the NTSB and Boeing pointed out that the piston design strength negated the possibility of this failure and that it was claimed no record of such an occurrence existed, it was still possible, especially since the number seven slat mechanism had been damaged previous to the upset. The aileron hinge bolt that had also failed had a design strength which, likewise, negated the possibility of its shearing, and there was also no record of such a fracture existing, yet the bolt had failed. If one component which was deemed impossible to fail had done so, it was surely possible that another on the same aircraft could have suffered a similar fate.

The passengers were also asked if they had heard any unusual noises during the flight. The investigators hoped that some might have mentioned the unmistakable, high pitched, shrill scream of the hydraulic motors operating which would have indicated extension of the trailing edge flaps. In the quietness of the cruise at 39,000 ft (11,890 m) the hydraulic motor noise would have been quite piercing, much louder than at the lower altitudes when flaps are normally operated after take-off or before landing. No-one, however, noticed any high pitched scream although many mentioned the light buffet and hearing the bangs of the compressor stall. This, alone, seemed conclusive proof that the crew did *not* operate the flaps.

The NTSB's accusations regarding untimely control inputs at the moment of the upset, disagreed by Francis McAdams, the only pilot on the Board, were also disputed. As early as 1975, Boeing had conducted a 727-100 flight test 'to investigate stability and control characteristics at high speed with one leading edge slat extended.' It is interesting to note that the test was ordered two years after the slat problem was assumed to have been solved by the issuing of the Airworthiness Directive in 1973, which required strengthening of the actuator piston rod ends. The aircraft took off with the number two slat bolted in the extended position and climbed in that condition to the required altitude. The test was abandoned at 33,400 ft and Mach 0.80 because of severe control difficulties resulting from heavy outboard aileron vibration. The test engineer at the time, D. L. Mahon, wrote on the flight log, as much in alarm as in jest, that the trial was curtailed 'due to several cases of extreme cowardice'. The aircraft, therefore, was never tested in this condition throughout its entire flight envelope. Captain Gibson's upset had occurred at 39,000 ft and Mach 0.81. Later it was demonstrated that the control forces at high altitude were not sufficiently effective to recover from a gross upset and that the more dense air at lower levels was required for full control effectiveness. The performance of the simulator during the trials was based on wind tunnel testing which had not been verified by flight test at high altitude. In fact, the analytical data for flap extension at lower altitudes had had to be adjusted significantly to match flight test results. As a note, the severe roll to the left experienced by Gibson after selection of the remaining slats at about 13,000 ft

(3,960 m) on the approach to Detroit could not be explained by Boeing. At slow speeds with the number seven slat missing and all other slats set, the aircraft should have rolled to the right.

During the Boeing 727-100 flight tests in October, 1980, while checking performance in cruise with slats retracted and 2° of trailing edge flap extended, it was discovered, to everyone's surprise, that the configuration rapidly deteriorated performance. If the practice of selecting this configuration had been as widespread as alleged, crews would have been aware of the poorer performance; the corollary being, of course, that once the crews had discovered the deteriorated performance the practice would not have been widespread. Something, somewhere was wrong.

The entire investigation was conducted without the assistance of Captain Gibson and his colleagues and, after the initial testimony taken under the glare of publicity about one week after the incident, the flight crew were denied due process. At no time after the first deposition, during the whole two and a half years of the investigation, the longest ever, were the flight crew consulted, cross examined or questioned by any investigator, except to answer 'no' to the question of whether or not Banks left the flight deck near the time of the incident. All of Gibson's phone calls to the NTSB were either refused, or never returned, and all notorised statements were ignored. The flight crew were also denied admission to the first hearing, which ended inconclusively, although the press and other interested parties were permitted to attend. Captain Gibson and First Officer Kennedy were admitted to the second hearing but they were not allowed to take part and they weren't even recognized by the investigators.

An appeal against the verdict was launched but it was to be eight years later before it reached the Supreme Court, the highest authority in the land. Unfortunately, Gibson's case was not one of those selected for review and the situation was never resolved.

In 1983, a few of the passengers who had suffered minor injuries in the upset took TWA and Boeing to court in pursuit of compensation. Judgement was found in the plaintiffs' favour and sums of money were paid by both companies.

In spite of the result, however, Gibson and his crew, according to Donald Mark Chance Jnr., the lawyer representing TWA, came out of the proceedings well.

Captain Gibson and his colleagues attempted to take their case against the NTSB and Boeing to court but were denied due process of the law. A federal judge declared Gibson a 'public person' and, because of other legal difficulties at the time, action by the flight crew through the civil courts was refused. A documentary, entitled 'The Plane that Fell from the Sky, was made by Paul and Holly Fine for CBS's 60 Minutes programme and was sympathetic to Captain Gibson's case, but was all to no avail.

The crux of the matter in this entire case lay in the statement of the original hearing officer, Les Kampschror, that if the crew were not found in error then the airworthiness of the 727 would be cast in doubt. Since, in the minds of the investigators, no mechanical fault of the 727 appeared feasible, they left themselves little choice but to blame the flight crew. The majority of the NTSB were satisfied with the decision. The publicity surrounding the case and the verdict against the crew, it was reasoned, would eradicate the alleged widespread practice of tampering with the flaps in the cruise and further inadvertent slat extensions would be prevented.

On 28 August 1982, a 727-100 of International Air Service Company Ltd experienced an uncommanded number seven leading edge slat extension, after slat and flap retraction, while climbing through 4,000 ft (1,220 m) *en route* from Tulsa to St. Louis. Substantial aileron inputs were required to maintain control. Inspection revealed that the number seven slat actuator had malfunctioned and extended the slat.

On 17 November 1984, once again on the climb out with flaps and slats retracted, an unwanted extension of the number eight leading edge slat occurred to an American Airlines 727-200 whilst flying from Las Vegas to Los Angeles. A large lateral control input was required to control the aircraft.

In the cruise, on leaving 31,000 ft (9,450 m) for 29,000 ft

(8,840 m), a partial and momentary extension of number two leading edge slat occured to a United Airlines 727 on 10 September 1985. The aircraft pitched and yawed moderately. No cause of the unscheduled flap extension was discovered. Remarked the captain later, 'I know Boeing says it can't happen, but it did.'

On 8 January 1986, a TWA 727-100 outbound from St. Louis with flaps and slats retracted, experienced an unscheduled extension of the number seven leading edge slat whilst climbing through 6,000 ft (1,830 m). The aircraft rolled to the left with moderate buffeting and the captain decided to return to St. Louis. On the ground the number seven slat remained extended when the flaps and slats were selected up. Inspection revealed that the inboard number seven slat track extend stops were missing.

Whilst climbing out from Detroit on 22 February 1987, with flaps and slats retracted, an unscheduled extension of the number two leading edge slat occurred to a North West 727. The captain reduced the speed to 210 knots, selected two degrees of flap (which also extended slats 3, 6 and 7 giving a symmetrical slat condition) and continued to destination, Cleveland.

Another incident occurred on 24 August 1987 to a TWA 727 *en route* to St. Louis. At a speed of 300 knots, and climbing through 10,000 ft (3,050 m), with the flaps and slats retracted, an unscheduled extension of the number seven slat began. The captain reduced speed and the number seven slat retracted of its own accord. Later, the number seven slat actuator was replaced.

In none of these incidents were the crew accused of tampering with the flaps. To this day, over a decade later, the NTSB's verdict of guilty on Captain Gibson, First Officer Kennedy and Second Officer Banks, still stands. It is time the judgement was reviewed.

Note: 'Hoot' Gibson flew as captain on L1011s for three years after the upset, then transferred to the Boeing 747 as a first officer in 1982. In 1983, the litigation in which he became involved took its toll and in January 1984 Gibson retired early on medical grounds, with review after five years, and went back to his farm in Costa Rica. After only two years of farming 'Hoot' Gibson returned to flying duties with TWA in January 1986 and is now a 747 captain. Jess Kennedy remained in service with TWA and gained his command, but Gary Banks never returned to flying.

Chapter 9
Strange Encounter

The Islamic Festival of Ramadan falls in the ninth month of the lunar calendar and celebrates the first revelation of the Koran to Mohammed. The event begins with the new moon and ends with the old, and marks a month of daylight fasting for the Muslim faithful.

In 1982, 24 June fell on the second day of Ramadan and the night was moonless and dark. At just after 20:00 local time a British Airways Boeing 747, en route from the United Kingdom to New Zealand, took off from the Malaysian capital of Kuala Lumpur into the clear, black night. A quick five-hour hop to Perth in Western Australia lay ahead. For the passengers enduring the twenty-nine-hour journey from London to Auckland, Kuala Lumpur marked the halfway stage of the flight. The travellers from Europe destined for Perth were joined by a tour group of about thirty returning to that city who had been on holiday in Malaysia. Of the 247 total on board, therefore, about a hundred were due to disembark in Perth, the first Australian stop. The flight was then scheduled to continue via Melbourne to Auckland.

In 1982, British Airways won the 'Airline of the Year' award, being voted number one by regular business travellers, but the accolade had not been achieved without difficulty. Industrial and security problems at Heathrow had disrupted many departures and British Airways' Flight BA 009 from London to Auckland on 23-25 June was no exception. As the Boeing 747, named *City of Edinburgh*, registration *G-BDXH*, known phonetically as X-ray Hotel, lifed off from Kuala Lumpur's runway 15, the flight was running about one-and-a-half hours late. The aircraft's weight at take-off was about 304 tonnes, which for the short journey was well inside the maximum structural take-off weight of 371 tonnes. The ninety tonnes of fuel carried was more than sufficient for the flight to Perth, which would consume about fifty-five tonnes, but with the Western Australian metropolis being one of the most isolated cities in the world more was needed in case of diversion. The designated second choice in case the destination airport was closed because of weather, or some other unforeseen circumstance, was the remote military base of Learmonth lying on the coast 600 nm north of Perth. With the high fuel load, however, BA 009 could, if necessary easily overfly Perth to Adelaide, a somewhat less isolated alternative. The briefing in Kuala Lumpur attended by the flight crew, Captain Eric Moody, Senior First Officer (SFO) Roger Greaves and Senior Engineer Officer (SEO) Barry Townley-Freeman, had given no indication of any adverse weather conditions en route, or in Perth, but the extra fuel carried was a sensible precaution. All expected a smooth, pleasant and routine flight.

On boarding in Kuala Lumpur, the passengers made themselves comfortable for the journey which lay ahead, but those travelling from London were becoming overtired and found it difficult to settle. Of the Australian tour group bound for Perth from Kuala Lumpur many were unhappy at being kept waiting for so long at the terminal before the late departure. Unconcerned with the problems at Heathrow the general feeling was that the Poms had let them down again. The big jet roared down the runway and after a long run because of the hot air became airborne at 20:09 local time (12:09 GMT). At 500 ft (150 m) in the darkness X-ray Hotel banked slightly left in the smooth air to turn towards Singapore and the route to Perth. Captain Moody established the aircraft on airway G79, flying towards the Johor Bahru VOR (very high frequency omnidirectional radio range beacon) which lay in the south of Malaysia at a distance of 150 nm. The climb continued steadily to the cruising altitude of 37,000 ft (11,280 m) and the aircraft was accelerated to the normal climb speed of 320 knots. The autopilot was now engaged and would remain so until approaching Perth. Passing 11,000 ft (3,350 m) the altimeters were adjusted to the standard pressure setting of 1013.2 millibars.

In the cabin drinks were served. At the request of the tired passengers the bar service was to be followed immediately by a light meal and then afterwards the lights were to be dimmed. In the quietness the weary travellers hoped to get some rest before arrival. As the journey progressed it soon became apparent to the passengers that they were being looked after by very able flight attendants, and they began to relax and make themselves at ease. Captain Moody's entire crew of sixteen, consisting of himself and the other two on the flight deck, plus thirteen in the cabin, led by Cabin Services Officer (CSO) Graham Skinner, had been together since they left London five days earlier. On this trip they had already been as far east as Jakarta, the Indonesian capital. The previous day the crew had then passengered out of Jakarta's International Airport, Halim, to position to Kuala Lumpur to pick up the BA 009 which they now operated down to Perth. The rapport enjoyed by the crew was reflected in the comments of their passengers, many of whom remarked that they seemed a 'happy band'. Little could any of these travellers have guessed at the time that before the flight was out they would be more than grateful for the professionalism of Captain Moody and his gregarious team.

As the climb continued the captain lifted the PA handset to say a few words to the passengers before they settled down for the night. He described the route the aircraft would follow as it flew over the Indonesian archipelago, explaining that from Singapore they would fly down the coast of Sumatra and across the Java Sea to overhead Jakarta,

A British Airways 747. (British Airways)

then from there they would proceed over the Indian Ocean to Carnarvon on the Australian coast, and on down to Perth. The remaining flight time of just under five hours gave them an estimated time of arrival of 01:30 local time in Perth, the time there being the same as in Kuala Lumpur. There was little or no cloud expected en route so flying conditions would be smooth, and the forecast for Perth was fine. The captain concluded by saying that he would leave them in peace now in the hope they might get some rest and he would talk to them again on the descent into Perth.

Approaching the Johor Bahru VOR beacon, X-ray Hotel levelled off at 37,000 ft (11,280 m) and accelerated to the cruise speed of Mach 0.85. The buffet speed of 265 knots the lowest cruise speed for that flight level and weight, was bugged on the airspeed indicator (ASI), and the three engine drift-down cruise level of 27,000 ft (8,230 m) was also noted. If an engine failed suddenly at 37,000 ft X-ray Hotel would be unable to maintain its altitude with reduced thrust and would have to descend to the more dense atmosphere at 27,000 ft to maintain level flight. As the aircraft weight reduced with fuel consumption the three engine drift-down level would, of course, rise and would eventually reach the actual cruise altitude if the aircraft remained level.

Over Johor Bahru, air traffic control (ATC) changed to Singapore, and BA 009 was cleared to proceed by airway B69 out of the Sinjon VOR beacon which lay just south of Singapore Airport. Airway B69 led all the way to Perth via Jakarta. Only the odd cumulonimbus (Cb) cloud could be seen flashing intermittently in the darkness below and the flight was proving to be as smooth and pleasant as expected. Captain Moody, SFO Greaves and SEO Townley-Freeman were being looked after by Stewardess Fiona Wright who had been assigned to duties in the first class upper deck area. The few first class passengers on board had all been accommodated in the main deck below and, with more than sufficient staff to cater for their needs, Stewardess Wright was free to attend to the flight crew. Normally they would have had to wait until after the passengers had been served but this evening they were able to eat right away. As the cockpit crew dined they remained strapped in their seats eating from trays on their laps. Duties were continued as usual with navigation and performance being monitored and communications being maintained en route.

Approaching Singkep radio beacon, situated on one of the outer Indonesian Islands, BA 009 was instructed by Singapore to call Jakarta on the HF long range radio frequency of 6556 KHz. Meanwhile, in the cabin, the trays were quickly cleared away and within one hour of departure, as the aircraft passed abeam Palembang in southern Sumatra, most were quietly asleep in the dimmed light. A few minutes later X-ray Hotel entered the Jakarta upper control area and SFO Greaves established VHF short range radio contact with Jakarta on 120.9 MHz. BA 009's progress was now monitored by radar, although position reports were still expected along the way. The crew were also monitoring their own radar sets, checking the small screens for any tell-tale signs of cloud activity lying in their path. A scanner in the nose was tilted down one degree to detect any weather ahead by reflecting transmitted signals from the large water droplets suspended in thunderclouds. The clouds would show up as small 'blips' on the pilots' radar displays. To avoid the Cb cloud it was simply a matter of engaging heading mode on the autopilot and steering round them. A faint outline of the northern coast of Java could be seen painted on the weather radar screen, but no large clouds were indicated on track.

An occasional flash of lightning appeared far to the east in the black night, and to the west lights on the coast of Sumatra were clearly visible. In the distance a low overcast obscured the brightness of the Indonesian capital. The Jakarta VOR was selected and the needles of the indicator pointed steadfastly ahead indicating that the inertial navigation system (INS) was accurately flying the south-east airway track of 157° magnetic towards the beacon. At 126 nm north of Jakarta X-ray Hotel crossed position Bidak, an important reporting point which lay on the Indonesian air defence identification zone. The wind blew from the north-east at about thirty knots giving a ground speed of 490 knots. The time was 13:17 GMT and it was estimated that X-ray Hotel would be overhead Jakarta in sixteen minutes' time.

'Jakarta, Speedbird 9,' called Greaves on the radio.

Captain Eric Moody.

'Go ahead.'

'Speedbird 9 was Bidak at 1317, level 370, Jakarta 1333.'

With the flight crew's meals finished the trays were removed and Stewardess Wright now took time to prepare her own fare. In the cockpit the relaxed progress of the flight left the crew at ease. In just under four hours they would be landing at Perth and after a short ride to the hotel some would be enjoying a few beers at a room party.

At 13:33 GMT BA 009 passed over the Jakarta VOR beacon, situated in a large bay to the north of the city, and from there the aircraft turned towards Halim VOR, lying only seventeen nm away. Its position six miles (9.6 kilometres) south-east of Jakarta marked the international airport from which the crew had departed northbound only the day before. At 13:35 GMT X-ray Hotel flew over Halim, still level at 37,000 ft, and Greaves once more radioed details of their progress to ATC.

'Speedbird 9, Halim at 35, level 370, Topar 52.'

The flight had been airborne now for almost one-and-a-half hours and Captain Moody felt it was time to stretch his legs. On long-haul journeys flight crew are encouraged to take short breaks from the cockpit and all emergency drills and procedures are designed to be performed by only two people. SFO Roger Greaves and SEO Barry Townley-Freeman were both very able and experienced airmen, with thirteen and eighteen years' flying experience respectively, so Captain

Eric Moody, himself with 9,000 flying hours and seventeen years' experience, had no reservations about leaving his copilot in charge. Flight conditions were satisfactory and a check of the weather radar indicated a clear path ahead. On exiting the flight deck Moody found the crew toilet occupied and seeing no sign of Fiona Wright assumed it was she who was having a wash and brush up before her own meal. He went below to the main deck but at the bottom of the stairs he bumped into the first class purser, Sara de Lane Lea, and stopped to talk to her for a while. He never did get to the toilet.

On the upper deck Stewardess Wright was at last able to begin her own meal, while on the flight deck all was calm. The quiet and routine progress of the aircraft reassured the two left in the cockpit that the flight was proceeding as planned, but it was an illusion of wellbeing that was soon to be shattered. In a few short moments the crew of BA 009 were to find themselves entangled in an emergency of dire proportions; caught up in an event unprecedented in the history of civil aviation.

At about 13:40 GMT Roger Greaves leaned forward in the right-hand seat to peer into the darkness beyond. His attention had been drawn by a strange visual effect on the windscreen and what appeared to be hazy conditions outside. A quick inspection of the radar screen showed no storm cells lying ahead and they were probably just catching some thin cirrus cloud or perhaps penetrating the tops of weak isolated cumulus clouds. He switched on a landing light to check the situation. At night it is sometimes difficult to tell if cloud penetration has occurred and such a procedure is an old pilot trick to illuminate the cloud if weather is encountered. A hazy effect seemed to be apparent and as a precaution the engine igniters and anti-icing systems were switched on. Combustion is normally self-sustaining once a jet engine is running, but heavy moisture can cause it to run down. In such circumstances igniters are employed which spark to maintain the engine lit without any adverse effect. Also, water droplets suspended in cloud can freeze on impact with engine sensors and nacelles and the anti-icing systems prevent disruption of the airflow through the engine. Flashing the landing light occasionally would indicate when the aircraft was in the clear and both kept a weather eye open for any change in the circumstances. It was not long before their vigilance was rewarded.

'Hey Barry, look at this.'

SEO Townley-Freeman had also noticed a glow and was already looking at the sight. Streaks of mini forked lightning flashed across the windscreen. The electrical effect is known as St Elmo's fire, which both men had witnessed before in their careers, but it is normally associated with electrical discharges in storm clouds. The flight remained smooth, nothing could be seen ahead in the form of flashing thunder clouds and no adverse weather returns appeared on the radar. All visual indications, however, were that BA 009 was on the edge of a storm cloud of intense electrical activity. Giant Cb cloud in the region can stretch up to 50-60,000 ft (15-18,000 m) and can give an aircraft quite a shaking, but it was most unusual for such thunderstorms not to appear on radar screens. If the flight was about to penetrate a storm cell it was as well to be prepared and SFO Greaves switched on the 'fasten seat belts' sign. Both flight crew already had lap and crutch straps fastened, but each now attached his shoulder harness.

The St Elmo's fire continued to develop and the two left on the flight deck had a front row view of a spectacular kaleidoscope of streaking light. The co-pilot called Stewardess Wright on the interphone.

'Fiona, come and see this. It really is quite a sight.'

'I've been flying too long to fall for anything like that, Roger. I'll be up in a minute.'

'No, no,' he insisted, 'come now. You won't believe your eyes.'

As BA 009 approached the south coast of Java, Fiona entered the flight deck and walked forward to stand behind the captain's empty seat. She stooped to look at the windscreen and was immediately taken by the beauty of the sight. The whole aircraft seemed to be enveloped in a glow and the scene ahead was like skiing at speed through a snowstorm of bright silver sparks. Amid the recognisable flashes of St Elmo's fire, streams of light like tracer bullets streaked up the windscreen. In the half-light of the darkened cockpit their faces almost shone in the irridescent light. But there was more. A strange ionised odour, like that associated with electrical sparking, could also be detected and a thin veil of mist with a bluish hue began to shroud the flight deck. Something very odd was happening to BA Flight 009.

'I don't like the look of this,' remarked Greaves. Quickly he turned to Stewardess Wright. 'I think you'd better get the skipper back up here.'

Fiona Wright made her way down the spiral staircase with some haste and from half way down called out to Captain Moody who was still standing chatting by the galley at the foot. He had been joined by CSO Skinner.

'Captain, you're wanted on the flight deck.'

He didn't need to hear her request twice to detect the urgency in her voice. Stewardess Wright turned and hurried back to the upper deck while Moody immediately dashed for the spiral staircase, mounting the steps two at a time. CSO Skinner followed behind. As the captain's head reached the floor level of the cabin above, however, he almost stopped in his tracks. In the dim light he could see from below the seats what appeared to be puffs of smoke emanating from the low level air conditioning ducts. It reminded him of the clouds of water vapour that billow from the air conditioning vents in hot and humid climates. He, too, could detect the acrid electrical odour. 'Smells like the London underground,' he thought to himself. It could also be an indication of an electrical fire. Any fire in flight is one of the greatest hazards to be encountered and he was now not at all surprised that he had been summoned back to his post.

He rushed into the flight deck and rapidly scanned the cockpit. The engine anti-icing and igniters were checked on and the fuel panel configuration was checked as satisfactory. Although the problem seemed to lie outside nothing could be seen on the radar.

'It's quite incredible,' said Greaves.

Captain Moody stared wide-eyed at the intense and spectacular display of dancing lights. For a moment the St Elmo's fire, tracer and glowing luminescence distracted his attention from the smoke enveloping the flight deck. The two cabin crew, Stewardess Wright and CSO Skinner, stood behind almost unable to believe the beauty of the sight.

'And look at the engines,' said Roger Greaves.

The captain turned to the left and could see what appeared to be the engines lit from within by an intense white hot light, like the illumination from a magnesium flare. Shafts of light extended forward from all four engines. The powerful brightness shining through the fan blades produced a stroboscopic effect which made the fans appear to turn backwards in the engine nacelles. Barry Townley-Freeman also rose briefly from the flight engineer's seat to peek from the side window behind Greaves while the flight attendants took turns to view from the captain's side.

In spite of the impressiveness of the display, however, there was mounting concern for the 'smoke' on the flight deck which was becoming more dense. If it became any worse they would have to don oxygen masks. CSO Skinner didn't need to be told something was amiss, and he quickly left the cockpit to warn the crew to prepare the cabin in case of problems. He was met by other flight attendants reporting evidence of smoke throughout the aircraft, and the instruction was given to switch off galley electrical power and to stow equipment safely.

To SEO Townley-Freeman, the thickening acrid smoke from the air-conditioning system suggested an electrical problem, and he carefully scanned his large instrument panel for any signs of trouble. The time was now 13:45 GMT and X-ray Hotel was crossing the Java coast and heading out over the Indian Ocean on Airway B69. As Townley-Freeman continued with his scrutiny, he carefully opened the check list at the drill for removal of air conditioning smoke. The procedure initially called for the donning of oxygen masks and goggles by the crew and for the switching off of gasper and recirculation fans. On the captain's command he would be ready for action. Suddenly the flicker of a light from an unexpected source caught his eye. The intermittent flashing was from a pneumatic valve light which was designed to illuminate when the valve was closed. The valve supplied compressed air direct from the engine compressors to feed air driven equipment and air conditioning and pressurisation systems. It was designed to close automatically when an engine ran down to prevent reverse flow, and its movement towards the shut position indicated more of an engine problem than trouble with the air conditioning system. Quickly Townley-Freeman glanced forward to check the main engine instruments situated in the centre of the pilot's panel. Number four engine on the outer right side was running down! Something had disturbed the flow of air through the compressors and the engine had surged, i.e. had suffered a giant hiccup, and the power was dropping. By contrast the exhaust gas temperature (EGT) was soaring and the indicator pointer was racing off the clock. The amber EGT warning light illuminated.

'Engine failure,' shouted Townley-Freeman, 'number four.'

The three flight crew: SFO Roger Greaves (left), Captain Eric Moody (centre) and SEO Barry Townley-Freeman (right) stand before *G-BDXH*'s number one engine.

The pilots also felt a slight swing of the nose as power dropped on one side and rudder was applied to hold the aircraft straight.

'Engine fire drill number four engine,' responded Captain Moody.

Greaves and Townley-Freeman carried out the fire drill from memory, closing the number four thrust lever, shutting off the fuel supply and pulling the fire handle to isolate services from the failed engine. As they completed the drill and confirmed their actions from the check list, the captain monitored the flying, holding on a little left rudder with his foot to balance the aircraft. The entire procedure had taken about thirty seconds. Some fairly quick thinking was now required, for on three engines X-ray Hotel was unable to maintain 37,000 ft (11,280 m). At about 280 tonnes they would need to descend down to 29,000 ft (8,840 m), the appropriate level for their weight and flight direction. Permission would be required from Jakarta and would have to be obtained promptly. It would also be unwise to commence a journey over the ocean on three engines, for if a second engine failed in mid-crossing BA 009 would be in a serious situation. Jakarta lay about ninety nm to the north and, with flight easily sustainable at the lower level on three engines, the sensible approach would be to turn back and land at Halim to have the engine checked. A rerouteing, however, would also require clearance over the radio. Before the crew could plan or even think procedures, events took a turn for the worse. Number two engine surged and began to run down. Once again the power drop was accompanied by a rising exhaust gas temperature.

'Number two's gone,' called the flight engineer.

Almost immediately Townley-Freeman saw to his horror that numbers one and three engines had also surged and were running down.

'They've all gone!' he shouted with disbelief.

If there was alarm on the flight deck at the situation, in the cabin the passengers, unaware of the problem, were becoming increasingly distressed. Although the main cabin lights were dimmed, the acrid 'smoke' in their midst was quite obvious and the thumping of the surging engines could be heard and felt through the structure. The glow from the engines could be clearly seen in the black night and huge sheets of flame could be seen shooting from the jet effluxes illuminating the interior of the cabin. To those by the windows it seemed as if the whole aircraft was on fire. Suddenly, for a reason they could not comprehend, those at the rear of the aircraft noticed a drop in the noise level and they could sense the cabin being engulfed in an eerie silence.

In the cockpit the initial reaction of the flight crew was one of total disbelief. Four engines don't just stop; what have we done wrong, was their first thought. Engine intakes can freeze up causing run down, but the engine anti-icing had been checked on. So, too, had the igniters to sustain combustion. Fuel mismanagement can also cause problems: it had been known on the rare occasion for all four engines to be run from one fuel tank for balancing purposes and for inattention to have caused the tank to run dry, stopping the engines. Moody and Townley-Freeman had both confirmed the fuel configuration as satisfactory. The engine indicator pointers were all over the place, causing confusion, but there was no doubt that, for whatever reason, all engines had failed.

Crew training, of course, covers every contingency, no matter how remote, and total engine failure was no exception. Only a few months earlier Captain Moody had successfully completed his biannual simulator check and had practised this very procedure. The simulator cockpit had quietened, the lights had dimmed to emergency level, the autopilot had disconnected and many instrument failure flags, including most of those on the co-pilot's side, had appeared. On X-ray Hotel's flight deck, however, the situation was quite different. The rush of air masked any reduction in engine noise, the cockpit lighting stayed on, the autopilot remained engaged and the instrumentation, apart from the engine indicators, displayed as normal. Whatever was happening to X-ray Hotel was a complete mystery. The strange storm producing the intense electrical activity also seemed to be affecting the power, and some unknown force had choked the engines into silence.

Immediately Captain Moody instructed Greaves to send

out a Mayday and Townley-Freeman commenced the 'loss of all generators' drill. The procedure assumed, of course, that primary services had been lost: pneumatics, including pressurisation and air conditioning, normally supplied by bleeds from the engine compressors, and electrics, normally supplied by the engine-driven generators. With total engine failure, however, no attempt to restart could begin without standby electrical power from the batteries to supply essential engine controls: fuel valves, start levers, standby ignition and some engine indicators. The 'loss of all generator' drill, therefore, began by checking that the battery switch was on and that the standby electrical busbars were powered. DC power from the battery was changed by an inverter into AC to supply the standby AC busbar. The combination of DC and AC buses also fed the captain's flight instruments, number one VHF radio, the interphone and the PA system. Having checked these items Townley-Freeman then opened the fuel cross-feed valves. In a matter of seconds he had reached the point at which the drill split: loss of electrical power caused by a massive short which had tripped all the generators, or the loss of all generators caused by total engine failure. In this case the course of action was obvious, and he prepared for in-flight starting of the engines. The air was too thin at cruising altitude for an engine start to be recommended, but under the circumstances there was little point in waiting and it seemed worth a try.

'Mayday, Mayday, Mayday,' shouted Greaves over the radio. He called on the Jakarta control frequency of 120.9 MHz, having set the emergency code of A7700 on the transponder. If there was no reply he would transmit on the emergency frequency of 121.5 MHz. 'Speedbird 9, our position is 100 miles south of Halim. We have lost all four engines. We're descending and we're out of level 370.'

Silence, and then a few moments later: 'Jakarta, Speedbird 9, have you got a problem?'

The electrical static produced by the storm was distorting transmissions and making communications difficult.

'Jakarta, Speedbird 9,' repeated Greaves carefully, 'we have lost all four engines. We are descending and we are now out of level 360.'

Another short pause followed, then Jakarta replied. 'Speedbird 9, understand you have lost number four engine?' Greaves turned in frustration to his Captain. 'The fuckwit doesn't understand.'

'Negative, Jakarta, Speedbird 9,' spoke Greaves slowly and distinctly, trying to control his frustration. 'We have lost all four engines, repeat all four engines. Now descending through level 350.'

The Jakarta controller was still unable to understand but fortunately a Garuda Indonesian Airways flight on the frequency interrupted the exchange.

'Jakarta, Garuda 875, Speedbird 9 has lost all four engines, he's lost *all* his engines, and he's out of level 370.

At last Jakarta got the message and understood the full force of the problem. Greaves requested radar assistance for return to Jakarta but the conditions prevented the controller from identifying X-ray Hotel. In the end the first officer simply told the controller they were turning back for a landing at Halim Airport. As Greaves battled with the radio, Townley-Freeman fought to bring the engines back to life. The number four engine had been fully shut down so the fire handle was reset and the thrust lever was placed in line with the others. Although it was not advisable to relight an engine after a non-recoverable stall unless a greater emergency occurred, this occasion surely qualified. It seemed common sense to Captain Moody and the crew to attempt to restart any one out of four engines instead of any one out of three. For the flight crew, however, the mental action of reading the check lists and performing the drills in such a desperate situation was, in Barry Townley-Freeman's own words, like trying to think through treacle. In such circumstances it is the man who can use his common sense who will win the day. The reinstating of number four engine may have seemed like a simple act but it was to prove significant at a later stage.

To begin the start sequence the start levers were selected to cut-off to reschedule the fuel control units. The standby ignition was switched on to power the igniters and the start levers were selected back to idle. The crew looked in hope to see the instrument readings rising indicating light up.

Nothing. Not even a hint that at least one engine was going to start.

In the meantime, the captain had exchanged height for speed and had commenced descent to stop the speed dropping further. Using the autopilot he had set 500 ft (150 m) per minute descent and the speed had settled on the airspeed indicator at 270 knots, which was close to the speed for minimum drag for the present all-up weight. He now turned the autopilot knob to bank left, clear of the airway, and although X-ray Hotel was descending over the sea he set the safety height for the return to Jakarta over Java of 11,500 ft (3,500 m) in the altitude select window. It is normal in such reports of aviation emergencies to read of aircraft plummetting to the ground, but in this case it was far from true. Normal descent procedures approaching destination include closure of the throttles to idle power at top of descent and the aircraft literally glides to the lower levels before power is re-applied for the approach to land. The aircraft's sleek aerodynamic lines reduce air resistances to a minimum and even slowing down without power at level flight takes time.

In X-ray Hotel's situation the lack of any residual thrust from idling engines and the drag from the big engine fans windmilling in the airstream would increase the rate of descent, but it would only be marginal. The windmilling engines, however, were a distinct advantage in this emergency, for even at speeds as low as 160 knots they could still turn the hydraulic pumps sufficiently to pressurise the system for operation of the aircraft's flying controls. Also, unknown to the crew at the time, in spite of complete engine failure number three generator had stayed on line, supplying electrical power. It was being driven by the windmilling engine and was the reason the autopilot had remained engaged and the cockpit lighting and instrumentation displays had stayed normal. The generator would not remain on line for long, however, but when it tripped electrical power would be supplied by the batteries for thirty minutes.

Meanwhile the captain could use the autopilot to control the aircraft in the descent and leave himself more free to concentrate on the problems in hand. The cabin was normally pressurised by the engine compressors to a pressure equivalent to 6,000 ft (1,830 m), but with power loss, even though the outflow valves had automatically run closed, the cabin pressure was beginning to drop.

In spite of total engine failure, therefore, the aircraft was under control. Hydraulic pressure was sufficient to operate the flying controls, electrical power, although fluctuating, was still available from number three generator, and the cabin pressure, although reducing, was still sufficient for normal breathing. X-ray Hotel was descending at around 2,000 ft (610 m) per minute and was now out of 35,000 ft (10,670 m). In its clean condition, i.e. no flaps or gear lowered, it could remain airborne without power for about a further twenty minutes, covering a ground distance of approximately 140 nm. Although the aircraft was in dire straits the problems were compounding gradually and the crew were able to gather their senses. The initial alarm had subsided and with the adrenalin flowing the three worked well together as a team. Even the early failures to start the engines did not leave the flight crew in complete despair. Engine start required conditions to be within certain parameters: indicated airspeed 250-270 knots, altitude below 28,000 ft (8,530 m), and windmilling engine compressor rotations within tolerance. Once they had descended to lower altitudes they were sure the engines would fire up again. After all, the same had happened in the simulator check; once below 28,000 ft the engines had relit. And X-ray Hotel was also a Rolls-Royce aircraft: American-built but fitted with good reliable British Rolls-Royce engines. Of course they would start.

Turning the aircraft back to Jakarta brought BA 009 closer to a suitable airport where the captain could land once the engines had been restored, but attempting a dead stick landing, i.e. touching down without power, could not be considered. Judging such a landing in the darkness would be almost impossible and the chances of success negligible. There was also a tall mountain range lying across Java which resulted in the safety height being as high as 11,500 ft. X-ray Hotel could not clear the high ground without power.

Many of the mountains were volcanoes and some of them were very active. In the Sunda Strait which separates the Indonesian islands of Java and Sumatra was once situated the then volcanic island of Krakatoa which erupted on 27 August 1883. The explosion was equivalent to a force of 50,000 H-bombs and blew the island completely off the map. The bang went on record as the loudest produced on earth and was heard halfway round the world. Almost one hundred years later, another mountain of the chain, Mount Galunggung, situated 100 miles (160 kilometres) south-east of Jakarta on the Java coast and one of sixty-seven currently active volcanoes in Indonesia, had recently been belching fumes into the atmosphere. It was considered the most dangerous and a major eruption was expected.

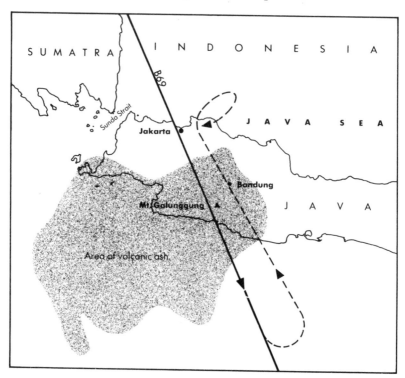

A map showing BA 009's routeing on B69. Also shown is the position of Mount Galunggung and the area of volcanic ash.

If the crew were unable to start the engines they would be forced to ditch in the black sea, but that was some way off. There was still about eighteen minutes to save the situation and all their efforts and thoughts could be concentrated on the single aim of getting the engines going again. They had no opportunity to think beyond their immediate tasks and they had not reached the point of assuming the worst. At that stage they were too busy to be afraid, but they were not to realise then that shortly the situation would turn much worse. Later they would know fear.

Once more the three scanned the flight deck for clues as to their predicament: engine anti-icing on; standby igniters on when required; fuel panel checked, all main boost pumps on; circuit breakers checked, all confirmed set. The thrust levers were simply left in the positions they were in at the time of failure. Again the start levers were selected to cut-off, fuel shut-off valves were checked open, standby ignition was selected and the start levers were set to idle. Still no response.

By now the crew had exhausted all their options and it was simply a matter of selecting the start levers to cut-off, switching on standby ignition, and reselecting the start levers to idle. Standard procedures in an emergency, requiring the captain to fly the aircraft and the other two to perform the drills, were abandoned in the circumstances. With the autopilot still engaged and the radio work completed for the moment, both pilots found themselves free to lend a hand and all three took turns at trying to start the engines. If persistence alone could have achieved results the four engines should have roared into life, but it was not to be. They remained ominously silent. In the cabin the initial alarm had subsided, but there was mounting apprehension at their plight. The captain had been too busy to make an announcement and the passengers were ignorant of the circumstances. Those at the rear of the cabin by the windows could witness, without realising what they saw, the crew's attempts to get the engines going. On each occasion when fuel was introduced and the engine failed to light-up, the kerosine ignited in a huge flame which shot from the jet efflux. The process shook and vibrated the aircraft.

The number three generator was still being driven by the slowly rotating windmilling engine via a constant speed drive which was unable properly to perform its task. The resulting fluctuating frequency made the cabin lights cycle on and off, brighter one moment then dim the next. At the back of the cabin the eerie silence also persisted and they could hear each other breathe. Under the circumstances it would not have been surprising if the passengers had panicked, but they remained remarkably calm. The cabin crew did a wonderful job in reassuring the travellers and moved about the cabin comforting those in greatest distress.

At this stage BA 009 was only a few minutes into the emergency and the crew seemed to have exhausted all their options except the hope of starting the engines at a lower altitude. Amazingly, most of the captain's and first officer's flight instruments remained powered and displayed normal indications and the cockpit lighting remained illuminated. Only occasionally did some engine instrument failure flags appear. The engine indicators were a mix of Smiths and General Electrics and the former gauges froze where they stood with power loss while the latter gauges ran to zero. It made it almost impossible to monitor properly the engine start sequence and difficult to know whether or not light-up was occurring. To add to the difficulties, the electrical storm still danced magnificently before their eyes and their headsets crackled loudly from static interference.

X-ray Hotel was now descending through 30,000 ft (9,144 m) out over the ocean at a rate of 2,000 ft per minute, in a gentle turn back towards Jakarta. If the engines failed to relight Captain Moody would simply turn back southbound away from the mountains and remain over the flat sea. Townley-Freeman continued with the start attempts and with the approach of 28,000 ft (8,530 m) all were confident of a recovery. Still nothing happened.

With time, the bluish mist in the cabin began to disperse as the inert compressors of the failed engines no longer pumped contaminated air, or any air at all for that matter, into the aircraft. Oxygen masks were not needed as a protection against the 'smoke', but they would shortly be required for another reason. As X-ray Hotel descended through 28,000 ft (8,535 m), an intermittent warning horn sounded in the cockpit: *beep, beep, beep*. Although the aircraft was descending the cabin was, in a sense, ascending, as the pressurised air within leaked to the atmosphere. On 'climbing' through 10,000 ft (3,050 m) the cabin pressurisation warning horn sounded and the flight crew were obliged to use oxygen. It was a moment that those in the cockpit did not relish for the masks were uncomfortable to wear, but what happened next took them all by surprise.

To begin with the flight engineer's mask stowage had been incorrectly installed and it was impossible to reach while seated. At a time when seconds were precious he had to undo his seat belt and stand up before stretching backwards to reach his mask. As the pilots pulled their oxygen mask harnesses over their heads, the supply hose of Greaves's mask came away in his hand and the connecting pieces fell apart in his lap. He was now without oxygen in the 'climbing' cabin and as the altitude increased he was in danger, if the situation continued, of suffering from anoxia. The effect induces a feeling of wellbeing in the victim although performance is akin to being drunk. Unconsciousness eventually results.

Standard procedure for the failure of pressurisation in the cruise is for a steep descent to be initiated immediately. The throttles are closed, speed brakes are deployed, the gear is lowered and the nose is pushed down to attain maximum speed. Descent rates of 10,000 to 12,000 ft (3-3,600 m) per minute can be achieved and the aircraft can quickly be brought to safety at a lower level. In X-ray Hotel's case it was pointless to throw away precious height if, for the time being, the crew could breathe normally, but with the threat of incapacitation to his co-pilot, the captain had little choice but to get down quickly. Captain Moody disconnected the autopilot to hand fly the aircraft and lowered the nose.

It was an agonising process for Moody to throw away height, the one commodity which could both buy them time and save everyone's life. The landing gear could be lowered by using the alternate system which simply released up-latches to allow the wheels to drop by gravity. Once lowered in such a manner, however, it could not be raised. If a ditching became necessary, touching down on the sea with the landing gear extended could place the aircraft in danger. Moody compromised on procedures by pulling only the speed brake lever, but even so the aircraft dropped at an alarming 6,000 ft (1,830 m) per minute. The passengers could feel a rumble through the cabin with the speed brakes deployed. As height tumbled from the altimeters, Greaves fought desperately to refit the hose attachment. With nimble fingers and cool nerves he managed to fit the pieces together and to place the mask on his face. By now the aircraft had descended to about 20,000 ft (6,100 m) and Moody retracted the speed brakes to resume the glide.

'I've got 320 knots,' said Greaves. 'We are going too fast.'

A speed of 250-270 knots was required for engine start but his airspeed indicator displayed a higher speed. The captain quickly glanced across.

'That's strange, I've got 270 knots.'

The discrepancy of fifty knots was a mystery, but could be the cause of the engines failing to relight at the lower altitude. Captain Moody varied the speed using the co-pilot's airspeed indicator and then his own while the flight engineer repeated the start sequence, but to no avail. In the cabin, to add to the passengers' discomfort, the aircraft now seemed inexplicably to climb and descend at varying rates and they could feel the heaving in the pits of their stomachs. There was still no time for the captain to explain the predicament, even if he knew what was happening, so in more ways than one the passengers were left in the dark.

The cabin continued to 'climb' in spite of the aircraft descending and, as X-ray Hotel passed 18,000 ft (5,485 m), the cabin height rose through 14,000 ft (4,270 m). Passenger oxygen masks dropped from their overhead stowages. The appearance of the masks added to the distress of the travellers who had just felt what seemed to be the aircraft plummetting out of control. The cabin crew donned portable oxygen equipment and performed a marvellous job in calming the more disturbed passengers and helping to fit masks as they walked round the cabin. Unbelievable as it may seem some travellers were still asleep. The emergency announcement explaining how to use the masks had failed to function and CSO Skinner found he was unable to operate the PA system from the cabin stations. He pulled a megaphone from one of the overhead lockers and attracted the attention of the passengers to give them instructions.

'Can you hear me, mother?' he called, imitating a northern English accent. It brought a moment of light relief that was greatly needed.

On the flight deck attempts to start the engines continued without success as Moody hand-flew the aircraft. The fifty knot discrepancy in the airspeed indicators was still evident and once again he varied the speed. X-ray Hotel was now running out of height although mysteriously descent was not as fast as expected. The outside air temperature was much higher than normal and the hot rising air was also producing an uplift to help keep BA 009 airborne, not unlike that experienced by a glider in an updraught. X-ray Hotel was still losing the race against time, however, and the flight crew were under a lot of pressure. The dim cockpit lighting, the wearing of uncomfortable oxygen masks, the flashing electrical discharges, and the loud headphone static did not make their tasks any easier. The electrical activity in the atmosphere also seemed to be playing some very strange tricks on the aircraft's electrical systems. The inertial navigation system displayed random digits and patterns, with nothing making sense, the VOR needles spun round in circles and the distance measuring equipment (DME) was blank. The dark night made visual flying impossible and the captain had to fly solely on instruments.

It was now that nagging doubts began to invade their minds and thoughts that the engines might not start entered their heads. Townley-Freeman surmised that water contamination in the fuel tanks could be causing the problem and if that was so there was no way the engines were going to light up again. There would be no choice but to ditch. Captain Moody had already decided that he would fly towards Jakarta until approaching 12,000 ft (3,660 m) just above the Java safety height, and then he would turn southbound to land on the sea. It was not a prospect any of them relished. The pilots understood the principles of ditching: landing along the line of the primary or predominant swell and upwind and into the secondary swell, or downwind and down the secondary swell. That was the theory, of course, but landing on a shark-infested ocean in a black night with a heavy sea running was a different matter.

The captain had watched flying boats as a boy and knew that they didn't operate at night because of the difficulty in judging height in the darkness. On X-ray Hotel, if only battery power was available for ditching, there would be no radio altimeter for precise height indication and there would be no landing lights available. The landing gear would remain retracted and the underside of the aircraft would present a smooth surface to the sea on touchdown, but it would not be possible to lower flaps and the ditching would be fast. The aircraft stall speed at sea level was 179 knots, so to achieve a safe margin a speed of 190-200 knots would be required. The engines would almost certainly break off on impact and the wings and structure might be damaged. Once stopped the aircraft would probably remain afloat for a reasonable time, but launching the life rafts in the heavy swell would be extremely difficult. There would also be a very real danger of sharks. X-ray Hotel was now descending through 16,000 ft (4,880 m) and if the engines didn't start soon they were going to be in trouble. Strangely, the flight crew as yet felt no great fear, but there was certainly growing apprehension.

In the cockpit the captain now felt they had caught up with events but there seemed little more they could do. Townley-Freeman had not rested in his attempts to start the engines and he continued to try. He would still be having a go when they touched the sea, if it came to that. Greaves repeated requests for radar assistance from Jakarta, but the electrical storm was making contact difficult. There was still no range or bearing information from Halim VOR, but one of the three inertial navigation sets suddenly made some sense and true track and distance were observed and transmitted to Jakarta.

'We're out of level 150 now, descending,' radioed Greaves, giving the position information at the same time.

'Roger, Speedbird 9, if possible maintain not lower than 12,000.'

It was a useful reminder of the safety height.

In these dire circumstances the captain felt the need to speak to CSO Skinner and to ensure that he understood the predicament, but Moody was unable to contact him on the interphone. The captain also felt that, in spite of the situation, he had some spare thinking capacity and now seemed an appropriate moment to talk to the passengers. He could also request Skinner's presence at the same time. Moody took a deep breath and spoke into the oxygen mask: there was little he could do but tell the truth.

'Ladies and gentlemen, this is your captain speaking. We have a small problem. All four engines have stopped. We are doing our damndest to get them going again. I trust you are not in too much distress.' There was a short pause in which the message took effect, then he added, 'Would the CSO come to the flight deck immediately, please.'

The last statement was not only a summons to Skinner, but was also a signal to the cabin crew in an emergency to stow their equipment and to move to their respective stations. No-one was now left in any doubt as to their predicament. As CSO Skinner made his way towards the flight deck, Captain Moody's mind flashed to a scene from a training film depicting a moment of great emergency. He could recall the amateurish screen captain delivering his instructions to his senior flight attendant in stilted but impeccable English.

'It seems it's not our day. I'm afraid we're going to have to ditch.'

It was not Moody's way. As Skinner entered the flight deck carrying his portable oxygen set the captain turned briefly to glance at his CSO.

'Got the picture?' he shouted with difficulty through his mask.

Skinner only heard a muffled sound, but didn't press the matter. He could guess from the activity on the flight deck that a forced landing, probably on the sea, was imminent. He gave a quick thumbs up to his captain and when no further direction was forthcoming he returned to the cabin.

The flight crew now began to feel more than concern for the height being lost. As the aircraft descended through

14,000 ft (4,270 m), about one minute remained before Captain Moody planned to turn southbound again and back out to sea. Once heading away from Java, approximately five minutes — no more — would be left for some further endeavour at starting the engines, then the flight would be over. The time was now 13:57 GMT, and X-ray Hotel had been without engine power for over twelve minutes. The crew had lost count of the number of failed attempts and it was unlikely at this stage that their efforts would bear fruit. It did not stop them trying, but a ditching seemed inevitable. Now they knew fear.

At that moment X-ray Hotel appeared to break from the hazy cloud of the electrical storm, and the aircraft flew into clear air. The dancing lights vanished from the windscreen and the glow from the engines dimmed. Suddenly Townley-Freeman let out a cry.

'Number four's started!'

The three watched with bated breath as the engine gauges rose steadily and the power settled. Gingerly Moody advanced the thrust lever and the engine ran successfully at normal power. It was only one relit, but they were on the road to recovery. In the cabin the roar of the engine was music to the passengers' ears. If they now turned out to sea and descended to a lower altitude, quickly dumping fuel at the same time to lose weight, they may just be able to remain airborne. If they couldn't hold height the power would at least slow the rate of descent. More importantly, the vital services of electrics, hydraulics and pneumatics could now be supplied from this one engine, and, if necessary, compressed air could be tapped from the engine to help turn the others and give a better chance of starting them. Number four engine had been shut down early in the emergency and it had been the least affected by whatever had caused the damage. It now seemed to have saved the day. The crew could also now remove their rather uncomfortable oxygen masks. X-ray Hotel's rate of descent eased to 300 ft (ninety m) per minute, but the aircraft continued inexorably downwards. Approaching 12,500 ft (3,810 m) there was little margin to clear the mountains for an emergency landing at Jakarta. As the seconds ticked by the initial relief abated, for all efforts to get the others going proved to be in vain. One minute passed in what seemed like an eternity, and still the moments slipped away. A further twenty seconds went by without success, then Townley-Freeman called out again.

'Number three's lighting up.'

Soon number three was restored to life and with two engines running at sufficient power to arrest the rate of descent, the aircraft was held level at 12,000 ft (3,660 m). X-ray Hotel could now remain safely at altitude if only the engines, which were probably damaged, could sustain power. They could then cross the mountains and execute a two-engine landing at Jakarta. Almost immediately numbers one and two engines relit in sequence and, from a situation of great danger only a few moments earlier, they now found themselves with all four engines running normally.

'Jakarta, Speedbird 9,' called a relieved Greaves, 'we're back in business, all four engines running, back at 12,000.'

Unfortunately the controllers were still unable to establish radar contact because of X-ray Hotel's proximity to the mountains, and a request was made for climb to 15,000 ft (4,570 m). As the aircraft levelled off at the new height, Greaves relayed their position giving bearing and distance from Halim VOR and radar contact was confirmed. The airspeed indicators had settled down and were reading in agreement but 'A' autopilot would not re-engage. 'B' was selected instead. They now seemed home and dry with their troubles over, but there was more to come. In the meantime the captain took advantage of this quiet moment to speak to the passengers.

'Ladies and gentlemen, we seem to have overcome our problems and have managed to start all the engines. We are diverting to Jakarta and expect to land in about fifteen minutes.'

Captain Moody had no sooner replaced the PA handset when they found themselves in the hazy conditions again. Once more the St Elmo's fire danced on the windscreen and their headsets crackled loudly with the static. There was no doubt that X-ray Hotel had climbed back into danger and Moody's first thought was to descend out of it as quickly as possible.

'Christ, we're not staying here.'

Moody slammed the throttles closed and lowered the nose for descent to 12,000 ft. Suddenly number two engine surged violently. It auto-recovered and surged again. The bangs from the engine were so loud they could be heard on the flight deck and the entire aircraft shook. Having lost all four earlier the crew were reluctant to shut one down again themselves, but they had little choice.

'Shut down drill number two engine,' shouted Moody.

The other two performed the drill and, with the fuel cut, engine power ceased. X-ray Hotel was now back on three engines. The failure of the number two engine for a second time gave the crew quite a shock, for it indicated that the damage sustained was worse than previously realised. Would another engine fail again?

'Speedbird 9,' radioed Greaves, 'we've lost another engine. We're now on three and we're descending to 12,000 feet.

Jakarta Control approved the lower level and issued radar vectors for their prompt return to Halim.

'Ladies and Gentlemen,' spoke the captain again on the PA, 'I have had to shut down number two engine as it was rough running. We shall be landing in Jakarta in about ten minutes.'

The general feeling aboard BA 009 was the sooner on the ground the better, but it was now obvious that some gentle nursing of the aircraft was going to be required to ensure a safe landing. Rather than adjust the power and risk further upsets, Moody used the speedbrakes to control aircraft speed, and once over the mountains began a shallow descent with speedbrakes deployed. Fortunately, Jakarta weather was fine with a calm wind, little cloud and good visibility. Runway 24 was in use and the temperature was 26°C. A few minutes later, Control instructed Greaves to call Jakarta Approach on 119.7 MHz.

X-ray Hotel now approached Halim Airport from the south-east but the aircraft was too high for landing and some height needed to be lost. Moody decided to overfly the 'AL' non-directional beacon, situated on the approach to the landing runway, at about 10,000 ft (3,050 m) and then to proceed initially to the north-west before banking right in a shallow descending turn out over the sea. He would complete the let-down by flying all the way round to land to the south-west. It was a non-standard approach which suited their purposes, and they would be guided on the way by radar which would direct them onto the instrument landing system (ILS). Fortunately, the crew had been into Halim in the daylight only the previous day, so they had recent experience of the airport. They knew there was a lot of high ground in the area.

X-ray Hotel was cleared initially to 2,500 ft (760 m), but as descent continued Captain Moody was informed of another difficulty. Normally the ILS displays both runway centre line and descent path guidance, but at Halim only runway direction guidance was available. On a big jet, judging the correct descent profile for the approach to land without radio aids is difficult at the best of times, but after a major emergency it was the last thing the captain wanted. Fortunately, visual approach slope indicators (VASIs) were available, if they could see them, for in spite of the clear visibility indicated by the controller, the conditions were very misty. It was more like dust haze, in fact, and there was a sulphurous odour in the air.

As X-ray Hotel completed the turn and approached the airport from the north, heading 210° under radar control, the flight was further instructed to descend to 1,500 ft (457 m) and was cleared to pick up the ILS extended runway centre line. Halim lay to the crew's right, about eight miles (13 kilometres) away, but the view through the first officer's number two window still appeared hazy.

'Can you turn up the runway lights?' requested Greaves over the radio.

Moody turned the 747 on to final approach and suddenly, from a dark area on the ground ahead, the brightened runway lights could be seen, but the view was very blurred. It was as if there was oil on the windscreen but the wipers and washers had no effect. Unknown to the crew, the 'cloud' which had choked the power from the engines had also sandblasted the windscreens. As BA 009 got nearer the brightness of the runway lights on the opaque windscreen blinded their vision and they had to be turned down a little.

A landing light window showing the opaque effect of the sandblasting. The 747 windshields were similarly affected. (British Airways)

The central and inboard sections of the pilot's windows were worst affected, but Moody managed to catch sight of the left side of the runway through a relatively clear three or four inch wide outer strip. Leaning sideways he established on the runway centre line and gingerly commenced descent on the final approach.

Greaves retained the distance measuring equipment on his side and called out the recommended descent profile of 300 ft per mile (fifty-seven m/kilometre): distance four miles, height 1,200 ft; distance three miles, height 900 ft; etc. He switched on the landing lights but their covers were also opaque and they gave almost no illumination. Greaves then called the Tower on 118.3 MHz and was given the clearance to land.

Captain Moody's view was not very clear but at least he could pick out the VASIs and had some descent slope indication. Sitting in this bent position, however, made it difficult to scane the instruments at the same time, an essential part of flying a large aircraft. As an aid, Greaves called out the heights every twenty-five feet (7.6 metres) from the radio altimeter. Townley-Freeman called the speed and the engine power settings. Cautiously X-ray Hotel groped towards the runway. On three engines, with visibility curtailed, difficulty in reading the flight instruments and no radio descent path indication, Moody had his work cut out. It was, in the immortal words of the Hampshire captain, like trying to negotiate your way up a badger's arse. The cabin crew had been instructed to prepare for an emergency landing and they waited anxiously at their stations.

BA 009 finally crossed the threshold of runway 24 at 150 knots but, as Moody prepared for touchdown, he still found his forward vision severely reduced. The lack of landing lights did not help. Normally pilots look far down the runway to judge height on landing but Moody could view only the hazy outline of the left-hand runway lights closest to him. Using all his experience he eased back gently on the control column and the wheels kissed the ground. It was a perfect touchdown. In the cabin the passengers spontaneously applauded. The time was 14:10 GMT and the emergency had lasted twenty-five minutes, thirteen of which had been totally without engine power. In a period of less than a quarter of an hour they had been to the edge of great danger and back again. Moody ran the aircraft right to the end of the 9,800 ft (2,990 m) runway then, with Greaves's help, managed to feel his way round the turning pad and to back-track towards the terminal. Eventually they groped their way to a turn off but as they approached the stand area the bright lights completely obscured their view.

'That's it,' said Moody. 'We're stopping here.'

Greaves called the tower to say they were unable to taxy further and asked for a towing truck to pull them to the arrival gate. Townley-Freeman then fired up the auxiliary power unit and shut down the engines. It was all over. The tension drained from their bodies and they were flooded with relief. There was no doubt it had been a close-run affair, but the cause of their predicament remained a mystery. As the arrival of the tow truck was awaited the flight crew had time to reflect on their experience. Black dust seemed to lie everywhere on the flight deck. Barry Townley-Freeman wiped his hand across a surface and rubbed the substance he picked up between his thumb and fingers. It was gritty, sooty and had a sulphurous smell.

'I reckon that's volcanic ash, you know.'

No-one believed him at the time, but the next day he was proved right.

On the evening of 24 June, Mount Galunggung, the active volcano situated on the south Java coast 100 miles (160 kilometres) south-east of Jakarta, had erupted violently. Giant plumes of ash and grit had been hurled eight miles (thirteen kilometres) into the air. The explosion had created a vast volcanic storm of thick, hot sulphurous gases and high electrical activity. No aviation warning was given. North-easterly winds aloft at twenty-five to thirty knots had blown the plume across the path of BA 009 and it had been engulfed in ash. Flying through the hot grit at speed had the effect of sandblasting X-ray Hotel's leading edges and had stripped paint and caused the opaque surface on the windscreens and landing light covers. The dust had penetrated aircraft sensors such as pitot tubes which sense dynamic pressure to measure airspeed. The discrepancy in the airspeeds was the result. The engine nacelles, intakes and fans were shot-blasted and stripped clean. Erosion of the compressor blades had occurred and the ash, which was of a silicate material, had fused in contact with the hot metal of the combustion chambers and turbines. Deposits of fused volcanic ash up to half an inch in diameter were later discovered in all the engine tailpipes. The effect was sufficient to disrupt the airflow and was similar to dampening a blaze with sand. As a result the engines had flamed-out and all power had been lost. At 13,500 ft (4,115 m), X-ray Hotel had broken into clean air and the least damaged engine, number four, had roared into life. Subsequent climb to 15,000 ft (4,570 m) had taken them back into the ash cloud. It was a close escape which might have ended

Fused volcanic ash on the number two engine high pressure nozzle guide vane. (Rolls-Royce Ltd)

tragically had not the skill, coolness and persistence of the flight crew won the day. But the passengers and crew of BA 009 were lucky, too, for the aircraft only just emerged from the volcanic storm in the nick of time.

'If the base of the ash cloud had dropped to the sea,' said Moody, 'so would we.'

Captain Moody was awarded the Queen's Commendation for Valuable Service in the Air, plus the British Airline Pilots Association Gold Medal, the Guild of Air Pilots and Navigators Hugh Gordon Burge Memorial Air Safety Award, British Airways Board Certificate of Commendation, Award as one of the twelve 'Men of the Year' in Britain, and Lloyds of London presentation set of crystal decanters. SFO Roger Greaves was awarded the British Airline Pilots Association Gold Medal and Lloyds of London presentation silver tray. SEO Barry Townley-Freeman was awarded the Flight Engineers International Association's Frank Durkin Award and Lloyds of London presentation silver tray. CSO Skinner

The entire crew of BA 009, with the then deputy chairman, Mr Roy Watts, at the British Airways Award presentation ceremony. (British Airways)

was awarded the Queen's Commendation for Valuable Service in the Air, the British Airways Board Certificate of Commendation and Lloyds of London presentation silver tray. Those mentioned above also received, together with the remainder of the crew, the British Airways Customer Service Award.

Of all the awards, accolades and applause received by the captain and crew of BA 009, however, no praise was more heartfelt than that offered by one of the Australian passengers. With great emotion she sobbed, 'I'll never rubbish the Poms again.'

Epilogue

Emergency proved to be a most enjoyable book to write and during its production many interesting and gracious people were met along the way. Much kindness and help was willingly given. The fact that most participants were quite humble about their experiences, in spite of the heroic endeavours involved, served only to increase their stature. On some occasions help was forthcoming under the most extraordinary pressure, as in the case of Captain Pat Levix, who still found the strength to offer assistance only shortly after his son Jim was discovered murdered by persons unknown in 1987. Such men are gracious indeed.

The author managed to contact, sometimes with great difficulty, at least one flight crew member from each of the major events outlined in the chapters of the book. The search was conducted from Anchorage to Auckland and from Ottawa to Miami and involved some time-consuming 'detective' work. Fortunately, some lucky breaks during the process contributed to success. A few gentlemen, such as Harry Orlady of the San Francisco office of NASA's Aviation Safety Reporting System, were able to furnish information which proved invaluable to the completion of the book and the tracing of individuals. Without such help *Emergency* could not have been written.

All the participants contacted were interviewed either by letter, telephone or in person and each was given the opportunity to read the manuscript. Advice was freely given on improving and correcting the texts and the assistance was gratefully received. The final drafts, without exception, were accepted and approved by all concerned, giving *Emergency* an accuracy and authenticity not always found in aviation publications. It is hoped that the reader has been given a refreshing insight into the drills and procedures of modern flight crew members, not only during difficult and unusual circumstances, but also in the day-to-day operation of an aircraft. It is seldom as simple a task as it seems.

In all the incidents recounted in this book, in spite of the seriousness of some of the situations, not a single life was lost, and this in itself is a great credit to the aircrew involved. The accolades received by the participants in these events are not only a testament to their skills, but also to the abilities of pilots and flight engineers everywhere. The problems outlined in *Emergency* are but a sample of the difficulties, some more critical than others, faced on rare occasions by crew members in the execution of their duties. Few of these incidents are heard outside the profession and, of those who perform their tasks so successfully, most remain unknown and unsung heroes.

The men and women of the airlines who fly their machines throughout the world give of their best at all times and are ready for any contingency. When the completely unexpected arises, as in some of the stories of this book, crews are known to respond with ingenuity and resourcefulness. And, in the everyday operation of an aircraft, the expertise and professionalism exercised daily by crews during flights contribute significantly to making air travel routine, commonplace and, most importantly, safe. The travelling public can feel comfortable with the knowledge that aircrews, like all other personnel within the aviation industry, are doing their utmost to maintain standards. It is hoped that *Emergency* has increased the general awareness of the capability of aircrews and that it inspires confidence. Above all, it is intended to reassure.

Abbreviations and Glossary

AC	alternating current
ADF	automatic direction finder
Aileron	wing turn control
AINS	area inertial navigation system
ALPA	Airline Pilots' Association (US)
Alternate	diversion airport
APU	auxiliary power unit
ASI	airspeed indicator
ATC	air traffic control
ATCC	air traffic control centre
ATIS	automatic terminal information service
Attitude	aircraft orientation relative to the horizontal e.g. nose up attitude
CAT	clear air turbulence
Cb	cumulonimbus cloud
circuit breaker	breaks electrical circuit
CSD	cabin services director
CVR	cockpit voice recorder
DC	direct current
DF	direction finding
DME	distance measuring equipment
Drag	air resistance to motion
Drift	wind effect
EGT	exhaust gas temperature
Elevators	tailplane climb and descent control
EPR	engine pressure ratio
ETA	estimated time of arrival

FAA	Federal Aviation Administration (US)
FDR	flight data recorder
F/E	flight engineer
FIR	flight information region
Flaps	wing trailing edge lift devices
Flare	arrest of descent on landing
Flight level	level expressed in hundreds of feet, e.g. FL 390 = 39,000 ft
FMC	flight management computer
FMS	flight management system
F/O	first officer
'g'	effect of gravity
Gear	undercarriage
Glide path	approach descent profile
GMT	Greenwich Mean Time
Go-around	missed approach procedure
Graduated take-off	reduced power take-off
HF	high frequency (long range radio)
Hold	holding area before approach to land
IAS	indicated airspeed
ILS	instrument landing system (runway centre line and glide path radio guidance)
INS	inertial navigation system
KHz	kiloHertz
Knot	nautical miles per hour
Mach number	aircraft speed relative to local speed of sound
Mayday	radio distress call
mb	millibar
MEL	minimum equipment list
MHz	megaHertz
NDB	non-directional radio beacon
nm	nautical miles
NTSB	National Transport Safety Board (US)

PA	public address system
psi	pounds per square inch
RMI	radio magnetic indicator
rpm	revolutions per minute
Rudder	tail fin yaw control
RVR	runway visual range
SEO	senior engineer officer
SFO	senior first officer
Slats	wing leading edge lift devices
sm	statute mile
S/O	second officer
Spoilers	speed brakes used to spoil lift
Squawk	radar transponder coded transmission
Squawk code	aircraft transponder code
Squawk ident	press aircraft identification button
Stabiliser	variable incidence tailplane
Stack	hold
Stall	loss of lift
TAS	true airspeed
Transponder	airborne equipment which receives a ground radar signal and responds by transmitting back a coded signal
V1	take-off go or no-go decision speed
VR	rotation or lift-off speed
V2	initial safe climb out speed
VASI	visual approach slope indicator
VHF	very high frequency (short range radio)
VOR	VHF omni-directional range radio beacon
Yaw	aircraft nose movement, left or right, in horizontal plane

Bibliography

Chapter One — Pacific Search
Air New Zealand Staff Magazine

Chapter Two — Bermuda Tangle
National Transport Safety Board Report, NTSB-AA4-84/04

Chapter Three — To Take-off or Not to Take-off
National Transport Safety Board Report, NTSB-AAR-72/17

Chapter Four — The Windsor Incident
National Transport Safety Board Report, NTSB-AAR/73/2
Eddy, Paul; Potter, Elaine and Page, Bruce, *Destination Disaster,*
Hart, Davies, McGibbon, 1976
Godson, John, *The Rise and Fall of the DC-10,* New English Library,
1975

Chapter Five — Don't be Fuelish
Pan Am Incident Report
Canadian Air Safety Board Report, T22-64/1985E

Chapter Six — The Blackest Day
Arey, James A., *The Sky Pirates,* Charles, Scribners' Sons, 1972,
Ian Allen, UK, 1973
Redfield, Holland L., *Thirty-five years at the outer marker,* Pitot
Publishing Co., 1981

Chapter Seven — Ice Cool
National Transport Safety Board Report, NTSB-AAR-82/4

Chapter Eight — Roll Out the Barrel
National Transport Safety Board Report, NTSB-AAB-86/3
National Transport Safety Board Report, NTSB-AAR-81/8
AL PA Report, 1981

Chapter Nine — Strange Encounter
Tootell, Betty, *All Four Engines Have Failed,* André Deutsch, 1985
Diamond, Captain Jack, 'Down to a sunless sea', BALPA 'Log'
article, April, 1986

CRASH COURSE

Third Edition

KU-277-041

History and
Examination

Books are to be returned on or before
the last date below.

7 → DAY
LOAN

LIBREX—

First edition author:

James Marsh BA, MBBS, MRCP

Second edition author:

Maxwell A Allan MBBS, BMedSci (Hons)
Anaesthetics Department, Sunderland Royal Hospital,
Sunderland

CRASH COURSE

Third Edition

History and Examination

Series editor

Dan Horton-Szar
BSc (Hons), MBBS (Hons), MRCGP

Northgate Medical Practice
Canterbury
Kent, UK

Faculty advisor

Professor John Spencer
Medical Education and Primary
Health Care
School of Medical Education
Development
University of Newcastle upon
Tyne
Newcastle upon Tyne, UK

John Brain
MBBS/BMedSci (Hons)
Academic Foundation Trainee in Transplantation
and Honorary Research Associate
Institute of Cellular Medicine/Newcastle Hospitals
NHS Foundation Trust
Newcastle Medical School
Newcastle upon Tyne, UK

MOSBY

ELSEVIER

Edinburgh • London • New York • Oxford • Philadelphia • St Louis • Sydney • Toronto 2008

MOSBY
ELSEVIER

Commissioning Editor	Alison Taylor
Development Editor	Lulu Stader
Project Manager	Joannah Duncan
Page Design	Sarah Russell
Icon Illustrations	Geo Parkin
Cover Design	Stewart Larking
Illustration Management	Merlyn Harvey
Illustrator	Cactus

First edition 1999
Second edition 2004
        Reprinted 2005, 2007
Third edition 2008

ISBN: 978-0-7234-3463-4

British Library Cataloguing in Publication Data
A catalogue record for this book is available from the British Library

Library of Congress Cataloging in Publication Data
A catalog record for this book is available from the Library of Congress

Note
Knowledge and best practice in this field are constantly changing. As new research and experience broaden our knowledge, changes in practice, treatment and drug therapy may become necessary or appropriate. Readers are advised to check the most current information provided (i) on procedures featured or (ii) by the manufacturer of each product to be administered, to verify the recommended dose or formula, the method and duration of administration, and contraindications. It is the responsibility of the practitioner, relying on their own experience and knowledge of the patient, to make diagnoses, to determine dosages and the best treatment for each individual patient, and to take all appropriate safety precautions. To the fullest extent of the law, neither the Publisher nor the author assumes any liability for any injury and/or damage to persons or property arising out of or related to any use of the material contained in this book.

The Publisher

The vast majority of medical students pass finals. Due to their intensely competitive nature, most medical students do a good job of convincing themselves that they are going to fail finals. This book is intended to act as a guide to the fundamentals of clinical medicine, the ability to take a history and perform a medical examination. As such, the focus is on learning what is necessary to pass the clinical exams, not on acquiring as much arcane knowledge as is physically possible. There is always something you don't know in medicine, and that is why consultants are specialists in one (or occasionally more) discipline. With a healthy attitude to revision and a pragmatic approach to learning, the book you're holding can help you hone your clinical skills sufficiently to pass the exams, with examples of the type of questions you are likely to face along the way. Good luck!

John Brain
John Spencer, Faculty Advisor

More than a decade has now passed since work began on the first editions of the Crash Course series, and over four years since the publication of the second editions. Medicine never stands still, and the work of keeping this series relevant for today's students is an ongoing process. These third editions build upon the success of the preceding books and incorporate a great deal of new and revised material, keeping the series up-to-date with the latest medical research and developments in pharmacology and current best practice.

As always, we listen to feedback from the thousands of students who use Crash Course and have made further improvements to the layout and structure of the books. Each chapter now starts with a set of learning objectives, and the self-assessment sections have been enhanced and brought up-to-date with modern exam formats. We have also worked to integrate material on communication skills and gems of clinical wisdom from practising doctors. This will not only add to the interest of the text but will reinforce the principles being described.

Despite fully revising the books, we hold fast to the principles on which we first developed the series: Crash Course will always bring you all the information you need to revise in compact, manageable volumes that integrate pathology and therapeutics with best clinical practice. The books still maintain the balance between clarity and conciseness, and provide sufficient depth for those aiming at distinction. The authors are junior doctors who have recent experience of the exams you are now facing, and the accuracy of the material is checked by senior clinicians and faculty members from across the UK.

I wish you all the best for your future careers!

Dr Dan Horton-Szar
Series Editor

Acknowledgements

Thanks to all my family, Mam, Dad, Blotty, Granddad and Martina. To Roger Searle for guiding me through the perils of six years at medical school, John Kirby for always making me think outside the box (or cell) and Graham Dark for stopping me learning ridiculous amounts of arcane knowledge for finals.

John Brain

Figure acknowledgements

Figure 1.1 adapted with kind permission from M A Stewart and D Roter. Communicating with Patients. Sage Publications, 1989

Figures 5.1, 5.2, 5.12, 14.6, 14.7, 14.12, 14.13, 14.15, 14.19, 21.16 redrawn with permission from O Epstein, G D Perkin, D P deBono and J Cookson. Clinical Examination, 2nd edition. Mosby, 1997

Figure 6.8 reproduced with permission from P M W Bath and K R Lees. ABC of arterial and venous disease. British Medical Journal 320, BMJ Publishing Group, 2000

Figures 6.9, 17.5, 17.7, 17.13, 17.14, 17.16, 17.20, 17.24, 22.1 reproduced with kind permission from D Lasserson, C Gabriel and B Sharrack. Crash Course: Nervous System and Special Senses. Mosby, 1998

Figure 7.3 adapted with kind permission from Advanced Life Support Group. Advanced Paediatric Life Support: The Practical Approach, 3rd edition. BMJ Books, 2000

Figure 8.4 courtesy of Dr Brian Lunn, Department of Psychiatry, University of Newcastle upon Tyne, Newcastle upon Tyne, UK

Figures 15.2, 15.3, 15.4 reproduced with permission from Panay et al. Obstetrics and Gynaecology. Mosby, 2004

Figure 21.4 reproduced with permission from S V Biswas and R Iqbal. Crash Course: Musculoskeletal System. Mosby, 1998

Figures 22.2, 22.3, 22.4, 22.5 courtesy of Myron Yanoff

Figure 24.8 reproduced with kind permission from P Kumar and M Clarke. Clinical Medicine, 5th edition. W B Saunders, 2002

Figure 24.9 reproduced with permission from J Weir and P H Abrahams. Imaging Atlas of Human Anatomy, 2nd edition. Mosby, 2003

Contents

Preface   v
Acknowledgements   vii

Part I: History taking   1

1.   **Introduction to history taking** **3**
      Overview   3
      Basic principles   3
      Obtaining and assessing
      information   4
      Relationship with the patient   5
      Difficult consultations   6

2.   **The history** **7**
      Introductory statement   7
      History of presenting complaint   7
      Past medical history   11
      Drug history   11
      History of drug allergy   12
      Social history   13
      Family history   15
      Review of symptoms   15
      Summary   18

3.   **Presenting problems: cardiovascular
      system** **19**
      Cardiac chest pain   19
      Non-cardiac chest pain   22
      Palpitations   22
      Heart failure   24
      Deep vein thrombosis   25
      Pulmonary embolism   25

4.   **Presenting problems: respiratory
      system** **27**
      Asthma   27
      Chronic obstructive pulmonary
      disease   29
      Chest infection   30

5.   **Presenting problems: abdominal** **33**
      Acute abdominal pain   33
      Acute diarrhoeal illness   36
      Jaundice   37
      Anaemia   40

Acute gastrointestinal bleeding   42
Change in bowel habit   43
Dysphagia   45
Acute renal failure   46
Chronic renal failure   48
Haematuria   50
Obstetric and gynaecological
histories   51

6.   **Presenting problems: nervous
      system** **55**
      The unconscious patient   55
      Blackouts   56
      Headache   58
      Epileptic seizure   60
      Stroke   62
      'Off legs'   64

7.   **Presenting problems in paediatric
      patients** **67**
      Taking a history   67
      Examination of the infant   72

8.   **Presenting problems in psychiatric
      patients** **75**
      Introduction   75
      A thumbnail sketch   75
      Referral method   75
      Presenting complaint   76
      Past medical history   77
      The mental state examination   77
      A case summary   80
      Deliberate self harm and suicide   80
      Depression   80
      Acute confusional state   80

9.   **Presenting problems: locomotor
      system** **83**
      Back pain   83
      Rheumatoid arthritis   84
      Osteoarthritis   87
      Perthes' disease   87
      Slipped upper femoral epiphysis   87
      Fractured neck of femur   87

Contents

10. **Presenting problems: endocrine system** 89
    The diabetic patient 89
    Thyroid disease 91

Part II: Examination 93

11. **Introduction to examination** 95
    Overview 95
    Setting 95
    Examination routine 96
    Examining a system 96
    Physical signs 96

12. **Cardiovascular examination** 97
    Examination routine 97
    Aortic stenosis 106
    Aortic regurgitation 107
    Mitral stenosis 110
    Mitral regurgitation 110
    Tricuspid regurgitation 111
    Infective endocarditis 112
    Heart failure 113
    Myocardial infarction 114

13. **Respiratory examination** 117
    Examination routine 117
    Pleural effusion 121
    Pneumothorax 121
    Lung collapse 124
    Consolidation 124
    Lung fibrosis 126
    Integrating physical signs to
    diagnose pathology 127
    Asthma 127
    Lung cancer 129

14. **Abdominal examination** 131
    Examination routine 131
    Hepatomegaly 138
    Splenomegaly 139
    Anaemia 140
    Acute gastrointestinal bleed 143
    Acute abdominal pain 144
    Chronic renal failure 148
    Palpable kidneys 148

15. **Obstetric and gynaecological examination** 151
    Obstetric examination 151
    Gynaecological examination 153
    Summary 156

16. **Surgical examination** 157
    Lumps in the groin 157
    Scrotal swellings 158
    Peripheral vascular disease 159
    Varicose veins 161
    Examination of an ulcer 161

17. **Neurological examination** 163
    Examination routine 163
    Motor system 170
    Sensory system 177
    Higher functions 178
    Patterns of neurological damage . . . 178
    Summary of neurological
    examination 180
    Stroke 181
    Epileptiform seizure 182
    Unconscious patient 183
    Multiple sclerosis 185

18. **Endocrine examination** 187
    Diabetes mellitus 187
    Hyperthyroidism 189
    Hypothyroidism 190

19. **Reticuloendothelial examination** 193
    Examination routine 193
    Skin examination 195

20. **Breast examination** 197
    Examination routine 197
    Breast mass 201
    Breast pain 201
    Gynaecomastia 201
    Investigations 202

21. **Locomotor examination** 203
    Examination routine 203
    Regional examination 206
    General examination 206

22. **Ophthalmic examination** 221
    Examination routine 221

23. **Writing medical notes** 225
    General points 225
    Structuring your thoughts 226
    Presenting your findings 227
    Sample clerking 228
    Ward round entries 228

24. Further investigations **233**
    Cardiovascular system 234
    Respiratory system 237
    Abdominal system 238
    Neurological system 240
    Locomotor system 240

Part III: Self-assessment **243**

    Examination technique **245**
    Multiple-choice questions (MCQs) **247**

Short-answer questions (SAQs) **255**
**Extended matching questions
(EMQs)** **257**
MCQ answers **273**
SAQ answers **283**
EMQ answers **287**

Index   295

HISTORY TAKING

1. Introduction to history taking   3

2. The history   7

3. Presenting problems: cardiovascular system   19

4. Presenting problems: respiratory system   27

5. Presenting problems: abdominal   33

6. Presenting problems: nervous system   55

7. Presenting problems in paediatric patients   67

8. Presenting problems in psychiatric patients   75

9. Presenting problems: locomotor system   83

10. Presenting problems: endocrine system   89

Introduction to history taking

Objectives

By the end of this chapter you should:

- Understand the principles of taking a medical history.
- Be able to recall the basic features of the Calgary–Cambridge framework.
- Be aware of the potential for history taking to be challenging.
- Understand the importance of an accurate history.

OVERVIEW

The first section of this book focuses on the history. After an introduction, the basic structure of a medical history is given, detailing the bare bones. It is important to adopt a systematic approach so that all relevant information is obtained. The chapters which follow illustrate the need to be flexible within a basic format, adapting your questions to different circumstances so that differential diagnoses can be explored. The examples of medical histories are not intended to provide a rigid checklist of questions. There are two kinds of history: those that you use in clinical practice and those that you use in examinations. Beware of any institutional preferences when exam time comes. Histories in clinical practice are context specific: what is appropriate for an 80-year-old woman unable to cope at home is different from a young man who has just been stabbed. The examples provided are only a framework, and should be adapted to suit your own preferences, the individual patient and, of course, the exam format, for example the Objective Structured Clinical Examination (OSCE) or long case.

BASIC PRINCIPLES

The principal aims of medical clerking are to establish:

- What is wrong with the patient today?
- How do these problems impact on the person's life?

The standard approach is to obtain the history before conducting a physical examination and requesting appropriate investigations. Despite being presented here as separate chapters, these components of the clerking interlock with each other in a dynamic fashion from the moment the doctor meets the patient.

The medical history has a traditional format, but it should not be considered as a rigid interrogation and checklist of questions. Effective history taking depends crucially on good communication skills. Although a structured approach is needed, it is important to adopt a flexible attitude, adapting your questions and differential diagnoses as information is received. The process of formulating a diagnosis starts as soon as the patient describes the presenting complaint(s). Each symptom should be explored in detail so that possible diagnoses can be identified and a list of probabilities weighed up. Once the patient has described the main symptoms, specific questions can be asked to refine the differential diagnosis, including or excluding specific diagnoses based on the patient's individual story. Thus the process of history taking is an active skill, and not one of passive listening. If you have been unable to ascertain a differential diagnosis by the end of the history, you will struggle to find signs on your examination to confirm your suspicions and will be unable to request appropriate investigations.

OBTAINING AND ASSESSING INFORMATION

Gleaning the important information is a fine art and takes years to master. Every medical student experiences the frustration of taking a garbled, incoherent history from a rambling patient, only to see a consultant or GP ask one or two seemingly simple questions, making the underlying diagnosis embarrassingly obvious.

The traditional approach is to start by asking 'open' questions:

- 'Why don't you tell me about your pain?'
- 'Why have you come to hospital today?'

This gives patients the opportunity to tell you how they perceive their problems before the agenda is taken out of their hands and your own prejudices take over. It is well recognized that doctors and patients often focus on different problems and may have conflicting agendas. By careful steering and gentle coaxing, even the most garrulous patients can usually give full, clear and reasonably concise descriptions of their current symptoms.

It is also necessary to ask 'closed' questions for further clarification, for example:

- 'Does your pain get worse after eating?'
- 'Did you black out completely or just feel lightheaded?'

It is a matter of judgement when to start interrupting and asking closed questions, but, as a general rule, think twice before interrupting a patient in full flow. Most patients won't go on for too long and this initial monologue may contain vital clues to their problem. If specific questions are introduced too early, vital information may never come to light.

It is also important to obtain information about the impact of the patient's problem, not only physically but also in a wider psychological and social context. The same pain or disability will restrict the activities of different people greatly (e.g. work, social interactions, hobbies). This information is vital for a full assessment and reassures patients that the doctor is taking a genuine interest in them, not just their chest pain or arthritis, etc. Information to explore includes:

- *Ideas.* What does the patient think is wrong?
- *Concerns.* What does the patient worry might be wrong?
- *Expectations.* What does the patient think is going to happen in the consultation and regarding his or her future health? What might happen following the consultation (e.g. investigations, operations)?

To say that effective history taking is underpinned by good communication may seem to be stating the obvious. Yet there is considerable evidence to show that doctors' communication skills – how they elicit a patient's story – are wanting, particularly in respect of finding out what the patient thinks is wrong and what concerns them.

A useful way of looking at the medical interview is the Calgary–Cambridge framework, which is increasingly used both in undergraduate and postgraduate education, including in the design of OSCE stations. It divides the interview into distinct phases, identifying the key tasks and 'microskills' associated with each one. In the context of history taking, as explored in this book, the 'explanation and planning' stage will not be addressed.

(Medical school curricula now place great emphasis on communication skills. For those looking to expand their communication horizons, Silverman et al (2005) is essential reading – see Further reading.)

Calgary–Cambridge framework

1. Initiating the session
   - Establishing initial rapport
   - Identifying the presenting problem(s)
2. Gathering information
   - Exploration of problem(s)
   - Understanding the patient's perspective
   - Providing structure to the consultation
3. Building the relationship
   - Developing rapport
   - Involving the patient
4. Explanation and planning
   - Providing the correct type and amount of information
   - Aiding accurate recall and understanding
   - Achieving a shared understanding through incorporating the patient's perspective
   - Planning through shared decision making
5. Closing the session

Key skills areas for the gathering of information include:

- Attentive listening.
- Eye contact, posture, nodding, facilitatory statements (e.g. 'Uh-huh' or 'Go on . . .').
- Appropriate use of questions. These can be classified as:

  closed questions – obviously necessary to ascertain facts, but generally invite a specific (often one-word) answer, and limit response to a narrow field

  open questions – less focused but still direct patients to a specific area whilst allowing them more discretion

  clarifying or probing questions – their purpose is self-evident (e.g. 'What do you mean by that?' or 'Can you tell me about the last time you had the pain?' or 'What makes you say that?')

  The key issue is to use an appropriate balance of question types – all have a place.

- Summarizing: summarizing periodically, and at the end of the interview, allows you to check whether you've heard the patient's story, review the details and deduce what else needs to be explored. This helps patients carry on with their accounts, by demonstrating that you have been listening. A summary closes the interview in an appropriate way.

Key skills for understanding the patient's perspective include:

- Asking about ideas, concerns and expectations.
- Responding to cues, both verbal and non-verbal: interpersonal exchanges are loaded with spoken and unspoken messages. These may give a clue as to what is really going on, and responding to them directly ('You mentioned your mother ended up in a wheelchair with her arthritis; are you worried that's what is going to happen to you?') or using reflection ('I sense that you are really very upset about this') will usually provide important extra information. Sometimes simply repeating back a 'cue' to the patient will suffice ('. . . something could be done . . .?' or '. . . very worried . . .?').

A patient-centred approach

The traditional medical history does not primarily aim to understand the meaning of the illness for the patient, focusing, as it does, on diagnosing disease in terms of an underlying pathology. The way that doctors have been taught to take a history in the past implied that if they simply asked the right number of (usually closed) questions about the functioning of an organ or system, they would gather all the information needed to make a diagnosis. Only lip service was paid to the impact the problem had on the patient's life beyond the symptoms themselves. This approach ignores the ways in which a disease can affect an individual – one person's 'illness experience' is very different from that of the next person, emotionally, psychologically and functionally. Remember that accurately eliciting and addressing the patient's concerns and expectations is not only important for diagnosis, but will also have a major influence on concordance with treatment, patient satisfaction, detection of hidden problems and some physical outcomes (e.g. glycaemic control in diabetics, pain relief after surgery).

The patient-centred approach aims to incorporate the patient's perspective and involves a parallel search of two frameworks – the 'illness framework', and the 'disease framework' (Fig. 1.1).

RELATIONSHIP WITH THE PATIENT

The atmosphere and setting is important when taking a history. Patients should feel free to express their fears and concerns without fear or embarrassment. An air of absolute confidentiality should be created. At the same time, take note of any non-verbal signs (e.g. hostility, embarrassment). Often the most crucial information needs to be coaxed out of the patient. Never appear to be in a rush. Patients expect and deserve full attention and sympathy for their problems. It is not unusual for patients to express their real concerns just as they are leaving the consulting room or you are leaving the bedside.

However tempting, do not ignore throwaway comments from the patient. They often carry the key to the whole problem. Never appear to be in a rush. Time invested in taking a good history pays dividends in the long term.

It is often useful (even if potentially time-consuming), to ask patients at the end of a consultation if there is anything else they wish to discuss. If patients feel at ease with you, they will talk more freely.

They should feel confident not only in your diagnostic abilities, but also in your empathy, understanding and motivation. After all, you are acting as their advocate. This process should start as soon as you greet patients. Remember that first impressions really do matter. Appear friendly, but professional. Patients must have confidence in your ability to act on their behalf.

> It is useful early in your career to start asking patients about their *Ideas* (what they think is happening; 'feels like indigestion or acid'), **Concerns** (what might be happening; 'it could be a heart attack') and **Expectations** (what they foresee happening in the future; 'I could end up in a wheelchair') – *ICE*. These questions often provide valuable information about patients and can influence how you interpret and relay information about their conditions.

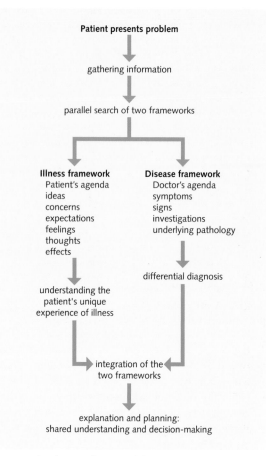

Fig. 1.1 The disease–illness model.

DIFFICULT CONSULTATIONS

In certain circumstances, the history can be considerably more challenging to obtain; for example if:

- There is a language barrier.
- The patient is confused, hostile or unconscious.

Remember that the history is the most important part of the examination. If there is a language barrier, help the patient to relax, do not rush, explain everything clearly and, if necessary, obtain an interpreter. If the patient is confused or unconscious, history taking is still vital. A relative, carer, nursing home worker or other witness is usually available to provide information. This process is time-consuming, but worthwhile.

There is no reason for you or any other member of staff to be harmed by a patient. Remember, there is no medical condition that cannot be managed in a police cell! There is, however, rarely any justification for resorting to a 'veterinary' approach. There

will be times when you will hear stories that are difficult to cope with emotionally. Be professional with the patient but do not be afraid to discuss this with others. We all need to share difficult experiences with one another as a part of becoming a healthy, reflective practitioner.

> Remember, the history is the most important part of the patient's assessment and provides the information required to make a diagnosis in up to 80% of cases.

Further reading

Silverman J, Kurtz S, Draper J 2005 *Skills for communicating with patients*, 2nd edn. Oxford: Radcliffe Medical Press

The history

Objectives

By the end of this chapter you should:

- Be able to take a basic medical history.
- Understand and apply the 'pain history' questions.
- Be able to summarize a medical history.
- Begin to develop your own skills of interpreting information from the patient and the medical notes.

INTRODUCTORY STATEMENT

Before taking a detailed history, it is essential to obtain some background information from the patient. This should include the patient's:

- Name.
- Age.
- Sex.
- What the person is like at their best.
- Occupation.
- Presenting complaint.

Ideally, try to use the words that the patient has used (e.g. people never complain of dyspnoea, but will say that they feel 'short of breath'). This statement should be short and pithy, for example 'John Smith is a 56-year-old electrician complaining of chest pain'.

This information is vital as it helps you (and any listeners) form a thumbnail sketch of the person in front of you, so that appropriate questions can be anticipated if the patient does not volunteer the information. The process of forming a differential diagnosis should have already begun, but at this stage it will, by necessity, be broad.

HISTORY OF PRESENTING COMPLAINT

This is the main component of the history. A detailed, thorough investigation into the current illness is performed. This is usually composed of two sequential (but often overlapping) stages:

1. The patient's account of the symptoms.
2. Specific, detailed questions by the doctor.

The relative proportion of each component depends upon the underlying problem, the communication skills of the patient and the listening skills of the doctor. Listening should be an active process. Ideally, the patient should be given every opportunity to talk freely at the start of the consultation with minimal interruption. A common mistake made by most students (and doctors) is to interrupt the patient and to intervene with closed questions too early.

Patients must feel that they are getting a fair hearing and have had an adequate opportunity to express themselves. They should be made to feel that the doctor is listening carefully and has a genuine concern for their problems. Subtle nuances are often missed if the doctor seizes the agenda too early.

A combination of art, experience and patience determines when and how to interrupt a patient in full flow, but it is prudent to err on the side of caution and allow the patient to drift a little (especially when you are first taking histories), while making a mental note of the most important features of the narrative and issues that require clarification.

The full circumstances surrounding a single event or symptom need to be explored in a systematic manner so that a complete picture can be obtained.

A patient complains of pain

Explore in detail the circumstances surrounding the episode so that a complete picture can be obtained.

What was the patient doing immediately before the episode?

Ascertain exactly what the patient was doing at the time of the onset of the pain (e.g. running, arguing with wife, sitting in chair).

Speed of onset

Determine the rate of development of the pain (e.g. seconds, minutes, hours, days). It may be helpful to draw a graph of symptoms versus time (Figs 2.1 and 2.2).

Time of onset

Try to obtain exact times and dates if appropriate.

Subsequent time course

Map out the fluctuations in symptoms with time.

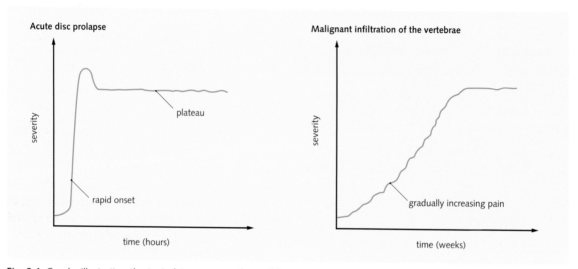

Fig. 2.1 Graphs illustrating the typical time course of two different types of back pain, which are of the same severity at presentation but of different aetiology.

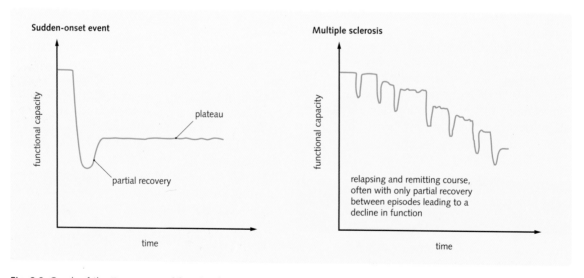

Fig. 2.2 Graph of the time course of functional capacity against time for a patient who has had a stroke or has multiple sclerosis. It is important to consider not only the speed of onset of a particular symptom, but its subsequent time course.

Duration

How long did the pain last? Patients will often over-estimate the duration of symptoms.

Nature of pain

It is particularly useful to record direct quotations from the patient, such as:

- 'Like a stabbing knife.'
- 'Like a tight band around my chest.'

They are often very descriptive, but some patients are less colourful in their language and may need help to describe a symptom. You could then provide examples, such as 'Would you describe the pain as burning, stabbing, crushing, throbbing or something else?' It can sometimes be useful to quantify the severity of the pain. The visual analogue pain score is a well established method (see Fig. 2.3).

Radiation

Radiation of pain is important and often gives clues to the aetiology. For example, the pain of a prolapsed intervertebral disc usually radiates down the back of one leg, whereas muscular strain is usually well localized.

Associated symptoms

Pain rarely occurs in isolation. Associated symptoms are often characteristic of certain pathological processes. For example, if a patient describes pleuritic chest pain, ask about the presence of cough, dyspnoea, fever, haemoptysis, etc.

Aggravating features

Different types of pain have different aggravating factors. For example, mechanical back pain is often exacerbated by exercise, but inflammatory pain may be worse after a period of prolonged rest.

Relieving factors

Factors producing relief of symptoms may give clues to the aetiology or severity of the pain (e.g. the pain of intermittent claudication resolves rapidly on rest compared with the pain of critical limb ischaemia).

Recovery

Note the time and speed of recovery.

Residual symptoms

Once the pain has resolved, there are often ongoing symptoms; for example, after the pain of myocardial infarction, the patient often reports ongoing fatigue, dyspnoea or palpitations.

Effect of any interventions

Ask about the effect of interventions, for example:

- 'Did your chest pain get any better after taking a glyceryl trinitrate tablet?'
- 'Did the pain resolve when you stopped running?'

In the event of repeated symptoms, additional information is also needed, for example frequency and pattern (e.g. getting progressively better or worse, more or less easily provoked).

Previous episodes

Ask the patient if he or she has had a similar episode of pain before.

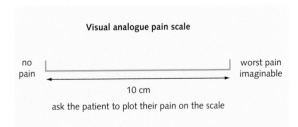

Fig. 2.3 Visual analogue pain score.

In remembering the features of pain, it is useful to consider the mnemonic
SOCRATES:
Site
Origin
Character
Radiation
Alleviating factors
Timing (duration and frequency)
Exacerbating factors
Severity
(www.valuemd.com)

Interpreting information

Do not accept the patient's (or the doctor's) interpretation of symptoms at face value. At this stage, you are most interested in what the patient was actually experiencing. If the patient has been given a label for the symptoms, it may be helpful to know by whom and on what evidence. For example, if the patient says 'My doctor tells me I have angina!', find out whether the patient has had investigations such as an exercise test, angiography or electrocardiography and, if so, what the results were.

Often, different patients mean different things by the same phrase. Be wary! For example, 'dizziness' may mean light-headed, muzziness or vertigo. This clearly has different implications for the underlying diagnosis. In addition, try to make your questions as clear and unambiguous as possible.

When recording the history, try to keep it as brief and 'punchy' as possible. An example of a history obtained for chest pain could be:

'He describes a sudden-onset retrosternal chest tightness "like a lorry parked on my chest". This radiates to the jaw and left arm and comes on every time he climbs the stairs at the underground station (approximately 100 steps). He has a sensation that he needs to stop and rest immediately, and also feels mildly nauseated, short of breath and occasionally sweaty. He never has palpitations, lightheadedness, pain at rest or syncopal episodes. The pain is relieved within 2 minutes of resting or seconds after placing a GTN tablet under his tongue, and he can then continue his daily activities. He has noticed that on a cold or windy day, the pain comes on earlier, but otherwise has been stable over the past two months.'

This illustrates the need to play an active role in the discussion, always being aware of:

- The differential diagnosis.
- How the symptoms interfere with functional activities.

Wherever possible, try to quantify the symptoms objectively. In addition, the presence of relevant 'negative' symptoms creates a more complete picture.

A long list of 'negatives' is very dull, and judgement is needed when deciding which ones to include in the narrative. This is where experience helps when presenting the history. Consider each diagnosis separately, and think of what information you would want to know to exclude each one.

Rounding off the history

Asking patients what they think the cause of their symptoms might be, and what they are worried it could be, always provides an important insight (ICE! – see Ch. 1). They are often worried about the consequences of what appears to be a trivial symptom to an objective observer. It is impossible to offer appropriate reassurance or counselling without this knowledge. Furthermore, if the patient's own judgement differs significantly from the doctor's, the doctor must reassess how he or she arrived at the differential diagnosis.

Often a history is very complicated or the patient appears to have given a contradictory account. It is helpful to read back your recollection of the history to the patient; that is, to summarize, so that he or she can verify its accuracy – you will have an opportunity to place your own interpretation at a later stage! You should both be in agreement about the facts before you introduce your own bias. It provides a message to the patient that you have been listening to his or her concerns. This is a useful way of focusing patients, especially if they are rambling. Finally, it is a good way of ending the history.

Summary

When taking the history of the presenting complaint:

- The primary objective is to obtain a comprehensive, but succinct, account of the presenting symptom(s).
- Remember, communication skills matter!
- Allow patients sufficient opportunity to describe their symptoms in the beginning.
- Resist the temptation to interrupt too early and too frequently.
- If patients drift, gently coax and steer them back on course to the main focus.

- Ask specific questions to clarify, obtain more detail or to investigate potential diagnoses only after the patient has described his or her symptoms.
- Sometimes negative answers are more important than positive answers.
- Ask patients what they think may be causing the symptoms.
- If the history is complicated, recount your interpretation of events back to the patient to ensure that your versions are concordant.
- Attempt to be as systematic and objective as possible. Look for collateral evidence to support any statement, especially if it is related to someone's interpretation of symptoms.
- Ask easily understood, unambiguous questions.
- If the patient presents with more than one complaint, this process may need repeating for each complaint.

PAST MEDICAL HISTORY

This is important for placing the current illness in the context of past events, as they are often related. A review of the past medical history should include the points addressed below (some students find a mnemonic useful in remembering the salient features; try an Internet search engine for medical mnemonics).

Previous hospital admissions

Outline essential information such as:

- Diagnoses/problems.
- Dates and places.
- Treatment and investigations.

Occasionally, verification of previous events may be necessary, particularly if the patient has had a complicated hospital admission. This may prevent repetition of unnecessary (and sometimes harmful) investigations. The patient's recollection may be hazy, especially of events occurring within hospital (e.g. investigations for acute renal failure). The patient's clinic letters are often an invaluable source of information and expert opinion.

Operations

List procedures chronologically in a similar way.

Known medical or psychiatric conditions

Remember to have a healthy scepticism about diagnostic labels used by the patient. If necessary, explore how a diagnosis has been established.

Problems related to underlying present illness

Ask about particular risk factors or associated diseases related to the primary complaint. For example, if a patient presents with chest pain, ask specifically about:

- Previous episodes of angina and myocardial infarction.
- Strokes and transient ischaemic attacks.
- Diabetes mellitus.
- Hypertension, etc.
- Hypercholesterolaemia.

This is the component of the past medical history that takes the most skill. An awareness of the likely differential diagnosis is needed, illustrating that the history should be adapted to different circumstances. Listening should be active, and relevant negative information often provides as much information as positive facts.

DRUG HISTORY

A complete list of current medication with doses is essential. Taking a drug history is difficult and it may be necessary to phone the GP and find out what medications the patient should be taking. Remember that what we consider a drug and what the patient may consider a drug are two separate things. Commonly, patients don't consider their inhalers as drugs or many feel that the term drug refers only to illegal substances. One way to get round this is to ask the patient, 'what pills, potions and puffers do you use?'

Why is drug history important?

A detailed drug history is important for the following reasons:

- It may give an indication of disease processes that the patient was either unaware of or has failed to mention (e.g. thyroxine indicating hypothyroidism).

- The drugs may be the cause of the present symptoms due to adverse effects or drug interactions (e.g. headache induced by nitrates used to treat angina). Try to establish a temporal relationship between the initiation or change of medication and the onset of new symptoms.
- Conversely, withdrawal of therapy may be responsible for the current symptoms (e.g. withdrawal of diuretics may lead to swollen ankles and orthopnoea).
- It provides an opportunity to explore the patient's understanding of the disease, often highlighting a need for further education (e.g. the controlled use and technique of inhaling steroids and bronchodilators by the asthmatic patient).

Finally, remember that adverse reactions and medication error are very common, occurring in at least 10% of all admissions to hospital, yet a significant proportion of these mistakes are preventable and avoidable.

The potential therapeutic options for the presenting illness can be explored.

Check concordance and explore the reasons for non-concordance, such as:

- Intolerable side effects.
- Perceived lack of efficacy.
- Ignorance.
- Poor communication from the prescribing physician.
- Four-times daily medication instead of once daily.
- Patients who think they don't need medication or are not 'the pill-taking' kind.

This exploration of concordance must be done very sensitively in an effort not to appear to be judgemental. Patients do not like admitting that they have not taken medication prescribed on their behalf. In a recent systematic review, one of the main factors contributing to non-concordance was whether or not the patient's beliefs about the medicine had been elucidated and discussed. Use statements such as 'Do you ever have difficulty taking

your tablets?' or 'Do you ever forget to take your tablets?' Even when approached sensitively, few patients admit poor concordance. This is important because there is no point in inexorably increasing the dose of medication if it is not being taken. Frequent culprits include those taking antihypertensive agents.

Ask specifically about the use of over-the-counter (OTC) medications, herbal remedies and (in women) oral contraceptives. These are often not considered to be 'medications'.

In the unconscious patient, check for the use of steroids, anticoagulants, antiepileptics, insulin, etc. The patient may carry a card detailing this information, or wear a 'medicalert' bracelet.

Ask to see the patient's medicines. This provides extra insight into the patient's understanding as well as concordance.

Drug histories

It is well known that doctors often take poor drug histories, relying on the patient's memory, handwritten lists and, occasionally, the GP summary. In practice, hospital pharmacists are invaluable in ensuring that patients are taking the correct medications and are often very keen to talk to medical students about a patient's medications.

HISTORY OF DRUG ALLERGY

This follows naturally from the drug history. As with other sections of the history, a systematic approach is rewarded.

Many people think that they have an allergy to medication, but a healthy scepticism is needed when assessing this. Always enquire how the 'allergy' was manifested; for example, 'stomach upset' after the use of antibiotics is common, but very rarely an allergic phenomenon. Skin rashes are much more likely to represent a true allergy.

It is important to clarify the circumstances of a suspected allergic reaction. Attempt to establish a 'temporal relationship' between the drug administration and the allergic manifestation. For example, it is not uncommon for a rash to be part of a viral illness or to be an epiphenomenon – patients with glandular fever who are prescribed amoxicillin will develop a macular erythematous rash. This is not an allergic phenomenon.

Enquire whether the patient has ever been 'rechallenged' with the same drug and, if so, whether any reaction occurred. However, in most cases of uncertainty, it is usually safest to assume that the patient does have an allergy, and to avoid the suspected drug.

Finally, enquire about the presence of atopic conditions (e.g. eczema, hay fever, asthma, urticaria, etc.).

SOCIAL HISTORY

The social history is crucial for every patient as it provides information on how the disease and patient interact at a functional level. It is particularly important that it is detailed in elderly, frail or socially isolated patients. Try to obtain a picture of daily activities, and consider the impact of the disease at each stage.

An elderly person presents with a hemiplegia

What is the structure of the home?

Consider the physical characteristics of the home, as these will affect mobility and the ability to function independently:

- Is it a flat or a house?
- If it is a flat, what floor is it on and is there a lift?
- How many bedrooms are there?
- What access is there to the house, bedroom, kitchen, toilet, etc. (e.g. stairs, ramps)?

Who are the potential carers?

Enquire about the structure of the household and whether other members of the household are out of the house during the day and physically fit. Find out whether any family members live nearby. Are there any other potential local carers?

Can the patient perform the usual activities of daily living?

Assess the practicalities of routine activities such as:

- Getting in and out of bed.
- Dressing.
- Toileting (mobility and continence).
- Cooking and eating.
- Bathing.
- Shopping.

What level of social support is already provided for the patient?

The patient may already be known to the social services. Enquire specifically about community nursing, 'meals on wheels', home help, social worker, benefits, etc.

In a hospital setting, discharge planning should begin from the first assessment so that potential problems can be anticipated before they arise. Some departments have their own early discharge teams.

Other patients may require the help of occupational therapists (e.g. patients with rheumatoid arthritis may need assessment for aids for eating, ramps for access to the home, stair rails, stair lifts, bath seats, etc.).

The young patient

Younger patients present different problems, and an illness may interfere with their lifestyle or be a direct result of it. It is important to find out the following information.

Occupation

Consider whether the occupation may have predisposed to the current illness, for example:

- Backache in manual workers.
- Repetitive strain injury in keyboard operators.
- Non-specific chest pain in people presenting with stress and people who are stressed by their jobs.

Consider whether the present illness may interfere with the patient's ability to continue in his or her occupation, for example:

- Epilepsy or myocardial infarction in a heavy goods vehicle driver.
- Ischaemic heart disease or rheumatoid arthritis in a manual worker.

Do not forget that lack of employment may contribute to the presenting illness. There is a well-documented association with the onset of morbidity (both physical and psychological) and unemployment.

Hobbies

These may be the cause of the illness (e.g. pigeon-fancier's lung) or be precluded by the current illness (e.g. squash, if the patient has newly diagnosed angina).

Travel

World travel is common and it is therefore important to be aware of recent travel. For example, you should consider:

- Malaria in a patient who has returned from an endemic area and presents with fever.
- Hepatitis A in a person with jaundice who has been to an endemic area.
- Schistosomiasis in a patient presenting with haematuria.

Other sources of stress

Common causes include financial difficulties, job insecurity or strained relationships at home. The list, however, is endless and recognition depends upon great sensitivity by the doctor.

Recreational drugs

It is always appropriate to enquire specifically about the use of recreational drugs. The following should raise your suspicions:

- If immunodeficiency is suspected.
- If needle marks are spotted.
- If the patient presents with unusual behaviour or a decreased level of consciousness.

Risk factors for human immunodeficiency virus (HIV) and hepatitis B and C

These diseases are becoming more common and an increasing index of suspicion is needed,

especially for those patients with risk factors, for example:

- Intravenous drug users.
- Individuals partaking in unprotected sexual intercourse.
- Sex workers.
- Haemophiliacs.
- Recipients of blood transfusion in Africa since 1977.
- Partners of the above groups.

Smoking

It is obligatory to ask every patient about cigarette smoking. A useful concept is 'pack years' (i.e. the number of packs of cigarettes smoked daily multiplied by the years of smoking). If the person tells you they are an ex-smoker, always ask them when they stopped. Do not forget other forms of smoking. Don't fall for the '... but I don't inhale, Doctor' explanation – nictotine and other toxins are still absorbed through the buccal mucosa.

Alcohol

Alcohol plays a contributory role in many illnesses. An attempt should be made to estimate consumption for every patient. This is usually quoted in units/week. Be cautious of the patient who offers their consumption in units – they may mean something quite different to you! One unit equals:

- 1/2 pint of standard strength beer/lager.
- 1/2 bottle premium strength beer/lager.
- 1/2 × 175 mL glass of wine.
- 1 single measure of spirit.

In certain circumstances, a more detailed history is appropriate. It is sometimes helpful to work through a typical drinking day for the 'heavier' drinker. Remember that patients often think that a doctor is likely to disapprove of heavy drinking, so denial is common. At this stage in the consultation it is not your job to be judgemental!

Nutrition

An attempt should be made to assess nutritional status. A detailed history is needed for certain patients (such as those presenting with weight loss or iron-, folate-, or vitamin B_{12}-deficient anaemia).

FAMILY HISTORY

Information about the health and age of other family members is often instructive, particularly for young patients or those with a suspected inherited disease. It is helpful to draw a family tree (Fig. 2.4).

Some diseases have a predictable mode of inheritance (Fig. 2.5):

- Autosomal dominant (e.g. Huntington's disease, adult polycystic kidneys).
- Autosomal recessive (e.g. cystic fibrosis).
- X-linked recessive (e.g. colour blindness).

Other disorders do not have a Mendelian mode of inheritance, but there is often a discernible genetic component, and family history is still significant. Such disorders include:

- Hypertension.
- Ischaemic heart disease.
- Schizophrenia.
- Breast cancer.

Even if the suspected disease has no recognized inheritable factor, it is useful to record a family tree with causes of death or major diseases. Doing this with the patient may open the door to some of the patient's concerns and may provide a useful insight into relationships within the family.

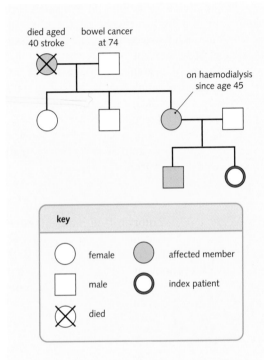

Fig. 2.4 A typical family tree written in hospital notes demonstrating family members affected by autosomal dominant polycystic kidneys.

REVIEW OF SYMPTOMS

A brief review of symptoms in a systems enquiry is essential in a detailed history. It may suggest other disorders that have not been considered, or highlight potentially serious complications that a patient may have considered to be trivial. In addition, it may be used as a screening process to highlight aspects of the history that require more detailed attention.

With practice, it is possible to cover this aspect of the history in a few minutes, and this is time well invested.

It is helpful to consider this section in systems. When the presenting complaint appears to relate to

Examples of some inherited disorders	
Pattern of inheritance	**Examples**
autosomal dominant	adult-onset polycystic kidney disease; neurofibromatosis; familial adenomatous polyposis
autosomal recessive	cystic fibrosis; sickle-cell anaemia; infantile polycystic kidney disease
X-linked recessive	colour blindness; haemophilia A
X-linked dominant	vitamin D-resistant rickets

Fig. 2.5 Examples of some inherited disorders.

one system, this system should be promoted to the detailed history of the presenting complaint, and a more detailed enquiry performed. The absence of particular symptoms attains a greater significance; for example when a patient describes chest pain, the presence or absence of dyspnoea or palpitations is clearly more relevant than when the presenting complaint is headache.

Since you have just spent some time exploring the history of the presenting complaint – usually the problem causing most discomfort or distress for the patient – it may seem odd to them when you embark on a detailed enquiry about other parts of the body. 'Signpost' your intention; a sentence like 'OK, Mrs Thompson, I'm now going to ask you some questions about other symptoms' will suffice.

The systems enquiry is often the most difficult component of the history to interpret. The significance of each symptom and its relevance to the primary complaint need to be analysed. This may lead to a review of the differential diagnosis and new questions to either confirm or refute new suspicions.

With a little practice, it is possible to draw up your own list of questions that can be asked and answered rapidly. It is helpful to consider each system of the body in turn. Some important symptoms and associated screening questions are listed in Figs 2.6–2.10.

Some screening questions for the cardiovascular system in the review of symptoms	
Symptom	Questions
chest pain	explore in detail; in particular note the characteristics of the pain (e.g. cardiac, musculoskeletal, pleuritic, pericarditic, oesophageal, etc.)
short of breath	quantify exercise tolerance; ascertain whether predominant pathology is cardiovascular or respiratory
short of breath when lying flat	quantify number of pillows needed ('what would happen if you were forced to sleep with no pillows?'); enquire specifically about paroxysmal nocturnal dyspnoea ('do you ever wake up in the night gasping for breath and need to sit on the side of your bed?')
ankle swelling	duration; degree; presence of facial or genital oedema, ascites
fatigue	'when did you last feel completely well?'; 'if I asked you to walk to the train station, what would make you stop?' (e.g. dyspnoea, chest pain, claudication, fatigue, etc.)

Fig. 2.6 Some screening questions for the cardiovascular system in the review of symptoms.

Some screening questions for the respiratory system in the review of symptoms	
Symptom	Questions
short of breath	see Fig. 2.6
cough	duration; sputum production; constitutional symptoms (e.g. coryza, fever, malaise, weight loss); time of day (e.g. left heart failure or asthma may present with night-time cough)
sputum production	amount per day (e.g. teaspoonful, eggcupful, etc.); characteristics (colour, tenacity, etc.), foul taste (e.g. anaerobic infection); associated haemoptysis, chest pain
wheeze	provoking factors (e.g. exercise, cold weather, house dust, etc.)

Fig. 2.7 Some screening questions for the respiratory system in the review of symptoms.

Some screening questions for the gastrointestinal system in the review of symptoms

Symptom	Questions
weight loss	'is your weight loss intentional?'; quantify; appetite; constitutional symptoms (e.g. fever, malaise, fatigue, etc.); diarrhoea, vomiting
nausea, vomiting	duration, frequency; time of day (e.g. early morning symptoms may indicate raised intracranial pressure); obvious precipitating factors (e.g. drugs, alcohol, pregnancy, food poisoning); presence of haemoptysis
dysphagia	'where does the food appear to get stuck?'
abdominal pain	needs to be explored in detail; establish site, acute versus chronic; characteristics; relationship to food; indigestion; relieving factors, etc.
stool frequency	remember that different people have very different perceptions about the terms 'constipation' and 'diarrhoea'; establish the patient's usual bowel habit and what changes have occurred for both frequency and stool consistency; rectal bleeding; duration of symptoms; it is the change in frequency that is crucial

Fig. 2.8 Some screening questions for the gastrointestinal system in the review of symptoms.

Some screening questions for the genitourinary system in the review of symptoms

Symptom	Questions
urinary frequency	if abnormal, quantify the number of times that urine is passed during the day and night (e.g. day/night = 6–8/2)
poor stream dysuria	enquire about other features of prostatism (nocturia, hesitancy, terminal dribbling, etc.)
haematuria	timing during the urinary stream; constitutional symptoms; urinary frequency; appearance of urine (e.g. cloudy, blood, offensive smell); degree of blood clots, 'like claret', bloodstained, etc.)
menstruation	cycle length, duration of menstruation, pain, menorrhagia, etc.
sexual activity	many patients are not willing to discuss their sexual history and often it is not relevant; relevant questions may cover number of sexual partners, 'do you practise safe sex?', homosexual encounters, libido, impotence, etc.

Fig. 2.9 Some screening questions for the genitourinary system in the review of symptoms.

Some screening questions for the nervous system

Symptom	Questions
headache	explore features in depth including precipitating factors, frequency, nature and location of pain, associated symptoms, timing during day, etc.
blackouts	if the patient describes a blackout, it is essential to devote time to exploring the event as this warrants investigation in its own right, regardless of other symptoms
fits	is the patient known to have epilepsy?; frequency and control, type of fits, duration of epilepsy, etc.
muscle weakness	duration, pattern of weakness, precipitating events
paraesthesia	distribution (e.g. dermatome, peripheral nerve, etc.)
change in vision	speed of onset, clarify visual acuity (e.g. 'can you read newspapers, watch television?' etc.); diplopia
dizziness	clarify exactly what the patient means by dizziness (e.g. vertigo, lightheadedness, muzzy feeling, etc.)

Fig. 2.10 Some screening questions for the nervous system in the review of symptoms.

Other general symptoms to consider include:

- Early morning stiffness
- Mobility
- Fevers
- Sweats
- Weight loss
- Tiredness.

SUMMARY

At the end of presenting the history, it is useful to provide a short summary of two or three sentences encompassing the most salient features. This will help you and the listener focus on the most relevant parts of the ensuing examination.

Presenting problems: cardiovascular system

3

Objectives

After reading this chapter you should be able to:

- Confidently take a history from a patient with chest pain, palpitations, heart failure or venous thromboembolism.
- Recall the key features of each presenting complaint.
- Identify cardiovascular risk factors.

CARDIAC CHEST PAIN

Detailed history

Obtain a detailed account of the pain in a systematic manner, as described in Chapter 2, asking specifically about the features discussed below.

Site

Ask patients to indicate on themselves where exactly they experience the pain. Cardiac chest pain is typically retrosternal, but may only be present in the neck, throat or arms (especially the left arm) (Fig. 3.1).

Radiation

The pain may not radiate, but classically goes to the throat and left arm (see Fig. 3.1).

Nature of pain

It is helpful to write down the exact words used by the patient. Cardiac chest pain is usually described as 'tight', 'crushing', 'gripping', 'like a band across my chest', 'a dull ache'. Patients often have difficulty finding words to describe abstract sensations such as pain, but it is important to try and ascertain its nature. You could give alternatives, for example 'Would you describe the pain as burning, stabbing, tightness, a tearing sensation?' Remember to try and avoid leading the patient too much!

Often, patients have had angina for a long time and may describe a pain as 'like my angina, only worse', when experiencing a myocardial infarction.

Your thumbnail sketch of the patient should include any risk factors for cardiovascular disease and, at best, what the patient's exercise tolerance is.

Make an attempt to distinguish the pain from other types of chest pain (Fig. 3.2). The main types of pain are:

- Cardiac.
- Pleuritic – sharp, stabbing, aggravated by coughing, deep breathing (especially inspiration) or, occasionally, posture.
- Gastrointestinal – often related to food ingestion, may be vague, described as burning or may be associated with an acid taste in the mouth.
- Musculoskeletal – usually easily recognized; the pain is often exaggerated by movement and there is often a good mechanical explanation, for example trauma or strain.
- Atypical diagnosis is partly by exclusion, partly by its atypical characteristics, for example sharp, lateral and may be precipitated by stress or anxiety.

Precipitating factors

Angina is typically provoked by exertion. If the pain is reproducible, try to find out the level of exertion necessary to induce it (e.g. walking up one flight of

Fig. 3.1 Typical location of cardiac chest pain and non-cardiac chest pain (e.g. medically unexplained symptoms, anxiety) in an anxious patient.

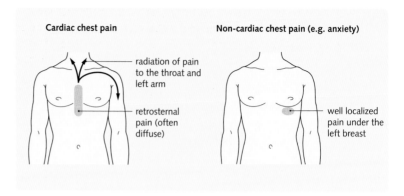

Cardiac chest pain

— radiation of pain to the throat and left arm

— retrosternal pain (often diffuse)

Non-cardiac chest pain (e.g. anxiety)

— well localized pain under the left breast

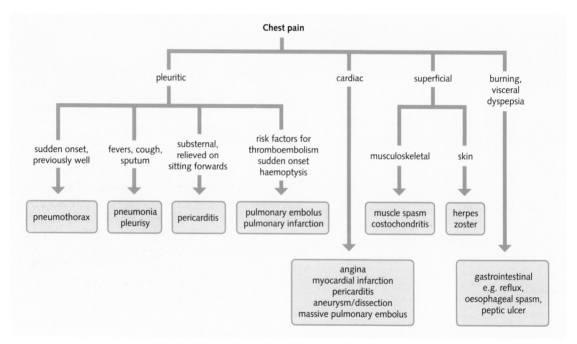

Fig. 3.2 Differential diagnosis of chest pain.

stairs, 200 metres on the flat). Note if this level of exertion has changed recently. Ask specifically about other precipitating causes (e.g. stress, excitement, sexual intercourse, meals).

If this is the first episode of pain or its nature has changed, it is important to know what the patient was doing *immediately* before the onset of pain (e.g. stable angina is provoked by a predictable stress, but unstable angina and myocardial infarction often occur at rest).

Time course and relieving factors

Angina usually lasts for only a few minutes and is typically relieved by rest. Patients often describe an urge to slow down or stop if they are walking at the onset of pain. Ask the patient exactly how long the pain takes to subside on rest – angina usually resolves within seconds or, at most, a few minutes. If the patient takes glyceryl trinitrate (GTN) tablets, enquire how quickly they seem to work. (Beware –

some patients may have a false label of angina, and just because they are taking GTN tablets, it does not mean that the pain they are describing to you must be angina – keeping an open mind is useful.) A myocardial infarction usually causes pain lasting for longer than 20 minutes. It would be rash to ascribe such pain to stable angina without other good evidence.

Associated features

Enquire specifically about any associated nausea, vomiting, sweating, shortness of breath, blackout or collapse during the pain. If the patient describes palpitations, it is crucial to know whether they preceded the onset of pain as, occasionally, a tachyarrhythmia may cause angina. Establish what the patient means by the term palpitation (ask them to tap out the rhythm). Make an attempt to distinguish between angina and myocardial infarction (Fig. 3.3).

Past medical history

Enquire specifically about the major risk factors for ischaemic heart disease:

- Previous episodes of angina or myocardial infarction. Record dates, events and how the diagnosis was established (e.g. exercise test, hospital admission, angiography).
- What is the patient like at their best?
- Cigarette smoking. 'Pack years' is a useful concept (see p. 14). Current smokers have a significantly increased risk compared with ex-smokers.

- Hypertension.
- Diabetes mellitus.
- Hypercholesterolaemia.
- Positive family history of ischaemic heart disease.
- Other vascular disease (e.g. stroke, peripheral vascular disease).

The major risk factors interact and the probability of disease is greatly increased if more than one is present. You should be suspicious of any chest pain in a patient with risk factors, even atypical chest pain.

Consider factors other than coronary artery disease that can cause angina (e.g. anaemia, arrhythmia, previous valvular pathology, rheumatic fever).

Drug history

A systematic review of the literature on accuracy of bedside findings concluded that with stable, intermittent chest pain, the patient's description of the pain was the most important predictor of underlying coronary disease. With acute chest pain, the ECG (electrocardiogram) was the most useful predictor for diagnosis of myocardial infarction (see Chun & McGee (2004), Further reading).

Characteristics of angina and myocardial infarction		
Feature	**Angina**	**Myocardial infarction**
site	retrosternal, throat, left arm	retrosternal, throat, left arm
radiation	typically to the throat or left arm	typically to the throat or left arm
nature	'tight', 'gripping', 'a dull ache'	similar, but usually recognized as more severe
duration	short, usually a few minutes	usually greater than 20 minutes and only terminated by opiate analgesia
precipitation	exertion, stress, cold, emotion	usually none, but may have similar precipitants
relief	rest, GTN (rapid)	often none (opiates)
associated features	usually none	sweating, lightheadedness, palpitations, nausea, vomiting, sense of foreboding

Fig. 3.3 Characteristics of angina and myocardial infarction.

A full list of the medications the patient is currently taking is essential; however, it is important to also note the following:

- Have there been any recent changes?
- The effect of antianginal drugs on symptoms as well as side effects. In particular, does the pain resolve rapidly with sublingual GTN?
- Has concordance been good?
- Is the patient taking aspirin? Check that there are no contraindications (e.g. active ulceration, asthma provoked by aspirin).
- Consider the role of other drugs that might aggravate angina (e.g. theophylline, tricyclic antidepressants, wrong dose of thyroxine).

Social history

Enquire whether there have been any recent changes in lifestyle (e.g. financial difficulties, stress at home or work). These outside influences may be the precipitant for angina or a reason for developing a non-cardiac chest pain. Ask how the chest pain has interfered with normal lifestyle.

Review of symptoms

A brief review of symptoms is important for several reasons, for example:

- Neurological symptoms may be provoked by decreased perfusion.
- To assess potential risks when considering invasive investigations or treatment (e.g. angiography or thrombolysis).
- To assess whether the patient has the mobility to tolerate an exercise ECG.
- To assess whether activity is limited by cardiac status or other factors such as poor mobility, obesity or chronic lung disease.

NON-CARDIAC CHEST PAIN

Chest pain is one of the commonest presenting symptoms in both primary care and hospital settings. Less than half of such patients have heart disease. Since many will go on to experience chronic symptoms, an early diagnosis of 'non-cardiac' chest pain is important. The key is a thorough history, first to establish the pattern of the pain, and second to search for potential causes, for example reflux oesophagitis. A few simple questions can be effec-

tive at establishing that the pain is unlikely to be cardiac in origin.

However, in many instances no obvious 'organic' cause is found. Chest pain is one of the commonest so-called 'medically unexplained physical symptoms' (formerly referred to as 'functional', 'psychosomatic' or, somewhat pejoratively, 'hypochondriac'). The pain is often accompanied by psychological distress and may be related to events in the patient's life. Thus it is important to explore these areas, and to tap into the patient's ideas, concerns and expectations (see Bass & Mayou 2002, Further reading).

PALPITATIONS

Presenting complaint

Palpitation is an awareness of the heart beating. Different people mean different things when they say they have experienced palpitations.

Detailed history

It is essential to explore the event in great detail so that the underlying rhythm disturbance and functional consequences can be appreciated.

Nature of the palpitation

It is often possible to make a reasonable estimate of the underlying rhythm from the patient's description (Fig. 3.4), for example in response to questions such as 'Can you describe what you experienced?' or 'Can you tap out the heart beat on the table?' The rate of the heart during the palpitation often provides a clue to the primary electrical disturbance (Fig. 3.5).

Duration and frequency of episodes

The functional impact on the patient may be revealed, as well as the likelihood of being able to 'capture' the event on a 24-hour tape (ECG) or event recorder.

Associated symptoms

Patients may have symptoms of cardiac decompensation, for example:

- Lightheadedness, fainting (syncope is due to poor cerebral perfusion and hypotension).
- Chest pain (angina).
- Sweating.

Fig. 3.4 Characteristics of common arrhythmias causing palpitations.

Characteristics of common arrhythmias causing palpitations	
Rhythm	**Typical features**
ectopic beats	'I felt as though my heart missed a beat'; 'a heavy thud'; usually due to awareness of post-extrasystolic beat
atrial fibrillation	'fast, irregular beating'; may be associated with dyspnoea or chest pain, especially if fast rate
supraventricular tachycardia	rapid palpitation, often abrupt onset; may be associated with polyuria; may have rapid termination; patient may have learnt to perform vagal manoeuvres to terminate episode
ventricular tachycardia	often associated with shock, collapse, dyspnoea or progression to cardiac arrest; can be hard to distinguish from supraventricular arrhythmias as features overlap, and may even be asymptomatic; conversely, supraventricular tachycardia can cause shock, especially if rapid

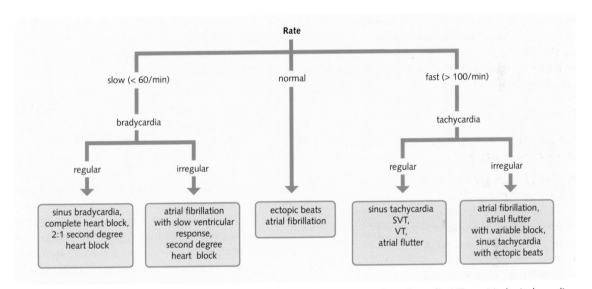

Fig. 3.5 Differential diagnosis of palpitations from the history. SVT, supraventricular tachycardia; VT, ventricular tachycardia.

The presence of these symptoms should alert the physician and prompt more detailed investigation.

Events immediately preceding palpitation

It may be physiological to experience some palpitations after exertion or emotional stress, and this may be evident from the response to 'What were you doing immediately before the palpitations started?' If there was chest pain, find out whether the pain preceded the palpitation or coincided with its onset.

Past medical history

Review the possible underlying diseases that cause palpitations, including:

- Risk factors for ischaemic heart disease (see Cardiac chest pain, above).
- Thyroid disease (especially atrial fibrillation).
- Rheumatic fever.

Drug history

This is very important. Particular attention should be paid to drugs with proarrhythmic effects, such as tricyclic antidepressants, digoxin, β-blockers (and other antiarrhythmic agents) and theophylline. Review the response of the palpitations to therapy.

Social history

Particular attention should be paid to alcohol consumption and caffeine-containing drinks. Use of recreational drugs (e.g. cannabis, ecstasy, amfetamines) may precipitate arrhythmias.

Summary of aims

The aims of the history for palpitations are:

- To determine whether the rhythm is slow or fast, regular or irregular.
- To note the presence of associated symptoms during the arrhythmia.
- To narrow down the differential diagnosis of the arrhythmia. This is difficult from the history alone, and ultimately a diagnosis can only be made by an ECG recording at the time of symptoms.
- To assess whether further investigation is appropriate and, if so, whether the episodes are long enough and frequent enough for the patient to use an event recorder or whether a 24-hour ECG is more appropriate.

HEART FAILURE

Presenting complaint

Acute

Acute heart failure classically presents with severe shortness of breath, severe distress, production of copious pink, frothy sputum and collapse (known as 'pulmonary oedema').

Chronic

This usually presents with shortness of breath, limitation of exercise tolerance, ankle swelling and fatigue.

Detailed history

The features of heart failure are usually distinctive enough to be recognized from the history alone, but airway obstruction can sometimes be confused with heart failure (indeed pulmonary oedema used to be known as 'cardiac asthma') and the two may coexist. A detailed history will clarify the presence of heart failure and establish its severity and possible aetiology.

Chronicity of symptoms

Has the patient had a recent sudden decline suggestive of an ischaemic event? Ask the patient 'What are you like normally?', 'How many hospital admissions for this have you had in the past year?' and/or 'Have you had to go to your GP with worsening symptoms?'

Severity of symptoms

Attempt to quantify the patient's impairment so that a reproducible assessment can be made. Focus on tolerance to exercise and the limiting factor for exercise: is it the breathlessness, osteoarthritic knees or chest pain that limits the patient?

Exercise tolerance

It is often difficult to be precise, but patients should, with assistance, be able to give quantitative answers to questions such as 'How many flights of stairs can you climb?', 'How far can you walk on the flat and uphill?' or 'Do you need help for any activities at home?' Be patient: the answers to these questions are important and it may take all of your communication skills to obtain an accurate answer. Try to quantify what factor limits exercise capacity (e.g. fatigue, coexisting lung disease, claudication). Severity of heart failure may be graded according to the New York Heart Association classification (Fig. 3.6).

NYHA grading of severity of heart failure	
Grade	Severity of symptoms
I	unlimited exercise tolerance
II	symptomatic on extra exertion (e.g. stairs)
III	symptomatic on mild exertion (e.g. walking)
IV	symptomatic on minimal exertion or rest (e.g. washing)

Fig. 3.6 The New York Heart Association (NYHA) grading of severity of heart failure provides a simple but reproducible assessment with interobserver agreement.

Limiting factors

Try to establish the limiting factor for exercise (e.g. dyspnoea, fatigue, chest pain).

Evidence of left heart failure

Enquire about features of pulmonary oedema, for example:

- Paroxysmal nocturnal dyspnoea. This is a feature of acute pulmonary oedema. Ask 'Do you ever wake up in the night fighting for breath?' Patients often describe having to sit upright on the edge of the bed and/or throwing open the windows.
- Orthopnoea. People may sleep with a few pillows for simple comfort or out of habit so it is important to find out if there has been any change. Ask 'How many pillows do you need to sleep with, and is this normal for you?'

Evidence of right heart failure

Enquire about symptoms related to fluid overload, which may result in:

- Ascites.
- Peripheral oedema (in severe cases, male patients may have scrotal oedema).
- Right upper quadrant discomfort (due to hepatic congestion).
- Nausea and poor appetite (due to bowel oedema).

Past medical history

The most relevant features include:

- Risk factors for ischaemic heart disease (see Cardiac chest pain, above).
- Previous cardiac investigations (e.g. echocardiography, angiography, exercise test).
- Other causes of left heart failure (e.g. rheumatic fever, valvular disease, cardiomyopathy), high output states (rare) (e.g. thyroid disease, Paget's disease, arteriovenous shunt, anaemia).
- Other causes of right heart disease (e.g. chronic lung disease, pulmonary embolus).

Drug history

A full list of medication is needed, but focus on:

- Current therapy for heart failure, for example angiotensin-converting enzyme (ACE) inhibitors (cough may be a side effect or due to mild pulmonary oedema) and diuretics (assess concordance and find out whether there has been a recent change in dose).
- Negatively inotropic drugs (e.g. β-blockers, verapamil, class I antiarrhythmic agents).
- Ask if the patient is on digoxin and, if they are, consider checking a level.

Social history

This section is very important. Assess daily activities, social support, mobility, etc. Review the patient's diet and appetite. Consider salt intake in oedematous states. Does the patient have sufficient mobility to cope with an increased diuresis whilst avoiding incontinence?

DEEP VEIN THROMBOSIS

Presenting complaint

Deep vein thrombosis (DVT) is a common condition. It is often asymptomatic; however, the most common features of presentation include (see box on p. 26):

- Calf pain.
- Leg swelling.
- Increased temperature of the leg.

Red tender leg

The history should be directed at finding risk factors for developing a DVT, which include (asterisks denote the more important ones):

- Pregnancy or puerperium.*
- Prolonged immobility (e.g. long-haul air travel).
- Contraceptive pill.
- Recent surgery.*
- Malignancy.*
- Lower limb fractures.*
- Heart failure.
- Dehydration.

PULMONARY EMBOLISM

Pulmonary embolus (PE) may be very difficult to diagnose. Its presentation can vary from asymptom-

Fig. 3.7 Differential diagnosis of leg swelling or inflammation other than venous thrombosis.

Differential diagnosis of leg swelling or inflammation other than venous thrombosis	
Condition	**Features**
infection (cellulitis)	subacute onset; fever; lymphangitis may be present; ask about portal of entry for infecting organism
ruptured Baker's cyst	preceding arthritis or swelling of knee; acute onset
torn calf muscle	acute onset, often during exercise
congestive cardiac failure	dyspnoea, fatigue, orthopnoea; risk factors for ischaemic heart disease; usually bilateral leg swelling
lymphatic obstruction	chronic; may be unilateral or bilateral
nephrotic syndrome	subacute or chronic leg swelling; bilateral; usually no features of inflammation

atic microemboli to sudden death caused by saddle embolism. The most common presentations are:

- Pleuritic chest pain.
- Shortness of breath.
- Haemoptysis.
- Collapse.

Thromboembolism is treatable but potentially fatal. It is under-recognized. A high index of suspicion is crucial.

Almost 50% of DVTs do *not* produce local symptoms, so PE may be the presenting feature. Presentation may be non-specific, and the differential diagnosis is wide.

In the presence of pleuritic chest pain of undetermined aetiology, do perform arterial blood gases and obtain a chest radiograph and an ECG. In determining if the patient has a DVT, it is useful to use the Hamilton or modified Wells criteria. These are evidence-based scoring systems that help predict the likelihood of a DVT (see Subramaniam et al (2006), Further reading).

Detailed history

If a PE is suspected, always ask specifically about risk factors suggestive of a DVT. In the presence of calf pain, investigate its features systematically (see Ch. 2). In particular, note any preceding symptoms, the speed of onset, any associated symptoms and whether the pain is unilateral or bilateral. Figure 3.7 highlights some of the more discriminatory features in the history.

Further reading

Bass C, Mayou R 2002 ABC of psychological medicine. Chest pain. *British Medical Journal* **325**: 588–591

Chun AA, McGee SR 2004 Bedside diagnosis of coronary artery disease. A systematic review. *American Journal of Medicine* **117**: 334–343

Subramaniam RM, Snyder B, Heath R et al 2006 Diagnosis of lower limb deep venous thrombosis in emergency department patients: performance of Hamilton and modified Wells scores. *Annals of Emergency Medicine* **48**(6): 678–685

Presenting problems: respiratory system

Objectives

By the end of this chapter you should:

- Be confident in taking a respiratory history.
- Recall the presenting features of asthma, chronic obstructive pulmonary disease and pneumonia.

ASTHMA

Presenting complaint

Can you actually obtain a history from the patient? If the patient cannot talk in sentences you have identified a medical emergency and you must seek senior help. However, the most common presentations include episodic wheeze, shortness of breath or a cough which is often nocturnal.

An acute asthma attack is often frightening both for the patient and the attending physician. The patient is often too dyspnoeic to provide much history. The priority is to make a rapid assessment and institute effective therapy. A more detailed history can be obtained once the patient is stable.

Detailed history

If the patient presents with an acute attack, investigate this attack in detail. Obtain a systematic, chronological account of the recent deterioration, focusing on:

- Severity – try to quantify in simple terms (e.g. unable to perform vigorous exercise, difficulty climbing stairs, unable to speak a complete sentence, being kept awake at night).
- Symptoms (e.g. wheeze, cough, dyspnoea).
- Time course (hours or days).

- Onset and precipitating events (e.g. exercise, emotional stress, viral illness, house dust, pets).
- Intervention during present attack, and response (e.g. nebulized bronchodilators, steroids).
- Reason for seeking medical attention at this stage.
- Whether the patient has been seeing their GP about their asthma.

Often asthma control has deteriorated chronically and insidiously. Ask either:

- 'Is there anything that you could do six months ago that you could no longer manage before this attack?'

or:

- 'Have you reduced your exercise over the past few months?'

Past medical history

Baseline asthma control

It is helpful to gain an awareness of the background control – in addition to allowing an assessment of disease severity, it may reveal information about the patient's understanding of the disease. Ask about:

- Usual exercise tolerance. Try to quantify as described above. (Young patients should have unlimited exercise capacity. Older patients may have coexisting morbidity).
- Frequency of attacks.
- Best recorded peak expiratory flow rate (PEFR). Ideally all asthma patients should have their

Fig. 4.1 An example of a peak flow recording from an adolescent with poor asthma control. Note the presence of cough at night and 'morning dips'.

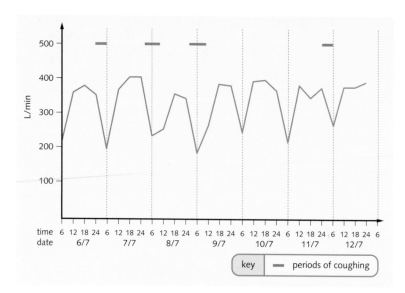

time 6 12 18 24 6 12 18 24 6 12 18 24 6 12 18 24 6 12 18 24 6 12 18 24 6 12 18 24 6
date   6/7   7/7   8/7   9/7   10/7   11/7   12/7

| key | ▬ periods of coughing |

own peak flow meter and know their baseline PEFR.

- Usual precipitating factors (e.g. pollen, stress, exercise, dust, pollution).
- Usual medication (see below).
- Usual response to therapy during exacerbations. For example, ask 'Is this the worst attack you've ever had?' or 'Would you normally expect your asthma attacks to get better after using a nebulizer?'
- Previous hospital admissions. Ask separately about intensive treatment unit admissions. For example, enquire 'Have you ever been admitted to hospital with asthma?' or 'Have you ever needed to be put on a ventilator?'
- Attendance at the GP's asthma clinic or for asthma review, if relevant.
- Symptoms suggestive of poor baseline control. This is very important and under-recognized. (e.g. 'morning dips', poor sleep, nocturnal cough, time off work or school). An example of a peak flow chart from a child with poor control is illustrated in Figure 4.1.

Other atopic conditions

Ask about other atopic conditions such as eczema, hay fever, urticaria.

Coexisting respiratory disease

Coexisting respiratory disease is particularly important in patients who present later in life, as it may

Inhaler technique scoring	
prepares device (e.g. shakes inhaler)	1
exhales fully	1
activates and inhales	1
holds breath for several seconds	1

Fig. 4.2 Inhaler technique scoring (total out of 4).

be hard to distinguish asthma from chronic obstructive pulmonary disease (COPD) from the history.

Drug history

Obtain a full list of medication. Ask specifically whether the patient:

- Has a nebulizer at home.
- Uses a bronchodilator.
- Takes theophylline or aminophylline (phosphodiesterase inhibitors). Have the drug levels ever been measured?
- Takes steroids: inhaled, nebulized or oral.

Ask patients to demonstrate their inhaler technique. It is possible to quantify inhaler techniques as in Figure 4.2. Find out whether function tests have been performed to assess airway reversibility, and responses to different agents, especially for older patients for whom it may be difficult to define the relative components of asthma and COPD to the overall morbidity. Consider medication that may aggravate the symptoms (e.g. β-blockers, aspirin).

Social history

Review how the asthma is interfering with lifestyle for both older and younger patients (e.g. school activities, absenteeism from work, limitation in sports, difficulty walking to the shops).

Always specifically enquire about smoking. During an exacerbation, it is timely to offer sensitive advice about smoking! Be non-judgemental in your approach. Try 'Have you tried to give up?' and ask about available support. You must let the patient make up his or her own mind. Enquire whether anyone in the patient's household is a smoker.

Review of systems enquiry

Focus on other diseases that may limit exercise tolerance, especially cardiovascular, respiratory pathology and arthritis.

Remember that asthma can be fatal. The morbidity and mortality are high but can be overcome by a holistic approach to management, including better supervision, objective assessment, improved patient understanding, active participation in management and the appropriate use of steroids. People with asthma should not smoke!

CHRONIC OBSTRUCTIVE PULMONARY DISEASE

Detailed history

Obtain a detailed history of chest symptoms. In an acute exacerbation, patients usually present following a cold with a deterioration of dyspnoea in association with a productive cough and discoloured sputum. Outline a detailed history of the present attack, following the usual systematic approach, to explore:

- Time course.
- Treatment given and effects.
- Functional impact on lifestyle.

- Any hospital admissions in the last year for COPD.
- Whether the patient has been seeing his or her GP with the problem.

Obtain a thorough history of baseline function, trying to get as objective answers as possible. For example, ask:

- 'How far can you walk?'
- 'Can you climb one flight of stairs easily?'
- 'Do you get short of breath while dressing?'
- 'Did you manage to walk to the outpatients' department without stopping?'

It is typical for a patient with COPD to have a pattern of chronically deteriorating exercise tolerance punctuated with acute declines during an infective exacerbation (Fig. 4.3). These may be seasonal, with an increased frequency in the winter months.

Sputum production and cough are characteristic. Try to quantify the usual amount per day and its characteristics (e.g. a teaspoonful, an eggcupful).

- Chronic bronchitis is defined on the basis of a history of cough, productive of sputum on most days, for 3 consecutive months, for at least 2 years. Emphysema is a pathological diagnosis of dilatation and destruction of the lungs distal to the terminal bronchioles. In practice, these conditions coexist and are labelled chronic obstructive pulmonary disease (COPD).
- In COPD patients, meticulous and realistic assessment of baseline lung function is essential – without this, it is impossible to make decisions regarding appropriate treatment and to set realistic goals of therapy.
- Many patients with COPD have a reversible component to their disease. This is under-recognized, but can be uncovered by a formal trial of steroids.
- Blood gases for assessment of COPD should be taken 3 months after any acute illness, otherwise a false result may be obtained.

Consider the possibility of cor pulmonale (right sided heart failure secondary to lung disease; see Ch. 3 for symptoms of right heart failure) in a patient with severe disease who describes ankle swelling.

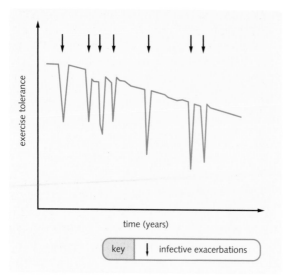

Fig. 4.3 Exercise tolerance in a patient with COPD.

Ascertain aggravating factors (e.g. cold weather, pollution, exertion).

Find out whether a satisfactory attempt has been made to establish the diagnosis, for example:

- Have lung function tests been performed to assess airway reversibility?
- Have arterial blood gases been analysed when the patient is well?

Past medical history

These patients may have multiple medical problems, which should be recorded, but specifically ask about:

- Previous admissions to hospital with acute exacerbations of COPD. Record the frequency, especially within the last year.
- Other smoking-related diseases (e.g. ischaemic heart disease, peripheral vascular disease, strokes, hypertension).
- Other causes of lung disease (e.g. occupational exposure to dust, bronchiectasis due to, for example, previous tuberculosis, childhood whooping cough).
- Asthma. There may be a reversible component to the disease.

Drug history

Review medication prescribed for COPD, for example:

- Bronchodilators (inhalers and nebulizers).
- Home oxygen. Who initiated therapy and on what evidence? For how many hours per day is it being used? Long term oxygen therapy (LTOT) should be used for >15 hours per day and its purpose is to prevent cor pulmonale. It is not for improving oxygen saturations per se (see Cullen (2006), Further reading).
- Theophylline. Have levels been measured recently?
- Steroids. Does the patient have a steroid card?
- Review inhaler technique.

Social history

This is particularly important for these patients as they often have significant limitation of exercise tolerance and rely heavily upon support from family, friends and state (e.g. are they receiving any benefits?). Consider all aspects of daily living.

A detailed occupational history may be important if there is any doubt about the patient's ability to continue working or the aetiology of the lung disease, for example:

- Exposure to inorganic dusts (coal-miner's lung, silicosis, asbestosis).
- Occupational asthma (isocyanates, colophony fumes).
- Extrinsic allergic bronchiolar alveolitis (farm workers, 'bird fanciers').

Obtain a detailed smoking history, because this is undoubtedly a smoking-related disease in the majority of patients. Remember that the patient must not smoke if they are using home oxygen!

Review of systems enquiry

Many patients with COPD have multiple pathologies related to their smoking, so a thorough trawl of their symptoms may raise suspicions of previously unrecognized conditions (e.g. ischaemic heart disease, malignancy, renal disease, peripheral vascular disease).

CHEST INFECTION

Detailed history

Perform a detailed enquiry about presenting symptoms, adopting a methodical approach. Ask specifically about symptoms referable to the respiratory tract, as follows:

- Cough – duration, whether productive or dry.
- Sputum production – quantity, colour, recent changes if the patient has a productive cough, any haemoptysis.
- Dyspnoea – obtain a quantitative account of exercise tolerance at baseline and during the current illness.
- Wheeze.
- Pleuritic chest pain – a common feature of pneumonia, but be aware of the possibility of a pulmonary embolus.
- Fever.

If symptoms are prolonged, recurrent or associated with weight loss (remember to ask!), consider the possibility of an underlying malignancy, especially in a smoker.

Ask about associated symptoms that have immediately preceded or coincided with the illness (especially gastrointestinal). These may give additional clues to the infecting organism causing pneumonia. Figure 4.4 illustrates how a detailed history may help to identify the microbiological cause of a pneumonic illness.

Drug history

Ask specifically about antibiotics used to treat this and any recent episode, and the duration of use, as the response to therapy may give a clue to the infecting agent as well as the likelihood of obtaining a positive blood culture, for example:

- Resistance of *Mycoplasma* to penicillin.
- Resistance of tuberculosis or *Pneumocystis* to repeated courses of antibiotics.

Find out if the patient is taking immunosuppressive medication (e.g. those taking steroids, transplant recipients) (see Fig. 4.4).

Clues to the underlying cause of pneumonia	
Organism	**Features from history**
*Streptococcus pneumoniae**	most frequent identifiable infecting organism in community-acquired pneumonia; classically associated with herpes labialis, commonly prominent fever and pleuritic pain; often abrupt onset in previously fit individual
*Mycoplasma pneumoniae**	occurs in epidemics with a 3–4-year periodicity; usually occurs in previously fit people, often young adults; may be preceded by a prodromal illness with headache and malaise; may be prominent extrapulmonary features (e.g. nausea, vomiting, myalgia, rash)
*Haemophilus influenzae**	most common bacterial pneumonia following influenza; associated with underlying lung disease (especially COPD)
Legionella pneumophila	associated with institutional outbreaks (e.g. hospitals, hotels); may be associated with mental confusion or gastrointestinal symptoms; typically causes a dry cough
Coxiella burnetii	contact with farm animals
Chlamydia psittaci	contact with infected birds ('Do you have a sick parrot?')
Staphylococcus aureus	associated with preceding influenza, intravenous drug abusers, patient is often very ill
Gram-negative organisms	hospitalized patients; may be community-acquired in elderly or diabetics; *Branhamella catarrhalis* is associated with exacerbations of COPD
Pneumocystis carinii, cytomegalovirus, *Nocardia asteroides*, *Mycobacterium avium intracellulare*	acquired immunodeficiency syndrome (AIDS); transplant recipients; chemotherapy
Mycobacterium tuberculosis	weight loss, chronic cough, foreign travel, infected family member

Fig. 4.4 Clues to the underlying cause of pneumonia. Asterisks denote the more common organisms.

Social history

Relevant clues may be provided by a travel history and details of hobbies (e.g. involving pets), occupation and risk factors (e.g. for HIV infection). Smokers are more likely to decompensate earlier in the course of the illness. Clearly it is important to assess the functional impact of the disease on patients and their families so that appropriate therapeutic and management decisions can be made.

Further reading

Cullen DL 2006 Long term oxygen therapy adherence and COPD: what we don't know. *Chronic Respiratory Disease* 3(4): 217–222

Objectives

By the end of this chapter you should:

- Be able to take a medical history focusing on common gastrointestinal and urological presentations.
- Understand the basic principles of taking an obstetric or gynaecological history.
- Be able to recall the common causes of jaundice, abdominal distension, anaemia and renal failure.
- Be able to state the initial management of a gastrointestinal bleed.
- Recall the presenting features of common abdominal malignancies.

The abdomen contains a number of organ systems that relate to different medical specialties. This chapter covers gastrointestinal (GI) and urological presenting complaints and has a separate section on taking obstetric and gynaecological histories. (For more detail, try Panay et al (2004), Further reading.)

ACUTE ABDOMINAL PAIN

Detailed history

A very careful history needs to be elicited as it will form the foundation for a working hypothesis and differential diagnosis, and rational subsequent investigation.

On the basis of the history, abdominal pain can be divided into three types:

1. Visceral.
2. Somatic.
3. Referred.

Visceral (deep) pain

This is dull, poorly localized pain referred to the midline. The site of pain is derived from its embryological origin (foregut, midgut, hindgut) (Fig. 5.1).

Somatic (peritoneal) pain

This is sharp, severe and more precisely localized pain. It occurs when the disease process involves the surrounding peritoneum and mesentery.

Referred pain

This is the perception of sensory stimuli at a distance from the source (e.g. acute cholecystitis causing diaphragmatic irritation, with the patient feeling pain over the right shoulder tip). The characteristics of the pain should be reviewed in a systematic manner as for other forms of pain (e.g. see Cardiac chest pain, Ch.3). There are several key areas which should always be investigated and the outline of how to take a pain history is outlined in Chapter 2.

Site of the pain

Define the initial location of the pain and whether it has subsequently moved (e.g. acute appendicitis). This is of great importance as certain disease processes tend to cause pain localized to a defined region of the abdomen (Fig. 5.2).

Time and mode of onset

Sudden-onset pain suggests a vascular event (e.g. rupture of an abdominal aortic aneurysm) or perforation of a viscus: 'One moment I was feeling fine, the next I was doubled up with pain!'

Frequency and duration

Colicky pain occurs when there is a pathological process (usually an obstruction) in a smooth muscular tube (e.g. small and large bowel, ureter, fallopian tube). It may be described as 'spasms'. Ask whether the patient has had previous similar episodes.

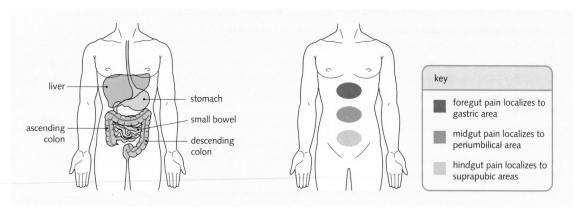

Fig. 5.1 The site of abdominal pain is related to the embryological development of the foregut, midgut and hindgut.

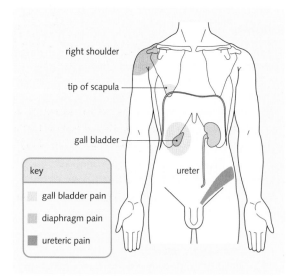

Fig. 5.2 Typical sites of radiation pain for pain originating in the gall bladder, diaphragm and ureter.

Character

The pain may change character, indicating progression of the pathology (e.g. transition of colicky pain to constant pain suggests transition of visceral to peritoneal involvement in acute appendicitis).

Severity

Is the pain getting worse, better or staying at the same intensity? 'If 0 equals no pain and 10 is the worst pain you have ever experienced, what value would you give this pain? Has it always been the same?'

Radiation

Loin pain, for example, radiates to the groin in renal or ureteric colic.

Aggravating factors

It is often apparent from first seeing patients what type of pain they have (e.g. patients with peritonitis lie still, patients with ureteric colic are often very restless). Certain foods may aggravate pain (e.g. classically, fatty foods may aggravate abdominal pain due to gallstones; however, this is unreliable).

Relieving factors

The pain of pancreatitis is, for example, characteristically relieved by sitting forward (the mesentery and other organs are then not pressing on the pancreas); duodenal ulcer pain may be relieved by eating, and antacids or sleeping upright may relieve the pain of reflux oesophagitis.

Cause

Ask the patient what he or she thinks is the cause of the pain (remember Ideas, Concerns, Expectations (ICE) from Ch. 1!).

Other symptoms

Review other symptoms referable to the abdomen:

- When were the bowels last opened and when was flatus last passed? This is particularly relevant if partial or complete obstruction is suspected.

- Change in bowel habit. This is likely to reflect a large bowel pathology (e.g. carcinoma or inflammatory bowel disease).
- Vomiting. Establish the nature of the vomitus (i.e. blood, bile (clarify what the patient means by bile), 'coffee grounds', faeculent) and when it occurs in relation to eating.
- 'Do you still feel hungry?' This is useful for discriminating non-serious pathology as the majority of patients with serious intra-abdominal disease have anorexia.
- Abdominal distension.
- Appetite and weight loss. Chronic weight loss is suggestive of an underlying malignancy. It is useful to ask if the patient's clothes still fit and if jewellery or false teeth have become loose.
- Dysphagia. Ask the patient to point to where the food appears to stick. Establish whether the dysphagia is for food, or food and drink. Enquire whether there is associated pain on swallowing.
- Are there any foods that are particularly associated with pain, e.g. fatty foods?
- Regurgitation, flatulence, heartburn, dyspepsia. Ask about these symptoms if there is a suspected peptic ulcer, gastro-oesophageal reflux or gallstone disease.
- Urinary symptoms. Frequency and dysuria may suggest a urinary tract infection. Nocturia, urgency and hesitancy are consistent with prostatic enlargement.

- History of trauma. Have a low index of suspicion for a splenic or hepatic tear.

Remember the five Fs as the causes of abdominal distension:
- Fat
- Fluid
- Faeces
- Fetus
- Flatus

Obstetric and gynaecological history taking are covered at the end of this chapter.

Consider the possibility of pregnancy in all women of childbearing age.

It is particularly important to include a cardiovascular and respiratory history, as several medical conditions can cause acute abdominal pain (Fig. 5.3).

Past medical history

Obtain a detailed history, paying particular attention to:

Medical conditions that can mimic a 'surgical abdomen'	
Medical condition presenting as abdominal pain	**Features**
myocardial infarction (MI)*	especially inferior MI; may have paradoxical bradycardia; risk factors for ischaemic heart disease
angina*	usually epigastric
chest infection*	especially lower lobe pneumonia; previous respiratory symptoms; pleurisy
diabetic ketoacidosis*	especially young patient; decreased level of consciousness; preceding polyuria, polydipsia, weight loss; positive family history
acute pyelonephritis*	dysuria, haematuria, frequency; loin pain versus central abdominal pain; history of renal stones
hypercalcaemia	often elderly; 'bones, stones, moans and groans'
sickle cell crisis	ethnic origin, usually known history

Fig. 5.3 Medical conditions that can mimic a 'surgical abdomen'. The history may distinguish medical conditions masquerading as surgical problems. Asterisks indicate the more common conditions.

- Previous operations – adhesions, recurrent pathology, etc.
- Recent myocardial infarction or cardiac arrhythmias – mesenteric embolus, especially in association with atrial fibrillation.
- Psychiatric history – patients will not volunteer this and a high index of suspicion together with old notes are needed for a diagnosis.
- Hypothyroidism.
- Constipation.

Drug history

Obtain a full list of medication. Pay attention to non-steroidal anti-inflammatory drugs (NSAIDs) and steroids. Also consider drugs that may provoke constipation (e.g. opiates, tricyclic antidepressants, antimuscarinic agents, antiparkinsonism therapy).

Family history

There may be a positive family history of inflammatory bowel disease or bowel carcinoma.

Social history

Alcohol history is extremely important (e.g. for peptic ulcer, pancreatitis). For diarrhoeal illnesses, consider foreign travel (e.g. amoebiasis, typhoid, giardiasis) or food poisoning (ask 'Are any of your friends or family also affected?'). The patient often has a strong inkling that symptoms have been caused by food and may be able to pinpoint exactly the suspect meal.

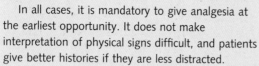

If the patient may need an operation, do not forget to ask when he or she last had food or drink.

In all cases, it is mandatory to give analgesia at the earliest opportunity. It does not make interpretation of physical signs difficult, and patients give better histories if they are less distracted.

ACUTE DIARRHOEAL ILLNESS

Detailed history

Diarrhoea is a symptom and not a disease. Therefore it is important to establish the underlying cause. Ask about the nature of the stools, frequency and events surrounding the episode. Important features to ask about include:

- Recent ingestion of undercooked meat, shellfish, unpasteurized milk, stream water (i.e. food poisoning).
- Associated abdominal pain and vomiting. Is the patient likely to need intravenous fluids or electrolyte replacement?
- Is the pain relieved by defaecation?
- Is there blood, pus or mucus in the stool?
- Are the stools pale and frothy (i.e. steatorrhoea)?
- Duration of symptoms (e.g. hours, weeks, days – different illnesses may present acutely or subacutely).
- Weight loss or anorexia.
- Recent return from a foreign country (e.g. amoebiasis, giardiasis).
- Allergy to gluten products.
- Symptoms of thyrotoxicosis (e.g. heat intolerance, agitation, palpitations); thyrotoxicosis occasionally presents with diarrhoea.

Past medical history

Obtain a detailed history, paying particular attention to:

- Previous operations (e.g. short bowel syndrome (almost exclusively Crohn's disease patients), gastrectomy and vagotomy leading to gastric dumping syndrome).
- Inflammatory bowel disease.

Drug history

The drug history is particularly important as drugs commonly contribute to a diarrhoeal state. Common culprits include antibiotics, laxative abuse (may be surreptitious) and magnesium-containing antacids.

Family history

Enquire about inflammatory bowel disease, carcinoma of the bowel and coeliac disease.

Social history

An infective aetiology is suggested if friends or relatives have a similar illness. If the patient is dehydrated, frail or responsible for child care, consider whether admission is indicated.

Causes of diarrhoea
infective
Clostridium difficile (if recent use of broad-spectrum antibiotics)
viral
Salmonella
Shigella
Campylobacter
enterotoxic Escherichia coli
inflammatory bowel disease
colorectal carcinoma
coeliac disease
drugs
anxiety states
miscellaneous
thyrotoxicosis

Fig. 5.4 Causes of diarrhoea.

The more common causes of diarrhoea are listed in Figure 5.4.

JAUNDICE

Presenting complaint

Jaundice presents with yellow discoloration, which is initially often not noticed by the patient.

The differential diagnosis of jaundice is broad. A detailed history is needed to focus further investigations. Painless jaundice in a patient over 55 years should be considered to be a cancer of the head of the pancreas until proven otherwise. In those under 55 years it is most likely to be hepatitis A. It is wise to consider the patient's socioeconomic circumstances while formulating a diagnosis; for example alcoholic liver disease is commonest in socioeconomic classes 1 and 5. Foreign travel may mean that more unusual causes of jaundice should be considered and this may include refugees and immigrants as well as the more well off. Remember: why *this* patient with *this* problem at *this* time?

Detailed history

The history of a jaundiced patient is very challenging as the pathophysiology is so varied. It is helpful to review the pathophysiology of jaundice (Figs 5.5 and 5.6). Focus on the major features.

Onset

Who noticed the jaundice (e.g. patient, family, abnormal blood test)? Establish the time course (e.g. acute onset in fulminant hepatitis A, insidious progression in biliary stricture).

Associated symptoms

It is often possible to narrow down the differential diagnosis of jaundice by a detailed history. Many causes of jaundice have typical features. The usual classification is prehepatic, hepatocellular or post-hepatic. Often the features of hepatocellular and obstructive jaundice overlap.

Prehepatic jaundice (haemolytic)

Jaundice is usually a minor component. The illness is often dominated by symptoms of anaemia, for example fatigue, dyspnoea, angina and palpitations (in older patients). It may be associated with gall-stones (pigment stones). On specific questioning, patients report normal coloured stool and urine. Patients may be transfusion dependent. Consider the possible causes (i.e. abnormal red cells or immune-mediated haemolysis).

Examples of abnormal red cells occur in:

- Congenital spherocytosis (northern Europe).
- Glucose-6-phosphate dehydrogenase (G6PD) deficiency (west Africa, Mediterranean, Middle East, south-east Asia).
- Sickle cell anaemia (sub-Saharan Africa).
- Thalassaemia (Mediterranean, Middle East, India, south-east Asia).

Causes of immune-mediated haemolysis include:

- Drugs (e.g. methyldopa, penicillin).
- Incompatible blood transfusion (acute onset).
- Warm autoantibodies (e.g. systemic lupus erythematosus, lymphoproliferative disorders).
- Cold agglutinins (e.g. infectious mononucleosis, *Mycoplasma*).

Fig. 5.5 The production, circulation and clearance of bilirubin (see also Fig. 5.6).

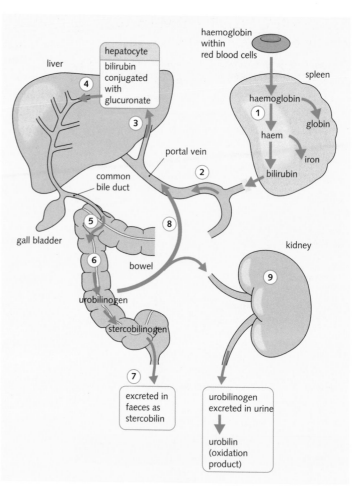

Production and clearance of bilirubin		
Stage	**Description**	**Example of pathology**
1	haemoglobin within the red cells broken down within the spleen producing non-water-soluble unconjugated bilirubin	excessive breakdown (e.g. haemolytic anaemia)
2	unconjugated bilirubin transported in the blood to the liver	
3	uptake of bilirubin by the hepatocytes and transfer to the smooth endoplasmic reticulum	drug toxicity; Gilbert's syndrome; Rotor's syndrome
4	conjugation with glucuronate	Crigler–Najjar syndrome
5	excretion of conjugated bilirubin in the bile into the small bowel	biliary obstruction (defect may occur at level of hepatocyte, bile canaliculi, or bile duct)
6	breakdown within bowel to stercobilinogen (urobilinogen)	
7	oxidation of stercobilinogen to stercobilin (causes brown colouration of faeces) and excretion	white stool in cholestatic jaundice
8	absorption of urobilinogen; most goes through enterohepatic recirculation	
9	small amount of urobilinogen (water soluble) reaches systemic circulation and excreted via the kidney	large amounts of urinary urobilinogen detectable if severe haemolysis or liver damage saturates the liver's capacity for enterohepatic recirculation

Fig. 5.6 Production and clearance of bilirubin.

Hepatocellular jaundice (inability to excrete bilirubin into the bile)

This is often dominated by symptoms of liver dysfunction, such as malaise, anorexia, right upper quadrant discomfort, abdominal distension, loss of libido, confusion. The list of diseases that may be responsible is vast, but the more important causes are illustrated in Figure 5.7.

Posthepatic jaundice (cholestatic)

The patient may complain of pruritus due to the deposition of bile salts. It is usually, but not always, relentlessly progressive rather than episodic. There is often a history of pale stools and dark urine due to a lack of stercobilinogen in the stool and retention of conjugated bilirubin. It is important to recognize extrahepatic causes of obstructive jaundice as these are often amenable to surgical intervention (Fig. 5.8).

Causes of hepatocellular jaundice	
Cause	**Examples**
viral*	hepatitis A* (common, especially in endemic areas may occur in epidemics; may present acutely); hepatitis B* (common in endemic areas, e.g. south-east Asia; ask specifically about risk factors for blood-borne infections); hepatitis C* (becoming more common; ask about blood transfusions, shared needles in drug addicts; usually chronic insidious illness)
alcoholic*	common; often presents as acute hepatitic illness
drugs*	common in hospitalized patients (e.g. rifampicin, isoniazid, prolonged course of antibiotics, paracetamol overdose, etc.)
cirrhosis*	of any aetiology (e.g. alcohol, biliary, haemochromatosis, etc.)
malignant infiltration*	primary or secondary (especially bronchus, bowel, breast)
congenital	for example Gilbert's syndrome* (common and mild); Crigler–Najjar syndrome
acute fatty liver of pregnancy	rare
inherited disorders	for example α-1-antitrypsin deficiency, Wilson's disease, etc.

Fig. 5.7 Causes of hepatocellular jaundice. Asterisks indicate the more common causes.

Causes of posthepatic jaundice	
Cause	**Features from the history**
gallstones*	common; often intermittent history of biliary colic or rigors; 'fat, female, forty, fertile'
carcinoma of head of pancreas*	weight loss; pain; relentless progression
pancreatitis*	acute onset; patient often very ill
benign stricture of common bile duct	may mimic carcinoma of the pancreas
sclerosing cholangitis	associated ulcerative colitis
cholangiocarcinoma	

Fig. 5.8 Causes of posthepatic jaundice. Asterisks indicate the more common causes.

Past medical history

Obtain a detailed history, paying particular attention to more recent events, for example:

- Alcohol abuse – recent bingeing.
- Ulcerative colitis – may suggest the presence of sclerosing cholangitis.
- Recent viral illness – Gilbert's syndrome, hepatitis A or B.
- Gallstones – either a cause of jaundice or a consequence of chronic haemolysis.

Drug history

An extremely careful drug history should be taken, as drugs may have precipitated the jaundice. In addition, certain drugs need to be avoided or used with care in liver disease. For example:

- Drugs causing haemolysis – acting as haptens (e.g. penicillin, sulphonamides), direct autoimmune effects (e.g. methyldopa), precipitating haemolysis in G6PD deficiency (e.g. primaquine, nitrofurantoin).
- Drugs causing hepatocellular damage (e.g. paracetamol overdose, alcohol, isoniazid).
- Drugs causing intrahepatic cholestasis (e.g. oestrogens, phenothiazines).
- Drugs causing gallstones (e.g. oral contraceptives, clofibrate).

Social history

Reviewing the patient's lifestyle may provide many clues to the aetiology. A detailed alcohol history is essential for acute alcoholic hepatitis and cirrhosis (see Ch. 2 for more on alcohol consumption). A travel history is particularly pertinent (e.g. to an area where hepatitis A is endemic). Risk factors for blood-borne infections, such as intravenous drug abuse, unprotected sexual intercourse, multiple blood transfusions, should be considered (e.g. for hepatitis B or C, or HIV infection). Cigarette smoking may point to malignant disease. Social contacts with hepatitis A may be apparent if there has been an epidemic.

Finally, review the patient's occupation and hobbies (e.g. leptospirosis in sewage workers or farmers, exposure to toxins by workers with organic solvents, hepatitis B in healthcare workers on dialysis units).

Family history

A family history is particularly relevant for younger patients (e.g. for Gilbert's syndrome, haemoglobinopathies, Wilson's disease).

Review of systems enquiry

The differential diagnosis is broad, so a complete systems enquiry is needed.

ANAEMIA

Presenting complaint

May be insidious, with lethargy, pallor, feeling tired all the time or an incidental finding on blood test.

Detailed history

Review the symptoms related to anaemia. Ask specifically about:

- Lethargy.
- Exercise tolerance.
- Palpitations.
- Angina and intermittent claudication (mainly older patients).
- Dyspnoea.

Try to establish the duration of symptoms; the causes of acute and chronic anaemia are different. For example, ask 'When did you last feel completely well?' Obtain the results of previous laboratory investigations – they may help in differentiating acute and chronic causes (Fig. 5.9).

Usually the result of the blood film will be available. It is helpful to categorize anaemia according to the mean corpuscular volume (MCV) (Fig. 5.10) as:

- Hypochromic (mean corpuscular haemoglobin concentratin (MCHC) <30%) microcytic (MCV <75 fl).
- Normochromic normocytic.
- Macrocytic (MCV >105 fl).

Past medical history

Take an extensive history. Ask specifically about the following conditions, including symptoms relating to:

- Peptic ulceration or indigestion – blood loss causing iron deficiency.

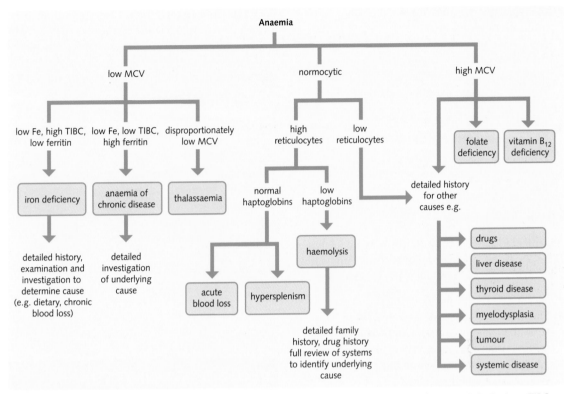

Fig. 5.9 Algorithm for the differential diagnosis of anaemia based on the mean corpuscular volume (MCV). Fe, iron; TIBC, total iron binding capacity.

Causes of anaemia	
Causes	**Features**
hypochromic	
iron deficiency*	overwhelmingly the most common cause of anaemia, usually a chronic insidious pattern (e.g. dietary, chronic blood loss)
thalassaemia trait and disease	disproportionately low MCV; Mediterranean; family history
anaemia of chronic disease*	may have normal MCV
congenital sideroblastic anaemia	rare
normochromic normocytic	
haemolytic anaemia*	often variable red cell indices
aplastic anaemia	may have variable MCV due to reticulocytosis (e.g. G6PD deficiency, drug-induced, etc.)
anaemia of chronic disease	usually multifactorial causes (e.g. malignancy, chronic renal failure, connective tissue disease, etc.)
macrocytic	
vitamin B_{12} deficiency*	common, megaloblastic (e.g. pernicious anaemia, veganism)
folate deficiency*	common, megaloblastic (e.g. nutritional, malabsorption, pregnancy)
alcohol*	the most common cause of an elevated MCV
hypothyroidism	
liver disease	
reticulocytosis	
myelodysplasia	rare
acquired sideroblastic anaemia	rare

Fig. 5.10 Causes of anaemia. Asterisks indicate the more common causes.

- Malignancy – chronic disease, marrow infiltration, blood loss.
- Renal disease – chronic disease, blood loss, haemolysis, erythropoietin deficiency.
- Connective tissue diseases.
- Thyroid disease – previous treatment with radioiodine.
- Diseases associated with pernicious anaemia – for example, vitiligo, diabetes mellitus, thyroiditis.
- Jaundice – alcohol abuse, chronic liver disease, haemolysis.

Drug history

A particularly detailed drug history is important as often drugs can cause or exacerbate anaemia. Drugs can cause anaemia in many ways, for example:

- Blood loss – aspirin or NSAIDs.
- Haemolysis – immune mediated (e.g. quinidine, methyldopa), G6PD deficiency (e.g. antimalarials, dapsone, favism).
- Aplasia – cytotoxic chemotherapy, idiopathic (e.g. sulphonamides).
- Megaloblastic anaemia – phenytoin, dihydrofolate reductase inhibitors (trimethoprim, methotrexate).
- Sideroblastic anaemia – isoniazid.

Family history

Consider the possibility of an inherited haemolytic anaemia (especially in the appropriate ethnic group), for example:

- Sickle cell anaemia – especially in sub-Saharan Africans and malarial areas.
- Thalassaemia – especially in those from the Mediterranean, Middle East, India, south-east Asia.
- Hereditary spherocytosis – northern Europeans.
- G6PD deficiency – in people from west Africa, the Mediterranean, Middle East, south-east Asia.

Social history

Focus on the diet, especially if there is iron, folate or vitamin B_{12} deficiency. In addition, alcohol can cause anaemia in many ways.

Review of systems enquiry

As the cause of anaemia is often multifactorial, the systems enquiry is often fruitful. In particular, con-sider causes of chronic blood loss (e.g. dyspepsia, melaena, menorrhagia) and symptoms suggestive of systemic disease (e.g. weight loss, fevers, sweats).

If a particular cause of anaemia is suspected, specific questions relating to that system should be asked in detail.

Anaemia is often multifactorial, so a detailed history is essential to elucidate different components. A diagnosis of iron deficiency is inadequate. The underlying cause for the deficiency must be found. Always ask about the use of aspirin or NSAIDs. Colorectal carcinoma is a common cause of chronic iron deficiency anaemia in the elderly so a high degree of suspicion is required.

ACUTE GASTROINTESTINAL BLEEDING

Presenting complaint

Typical presentations include vomiting blood (haematemesis), dyspepsia, abdominal pain and 'tarry black stools', which are very often smelly (melaena).

Detailed history

The most common causes of acute upper GI bleeding include the following (those with an asterisk being the most common):

- Gastric ulcer.*
- Duodenal ulcer.*
- Gastric erosions and gastritis.*
- Mallory–Weiss tear.
- Oesophageal varices.
- Haemorrhagic peptic oesophagitis.
- Gastric carcinoma (rarely presents with an acute GI bleed).
- Hereditary haemorrhagic telangiectasia (rare).

Acute lower GI bleeding may be due to:

- Bleeding piles.
- Diverticulosis.

Acute GI bleeding is a medical emergency and should be initially assessed via the ABC approach. Placement of two large-bore cannulae and a 'group and save' are mandatory management.

The presence of melaena indicates that the source of blood loss is probably proximal to and including the caecum. It is not enough to accept a history of melaena. A digital per rectum (PR) examination must be performed to positively confirm or refute this. A useful guide to the seriousness of the bleed is given by the Rockall scoring system. This uses a variety of factors to produce a mortality estimate for GI bleeding and is also used as a guide as to when emergency endoscopy is required (see Rockall et al (1996), Further reading).

You only have time to ask specifically about symptoms suggestive of haemodynamic instability if you are receiving the history second hand, for example as a triage telephone conversation in a GP surgery, including:

- Faintness and loss of consciousness.
- Sweating.
- Palpitations.
- Confusion.

Obtain a detailed history, focusing on symptoms referable to the GI tract. Ask specifically about abdominal pain, dyspepsia and heartburn, vomiting and nausea, weight loss and early satiety.

Ask about the duration of symptoms. It is worth enquiring whether the patient has experienced any symptoms suggestive of anaemia (e.g. lethargy, angina, palpitations, unexplained fatigue).

There may be a periodicity and relationship to food or identifiable precipitating events, for example an alcoholic binge, vomiting (e.g. Mallory–Weiss tear, pyloric stenosis).

Past medical history

Ask about pre-existing GI tract pathologies and investigations (e.g. endoscopy, barium meals). Liver disease or jaundice may suggest gastritis or oesophageal varices in the presence of portal hypertension.

Drug history

Ask specifically about:

- Aspirin and NSAIDs (common causes of gastritis).
- Steroids (may exacerbate pre-existing ulcer).
- Use of antacids, histamine H_2 blockers, proton pump inhibitors.

Family history

Patients with peptic ulceration often have a positive family history.

Social history

Cigarettes are associated with peptic ulceration. Alcohol is strongly associated with liver disease and gastritis. Binge drinkers may have been vomiting and have produced Mallory–Weiss tears.

The underlying cause of the bleed is often indicated from the history, but subsequent confirmation by endoscopy is almost invariably indicated.

CHANGE IN BOWEL HABIT

Presenting complaint

Patients may present with either a change in their normal stool frequency or a change in the nature of the stool.

Detailed history

The main conditions producing a change in bowel habit are illustrated in Figure 5.11. Find out what the normal pattern of bowel movements are for the patient. A normal pattern varies from one stool every three days to three stools a day. Enquire specifically about the frequency of stools and do not accept terms such as 'diarrhoea' or 'constipation' without clarification.

Ask about the duration of symptoms. A very short history of a few hours is likely to indicate an infective aetiology, whereas altered bowel habit for many years is more likely to indicate irritable bowel disorder in a young patient. Ask specifically about weight loss, anorexia, fatigue, etc., and their onset.

Associated abdominal pain may suggest an anatomical site of pathology (e.g. left iliac fossa pain is common with disease of the sigmoid colon).

Fig. 5.11 Causes of change in bowel habit.

Causes of change in bowel habit	
Condition	**Features**
colorectal carcinoma	weight loss; chronic history; blood in the stool
inflammatory bowel disease	Crohn's disease; ulcerative colitis; ask about systemic manifestations (e.g. arthropathy, oral ulcers, weight loss, etc.)
diverticular disease	very common; older patients; hard to diagnose from the history and examination alone
colonic polyps	may have mucoid discharge
infective colitis	usually acute, explosive history
irritable bowel syndrome	colicky abdominal pain; bloating; mucus; related to stress; absence of any sinister features in the history; very common

Enquire about the presence of blood in the stool (Fig. 5.12). The colour and relationship to the stool may reveal its origin, as follows:

- Bright red blood on the surface of the stool occurs with rectosigmoid lesions (e.g. polyp, carcinoma) or haemorrhoids.
- Red blood mixed with the stool is a feature of colorectal lesions (e.g. polyp, carcinoma, inflammatory bowel disease, diverticular disease).
- Altered blood or clots almost always imply significant pathology (e.g. colorectal lesion such as polyp, carcinoma, inflammatory bowel disease, diverticular disease).

Enquire about the presence of mucus or slime in the stool. If it is associated with blood, the most likely causes are inflammatory bowel disease or colorectal carcinoma. If mucus or slime occurs in isolation, irritable bowel syndrome may also be a cause.

Finally, ask about other characteristics of the stool. For example:

- Reduced calibre/diameter stools occur in low strictures.
- Fatty, floating, difficult to flush, offensive stools suggest steatorrhoea.
- Pellet-like or 'stringy' stools occur in diverticular disease or irritable bowel syndrome.

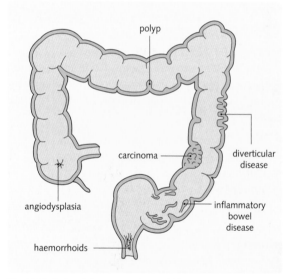

Fig. 5.12 Potential sources of rectal bleeding.

Past medical history

A detailed history is essential, but previous surgical or medical problems may elucidate the cause of the change in bowel habit, for example:

- Previous colonic polyps, abdominal surgery.
- Thyroid disease.
- Malabsorption syndromes (e.g. pancreatitis).
- Diabetes mellitus (autonomic neuropathy).

44

Drug history

Many drugs can cause a change in bowel habit, for example:

- Constipation – opiates, anticholinergic agents, tricyclic antidepressants.
- Diarrhoea – thyroxine, laxative abuse, magnesium salts, broad-spectrum antibiotics (specifically consider pseudomembranous colitis).

Family history

Some diseases causing a change in bowel habit have a genetic component, for example:

- Familial adenomatous polyposis.
- Inflammatory bowel disease.
- Carcinoma of the bowel.

Social history

Ask about foreign travel for amoebiasis, giardiasis and typhoid. If there is unexplained diarrhoea, consider the patient's risk factors for HIV infection.

DYSPHAGIA

Detailed history

Dysphagia refers either to difficulty in swallowing or to pain on swallowing. Although the cause usually requires specific investigations (e.g. barium swallow, endoscopy and biopsy), the history is important in directing these investigations. The main causes of dysphagia are indicated in Figure 5.13.

What does the patient mean by dysphagia?

It is important to clarify exactly what patients mean when they say that they have difficulty in swallowing. It is not acceptable to write 'Patient complains of dysphagia' in the medical notes.

True dysphagia almost always indicates the presence of an organic lesion. It is important to distinguish dysphagia from 'globus hystericus' (the sensation of a lump or fullness in the throat associated with chronic anxiety states).

How bad is the dysphagia?

Weight loss is a useful indicator of a serious underlying organic disorder and should always be asked about specifically.

Try to assess the functional impact. Dysphagia often progresses from solid food to soft food and liquid. Ask the patient to describe exactly which foods cause difficulty. Ascertain whether there is complete obstruction (e.g. regurgitation immediately after attempting to swallow food, vomiting).

Duration and time course of symptoms

Ask patients how long they have had difficulty swallowing.

- Malignancy often presents over weeks or months and is typically progressive.

Causes of dysphagia in the oesophagus	
Type of lesion	**Example**
obstruction within the lumen	carcinoma of the oesophagus; peptic stricture; foreign body; lower oesophageal ring
extrinsic compression of the oesophagus	mediastinal lymphadenopathy
motility disorder of the oesophagus	achalasia of the oesophagus; oesophageal spasm; scleroderma; Chagas' disease; diabetic autonomic neuropathy

Fig. 5.13 Causes of dysphagia in the oesophagus.

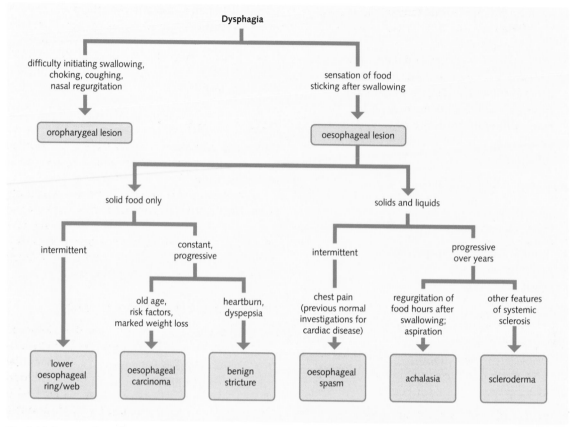

Fig. 5.14 Features from the history to aid the differential diagnosis of dysphagia.

- An oesophageal ring may present over a similar time course, but produces a more intermittent pattern.
- Other causes may be present for years without any obvious systemic disturbance (e.g. globus hystericus).

Clues to underlying pathology

Specifically enquire about previous dyspepsia, proven peptic ulcer disease or reflux. Ask about symptoms of heartburn such as acid taste in the mouth, retrosternal burning, relationship to posture. These symptoms suggest the presence of a benign oesophageal stricture.

Look for risk factors for oesophageal cancer, such as:

- Cigarette smoking.
- Barrett's oesophagus.
- Old age.

- Heavy alcohol use.
- Significant weight loss.

Dysphagia to solid foods alone is suggestive of a mechanical obstruction whereas dysphagia to liquids to a greater extent than solids suggests a neuromuscular cause.

Ask the patient where the food appears to get stuck; this is above the level of the lesion. Symptoms such as difficulty initiating swallowing, coughing, choking or nasal regurgitation suggest an oropharyngeal pathology. A sensation of food sticking after swallowing suggests an oesophageal lesion (Fig. 5.14).

ACUTE RENAL FAILURE

Presenting complaint

Patients may present in various ways, and these include the following:

- With symptoms directly referable to the renal tract (relatively rare presentation) (e.g. with haematuria, loin pain).
- With the consequences of renal failure (e.g. oedema, uraemic symptoms, hypertension).
- As an incidental finding from laboratory investigation (e.g. biochemical profile from investigation of other disease).

Detailed history

Ask specifically about uraemic symptoms as these may indicate the need for haemodialysis. Ask about:

- Nausea, vomiting.
- Anorexia.
- Malaise, lethargy.
- Pruritus.
- Hiccupping.

Note that many of these symptoms are non-specific.

Specifically enquire about symptoms referable to the urinary tract, as these may indicate the aetiology of the renal dysfunction, for example:

- Prostatism – may suggest outflow obstruction.
- Haematuria – enquire specifically about the colour if haematuria is present (often described as 'like cola' (or 'smoky' in textbooks) in glomerulonephritis; bright red usually implies lower urinary tract bleeding).
- Dysuria, frequency (may suggest infective aetiology).
- Oliguria or anuria (may suggest prerenal disease or severe renal failure).

Complications of renal failure

These may be present, for example:

- Peripheral oedema.
- Hypertension.
- Dyspnoea due to pulmonary oedema.

Duration of disease

It is often difficult to elicit the duration of the disease, as often the symptoms begin insidiously and are usually very non-specific. Clues may be obtained by asking specifically about, for example, change in weight or fatigue. Ask 'When did you last feel completely well?' or 'Have you been more tired than usual lately?' This may date the onset of renal failure, but usually renal pathology remains clinically silent until decompensation occurs or it is discovered incidentally. However, a meticulous history may help date the original renal insult in different circumstances.

Hospitalized patients

Most cases of acute renal failure occur in hospital. Create a flow chart of the blood results (especially biochemical profile) dating back to the decline in renal function. It is usually possible to identify within one or two days when the creatinine started to rise. At this point, focus on events that might have provided a critical insult to the kidneys, such as a period of hypotension or dehydration, toxic levels of aminoglycosides, coexisting infection.

New referral from the community

Almost invariably, there is less immediate information to chart, but it is still essential to obtain historical records of previous renal function. Clues may be obtained from various sources (e.g. previous blood test results, results of urinalysis for women who have previously been pregnant). The GP is a source of useful information. A quick telephone call can reveal all.

Past medical history

As the causes of renal failure are numerous, detailed past medical history is essential. In particular, you need to consider:

- Diabetes mellitus – duration, and presence of neuropathy or retinopathy, which are almost invariably associated with diabetic nephropathy.
- Hypertension – did it predate or postdate renal dysfunction?
- Risk factors for renovascular disease, for example claudication, aortic aneurysm, ischaemic heart disease, hypercholesterolaemia.
- Childhood enuresis or frequent urinary tract infections suggesting reflux nephropathy.
- Renal stones or colic.
- Autoimmune diseases, for example systemic lupus erythematosus (SLE), rheumatoid arthritis, scleroderma.
- Jaundice, for example hepatitis B, hepatitis C-associated glomerulopathy, leptospirosis, hepatorenal syndrome.

Mechanisms of drug-induced renal dysfunction	
Pathology	**Drugs**
decreased renal perfusion	diuretics* (hypovolaemia); NSAIDs* (also cause interstitial nephritis, hyperkalaemia, and rarely papillary necrosis)
decreased glomerular filtration pressure	ACE inhibitors*
nephrotic syndrome	gold; penicillamine
acute tubular necrosis	aminoglycosides* (especially if toxic drug levels); antibiotics (e.g. cephalosporins); contrast agents (especially in diabetics); chemotherapy (e.g. cisplatin)
interstitial nephritis	NSAIDs*; antibiotics* (e.g. penicillin, sulphonamides)
renal stones	cytotoxic agents (especially in lymphoma)
electrolyte disturbances	diuretics* (especially hypokalaemia); renal tubular acidosis (acetazolamide); inappropriate ADH secretion (carbamazepine, chlorpropamide); hyperkalaemia (NSAIDs, ACE inhibitors, diuretics acting on distal tubule)
retroperitoneal fibrosis	methysergide

Fig. 5.15 Mechanisms of drug-induced renal dysfunction. Asterisks indicate the more commonly implicated drugs. ACE, angiotensin-converting enzyme; ADH, antidiuretic hormone; NSAID, non-steroidal anti-inflamatory drug.

- Recent infections, for example postinfectious glomerulonephritis (rare), presentation of IgA nephropathy with haematuria following a sore throat.

Drug history

Again a detailed drug history is essential. Very often, drugs have precipitated the renal failure. Remember to ask about over-the-counter medication and herbal remedies.

Drugs may precipitate renal failure by various mechanisms (Fig. 5.15). Other drugs must be used with caution in renal failure, for example:

- Renally excreted drugs (aminoglycosides).
- If there is an accumulation of metabolites due to failure of clearance (opiates).

Family history

Consider inherited conditions such as polycystic kidneys, Alport's syndrome.

Social history

It is essential to obtain as much background information as possible to assess normal functional capacity. In addition, consider whether the patient's lifestyle may have contributed to the renal failure, for example:

- Cigarette smoking (renovascular disease).
- Alcohol (hepatorenal disease).
- Risk factors for HIV and hepatitis B and C (glomerulonephritis, hepatorenal disease).
- Travel and ethnic origin – many forms of glomerulonephritis demonstrate great geographical variation (e.g. IgA nephropathy is more common in Caucasians, SLE is more common in Afro-Caribbeans).

Review of systems enquiry

Vital information may be omitted if a systems enquiry is not performed. In particular, consider:

- Symptoms suggestive of autoimmune aetiology – skin rash, arthralgia, myalgia, alopecia, early morning stiffness.
- Fevers – any infective or inflammatory disease.

CHRONIC RENAL FAILURE

Presenting complaint

It is assumed that the patient will already be on dialysis or is being reviewed in the predialysis clinic, and that the cause of renal disease has already been investigated (see previous section).

Detailed history

Dialysis

Assess symptoms indicative of inadequate dialysis or need to commence dialysis, such as:

- Anorexia.
- Nausea, vomiting.
- Fatigue, malaise.
- Pruritus.
- Confusion, drowsiness.

Although many of these symptoms are non-specific, if no other cause is found, assume that they represent uraemia.

Dialysis-related problems

Dialysis is associated with a number of specific problems or issues which need to be considered, whether the form of dialysis is peritoneal dialysis or haemodialysis. Review the mechanics and complications of dialysis (Fig. 5.16).

Fluid balance

There are many common problems of chronic renal failure and these should always be addressed. Review fluid balance as this is central to the management. It is essential to ask about the following:

- Does the patient have a target 'dry' weight? Fluctuations from this weight in the short term usually indicate fluid shifts.
- Urine output and daily fluid restriction, which is usually 500 mL more than daily urine output.

- Interdialytic weight gains. Large gains may indicate poor compliance and understanding of self-management.

Anaemia

Review symptoms (e.g. dyspnoea, lethargy, decreased exercise tolerance). Many patients will be taking recombinant erythropoietin. Always ask specifically about and record:

- The dose.
- Side effects (e.g. hypertension, hyperkalaemia).
- Reasons for lack of response to erythropoietin. (e.g. iron deficiency, intercurrent infection, hyperparathyroidism).

Renal osteodystrophy

Renal osteodystrophy is a common problem of chronic renal failure and should be explored. From the notes and patient account, review:

- Calcium and phosphate balance.
- Diet.
- Calcium carbonate dose.
- Biochemical evidence for hyperparathyroidism.

Transplant status

Review plans for discharging the patient from this form of dialysis. In particular, consider (at every visit) the appropriateness for transplantation, taking into account the patient's wishes and knowledge. Review intercurrent medical issues that may preclude transplantation: for example, age, infection, malignancy, severe vascular disease, untreated ischaemic heart disease, active peptic ulceration.

Features of peritoneal dialysis and haemodialysis to elicit from the history		
Parameter	**Peritoneal dialysis**	**Haemodialysis**
mode of dialysis	continuous ambulatory peritoneal dialysis (CAPD); automated peritoneal dialysis (APD)	hospital haemodialysis; home haemodialysis
dialysis dose	number and type of bags (e.g. 'light/heavy'); volume of fluid (typically 2L)	hours on dialysis; frequency (typically 4 hours three times weekly)
access	PD exit site	arteriovenous (AV) fistula; temporary catheter (e.g. 'vascath'); AV shunt
complications of dialysis	peritonitis; exit site and tunnel infections	dialysis disequilibrium; hypotensive episodes; difficulty needing fistula; exit site infections; vascular stenosis

Fig. 5.16 Features of peritoneal dialysis (PD) and haemodialysis (HD).

Blood pressure control

Review the documentary evidence of blood pressure measurements, for example from dialysis charts or recordings made at home (before erythropoietin dosing) or by the GP.

Past medical history

This is usually well known. Do not forget the original cause of the renal failure!

Drug history

A meticulous drug history is essential. Very often, the patient will have a long list of medication. Enquire specifically about:

- Antihypertensive agents.
- Erythropoietin (see above).
- Phosphate binders and vitamin D.
- Iron supplements.
- Over-the-counter medication.

Consider drugs to be used with caution in renal failure (see Acute renal failure, above).

Social history

A detailed assessment should be made, especially in the predialysis patient, when considering whether dialysis would be appropriate and, if so, which form.

If considering haemodialysis:

- Will transport be needed to the hospital?
- Does the patient have motivation and space at home, and a partner for home haemodialysis?

If considering chronic ambulatory peritoneal dialysis (CAPD):

- Does the patient have space at home to store boxes containing peritoneal dialysis fluid?
- Does the patient have the manual dexterity needed to change bags?
- Does the patient have the motivation and understanding to manage the care so that there is not an undue risk for developing peritonitis or exit site infections?
- Is the patient obese?

Review the patient's nutrition and diet. This is often specialized, and referral to a renal dietician is indicated.

Review of systems enquiry

This is particularly important as often these patients are multisymptomatic and renal failure is associated with so many other diseases (e.g. ischaemic heart disease, arthritis, GI bleeding).

Chronic renal failure results in multisystem dysfunction. Assessment needs to be detailed. It is pointless to rush. Always allow enough time. In predialysis patients, check that adequate plans have been made for the initiation of dialysis so that the transition can be as smooth as possible. Consider patient education, mode of intended dialysis, access for dialysis and estimated time to dialysis initiation.

HAEMATURIA

Detailed history

Haematuria is a common symptom and may be due to a wide variety of pathologies (Fig. 5.17). Take a full history of the presenting symptom.

Ascertain that true haematuria is present

Some patients with uterine bleeding mistakenly believe that they have haematuria.

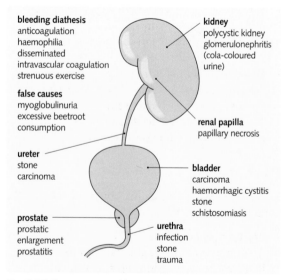

bleeding diathesis
anticoagulation
haemophilia
disseminated
intravascular coagulation
strenuous exercise

false causes
myoglobulinuria
excessive beetroot
consumption

ureter
stone
carcinoma

prostate
prostatic
enlargement
prostatitis

kidney
polycystic kidney
glomerulonephritis
(cola-coloured
urine)

renal papilla
papillary necrosis

bladder
carcinoma
haemorrhagic cystitis
stone
schistosomiasis

urethra
infection
stone
trauma

Fig. 5.17 Some causes of haematuria.

Duration

Note whether the haematuria is an acute presentation or has been present for many years or months.

Nature

A small amount of blood creates visible discoloration of the urine. Try to establish how much blood is present in the urine. It may be helpful to ask:

- 'Are there any blood clots in your urine?'
- 'Is your urine bright red or stained like blackcurrant juice?'
- 'Does your urine appear cloudy or like cola?' (glomerulonephritis).

The timing of blood during the urinary stream may provide a clue to the origin of bleeding. For example:

- Bleeding at the start of the urinary stream suggests a urethral lesion.
- Bleeding through the whole stream suggests a source in the bladder or higher in the urinary tract.
- Bleeding at the end of the urinary stream suggests a source in the lower bladder.

Associated urinary symptoms

Other symptoms referable to the urinary tract often provide useful clues to the cause of haematuria. Specifically ask about:

- Dysuria and frequency with small quantities of urine – urinary tract infection.
- The above symptoms in association with fever and loin pain – suggests pyelonephritis.
- Colicky loin pain–indicative of renal stones.
- Symptoms of renal disease such as ankle swelling or uraemic symptoms.
- Terminal dribbling, hesitancy and poor stream – common in prostatic obstruction.

Past medical history

A detailed past medical history is important. In particular, ask about:

- Previous renal disease.
- Abdominal trauma (e.g. renal capsular tear).
- Renal stones or previous episodes of colic.
- Previous cystoscopies.

- Prostatectomy (in men).
- Sickle cell anaemia (papillary necrosis).

Drug history

Some drugs may aggravate or cause haematuria. For example:

- Cyclophosphamide (haemorrhagic cystitis, carcinoma of the bladder).
- Warfarin (bleeding diathesis).
- Analgesic abuse (papillary necrosis).

Family history

Many renal diseases have a familial tendency. For example:

- Polycystic kidney disease (adult variety is autosomal dominant).
- Alport's syndrome (X-linked recessive). Ask about deafness.
- Immunglobulin A nephropathy. Ask about the relationship of macroscopic haematuria to infections.
- Sickle cell anaemia.

OBSTETRIC AND GYNAECOLOGICAL HISTORIES

Taking an obstetric or gynaecology history does not differ from the standard history format outlined in Chapter 2. However, there are some specialist questions that need to be asked. It is important to remember that this is a sensitive area for many women who may not feel comfortable discussing this area with medical students, especially male medical students.

The gynaecological history

This should include the following points.

Menstrual history

- Ask about the menarche and, if appropriate, the menopause.
- When was the first day of the last menstrual cycle?
- Ask about the pattern of bleeding and the length of time between cycles.

It is important to remember that the menstrual cycle is from the first day of bleeding to the next first day of bleeding. Try to establish the pattern of bleeding. How many days does this last for; what kind of protection is used; how often does protection require changing; does the woman ever flood through her protection; and does she ever pass clots?

- Ask the patient about menstrual pain; for example does the pain occur before menstruation starts and is relieved by it, or is menstruation itself painful? (Mittelschmerz (middle pain) is the name given to pain around the time of ovulation (day 14 of the cycle): this is unlikely to be pathological.)

- Try to determine the severity and what functional impact this has on the woman's life. What analgesics does she use and for how long? Does she have any other associated cyclical symptoms?

Try to remember at all times what is happening to your patient's hormone levels throughout the normal menstrual cycle (Fig. 5.18).

Contraception and sexual history

A contraceptive history should be taken. This should include which methods have been tried and how suitable the woman felt they were for her.

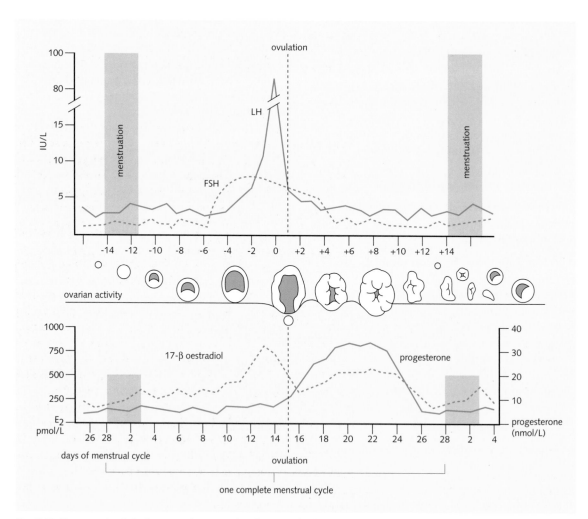

Fig. 5.18 Hormone levels in the normal menstrual cycle. Considerable variations can be found that are still compatible with normal menstrual function. FSH, follicle stimulating hormone; LH, luteinizing hormone.

If appropriate, you may need to ask about the woman's sexual history. Is she sexually active? Do not make assumptions about her sexuality or sexual orientation. This is a very sensitive area and great care should be taken when asking about sexual history. It is often a good idea to leave this until the end of the interview when you have had the most amount of time to build up a rapport. This should include the number of partners she has had over the past year and the sex of the partners, how often she has had intercourse and, if appropriate, whether or not she has had dyspareunia. You should also ask if she has had any sexually transmitted diseases.

Smear tests

You should also ask when the patient's last smear test was. Has she been attending a clinic and has she had any abnormal smears? What treatment did she require for these? The UK criteria for smear screening has recently changed: make sure you know when screening commences and how often this is repeated (www.cancerscreening.nhs.uk/cervical).

At present, the screening intervals are:

- age 25 years – first invitation
- age 25–49 years – 3 yearly
- age 50–64 years – 5 yearly
- age 65+ years – women in this age group are only screened if they haven't been seen since age 50 or have had recent abnormal tests.

Vaginal discharge and continence

Symptoms of any vaginal discharge should be recorded. When does it occur (when during the cycle, post coital or post menopausal), how much is there and what is its nature (consistency, colour, smell and is there any blood)?

Urinary abnormalities should be asked about, e.g. frequency, nocturia and dysuria. Are there any symptoms of stress incontinence (usually a consequence of lax pelvic musculature following childbirth) or urge incontinence (usually due to detrusor instability)?

The obstetric history

Pregnancy is usually a happy time for people, but remember there may also be a lot of anxiety, especially if previous pregnancies have been lost. You should find out about any possible current pregnancy first. Ask about the last menstrual period and,

if pregnancy is confirmed, use a date wheel to calculate the expected due date. Was there any difficulty in getting pregnant? Ask if she has had any of the common symptoms of pregnancy, for example early morning sickness. You may ask about any cravings. Has the woman felt any fetal activity yet (be careful to ensure this is an appropriate question or you may cause maternal anxiety)?

Previous obstetric history

- How many times has she been pregnant and how many children does she have? This is a delicate way of starting to find out about miscarriages and abortions.
- Are all the children fathered by the same man? If not, clarify the paternities.
- When the children were born, what was their gestation and did any of them have to go to a special care baby unit?
- Have there been any fetal abnormalities or any postnatal complications?
- During any previous pregnancies, has she had any medical problems, for example pre-eclampsia or diabetes?
- Were her previous labours difficult (what did she use for analgesia and was this satisfactory)?
- Did she require instrumental deliveries or caesarian sections?
- Has she had any postnatal psychological problems, for example postnatal depression.

You should also take a gynaecological history as outlined above.

When asking about medications the patient is on, ask whether or not she is taking folic acid.

It is important to take a family and a social history. Particular attention should be paid to any inherited diseases that run in the family, for example cystic fibrosis. It is also useful to know if there is a history of twins in the family. Ask the woman about her marital status. Ask if she is employed and how long she intends to work while pregnant. Is the home suitable for children or will she need to be re-housed? Does she still drink alcohol? If so, how much? Does she smoke? Again, if so, how much? The woman should be encouraged to stop or at least cut down on smoking and drinking. This is also a good opportunity to find out if she is considering breast feeding.

When dealing with pregnant women, remember that pregnancy is not a disease state. Most pregnancies are uncomplicated and don't require to be medicalized. At times, we can become biased in our views of pregnancy because we see and remember the problematic patients. You will need to decide for yourself what you feel about medical interventions in pregnancy and what an appropriate level of medical involvement is.

Further reading

Panay N, Dutta R, Ryan A et al 2004 *Crash course: obstetrics and gynaecology*. St Louis: Mosby

Rockall TA, Logan RF, Devlin HB et al 1996 Selection of patients for early discharge or outpatient care after acute upper gastrointestinal haemorrhage. National Audit of Acute Upper Gastrointestinal Haemorrhage. *Lancet* 347(9009): 1138–1140

Objectives

By the end of this chapter you should:

• Understand the potential problems in taking a neurological history.
• Demonstrate an understanding of some of the terminology patients may use and what this may mean.
• Realise that diagnosis of neurological disorders can be difficult and adopt a pragmatic approach.

The specialty of neurology is littered with epony-mous syndromes, but, as a student, your focus should be on the most common and serious conditions. With this in mind, this chapter has been divided into broad categories representing common presentations. For a more in-depth look at neurological conditions, investigation and how to find the level of the lesion, the reader is referred to Turner (2005); see Further reading.

THE UNCONSCIOUS PATIENT

Detailed history

The history is especially important in the evaluation of the unconscious patient, since it does not usually come from the patient. Obtain the history from relatives or friends, the ambulance crew or police, if appropriate. Try to establish the following.

Time of onset of unconsciousness

Who found the patient unconscious, and when? When was the patient last seen conscious? Where was the patient found?

Duration of illness preceding unconsciousness

Had the patient been well before being found unconscious? Was the illness sudden (e.g. minutes), gradual (hours) or chronic (days to weeks)?

Nature of the preceding illness

It is helpful to consider the differential diagnosis, so that questions can be more focused (Fig. 6.1).

Past medical history

Obtain a full history. There may have been previous episodes. Enquire specifically about:

• Diabetes mellitus (hypoglycaemia, hyperglycaemic coma).
• Risk factors for cerebrovascular disease (e.g. stroke; see Ch. 3 for vascular risk factors).
• Epilepsy and other neurological disorders.
• Head trauma – no matter how mild and how much in the past; a subdural haematoma may be preceded by a history of a trivial head injury, especially in the elderly.
• Preceding headaches – for example, meningitis, intracranial mass lesion, subarachnoid haemorrhage.
• Renal failure.
• Liver failure.
• Vomiting.

Drug history

Consider all drugs that may depress the conscious level if taken in therapeutic or toxic amounts. Remember to ask about analgesic agents and psychotropic medication.

Social history

This is particularly important, especially in younger patients. Ask specifically about:

• Alcohol (very important).
• Recreational drugs (very important).
• Possible reasons for deliberate self-harm.

Clues from the history on the underlying cause of unconsciousness	
Description	Indications

vascular*
  subarachnoid haemorrhage*
  intracerebral bleed*
  massive infarction*
  brainstem stroke*
	preceding headache; sudden onset; often young adult; may have had 'herald bleed'
	risk factors for cerebrovascular disease
	(hypertension, diabetes mellitus (DM), ischaemic heart disease, age, family history, etc.)

metabolic*
  DM*
 	
  drugs and toxins*
  hypoxia,
  hyponatraemia
  hypothyroidism*
  uraemia,
  hepatic encephalopathy
	hyperosmolar coma (type II DM); diabetic ketoacidosis
	(type I—may be presenting feature of disease); hypoglycaemia
	alcohol*; sedative drugs (opiates, benzodiazepines, barbiturates, etc.)
	often present non-specifically

sepsis
  generalized,
  meningoencephalitis
  brain abscess
	usually preceding illness; ask about rash, photophobia, fevers, headache, vomiting, etc.

subdural*/extradural haematoma	may be history of trauma (often absent)
postictal	may find bottle of anti-epileptic tablets
intracranial mass lesion	ask about features of raised intracranial pressure (e.g. increasing morning headache, vomiting, developing focal neurological problems)
factitious/hysteria	often unusual presentation; past psychiatric history

Fig. 6.1 Clues from the history on the underlying cause of unconsciousness. Asterisks indicate the more common causes.

Do not underestimate the importance of the history, even if the patient cannot provide one directly. The differential diagnosis is broad, but can usually be narrowed down with the aid of a well-taken history from a third party. Detailed history taking often needs to be deferred until appropriate resuscitation or stabilization has been carried out.

BLACKOUTS

Detailed history

A common problem for admitting medical teams is the investigation of a patient who has had a blackout. As usual, the history is central to the diagnosis. It is imperative to find out what the patient means by the term 'blackout'. Follow the usual systematic approach to investigate circumstances of the blackout. It is helpful to consider the differential diagnosis of blackouts (Fig. 6.2).

Investigate the episode chronologically; this will require excellent communication skills as the patient may be unsure or confused due to the cause of the blackout. Find out what the patient was doing immediately before blacking out, whether the patient had any warning symptoms and how the patient felt immediately after regaining consciousness (Fig. 6.3).

Did anyone witness the episode?

This is probably the most useful piece of information. If so, ask specifically about:

- How long the blackout lasted (seconds, minutes or hours).
- What the patient was doing during the episode (e.g. lying still, shaking, appearing confused, purposeful movements)?

Differential diagnosis of blackouts
epilepsy*
decreased cerebral perfusion   vasovagal episode*; cardiac disturbances* (e.g. arrhythmia, aortic stenosis, ischaemia); postural hypotension*; TIA (especially in posterior circulation); micturition syncope (decreased venous return during breath holding); cough syncope (decreased venous return); carotid sinus hypersensitivity
metabolic disturbances   hypoglycaemia; hypocalcaemia
psychological   panic attacks*; hyperventilation*; factitious
drugs   alcohol*; recreational drugs of abuse; prescribed medication (e.g. decreased threshold for epileptic fit, sedative, β-blockers provoking profound bradycardia, etc.)

Fig. 6.2 Differential diagnosis of blackouts. Asterisks indicate the more common causes. TIA, transient ischaemic attack.

Clues from the history on the underlying pathology responsible for an episode of loss of consciousness	
Clues from the history	**Possible underlying cause of blackout**

'What were you doing immediately before blacking out?'
    standing up quickly
    turning head sharply
    trauma
    completely at rest
    standing still in hot environment
	
	postural hypotension
	cervical spondylosis (occlusion of vertebral artery)
	subdural haematoma; extradural haematoma; contusion injury
	arrhythmia; cerebrovascular disease, etc.
	vasovagal

'Did you have any warning that you were going to black out?'
    aura
    palpitations, chest pain
    lightheadedness
    sweating, hunger
	
	epileptic fit
	cardiogenic; panic attack
	vasovagal, etc.
	hypoglycaemia

Fig. 6.3 Clues from the history on the underlying pathology responsible for an episode of loss of consciousness.

- The presence of any incontinence or shaking to suggest an epileptic fit.
- Whether anyone felt the pulse either during or immediately after the blackout. A normal pulse during the blackout could exclude an arrhythmia. Remember to evaluate the competence of the person who felt the pulse.

Try to establish whether the episode was a true syncopal attack (loss of consciousness and motor tone) or just a period of lightheadedness. Very often, people say that they have blacked out when they do not completely lose consciousness. Ask, for example, 'Did you lose any time?' or 'Were you out cold?'

Enquire about the immediate period following recovery of consciousness. Symptoms at this stage may give clues to the precipitating event, for example:

- Immediate recovery – vasovagal.
- Confusion and disorientation – for example, postictal.
- Weakness – Todd's paresis, transient ischaemic attack (TIA).

Past medical history

Ask about previous blackouts and investigations performed. Consider clues from the past history that may increase the probability of certain underlying problems, for example:

- Risk factors for epilepsy – for example, head injury, cerebrovascular disease, meningitis.
- Cardiac diseases.
- Diabetes mellitus – enquire about medication, diabetic control and previous episodes of hypoglycaemia.

Drug history

A full drug history is essential. In particular, consider:

- Recent changes to prescribed drugs.
- Negative chronotropic and inotropic agents (e.g. β-blockers).
- Drugs likely to cause arrhythmias (e.g. tricyclic antidepressants, theophylline).
- Insulin (hypoglycaemia).
- Antihypertensive agents (postural hypotension).
- Glyceryl trinitrate (GTN) syncope.
- Illicit drug use.

Social history

Investigate whether the home environment is safe for someone who may blackout unexpectedly (e.g. are there any carers at home?). The patient's lifestyle may suggest underlying risks for a blackout (e.g. alcohol consumption, unusual stresses at home or work).

Review of systems enquiry

Blackouts may result from a wide range of pathologies, so a review of systems may reveal unexpected pathology. Focus on the cardiovascular, neurological, metabolic and locomotor systems.

The key to diagnosing the cause of a blackout is a well-taken history. The most useful information comes from an eyewitness account.

HEADACHE

A detailed guide to diagnosis and management of headaches from the British Association for the Study of Headache can be found at www.bash.org.uk. An easier introduction can be found in Fuller & Kaye (2007); see Further reading.

Detailed history

Consider the following common causes of single and recurrent headaches:

- Tension-type headache (by far the most common).
- Migraine (common).
- Hangover (apparent from the history).
- Analgesic-induced headache.
- Subarachnoid haemorrhage (rare, but consider for any sudden-onset headache).
- Meningitis, encephalitis.
- Raised intracranial pressure (e.g. tumour, hydrocephalus).
- Temporal arteritis.

Headache is a universal condition. The principal aim is to distinguish non-serious, benign headaches from those that may represent serious underlying pathology and therefore require further investigation. Be wary that only appropriate reassurance is given to the patient.

Always consider headaches representing serious disease (see Fig. 6.5 for a diagnostic algorithm).

Red flags may include:

- Scalp tenderness, proximal limb stiffness – temporal arteritis
- Rash, fever, photophobia, neck stiffness – meningitis
- Sudden-onset occipital headache – subarachnoid haemorrhage
- Early morning headache, nausea – raised intracranial pressure
- 'Worse headache of my life'
- New-onset headache in patient over 50.

For headaches presenting in A&E, subarachnoid haemorrhage should be ruled out. The most common mistake is to diagnose them as migraine; a high degree of suspicion is therefore required.

For any new-onset headache, always ask specifically about photophobia, neck stiffness, rash, fever and vomiting.

Obtain a detailed history about the frequency of headaches; for example recurrent headaches are typically tension headaches or migraine, while a headache every morning can be due to raised intracranial pressure associated with a tumour.

Obtain information about the onset, nature and location of headache. It is often possible to identify the cause of a headache from these parameters (Figs 6.4 and 6.5).

Features of different types of headache	
Headache	Characteristics
tension headache	most common recurrent headache, typically described as 'throbbing', 'pressure', etc. often identifiable precipitating factor (e.g. stress, depression)
migraine	common cause of recurrent headache; usually presents in young adults; prodrome—often visual (e.g. scotomata, teichopsia), tingling, etc.; headache—often starts unilaterally; associated symptoms (e.g. nausea, vomiting, photophobia, etc.)
subarachnoid haemorrhage	may have had 'herald bleed' with milder 'subclinical' episodes; typically sudden onset 'like a hammer hitting the back of my head'; may have associated neurological deficit or decreased level of consciousness
hangover	preceding alcohol consumption; associated nausea
meningitis	photophobia; neck stiffness; fever; rash
raised intracranial pressure	usually subacute onset; relentless; present on waking; aggravated by coughing, sneezing, stooping, associated nausea
temporal arteritis	pain over superficial temporal arteries, especially on touching area (e.g. combing hair); may have associated malaise, proximal muscle weakness, and stiffness, and visual loss; usually older than 60 years

Fig. 6.4 Features of different types of headache.

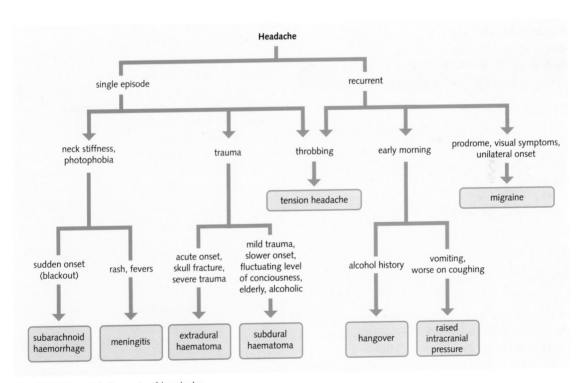

Fig. 6.5 Differential diagnosis of headache.

Past medical history

Ask specifically about previous malignancy if raised intracranial pressure is suspected.

Drug history

Ask about medication taken to relieve symptoms and efficacy. It is often hard to assess the severity of headache as people can rarely give a quantitative description. It may be worth asking 'If 0 is no pain, and 10 is the worst headache you have ever had, what score would you give this pain?' Ask girls and women about the use of oral contraceptives (may precipitate or worsen migraine, particularly in the pill-free week).

Social history

For recurrent headaches, it is essential to ask about stressors, for example at home and work, that may be precipitating tension-type headaches. The patient's alcohol history is also relevant. Eliciting the patient's ideas and concerns about the headache is especially important as many are worried about serious pathology, for example brain tumour or impending stroke.

EPILEPTIC SEIZURE

Presenting complaint

Epilepsy may present as an unwitnessed blackout (single or recurrent) or as an eyewitness account of a fit. More rarely it may present with behavioural changes.

Detailed history

Start with the patient's own recollection of the event, recording the following chronologically.

Prodrome

Ask 'Did you have any warning that you were going to black out?' Typical auras include sensations that the patient may find difficult to describe, including strange thoughts, emotions or hallucinations (e.g. smell, taste, vision). They often follow a predictable course for the individual patient.

Seizure

Ask 'Tell me what you remember about the attack' and 'Did you blackout and lose consciousness?' The answers to these questions may help distinguish between generalized (loss of conciousness) and partial (no loss of conciousness) seizures. If the patient was conscious during the episode, ask him or her to describe exactly what happened. If the patient has experienced a partial seizure, the symptoms are often characteristic of the site of epileptic focus, for example:

- 'Jacksonian' motor seizures.
- Temporal lobe epilepsy (déjà vu, jamais vu, hallucinations of smell or taste, etc.).

It is usually possible to distinguish partial seizures, generalized seizures and partial seizures with secondary generalization.

Postictal period

Ask 'How did you feel when you came around?' It is usual to experience a headache, 'muzziness', lethargy, confusion and non-specific malaise, etc. A focal weakness may be present for up to 24 hours (Todd's paresis).

Obtain an eyewitness account if possible. This is invaluable, and it can provide information that clinches the diagnosis.

If the episode was a single blackout, the most common difficulty is distinguishing between epilepsy and syncope. Look for an identifiable precipitant for syncope (e.g. emotional stress, prolonged standing, cough, micturition syncope), gradual onset and recovery, and pallor and flaccidity during the episode. Beware that a convulsion or urinary incontinence only rarely occur during syncopal episodes.

Generalized seizures imply widespread abnormal electrical activity in the brain while partial seizures imply a discrete area of abnormal electrical activity which may or may not spread. In common with many other neurological events, try to obtain a good eyewitness account of a typical episode, as very often the patient is asymptomatic and has no physical signs on presentation to a doctor. Ask about previous seizures, when they occurred, what happened and their frequency. It is important to build up a picture of a typical seizure and any change in symptoms, prodrome, etc.

Historical features of different types of epilepsy	
generalized seizures	
tonic–clonic seizure (grand mal fit)*	most commonly perceived form of epilepsy—'convulsions'
absence seizures (petit mal)*	especially in childhood; brief (few seconds) loss of consciousness; no fall; no convulsion
myoclonic seizures	especially in childhood; usually symmetrical
tonic seizures	especially in childhood; loss of consciousness; usually underlying organic brain disease
akinetic seizures	no prodrome ('drop attacks')
partial seizures	
simple partial seizures*	remains fully conscious during attack (e.g. Jacksonian fit)
complex partial seizures*	originates in the temporal lobe; often complex sensory hallucinations, deja vu, jamais vu, lip smacking, chewing, behavioural disturbances; etc.
partial seizure with secondary generalization*	as above, but progresses to tonic-clonic seizure

Fig. 6.6 The history often reveals the type of epilepsy. Asterisks indicate the more common causes.

Underlying pathology causing seizures
cerebrovascular disease*
most common cause in older age group; risk factors (e.g. hypertension, smoking, ischaemic heart disease, diabetes mellitus, etc.)
alcohol and drug withdrawal*
drugs*
for example tricyclic antidepressants, phenothiazines, amfetamines, etc.
head injury and neurosurgery
tumours
encephalitis
degenerative brain disease
metabolic disorders
for example hypoglycaemia, hyponatraemia, hypocalcaemia, uraemia, liver failure, etc.
fever
NB febrile convulsions are not epilepsy

Fig. 6.7 Underlying pathology responsible for epileptic fits can often be identified. Asterisks indicate the more common causes.

On the basis of the history, try to decide on the type of epilepsy (Fig. 6.6). If the patient has had multiple fits, ask about seizure control (i.e. frequency of fits, duration of fits). Ask about precipitating factors (e.g. flickering lights). It is important to ask a patient with epilepsy about their first seizure, last seizure and any change in the nature of the seizures.

Past medical history

Consider possible underlying causes of fits (Fig. 6.7).

Drug history

Ask about drugs used to control epilepsy, and their side effects (e.g. phenytoin, sodium valproate, carbamazepine). Ask specifically about symptoms suggestive of overdose (e.g. drowsiness, ataxia, slurred speech). Review concordance, especially if there is evidence of poor seizure control, and consider the need to check drug levels (phenytoin has a narrow therapeutic index).

Consider medication that may:

- Interact with anticonvulsants (e.g. oral contraceptives, warfarin; see the *British National Formulary* Appendix 3).

- Lower the seizure threshold (e.g. phenothiazines, tricyclic antidepressants, amfetamines).

Family history

In younger patients, there is often a positive family history.

Social history

There are multiple social issues surrounding epilepsy. Often, patients feel stigmatized and socially isolated. School performance is often poor. The reasons for this should be explored (e.g. poorly controlled seizures, drug intoxication, social isolation and bullying, underlying brain disease). Parents are often understandably overprotective. Consider sensible restrictions on activities for children while allowing a full social life (e.g. avoiding known precipitants such as strobe lighting at discos, bathing in a locked bathroom or dangerous sports such as rock climbing).

Review occupational problems (e.g. use of dangerous machinery, employers who do not understand the disease).

Consider the restrictions on driving. This may affect the decision to start withdrawing medication in patients with well-controlled disease if driving is particularly important to them. Consult the DVLA *Medical Aspects of Fitness to Drive* handbook for further information.

STROKE

Strokes are common and take up large amounts of NHS resources. They are the third most common cause of death in the developed world. Eighty per cent of all strokes are embolic events and 10% of people with infarcts will die within 30 days (see Gubitz & Sandercock (2000), Further reading).

Detailed history

Establish that the described event is a stroke. It is usually obvious from the history that a stroke has occurred. Try to work out where in the cerebral circulation the stroke has occurred as this has prognostic value (Figs 6.8 and 6.9).

Symptoms and signs of stroke	
Anterior circulation strokes	**Posterior circulation strokes**

• Unilateral weakness
• Unilateral sensory loss or inattention
• Isolated dysarthria
• Dysphasia
• Vision:
  Homonymous hemianopia
  Monocular blindness
  Visual inattention
	• Isolated homonymous hemianopia
	• Diplopia and disconjugate eyes
	• Nausea and vomiting
	• Incoordination and unsteadiness
	• Unilateral or bilateral weakness and/or sensory loss
	Non-specific signs
	• Dysphagia
	• Incontinence
	• Loss of consciousness

Characteristics of subtypes of stroke				
	Lacunar	Partial anterior circulation	Total anterior circulation	Posterior circulation
Signs	Motor or sensory only	2 of following: motor or sensory; cortical; hemianopia	All of: motor or sensory; cortical; hemianopia	Hemianopia; brain stem; cerebellar
% mortality at 1 year	10	20	60	20
% dependent at 1 year	25	30	35	20

Fig. 6.8 Symptoms and signs of stroke, and characteristics of subtype.

A stroke is an acute focal neurological deficit due to a vascular lesion, lasting longer than 24 hours. A TIA is a focal neurological deficit due to a vascular lesion that resolves within 24 hours.

Onset

Obtain a chronological account of the onset and progression of neurological disability. Ask 'What were you doing immediately before this happened?' Typically it occurs at rest, but patients may wake up to find that they cannot move a limb. The onset is typically immediate, but may evolve over a few hours – 'One moment I was fine, and the next, I was unable to move my right arm'.

Define the neurological disability; for example ask 'What can't you do now that you could manage before this happened?' This is important because:

- It provides a baseline to assess recovery or subsequent deterioration.
- The anatomical site of the lesion can be identified. The neurological disability typically corresponds to the vascular territory of the occluded or haemorrhaging artery (see Figs 6.8 and 6.9).
- It allows an assessment of whether the event is likely to be a stroke or a TIA and any evidence of recovery between the onset and presentation.

Past medical history

Consider the possible aetiology of the stroke as this will affect subsequent management (Fig. 6.10). Ask specifically about previous TIAs, including transient loss of sight and peripheral vascular disease (PVD).

Drug history

Ask specifically about warfarin, aspirin, clopidogrel and heparin (usually in inpatients, e.g. perioperative strokes) if an intracranial bleed is suspected. Reviewing the list of medication may highlight a risk factor that the patient or his or her relatives may have forgotten.

Social history

An extremely detailed social assessment is necessary to ascertain if and when the patient can be

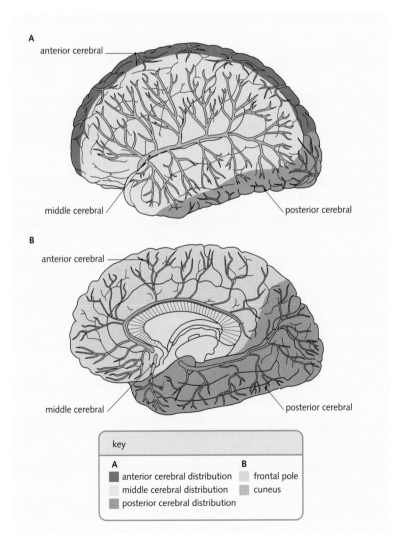

A
anterior cerebral
middle cerebral
posterior cerebral

B
anterior cerebral
middle cerebral
posterior cerebral

key

A		**B**	
■	anterior cerebral distribution	░	frontal pole
░	middle cerebral distribution	▨	cuneus
▨	posterior cerebral distribution		

Fig. 6.9 The vascular territories of the main cerebral arteries.

Common causes of stroke	
Pathology	**Features in the history**
ischaemic stroke*—thrombosis*, embolism*	by far the most common; ischaemic heart disease, diabetes mellitus, hypertension, smoking, valvular heart disease (especially mitral stenosis/prosthetic heart valve), atrial fibrillation, family history, older age; hyperviscosity (e.g. polycythaemia, Waldenstrom's macroglobulinaemia)
cerebral haemorrhage*	common; as above, but hypertension is often pronounced; headache is often a prominent feature as is a disturbed level of consciousness
extradural haematoma	severe head injury
subdural haematoma	history of head injury (often mild); alcoholic; old age
subarachnoid haemorrhage	'herald bleeds', sudden-onset headache; loss of consciousness common; meningism
vasculitis	giant cell arteritis, systemic lupus erythematosus (SLE), Wegener's granulomatosis, sarcoidosis, etc. (consider features of above diseases)
infections	syphilis, infective endocarditis (rare)

Fig. 6.10 Aetiology of strokes. Asterisks indicate the more common causes.

discharged home. Remember that discharge planning should begin at the point of admission to hospital.

The diagnosis of stroke is usually apparent on presentation, but an attempt should be made to define its aetiology and severity. There is as yet little proven medical therapy to limit the neurological deficit, though strokes presenting within three hours of onset need an emergency computed tomography (CT) and assessment for thrombolytic therapy. The bulk of treatment will be supportive to aid rehabilitation. This relies upon an adequate initial assessment and the detailed social history is essential for coordinating the various members of the multidisciplinary team (e.g. physiotherapist, occupational therapist, social worker, dietician, speech therapist).

When stroke is diagnosed, coordinate all members of the multidisciplinary team as soon as possible. Discharge planning should begin on admission to hospital.

'OFF LEGS'

Detailed history

Elderly people often present in a non-specific manner with difficulty functioning at home with their normal day-to-day activities. The ability to compensate for relatively minor physiological derangements is impaired in the elderly. The differential diagnosis is enormous, and represents one of the greatest (and most frequent) challenges in history taking. Consider some of the more common underlying disorders listed below that may present non-specifically as confusion or inability to cope at home (the more common are denoted with an asterisk):

- Infection* (e.g. urinary tract infection*, chest infection).
- Drugs* – very common (e.g. diuretics causing electrolyte disturbances, sedatives, antihypertensives).
- Metabolic – for example hypothyroidism*, hyperthyroidism, diabetes mellitus (new presentation or established disease), electrolyte disturbances* (e.g. hyponatraemia, hypernatraemia, hypercalcaemia), dehydration.

- Neurological (e.g. dementia*, Parkinson's disease, subdural haematoma*).
- Cardiovascular – for example cerebrovascular disease*, myocardial infarction* (e.g. heart failure*, hypotension*), arrhythmias* (especially atrial fibrillation*).
- Haematological (e.g. anaemia due to peptic ulceration, bowel cancer, vitamin B_{12} or folate deficiency).
- Liver disease.
- Psychiatric (e.g. depression*).

Often the history from the patient is vague or uninformative. A collateral history from friends, family or other carers is invaluable. In considering diagnosis in the elderly, it is helpful to remember RAMP: **R**educed physiological reserve, **A**typical presentation, **M**ulti-pathology and **P**oly-pharmacy. These factors often cloud diagnosis in the elderly patient resulting in a presentation of 'off legs' in a patient with multiple problems. There are usually several drugs/treatments already in place; it doesn't take much extra physiological stress to result in a non-specific presentation.

Almost the whole medical dictionary can be added to this list, including apparently mundane conditions (e.g. urinary retention, constipation). Always consider: why *this* patient with *this* problem at *this* time? Stepping back and taking a broad view of the patient's social situation may reveal the underlying problem.

Try to obtain a chronological, systematic history in the usual way, focusing on:

- Time course.
- Rate of decline (e.g. sudden decline may be suggestive of a vascular event, acute decline may suggest an infective or metabolic disorder and chronic decline may be due to dementia).
- Pattern of dysfunction (e.g. stepwise deterioration, sudden decline, relapsing and remitting).
- Functional skills. Ask 'What does the patient have difficulty doing now that he or she could manage last week?'

Past medical history

A detailed past medical history should be obtained. The dysfunctional state may result from deterioration of a pre-existing condition or a new event superimposed on a pathological process. In particular, ask about:

- Ischaemic heart disease and cerebrovascular disease (including risk factors).
- Diabetes mellitus.
- Psychiatric disorders (dementia, depression).
- Malignancy.

Drug history

The drug history is particularly important. Often the carer may not be aware of the detailed past medical history and the patient may not be in a state to provide one, so a look at the patient's medication may offer clues about pre-existing conditions. Alternatively, the drugs often contribute to the presentation. Concentrate on:

- Newly prescribed medication – why, when, effect?
- Psychotropic medication (e.g. sleeping tablets, antidepressants).
- Antihypertensive medication (especially if newly prescribed).
- Drugs that may cause electrolyte derangement (e.g. diuretics).

Social history

Obviously, the social history is central to a full assessment. It is important to understand the normal pattern of daily activities and how they are normally negotiated, such as dressing, bathing, cooking, toileting, shopping, managing medication, social interaction.

The social support usually provided, and the possibility that this could be improved, should be assessed (e.g. family, friends, home help, 'meals on wheels', community nurse, previous social worker involvement and benefits).

Consider other factors in the patient's lifestyle that may have contributed to decompensation, for example:

- Alcohol is a surprisingly common contributory factor in declining social functioning in elderly patients.

- Smoking history may highlight risk factors (e.g. for cardiovascular disease, chest disease, malignancy).
- Diet and nutrition are often inadequate in elderly patients.

Elderly patients often present in a non-specific manner with an inability to cope in their home environment. A thorough assessment should be made to elucidate the reason(s) behind this. Usually there are multiple pathologies. Try to diagnose each individual problem and consider which ones are active and which are responsible for the current presentation. Do not ignore apparently trivial illnesses! Review the home circumstances in detail and consider whether the home is safe in the short term or a more detailed assessment is needed (e.g. by occupational therapists and social workers).

Review of systems enquiry

There are few circumstances when it is more important to perform a thorough review of systems enquiry. Inability of elderly people to cope in their home environment is usually due to the interaction of multiple pathologies reaching a critical level and resulting in decompensation. If adequate time is allowed, it usually becomes apparent that the patient was polysymptomatic from different systems before this presentation. For example, you could have an elderly man with:

- Osteoarthritis affecting his hips and limiting mobility.
- Early Alzheimer's dementia.
- A peptic ulcer resulting in iron deficiency anaemia.
- Urinary retention due to prostatism.
- Recently prescribed tricyclic antidepressants.

It is sometimes difficult to distinguish symptoms relevant to the current presentation, but that is part of the skill and challenge of good history taking!

Further reading

Fuller G, Kaye C 2007 Headaches. *British Medical Journal* **334**: 254–256

Gubitz G, Sandercock P 2000 Prevention of ischaemic stroke. *British Medical Journal* **321**: 1455–1459

Turner C 2005 *Crash course: neurology*, 2nd edn. St Louis: Mosby

Presenting problems in paediatric patients

Objectives

By the end of this chapter you should:

- Be able to take a paediatric history.
- Understand that the dynamics of history taking and examination are different in children.
- Appreciate that examination in paediatrics relies upon an opportunistic approach.

Paediatrics covers a wide range of patients from the newly born to the adolescent verging on adulthood. Clearly, a uniform approach to clinical evaluation is not applicable. This chapter is not designed to be a comprehensive guide to paediatrics, merely an introduction. For a more detailed view, the reader is referred to Pang & Newson (2005); see Further reading.

It is useful to distinguish the following age groups:

- Fetus: in utero.
- Neonate: birth to 28 days.
- Infant: birth to 1 year.
- Toddler: 1–3 years.
- Preschool: 3–5 years.
- School child: 5–16 years.
- Adolescent: 12–18 years.

A clinical approach for two important categories of paediatric patient is set out here:

- The toddler and preschool child (aged 1–5 years).
- The newborn infant.

TAKING A HISTORY

For the majority of paediatric patients, the history will be mainly from a parent, usually the mother, but can be from a carer or guardian – be clear about who you are interviewing. The general format is the same as that in adult medicine, but with some very important differences in emphasis. Set out below is a scheme for paediatric history taking. During the history taking:

- Remember parents, especially mothers, observe their children very closely.
- Never ignore or dismiss parental observations.
- Listen carefully: the diagnosis and the parent's concerns are usually apparent in the history.
- Avoid leading questions.

On taking a history, be comprehensive: if you don't ask, they won't tell. Formulate a differential diagnosis by the end of the history. Examine the patient with your diagnosis in mind.

Communication skills are very important in paediatrics as you will usually be taking the history from two or more people. You need to develop a strategy for dealing with this, depending upon how old and communicative the child in question is. Remember that the child is the patient and is full of useful information that he or she may need help to express.

Introductions

On meeting the patient:

- Introduce yourself and get down to the same level as the patient (this makes you less intimidating).
- Identify the patient (find out name, age and sex in advance), for example 'Is this Billy? How old is he?'

- Confirm the relationship of the accompanying adult; for example 'Are you his mother?' (it could be the au pair, older sister, social worker, etc.).

Presenting complaint

Give a prompt to allow the parent to have their say; for example 'What has been your main worry?' Listen carefully and patiently and then follow up with specific questions to elicit the full details of presenting symptoms. Ask about any treatment, such as medicines from the GP or chemist.

Previous history

This should include:

- Birth history: 'Where was the child born? Were the pregnancy and delivery normal? Was the baby early or late? What did the baby weigh? Were there any problems in the newborn period?'
- Immunizations: 'Has the child had all immunizations?' (see Fig. 7.1).
- Medical: 'Any hospital admissions or operations?'

Developmental history

Have there been any problems or concerns (either from parents or doctors/health professionals) about

Immunization schedule
at 2,3 and 4 months DTP (diphtheria, tetanus and pertusis) HIB (*Haemophilus influenzae* B) Meningococcal group C Polio
at 12–15 months MMR (measles, mumps and rubella)
Before school or nursery DTP booster Polio booster MMR booster
at 10–14 years BCG
before leaving school Diphtheria and tetanus booster Polio booster

Fig. 7.1 Immunization schedule. These guidelines change, so check the *British National Formulary* close to exams for up-to-date information.

the child's development? See Figure 7.2 for developmental milestones.

Family history

- Age and sex of siblings? 'Do you have other children?' What is the relationship between the parents looking after the child? 'Are you John's Mum or step-Mum?'
- Family illness, for example tuberculosis, asthma, epilepsy? 'Do any illnesses run in the family?'

Social history

Build up a picture of the family circumstances. Identify socioeconomic status (sensitively). 'Are you or your partner working at present? Any problems with your housing? Does anyone help to look after your child?' If appropriate, 'As a lone parent, what support do you have?' or 'How is your child getting on at school?'

Systems review

This is a set of questions used to identify key symptoms; it is similar to the adult system, with the emphasis on the system implicated by the presenting complaint:

- Respiratory system: breathing difficulties or does the child have any difficulty breathing while feeding?
- Cardiovascular system: any fainting episodes (syncope), breathlessness or blueness about the mouth (cyanotic episodes)?
- Gastrointestinal system: appetite, vomiting, bowel habit, or abdominal pain, frank weight loss or crossing centiles?
- Genitourinary tract: excessive thirst, passing a lot of urine (polyuria) or pain (dysuria)?
- Central nervous system: headache, regression (loss of skills) or fits and funny turns?

Examination

The key word in a paediatric examination is 'opportunistic'. While in adult examination you can have a system to work through, with a child you must be flexible enough to examine what you can in the circumstances. Young infants and school-age children are relatively easy to examine, but the most commonly encountered patient in general paediatric

Developmental milestones				
Age	Gross motor	Fine motor and vision	Hearing and speech	Social behaviour
newborn	symmetrical movements in all four limbs, normal muscle tone	fixes on mother's face	cries	settles on being picked up
6 weeks	good head control, presence of the Moro reflex, transiently holds head in horizontal plane when held in ventral position	follows mother's face	loud noises will startle, makes contented noises	has started to smile
8 months	sits unsupported, starting to crawl	palmar grasp, moves objects from hand to hand, eyes follow a dropped toy	'Dadda' and 'Mamma', reacts to name, positive distraction test	stranger anxiety, separation anxiety, plays 'peek-a-boo'
18 months	walks, climbs onto chair	pincer grip, builds 3-brick tower	three-word sentence, comprehends simple commands	begins toilet training, uses spoon
3 years	runs and jumps, manages stairs, kicks a ball	builds 8-brick tower, copies lines and circles	short sentences	dry by day, Plays with other children, dresses under supervision
5 years	heel–toe walking, catches ball	draws man with features	comprehensive language skills	can play games, tells time

Fig. 7.2 Developmental milestones.

practice is an uncooperative toddler in the 1–3 year age group. In this group, particularly, the following 'dos and don'ts' apply.

Do:

- Be friendly and cheerful: try to smile and keep up some idle chatter (unless of course the child is acutely and severely ill).
- Be gentle: rapport is lost if you cause pain or discomfort.
- Be opportunistic: if asleep, auscultate the chest; if screaming, inspect the tonsils.
- Explain or demonstrate what you are about to do: auscultate a doll or teddy.
- If appropriate or feasible, try and distract the child while examining them, for example shining a pen torch while examining the lung fields.

- Leave unpleasant procedures until last: examination of the ears, nose and throat, rectal examination (rarely necessary), blood pressure measurement.

Don't:

- Tower over the child.
- Stare at the child: avoid looking too intently at toddlers.
- Separate the child from the mother: a toddler is best examined sitting on the mother's lap.
- Undress the child: ask the mother to take off outer layers while the history is taken, but don't strip the child naked – a certain way to make toddlers cry is to undress them and lie them on a cold couch.

In assessments, you may be expected to examine a child. This is more likely in an end of attachment exam or a long case or Objective Structured Long Examination Record (OSLER) in finals. In these settings, you will have relatively more time to put the patient at ease. Examiners should always take into account the individual situation in marking you (a screaming uncooperative child will yield much less information, for example). Observation is key in paediatric examination. Be opportunistic and try not to show your frustration, as this will only exacerbate the situation.

Paediatric vital signs			
Age	Respiratory rate	Heart rate	Systolic BP
<1	30–40	110–160	70–90
1–2	25–35	100–150	80–95
2–5	25–30	95–140	80–100
5–12	20–25	80–120	90–110
>12	15–20	60–100	100–120

Fig. 7.3 Paediatric vital signs.

The features of a 9-month to 5-year-old child unique to a paediatric examination are outlined below; the framework of 'inspection, palpation, percussion and auscultation' still hold true. For a detailed description of examinations, please see the specific system chapter. The following should be read in conjunction with the specific system chapter.

Inspection

Careful initial observation should be made to assess:

- Severity of illness: does this child appear well, unwell or ill?
- Growth: is the child well grown and well nourished? Height, weight and head circumference should be entered on the centile chart.
- Appearance: are there any dysmorphic features? Is the child clean and well kempt?
- Fever or rash: infection?
- Major signs relating to specific systems: level of consciousness, pallor or bruising, cyanosis, tachypnoea or jaundice.

Hands/neck/pulse

The first touch on the hands should be gentle, so as to be non-threatening, as should be palpation of the neck for cervical lymphadenopathy. The radial or brachial pulse can be palpated for rate, rhythm and volume.

Chest

This will include examination of the cardiovascular and respiratory systems. Important signs common to both will have been noted on initial inspection:

- Cyanosis.
- Tachypnoea (Fig. 7.3).
- Clubbing (rare).

Respiratory system

Look for:

- Intercostal or subcostal recession.
- Nasal flaring.
- Use of accessory muscles.
- Chest shape and movement, specifically Harrison sulcus and asymmetrical movements.

Percuss (but this is seldom helpful in very young infants).
Auscultate and note:

- Breath sounds.
- Added sounds, for example wheeze and crackles.

The cardiovascular system should be examined as described in Chapter 3. Detailed evaluation will include:

- Measuring blood pressure (this is quite unpleasant for a child and so should be left until the end, if possible).
- Checking for hepatomegaly: this is an important sign of cardiac failure in infants.

Additional features in the examination of the abdomen

Inspect for:

- Distension: this could be because of intestinal obstruction or ascites.
- Peristalsis, which is a useful sign in pyloric stenosis.
- Inguinal region and genitalia (hernia, hydrocele, testicular torsion).

Palpate the abdomen for tenderness or guarding.

Masses – organomegaly

Auscultate and listen to the bowel sounds, which are:

- Increased in obstruction and acute diarrhoea.
- Reduced in ileus.
- Absent in peritonitis.

Rectal examination may be indicated in suspected appendicitis or intussusception. Remember that this is an unpleasant procedure. Try to have a parent in the room with you for reassurance and use your little finger. **ALWAYS** have a chaperone present.

Nervous system

Important signs noted on inspection include:

- Level of consciousness.
- Which developmental milestones have been met (as in Fig. 7.2).

In infants, it is important to:

- Measure the occipitofrontal head circumference.
- Palpate the anterior fontanelle: this is a window in the skull that allows intracranial pressure and levels of dehydration to be assessed. It closes at about 12 months of age.

If indicated, more detailed evaluation may include the same features as would be examined in an adult: tone, power, coordination, sensation and reflexes (see Ch. 17).

Ears and throat

These are usually left to last, as their examination is not enjoyed by toddlers, mothers or doctors. The key to success is parental help in holding the child. The child should be seated on the mother's lap and held firmly by her with one hand on the forehead and one around the trunk and both arms.

Examine the ears first and the throat last; the occasional child will cooperate by 'opening wide'. In some cases, it is necessary to insert a wooden spatula between clenched teeth onto the tongue.

History in an infant

Before examining the infant, details of the mother's health, the pregnancy and labour should be ascertained.

Maternal health

Some maternal conditions may affect the baby. Ask about:

- Diabetes mellitus type I.
- Autoimmune disorders, for example hyperthyroidism.

Maternal drugs

Find out about drug use; for example which drugs has she taken during pregnancy (prescribed and illicit plus cigarettes and alcohol). The paediatric *British National Formulary* is a great reference book and is kept up to date. Check the drugs the mother mentions in it to find out if there are likely to be adverse effects.

Maternal infections

Infections that may affect the fetus include:

- Rubella.
- Cytomegalovirus.
- *Toxoplasma gondii*.
- Chickenpox.
- *Listeria monocytogenes*.
- *Treponema pallidum* (syphilis).
- HIV.

Pregnancy

Ask about:

- Length of gestation.
- Complications, for example pre-eclamptic toxaemia, intrauterine growth retardation.
- Antenatal diagnoses.

Birth

Ask about:

- Mode of delivery.
- Intrapartum complications.

The mother should be encouraged to express any concerns or questions about her infant. Be proactive and ask specifically, since people do not always express their concerns spontaneously (for example, they might feel that the doctor will think them silly or fussy). Enquire about the mode of feeding; for example, if breast fed, then for how long and when did weaning take place?

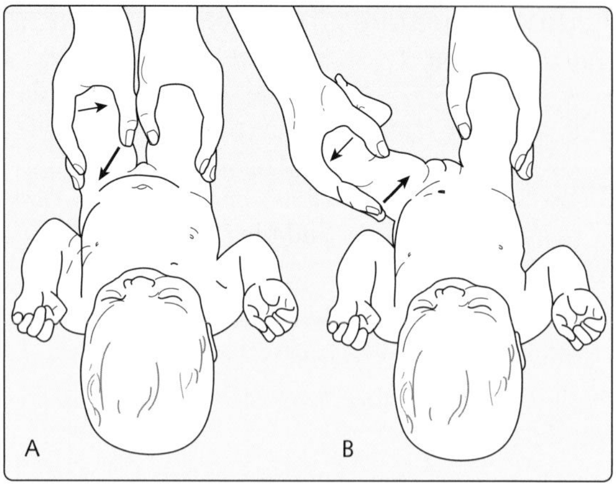

Fig. 7.4 Examination of the hip in newborn infants. (**A**) The Barlow manoeuvre; (**B**) the Ortolani manoeuvre.

EXAMINATION OF THE INFANT

Routine examination

As soon as a baby is born, the midwife (obstetrician or paediatrician, if present) will check that the baby is pink, breathing normally and has no major congenital malformations. Obviously, if the infant is of low birthweight (<2500 g) or ill (e.g. after-birth asphyxia), admission to a special care baby unit will be required. A useful measure of how well a newborn is can be shown by the APGAR scoring system (see Pang & Newson (2005), in Further reading).

About 95% of babies are born at term and appear healthy. However, they all need a full medical examination within the first 24 hours of life. The purpose of this is:

- To give the parents a chance to ask any questions about their baby.
- To identify any problems anticipated as a result of maternal disease or familial disorders, for example congenital infection, maternal diabetes mellitus.
- To detect congenital abnormalities, which may not be immediately obvious at birth, for example cataract, cleft palate, heart murmur, undescended testes, dislocatable hip.

A scheme for the routine examination of the normal term infant is outlined below.

Neonatal screening

On day 6 of life, at which time feeding has been established, all babies have a blood sample from a heel prick taken on to a card (the Guthrie test). This is analysed to detect two inborn errors of metabolism:

1. Phenylketonuria: 1 : 6000 births.
2. Hypothyroidism: 1 : 3000 births.

Look hard but unobtrusively before you touch. Careful observation is the key to success. Vital information is obtained just from looking. Upper airways noises are readily transmitted to the upper chest in infants. They can be difficult to distinguish from coarse rhonchi.

Watch a child's face for grimaces as you palpate the abdomen. Putting the child's hand under yours may enhance cooperation.

In the first 24 hours, many babies have a quiet systolic 'flow' murmur. Features suggesting a significant murmur include:

- Loud murmur.
- Diastolic murmur.
- Associated cardiac signs.

These should be investigated with chest X-ray, electrocardiogram and echocardiogram.

To test for congenital dislocation of the hip (Fig. 7.4), use the:

- Barlow manoeuvre – stabilize the pelvis with one hand, and with the other, abduct the hip to 45°. If the hip is dislocated, forward pressure with your middle finger will cause the femoral head to slip back into the acetabulum.

- Ortolani manoeuvre – with the child supine, flex his or her hips to 90° and also bend the knees to 90°. Place your middle finger over the greater trochanter and your thumb on the inner aspect of the thigh over the lesser trochanter. The child has a dislocated hip if slow abduction causes a palpable or audible jolt. A click is more commonly felt and isn't diagnostic of a dislocated hip.

After children have been discharged from hospital, they should have a 6–8 week check at their GP surgery. This will include:

- Weight and head circumference.
- Hips again.
- Testes and penis.
- Heart sounds.
- Primitive reflexes.
- Muscle tone.
- Smiling.
- And most importantly, any parental concerns.

Further reading

Pang D, Newson T 2005 *Crash course: paediatrics*, 2nd edn. St Louis: Mosby

Presenting problems in psychiatric patients

Objectives

By the end of this chapter you should:

- Understand the principles of taking a psychiatric history.
- Understand and recall the features of the mental state examination.
- Be able to assess suicide risk.

INTRODUCTION

Psychiatry can be a challenging area of medicine. It also makes up more of the workload in many non-psychiatric specialties than you may at first imagine. Around a quarter of the population may experience some form of psychiatric symptom in a year. At any given time, roughly 7% of the population will have depression significant enough to impair them, and the lifetime risk of schizophrenia is 1% (remember schizophrenia carries a 10% lifetime risk of suicide for those affected). Patients with physical disease may also (and often do) develop mental distress as part of their reaction to ill health.

Remember your safety. With a potentially aggressive patient it may be advisable to have a chaperone with you, but this is not always possible or appropriate: always make sure that the patient does not come between you and the door.

The format of a psychiatric history and the mental state examination are outlined in the following paragraphs.

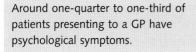

Around one-quarter to one-third of patients presenting to a GP have psychological symptoms.

With psychiatric problems, more so than in any other area, history taking is key. It can be very difficult as patients may be too distressed or unwilling to communicate. Developing a rapport is essential but can be very challenging.

A THUMBNAIL SKETCH

This should include the patient's name, age and occupation.

REFERRAL METHOD

Who referred the patient? Was it self-referral, via concerned family, the GP or even the police? Note the legal status of the patient, for example informal or under which part of the Mental Health Act they are being held, if appropriate.

PRESENTING COMPLAINT

Always ask open questions: 'What brought you to see a psychiatrist?' or 'Why do you think you're here?' As with all histories, it is important to record the words that the patient uses, for example 'Life just isn't worth living'.

History of the presenting complaint

As discussed in Chapter 2, it is important to formulate an idea of how long this has been going on. Is it gradual in onset or is it acute? Have there been any clear precipitants, for example drug induced psychosis (notably with amfetamines or cannabis)? Does anything make this better; does anything make it worse? How often does this happen? Is it weekly, daily or even monthly? It is important to get an idea of how severe an impact on the patient's life the problem is having. Ask if it is interfering with employment and relationships.

Past psychiatric history

Has this person previously had contact with psychiatric services? For example, have there been depressive episodes or deliberate self-harm? What was the form of this? Have they previously been sectioned under the Mental Health Act and, if so, how many times? Have they been admitted to hospital and have they ever had psychotherapy or counselling? Each episode should be investigated and detailed. Many psychiatric disorders are chronic and so there have often been multiple contacts with the patient. You might also ask about on-going contact with psychiatric services (e.g. community psychiatric nurses (CPNs), community mental health teams, partial hospitalization and/or regular attendance at the GP surgery).

Family history

This is generally split into two sections.

Relationships with parents and siblings

Who is in the immediate family? Does the patient get on with the family and what contact do they have with them? Is home a safe and supportive place to be? What have been the learnt coping mechanisms that they have developed; for example you might ask

'In the past, when you felt down, what actions have you taken?' Do any family members have criminal convictions and, if so, what for? The answer might be 'Dad has spent the last ten years in prison'.

Family psychiatric history

As many psychiatric illnesses have a genetic component, it is important to know if any family members have psychiatric illnesses.

Personal history

Early development

Were there any problems during pregnancy or at birth and did they meet the developmental milestones (see Ch. 7)? If the patient is young enough, you may have access to their parents to ask these questions, otherwise you will have to ask the patient if they can recall their mother discussing their early childhood.

Childhood behaviour

Did they play with other children or were they anti-social? You will need to explore the possibility that the person may have been abused at some point in their childhood. This is a difficult area to broach and not everyone will volunteer the information at the first time of asking. Direct questioning in such a sensitive and distressing area is often inappropriate and ineffective. A more oblique approach, for example using a statement instead of a question, may help: 'Sometimes distressing experiences in childhood can make people feel this way'.

School history

What kind of school did they go to, for example was it single sex, boarding, young offenders' institution, etc.? Were they bullied or did they do the bullying? Did they manage to make and maintain friendships at school, did they play truant and what was their disciplinary record like (e.g. suspensions and expulsions)? Were there any family upheavals during these years (e.g. divorce, deaths in the family, etc.)?

Occupational history

Ask how many jobs have they had, how long did they hold them down and why did they leave (e.g. the 16-year-old boy who went into mining and was then made redundant and has not found work in the past 20 years).

Sexual history

This includes sexual orientation, number of partners (including use of prostitutes) and whether or not the relationships have been successful.

Relationship/marriage

Often this is quite different from the above. Has this person been in a stable relationship that has recently ended and so they have lost their social support network?

Children

How many children do they have and with how many partners? What sort of contact do they have with the children and how do they feel about this?

Forensic history

Have they ever been in trouble with the police and, if so, how much contact was there? Have they been to prison and, if so, for how long and how many times? Do they have a case pending?

Current social circumstances

With whom are they living at present? Where are they living? Are they the owner/occupier or is it rented accommodation and is it private or council owned? How many people live there? The answers to these questions can be quite revealing.

Premorbid personality

It is important to ask the patient how they perceived themselves before they became ill. Ask them about the following topics.

Social relationships

Do they feel they get on with people? Did they have a social support network? How were their friends and how did they perceive the relationships?

Hobbies/interests

For example, find out if the patient likes sport or has a hobby or is a member of any clubs.

Predominant mood

Would they describe themselves as predominantly anxious, pessimistic, depressed, happy or optimistic? It's amazing the number of miserable and depressed people who will describe themselves as happy-go-lucky!

Character

Would they describe themselves generally as irritable, self-centred, obsessive or suspicious?

Habits

Do they drink alcohol and, if so, how much? Do they use recreational drugs? If so, which drugs do they use, for example amfetamines, ecstasy, cannabis, heroin and tobacco. It is important to quantify drug use. A full and comprehensive drug history is essential.

PAST MEDICAL HISTORY

This is no different from taking a history in any other setting, but remember particularly to ask about previous head injuries and epilepsy. Medical problems can have psychological presentations or sequelae (e.g. hypothyroidism). Ask about medication as this can also cause psychiatric problems. A full psychiatric history can take upwards of an hour to take. Do not worry if the patient chooses to terminate the interview. Very often, more than one interview is required to build a rapport, and thus obtain a full picture.

THE MENTAL STATE EXAMINATION

This is the psychiatrist's equivalent of the physical examination. It begins when you first meet the patient and continues through the interview. It is your assessment of their mental state at the time you see them and not the history of their illness. It includes a few specific tests, for example testing memory (Fig. 8.1). It is broken down into the following categories:

- Appearance.
- Hygiene.
- Clothes, signs of self-neglect (this may indicate depression, dementia or schizophrenia) or strange or inappropriate behaviour, for example in mania.

Mini-mental test
1. time of day (to nearest hour)
2. year
3. age
4. birthday
5. place
6. recognize two people
7. dates of Second World War
8. name of the monarch
9. count backwards from 20 to 1 with no mistakes
10. recall an address given 5 minutes earlier

Fig. 8.1 The mini-mental test. Score 1 point for each correct answer. Scores of less than 7 indicate well-established dementia.

- The state of nutrition – can be a sign of alcoholism or anorexia nervosa.
- Facial expression, for example smiling, Parkinsonism dyskinesia and posture.
- Tremors – whether they are at rest or on movement and whether they are fine or coarse.
- How much eye contact the patient has during the interview.

Behaviour

You should mention the quality of the rapport generated with the patient. You should also mention how the patient behaved. Most will be pleasant and cooperative, but they may also:

- Be aggressive.
- Be restless or agitated.
- Be slow/withdrawn.
- Be overly familiar or frankly sexually disinhibited.
- Be catatonic, for example when the patient is mute, stuperous or adopts bizarre postures.
- Have unusual mannerisms, for example repetitive hand actions.

Speech

You should consider the patient's speech with respect to:

- Amount; for example how much talking is the patient doing?
- Intonation; for example is the patient monotonous in their speech or does the pitch rise and fall naturally?
- Rate (pressure of speech or slow speech).

- Dysphasia/dysarthria.
- Quality: does the patient use proper words or are they using made-up words (neologism), attaching different meanings to proper words (paraphasia) or are they repeating words spoken by another person (echolalia)?

Mood

A subjective and objective assessment of the patient's mood should be made. If the two fail to match, the patient is said to be 'mood incongruent'. Patients may be:

- Blunted – the patient cannot externalize emotions fully.
- Flat – no emotional reactivity.
- Depressed – feels down.
- Irritable.
- Euphoric.
- Angry.
- Anxious or worried.

Thought

The patient's thoughts should be assessed for both form and content.

Form

Disorders of thought form reflect a disturbance in the process of formulating and expressing thoughts. Thoughts may be classified as:

- Loosening of associations – here there is a link between the patient's thoughts, but it can be difficult to see it.
- Disordered – one thought follows the next with no obvious connection.
- Knight's move thinking – here the patient's associations are tortuous, at best.
- Word salad – this is when the patient is speaking a jumble of unrelated words.
- Flight of ideas – this is common in manic patients where they jump from one idea to the next with some discernible association.
- Thought insertion/withdrawal/block/broadcast – here patients may feel that someone or something is putting thoughts into their head or taking them out or stopping them having thoughts. In thought broadcast, patients feel that everyone can read their thoughts.

Content

The content of a patient's thoughts must be assessed, although this will largely have occurred during the history taking. It is important to ask if the patient has any thoughts that are troubling them or that they do not think others would have. Thought content disorders include:

- Preoccupations. Does the patient have something that they cannot stop thinking about?
- Obsessions. An obsession is a recurrent thought or feeling which is unpleasant and cannot be got rid off.
- Delusions. This is a fixed and false belief abnormal to society.
- Over-valued idea. These are thoughts that the patient places an abnormal emphasis on.
- Depressive. Is the patient depressed? Do they have predominantly unhappy thoughts?
- Suicidal ideation. You must ask all patients if they are considering taking their own life. Do not beat around the bush, but be sensitive. It is a myth that asking directly will precipitate the thought in a depressed patient. Most patients will answer honestly. Beware of patients who are vague in their answers.

Perceptual abnormalities

Remember that hallucinations are perceptions without stimulus while illusions are the misinterpretation of a stimulus. You should identify which sensory modality is affected (visual, auditory, somatic, etc.). For auditory hallucinations, you should identify where the voice or voices come from (e.g. inside the patient's head or outside), how many

of them there are and whether the first, second or third person is used.

Visual hallucinations point towards an organic problem whereas auditory hallucinations are commonly associated with schizophrenia.

Cognitive functions

The patient's cognitive state is assessed under the following divisions:

- Orientation in time, person and place.
- Memory (short and long term).
- Concentration.
- Attention.
- Intellect.

Insight

This is the final section of the mental state examination and deals with the patient's understanding of whether they are ill or not. It also includes the patient's willingness to accept treatment, and which treatments he or she is prepared to accept. Note that it is not very useful or helpful to dismiss the patient's thoughts and perceptions – they are very real to the patient. You can be empathic without directly agreeing with the content of their thoughts and emotions.

Once you have taken the history and mental state, you need to pull it all together so it can be succinctly presented. Figure 8.2 shows how to do this. The case should be presented along the following lines.

Sample case summary for a psychiatric patient			
	Biological	Psychological	Social	Example
Predisposing	(???)	(???)	(???)	Family history, personal childhood
Precipitating	(???)	(???)	(???)	Bereavement, disaster, divorce
Perpetuating	(???)	(???)	(???)	Organic damage, self-esteem
Example	Genetic, biochemical, drugs, endocrine	Stress, coping strategy, mental mechanisms	Poverty, social class, culture, institutions	

Fig. 8.2 Sample case summary for a psychiatric patient. Try to fill in the 'question boxes' (???) table for every patient you see. It makes pulling the history together easier and will enable management plans to be formulated.

A CASE SUMMARY

- The differential diagnosis you have reached.
- The aetiology. It is often useful to divide into predisposing, precipitating, perpetuating and protective factors for each diagnosis.
- The management plan.

In this next section, we will consider some common psychiatric problems that, as a junior doctor, you will be exposed to and will be expected to deal with. They are deliberate self-harm and suicide risk assessment, depression and acute confusion.

DELIBERATE SELF HARM AND SUICIDE

You must ask directly about suicidal ideation at some point during any psychiatric history. Failure to do so is considered negligent. Suicidality is divided into ideation (thinking about suicide), planning (patients have thought about how they will do it) and intent (what has stopped the patient from doing it so far). The risk of suicide can be inferred from these.

Risk factors for completed suicide:

- Men > women.
- Older age.
- Single, divorced, widowed.
- Psychiatric illness.
- Chronic illness.
- Traumatic means.

Deliberate self-harm is a self-initiated act in which the individual person injures him or herself in a way that does not result in death. Suicide is a self-initiated act that leads directly to the person's death.

After the patient has been stabilized and any medical complications treated, an assessment of suicidal intent must be made.

Figure 8.3 shows the widely used Beck Suicide Intent Scale.

Key points to ascertain are:

1. A clear intention to die and remorse at having failed.
2. Detailed planning of the event.
3. Attempting to avoid detection.
4. Not seeking help after the event.
5. Using traumatic means.
6. Undertaking final acts, for example changing a will, paying off bills.

DEPRESSION

Depression as an isolated symptom is one of the commonest presenting complaints that GPs see. It often presents in both hospital and community settings with vague physical symptoms, for example tiredness. It is easily missed and accurate diagnosis depends upon the communication skills of the doctor. When taking a history from a depressed patient, open and closed questions are needed. It is important to elucidate any of the symptoms listed in Figure 8.4 and must include suicidal intent. It is a myth that asking a patient about suicidal ideation will precipitate the thought in their minds. It is negligent to fail to ascertain any such ideation. It is often possible to ask a direct question (e.g. 'Have you ever been so unhappy that you've thought about ending your life?'), but it may be more appropriate to ask in a more oblique fashion (e.g. 'Sometimes when people are so unhappy they feel that they can't go on?'). It is important to differentiate between real suicidal intent and the more common feeling that a patient wishes they were dead but has no intention of taking their life. This may require direct but sensitive probing.

ACUTE CONFUSIONAL STATE

Up to 20% of hospital inpatients will manifest some degree of acute confusion, and early on in your medical career as a newly qualified house officer you will be asked to come and sedate a patient (usually an elderly lady) because they are disturbing the night staff. Be warned that sedating such patients is a last resort and should only be done if they are presenting a danger to themselves and/or to others.

Suicide Intent Scale		Circumstances related to suicidal attempt
1. Isolation	0	Somebody present
	1	Somebody nearby or in contact (as by phone)
	2	No one nearby or in contact
2. Timing	0	Timed so that intervention is probable
	1	Timed so that intervention is not likely
	2	Timed so that intervention is highly unlikely
3. Precautions against discovery and/or intervention	0	No precautions
	1	Passive precautions e.g. avoiding others but doing nothing to prevent their intervention. (Alone in room, door unlocked)
	2	Active precautions, such as locking doors
4. Acting to gain help during or after the attempt	0	Notified potential helper regarding attempt
	1	Contacted but did not specifically notify potential helper regarding the attempt
	2	Did not contact or notify potential helper
5. Final acts in anticipation of death	0	None
	1	Partial preparation or ideation
	2	Definite plans made (e.g. changes in a will, taking out insurance)
Self report		
1. Patient's statement	0	Thought that what he had done would not kill him
	1	Unsure whether what he had done would kill him
	2	Believed that what he had done would kill him
2. Stated intent	0	Did not want to die
	1	Uncertain or did not care if he lived or died
	2	Did want to die
3. Premeditation	0	Impulsive, no premeditation
	1	Considered act for less than one hour
	2	Considered act for less than one day
	3	Considered act for more than one day
4. Reaction to the act	0	Patient glad he has recovered
	1	Patient uncertain whether he is glad or sorry
	2	Patient sorry he has recovered
Risk		
1. Predictable outcome in terms of lethality of patient's act and circumstances known to him.	0	Survival certain
	1	Death unlikely
	2	Death likely or certain
2. Would death have occurred without medical treatment?	0	No
	1	Uncertain
	2	Yes

Fig. 8.3 The Beck Suicide Intent Scale. Scores of 0–3 indicate a low risk of repeat, 4–10 indicate a moderate risk and those of 11 and over indicate high risk. Scores of 18 and over indicate a very high risk of repeat in the short term.

Ask your senior house officer for guidance before prescribing anything. Common clinical features of acute confusion are:

- Impaired consciousness: the level of this commonly fluctuates.
- Disorientation in time, person, place.
- Incoherent or rambling conversations.
- Development of perceptual disturbance, for example illusions or hallucinations.
- Diurnal variation in the symptoms and the sleep/wake cycle, if often disturbed.

- The patient's behaviour is quite out of character; for example a normally quiet elderly lady becomes abusive and violent.

Try to identify and treat the underlying problem (e.g. a urinary tract infection or an upper respiratory tract infection). The best method of managing the patient is with good nursing care. Nurse the patient in a well lit, single room. Provide plenty of reassurance and repeated explanations.

Symptoms of depression

Mood
Low mood for most of the time
Anhedonia
Anxiety (common)

Speech and thought
Suicidal ideas and/or intent
Slow speech
Poverty of thought
Pessimism, hopelessness
May develop delusions (mood congruent)

Biological function
Disturbed sleep (often with early morning wakening)
Anergia
Reduced appetite/weight loss
Reduced libido

Perception
May develop hallucinations if severe

Behaviour
Avoids social interaction
Self neglect
May show psychomotor retardation or agitation
Actions in preparation for suicide

Minimum time for diagnosis
2 weeks

Fig. 8.4 The symptoms of depression. (Courtesy of Dr Brian Lunn, Department of Psychiatry, University of Newcastle upon Tyne.)

High risk groups include the very young, very old, alcohol and drug abusers and those with dementia.

Presenting problems: locomotor system

By the end of this chapter you should:

- Be more confident in taking a musculoskeletal history.
- Be confident in taking a history from a patient with back pain.
- Demonstrate an understanding of the wide ranging biopsychosocial effects related to musculoskeletal disorders, with specific reference to back pain and rheumatoid arthritis.

BACK PAIN

Detailed history

Back pain is extremely common. It is the largest single cause of lost working hours in the developed world among both manual and sedentary workers: in the former, it is an important cause of long-term disability. It is also one of the commoner so-called 'medically unexplained physical symptoms'. The history is used to highlight potentially serious or treatable causes of the pain. Consider the differential diagnosis of back pain (Fig. 9.1).

Ask patients what *they* consider is causing the pain. This is always extremely informative, and may help pave the way to effective treatment. Take a detailed history in the usual systematic manner, focusing on the factors below.

Location of the pain

Most back pain is in the lumbar region. Thoracic pain is usually due to an organic cause (e.g. tuberculosis, osteoporotic crush fracture, myeloma).

Radiation of the pain

For example, radiation down the distribution of the sciatic nerve after lumbar disc prolapse.

Speed of onset

This may be acute (e.g. disc prolapse, crush fracture), insidious (e.g. malignancy, infection, inflammatory causes) or chronic for years (non-specific back pain).

Aggravating and relieving factors

Mechanical pain is often exacerbated by exercise. Inflammatory pain is often worse after a period of inactivity. Spinal stenosis may be worse on walking, but relieved by leaning forwards (e.g. when walking uphill) or resting.

Pattern of severity with time

For example, is the patient experiencing chronic unrelenting pain (e.g. inflammatory disease, psychogenic), intermittent or relapsing pain (e.g. disc disease)?

Associated symptoms

A full review of systems should be performed as the back pain may be part of a systemic disease such as ankylosing spondylosis (polyarthritis, dyspnoea, etc.), malignancy or renal failure (hypercalcaemia, etc.).

Past medical history

A detailed past medical history may elicit a potential underlying cause for the pain. Consider psychiatric disorders, especially depression, which may be a cause or result of the pain (e.g. somatization or lowered pain threshold).

Drug history

Ask patients what analgesics they have taken in an attempt to relieve the symptoms and their efficacy.

Fig. 9.1 Differential diagnosis of back pain. Asterisks indicate more common causes.

Differential diagnosis of back pain
Inflammatory
ankylosing spondylitis*; psoriatic arthropathy; enteropathic arthropathy
Bone disease
osteoporosis*; osteomalacia; renal bone disease; malignancy
Disc disease and osteoarthritis
spondylosis*; acute disc prolapse*; tuberculosis and septic discitis
Mechanical disease
posture* (pregnancy, obesity, scoliosis, etc.); spondylolisthesis; spinal stenosis
Soft tissue disease
'fibrositis'*; muscle strain*
Back pain is unlikely to have a serious cause when:
• the patient can get on and off the examination couch without discomfort
• there is no associated spasm of the spinal muscles and/or local tenderness
• the spine has a full range of movement
Non-specific back pain*
the most common cause; usually has a mechanical basis
Referred pain
chronic pancreatitis; posterior duodenal ulcer; abdominal aortic aneurysm, etc.

Social history

Explore how the pain limits functional activity. Ask specifically about time off work due to the pain. For suspected non-specific pain or psychogenic pain, explore current social pressures being experienced by the patient.

Review of systems enquiry

It is essential to consider systemic illnesses that may have precipitated the pain. In particular, consider weight loss, fevers, sweats, features suggestive of malignancy, polyarthritis, etc.

> Back pain is very common. A good history is the key to efficient diagnosis and can prevent unnecessary and occasionally expensive or unpleasant investigations, which can also reinforce illness behaviour. Try to distinguish between systemic disease and mechanical pain.

Red flag symptoms suggest a potentially serious cause of back pain. These are:

- Age under 20 or new symptoms over 55.
- Non-mechanical pain.
- Thoracic pain.

- History of cancer, steroid use or HIV.
- Unwell, weight loss.
- Widespread neurology, e.g. bilateral leg signs.
- Structural deformity.
- Persistent night pain (with or without night sweats).
- Saddle anaesthesia/sphincter disturbance (www. gp-training.net).

Patients with one or more red flags should be referred as a matter of urgency; where to and how quickly depends on the potential diagnosis (e.g. bilateral neurological signs in the lower limbs could suggest spinal cord compression and this requires magnetic resonance imaging and referral to a spinal or neurosurgeon as a surgical emergency). So-called psychogenic back pain (i.e. 'medically unexplained physical symptoms') is a diagnosis of exclusion. An attempt to make a positive diagnosis should be made as management can be tailored towards the diagnosis.

RHEUMATOID ARTHRITIS

Detailed history

Rheumatoid arthritis is a multisystem disease and the history should be taken in a systematic manner.

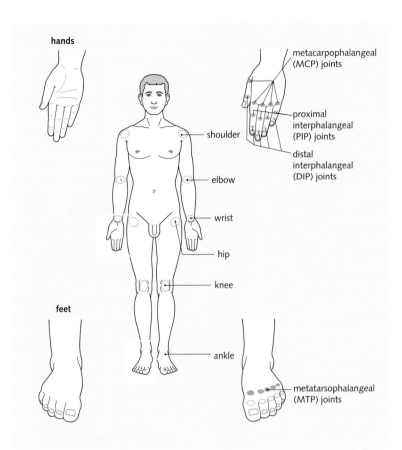

Fig. 9.2 Simple diagrams can be used to illustrate the distribution of active synovitis. Shaded circles represent inflamed joints.

Background disease

Ask about age of onset (typically 25–40 years) and usual pattern of arthritis, etc. About 5–10% of patients will have a positive family history.

Current disease activity

Attempt to assess whether the patient has active synovitis and try to distinguish it from secondary osteoarthritis due to burnt out rheumatoid disease. Enquire about the time pattern of disease activity (e.g. relentless progression, disease flares separated by periods of remission) and the presence of red or swollen joints. If so, note the response to analgesics and non-steroidal anti-inflammatory drugs (NSAIDs) and the duration of early morning stiffness (>30 minutes is significant). Map out the joints that the patient thinks are inflamed (Fig. 9.2).

Functional impact of the disease

Consider how current disease activity has altered functional ability. For example, ask 'Is there any-thing that you have difficulty doing now that you could manage a few weeks ago?' Consider mobility, grip, doing buttons, climbing stairs, etc. Ascertain whether activities are limited by pain, weakness or other factors.

Extra-articular features of the disease

Rheumatoid arthritis is a systemic disease and history taking should reflect this. For each system, consider how disease activity may be manifested:

- Lung – dyspnoea (e.g. due to rheumatoid nodules, pleural effusion, bronchiolitis obliterans).
- Skin – rash, vasculitic leg ulcers, rheumatoid nodules.
- Nervous system – paraesthesiae (especially carpal tunnel syndrome), symptoms of peripheral neuropathy.
- Eyes – dry eyes (Sjögren's syndrome) especially if associated with a dry mouth, red eye (scleritis).

- Renal – proteinuria or known dysfunction (e.g. due to amyloid, medication).
- Anaemia – many patients will have an anaemia of chronic disease; this may be exacerbated by anaemia due to blood loss if the patient is on NSAIDs.

Drug history

A particularly detailed drug history is absolutely essential. Concentrate on drugs currently being used to control the disease (e.g. steroids, analgesics, NSAIDs, disease-modifying agents). Often, a multitude of drug combinations has been used previously. Try to chart previous experiences objectively so that an assessment can be made about changing agents, if necessary, to improve control of disease activity. For each disease modifying agent, chart:

- Acceptability of the agent to the patient.
- Time period of use.
- Efficacy – use objective and subjective parameters if possible; for example what did the patient think of it, early morning stiffness, erythrocyte sedimentation rate (ESR), progression of joint erosions (radiologically).
- Reason for discontinuation – for example side effects, lack of response.
- Doses used – especially cumulative dose.
- If taking NSAIDs, whether the patient had any gastrointestinal side effects and whether the patient has tried selective cyclo-oxygenase-2 (COX2) antagonists (these agents can still be prescribed by rheumatologists).

All second-line agents have side effects (Fig. 9.3). Ask specifically about the use of steroids.

Social history

It is important to investigate the functional impact of the disease on daily life, exploring the home environment as well as occupation. Many aids to living are available, and specialist use of physiotherapists and occupational therapists may be invaluable. As with many long-term disabling conditions, it is easy to concentrate on symptom control and practical issues; enquire how the patient is feeling and coping, and what the patient is worried about. This is equally important.

Remember that rheumatoid arthritis is a chronic (and often disabling) condition. Patient education and motivation are crucial for optimal rehabilitation. It is particularly important that an air of mutual trust is fostered between the patient and doctor. Patients are likely to require extensive long-term follow up. Rheumatoid arthritis exemplifies the humanitarian aspect of being a doctor. Many treatments are focused around limiting disease progression or treatment of symptoms. Good rapport is essential for ensuring the correct treatment for the correct patient at the correct time.

Side effects of second-line agents used in the treatment of rheumatoid arthritis	
Drug	Side effects
gold	proteinuria (membranous nephropathy); thrombocytopenia and agranulocytosis; skin rash (approximately 25%); stomatitis
penicillamine	proteinuria (common); nephrotic syndrome; thrombocytopenia; agranulocytosis; anorexia; nausea (early in treatment); rash
methotrexate	bone marrow suppression; oral ulceration; gastrointestinal disturbances; hepatotoxicity (especially prolonged use); teratogenic
azathioprine	bone marrow suppression; increased risk of malignancy; infections; gastrointestinal disturbances
sulfasalazine	nausea; vomiting; skin rash; blood dyscrasia (rare)
hydroxychloroquine	retinopathy (especially long-term use)

Fig. 9.3 Side effects of second-line agents used in the treatment of rheumatoid arthritis.

OSTEOARTHRITIS

This is the commonest joint condition and is three times more common in women than men, and normally presents at around 50 years of age. It usually occurs as a primary feature, but may occur chronically after injury. The patient will normally complain about pain on exertion which is relieved by rest. The pain is often worse at the end of the day. This process begins insidiously and develops over years. With time, the relief with rest is less complete. The patient may also complain about stiffness after periods of rest. Unlike the inflammatory joint diseases, however, there are no systemic features of osteoarthritis.

Ask the patient what joints are particularly affected. (In order) it is the distal interphalangeal, first metacarpophalangeal and first metatarsophalangeal joints which are most commonly affected. The hips and knees are also commonly affected. Determine what the patient's current level of function is and what exactly stops the patient from doing more. It is useful to find out if they have had any joint replacements and whether they feel them to have been successful. As the patient will often complain about pain, it is important to take a drug history to determine what analgesics are used and which are most effective. Refer to Chapter 21 on locomotor investigations for detail about the X-ray changes associated with osteoarthritis.

PERTHES' DISEASE

This condition is a subject much loved by orthopaedic surgeons as it demonstrates the changing anatomy of the blood supply to the hip with age. It is a relatively rarer disease of childhood and has an incidence of 1 : 10 000. The patient is usually a boy between the ages of 4 and 8 with delayed skeletal maturity. (Boys are four times more affected than girls.) Initially they will complain of hip pain and then start to limp. The joint will be irritable; thus all movements are diminished and painful at the extremes. Abduction and internal rotation are the most commonly affected movements.

Between the ages of 4 and 7, the femoral head is dependent on the lateral epiphyseal vessels for its blood supply. Their course makes then susceptible to occlusion by pressure from any effusion around the hip. After the age of 7, the blood vessels in the ligamentum terres are developed and supply the femoral head.

The pathology is a three-stage process. Initially, there are one or more episodes of ischaemia causing bone death. Revascularization and repair then occur, but there is then distortion and remodelling of the femoral head and neck. This can lead to the incomplete covering of the femoral head by the acetabulum.

SLIPPED UPPER FEMORAL EPIPHYSIS

A slipped upper femoral epiphysis is basically a fracture through a hypertrophic zone of the cartilaginous growth plate. The patient is usually a pubertal boy who will present with referred hip pain (e.g. knee pain, groin pain, anterior thigh pain and limping). They can present with acute pain following trauma, chronically or with acute-on-chronic pain. These patients are very often unusually tall and thin, or overweight and sexually underdeveloped. On examination, the leg will be externally rotated and shorter and all movements will be painful.

FRACTURED NECK OF FEMUR

This usually happens to older women who have osteoporosis. It often follows the simplest slip or fall; however, care should be taken to ensure that there is no medical cause for the fall. The patient cannot normally weight bear but some will be able

Garden's classification of fractured neck of femur	
Type 1	inferior cortex not completely broken but trabecule lines are angulated
Type 2	inferior cortex clearly broken but trabecule lines are not angulated
Type 3	obvious fracture line and rotation of head in acetabulum
Type 4	fully displaced fracture

Fig. 9.4 Garden's classification of fractured neck of femur.

to walk, albeit with considerable pain. On examination, the affected limb will be shortened and externally rotated. This condition has a 1-year mortality of approximately 50%. The fractures are often classified according to Garden's system; this may be used to decide whether a hip fracture is fixed with cannulated screws or a hemiarthroplasty (Fig. 9.4; see also Ch. 21 on locomotor examination and Ch. 24 on further investigations). Patients are often frail and susceptible to hospital-acquired pneumonia, pressure sores, deep vein thromboses and pulmonary emboli. If the patient is to make a successful recovery, operative management is mandatory to facilitate early mobilization.

Presenting problems: endocrine system

10

Objectives

By the end of this chapter you should:

- Be able to take a history from a patient with diabetes.
- Know the potential complications of diabetes.
- Be able to take a history from a patient with thyroid disease.

THE DIABETIC PATIENT

Detailed history

Review presentation with diabetes mellitus

The most important features of diabetes mellitus are:

- Age of onset.
- Presenting symptoms (e.g. weight loss, polyuria, polydipsia, coma).

Form of diabetes mellitus

Note the class of disease, as complications and management strategies will vary, for example:

- Type I – due to lack of insulin.
- Type II – due to insulin resistance.
- Secondary (due to glucocorticosteroids, pancreatitis, Cushing's disease).

Diabetic control

This is the cornerstone of managing diabetes mellitus, so much of the time available at consultation should be devoted to this. Review the form of blood sugar control (e.g. diet alone, oral hypoglycaemic agents, insulin). Try and ascertain concordance, especially in teenage patients who may feel socially stigmatized by having to take insulin. Be sensitive in inquiring about this, for example 'Do you ever forget to take your tablets?' or 'Do you ever eat food you're not supposed to?'

Establish how the patient monitors glycaemic control. Check the form of monitoring (e.g. measuring blood glucose, urinalysis), frequency of measurements and the levels attained. Most patients will have a book charting the blood glucose level. This should be reviewed to assess general control and fluctuations at different times of the day. It also provides an additional insight into the patient's concordance. In an analogous way to checking inhaler technique, it is often informative to observe a patient performing a blood glucose estimate. While patients may try to improve their diabetic records, they cannot fake the glycosylated haemoglobin (Hb_{A1C}), which is a measure of long-term glucose control. This is produced by the attachment of glucose to Hb. The measure of the glycosylated fraction gives the average glucose concentration over the lifetime of the Hb molecule (120 days, normal range 4–8%).

Type I diabetes

In type I diabetes, patient education and motivation are crucial to the long-term prognosis. Good long-term control is pivotal for reducing the long-term complications of diabetes mellitus. Focus on risk factors for ischaemic heart disease (see Ch. 3) and always ask specifically about visual problems and foot care. Your goal should always be to empower your patient so that he or she feels in control of the disease, not the other way round. One method is the Dose Adjustment For Normal Eating (DAFNE) regime (http://medweb.bham.ac.uk/easdec/prevention/diabeteseducation.htm) which allows patients to adjust the dose of their insulin based on meal carbohydrate content, giving greater dietary freedom.

Complications of diabetes

Review the frequency of acute complications of therapy or hyperglycaemia, for example:

- Hypoglycaemic attacks.
- Diabetic ketoacidosis (DKA).
- Hyperosmolar non-ketotic coma (HONK).

Has the patient required hospital admissions for any of these?

Ask specifically about symptoms suggestive of poor control, for example:

- Weight loss, malaise, fatigue.
- Polyuria, polydipsia (due to osmotic diuresis).
- Blurred vision (refractive changes in the eye).
- Balanitis, pruritus vulvae, thrush.

Macrovascular complications

Diabetes mellitus is strongly associated with macrovascular disease. It is imperative that any coincident risk factors such as hypertension and hyperlipidaemia are identified and minimized. Check for symptoms related to the three main forms of macrovascular disease:

1. Ischaemic heart disease (ask specifically about chest pains, myocardial infarction).
2. Peripheral vascular disease (e.g. claudication).
3. Cerebrovascular disease.

Microvascular complications

Long-term disease is associated with relentless progression of microvascular disease. Good control is clearly associated with a decreased likelihood of microvascular complications. Early recognition may allow specific therapy to be instituted. Ask about the three main forms of microvascular disease:

1. Retinopathy – ask about visual symptoms, laser therapy.
2. Neuropathy – paraesthesiae and pain, especially in the lower limbs.
3. Nephropathy – proteinuria and hypertension.

Other complications

Diabetes mellitus is associated with other complications, which may be multifactorial. Specific enquiry should be made about:

- Impotence – rarely spontaneously volunteered by the patient, but common, so always ask (be sensitive).
- Staphylococcal skin infections – for example boils, carbuncles.
- Gastroparesis – nausea, vomiting, early satiety.
- Foot ulcers – neuropathy and vasculopathy. This normally manifests as numbness, pain and paraesthesia. 'Do you feel like you're walking on cotton wool?', 'Do you know where you're putting your feet?'

Past medical history

Review the other risk factors for ischaemic heart and cerebrovascular disease. Consider other possible autoimmune diseases (especially thyroiditis, pernicious anaemia, vitiligo, Addison's disease), renal failure and systemic infections.

Drug history

Review the agents used to control the diabetes and drugs that may be related to complications, for example antihypertensives, angiotensin-converting enzyme inhibitors, statins, etc.

Oral hypoglycaemic agents

Consider whether it is appropriate for the patient to be on an oral hypoglycaemic agent. Beware of metformin in the presence of renal dysfunction (metformin is contraindicated as it may cause lactic acidosis). Consider whether it is more appropriate to use agents that are hepatically metabolized or renally excreted.

Insulin

Review the form of insulin used (e.g. long or short acting, porcine or human). Assess whether the dose needs to be modified. Ask about problems at injection sites (e.g. lipohypertrophy, scarring).

Review the use of other medication that may aggravate diabetic control (e.g. glucocorticosteroids, thiazide diuretics). It may be possible to improve diabetic control by appropriate adjustment of other medication.

Family history

This is usually positive for patients with type II diabetes mellitus.

Social history

Review home circumstances, for example:

- Is the patient's eyesight good enough to read BM sticks?

- Does the patient have sufficient manual dexterity (especially if he or she had neuropathy) to monitor his or her own therapy?

Education and motivation are central to the long-term outlook for the diabetic patient. An appropriate diet is essential for improving the control of all diabetic patients, especially those who are obese. Help from a dietician may be indicated. It is particularly important that the patient understands the unacceptable risks of smoking.

Diabetes is a condition in which a 'patient-centred' approach has been shown to positively influence outcome.

In diabetes mellitus, patient education and motivation are crucial to the long-term prognosis. Good long-term control is pivotal for reducing the long-term complications of diabetes mellitus. Focus on risk factors for ischaemic heart disease and always ask specifically about visual problems and foot care.

THYROID DISEASE

Presenting complaint

Hypothyroidism and hyperthyroidism may present in many ways and form part of the differential diagnosis of a variety of presenting complaints. Some of the more common presentations are shown in Figure 10.1. A high index of suspicion may be needed to diagnose these conditions as presentation is often non-specific.

Detailed history

Hyperthyroidism

Ask about symptoms of hyperthyroidism. Enquire specifically about palpitations (see Fig. 10.1).

The common causes of atrial fibrillation are ischaemic heart disease, hyperthyroidism and mitral valve disease.

The following specific features may suggest underlying Graves' disease:

- Eye disease (diplopia, proptosis).
- Goitre (ask about difficulty swallowing or breathing).
- Pretibial myxoedema.

Hypothyroidism

Ask about symptoms of hypothyroidism (see Fig. 10.1): tiredness, general slowing down and deepening of the voice. In particular, enquire about weight gain, fertility problems, menstrual difficulties and cold intolerance. The symptoms are often insidious and are not noticed by patients or their immediate family, but may be observed by occasional visitors or general practitioners.

Presenting features of thyroid disease	
Hypothyroidism	**Hyperthyroidism**
weight gain*	weight loss despite good appetite*
general slowness	
mental slowing, poor memory	poor concentration
anorexia	
cold intolerance*	heat intolerance*, excessive sweating
depression	agitation*, restlessness*
coma	
constipation	diarrhoea
altered appearance	eye changes
	palpitations

Fig. 10.1 Presenting features of thyroid disease. Asterisks indicate the more common or discriminatory features.

Fig. 10.2 Secondary causes of thyroid disease. TSH, thyroid-stimulating hormone.

Secondary causes of thyroid disease	
Hypothyroidism	**Hyperthyroidism**
dietary iodine deficiency	
antithyroid drugs used at inappropriate dose or on resolution of hyperthyroid state	over-dosage with thyroxine
post-thyroid surgery	
radio-iodine therapy	thyroid carcinoma (rare)
tumour infiltration (rare)	TSH-secreting tumours (rare)
hypopituitarism	
post-subacute thyroiditis	acute thyroiditis (e.g. infective, autoimmune)

Past medical history

Ask about:

- Other autoimmune disorders (e.g. diabetes mellitus, Addison's disease, vitiligo, myasthenia gravis).
- Previous partial thyroidectomy or treatment with radio-iodine if there is suspected hypothyroidism.

Hypothyroidism and hyperthyroidism can produce a multitude of symptoms (see Fig.10.1). A high index of suspicion is needed, particularly in the context of mental changes, palpitations, changes in weight, altered conscious level or cardiac disease. The most discriminatory features from the history are cold and heat intolerance and weight changes despite contradictory dietary history. A relative may provide invaluable clues as the patient may not spontaneously report symptoms. The patient may report that family or colleagues have mentioned certain symptoms, which may be helpful if the patient attends alone.

In treated patients, thyroid status should be assessed clinically as biochemical tests lag behind therapeutic responses by several weeks.

Also consider treatable causes of thyroid disease (Fig. 10.2) and the presence of ischaemic heart disease in hypothyroid patients, as thyroxine therapy will then need to be more cautious.

Drug history

Review the use and dose of antithyroid medication and thyroxine. Consider drugs that may aggravate symptoms (e.g. β-blockers causing bradycardia in hypothyroidism). In addition, consider drugs that may interfere with interpretation of thyroid function tests (e.g. oestrogens increase thyroid-binding globulin concentration). Thyroid-stimulating hormone (TSH) levels are usually (if not always) needed in addition to total thyroxine concentration, when interpreting thyroid function tests.

Review of systems enquiry

A detailed screen of systemic symptoms is essential as many features are non-specific. It is often fruitful to concentrate on psychological changes, for example symptoms of depression.

The common causes of atrial fibrillation are ischaemic heart disease, hyperthyroidism and mitral valve disease.

EXAMINATION

11. Introduction to
    examination 95

12. Cardiovascular examination 97

13. Respiratory examination 117

14. Abdominal examination 131

15. Obstetric and gynaecological
    examination 151

16. Surgical examination 157

17. Neurological examination 163

18. Endocrine examination 187

19. Reticuloendothelial examination 193

20. Breast examination 197

21. Locomotor examination 203

22. Ophthalmic examination 221

23. Writing medical notes 225

24. Further investigations 233

Objectives

By the end of this chapter you should:

- Understand the principles of patient examination.
- Accept that examination routine may vary between individuals and situations.

OVERVIEW

This section of the book is divided into the individual systems of the body. The first part of each chapter describes the routine examination of that system. The second part illustrates the process described above in practice by providing examples of pathologies or presentations related to the system. These examples are intended to provide a skeleton for the student's examination, illustrating the important features. When it comes to exams, there is a set way in which your examiners will expect you to examine patients. Failure to use this approach will unsettle examiners and they may not appreciate the visionary methods you are using! This is especially true in the Objective Structured Clinical Examination (OSCE) environment where marks are awarded for doing a specified task. There is, however, an acceptable variation in examination technique: as long as you are comfortable, practised and don't hurt the patient, minor variations should be accepted by the examiners (e.g. using tactile vocal fremitus rather than vocal resonance in respiratory examination). It is important to get as much feedback as possible on your examination technique. Take every opportunity for someone more experienced to be present, to watch what you are doing and to give you constructive feedback. This can be unsettling but is the closest to an exam situation you can achieve (outside of an exam itself, of course!).

SETTING

A good physical examination relies upon a cooperative patient and a well-lit room. It is important to engender an atmosphere of trust and professionalism during the history taking and to explain the steps of the examination appropriately, and, if necessary, what information you hope to elicit from the process. This will help to avoid any misunderstanding and will reassure the patient, taking away some of the mysticism that may surround the doctor–patient relationship. It is good practice to start any examination with the hands. It is much more socially acceptable to have your hands examined than your abdomen. As the examination progresses up the patient's arm, you have time for the patient to relax and trust in your professionalism. It is important to be sensitive to the patient. It is always appropriate, and often mandatory (e.g. pelvic examination), for a chaperone to be present; it may also be inappropriate for the patient's partner to be in the room during the examination.

The patient must be appropriately positioned, comfortable and in a well-lit warm room (in an OSCE examination, this alone can make up a significant proportion of the marks). Ensure that there is adequate privacy. It is clearly unacceptable for a semi-clad patient to be examined on a couch without screens if the door to the examination room is potentially going to be opened without warning, or for the patient to be examined on a ward without curtains pulled round.

EXAMINATION ROUTINE

Although the examination is described as a separate process from the history, a good assessment should start as soon as the patient walks into the examination room (Fig. 11.1). There are countless other observations that can be made, and a rapid inspection of the patient will put the subsequently elicited history into context. For example, does the patient look ill or well? Have they apparently lost weight? Are there any obvious facial features (the staring eyes of thyrotoxicosis, yellow sclera of jaundice, etc.)?

The examination, like the history, should be performed in a systematic manner but, again, it is vital to be aware of the differential diagnosis as each sign is elicited. It is also important to think about what you expect to find from the history – it's easier to find something if you know what you're looking for. In many clinical examinations, including OSCEs and long cases, you may be asked to describe what you're doing and why, so it is a good idea to get into the habit. However, it is also necessary to keep an inquiring mind, as it is not difficult to convince yourself you've found the signs you expected, when in fact they're not actually present!

Once again, the process is active. The examination routine usually follows a set order. For most systems, the order of examination is as follows:

1. Inspection.
2. Palpation.
3. Percussion.
4. Auscultation.

Or, in orthopaedics:

1. Look.
2. Feel.
3. Move.

Points to consider when the patient walks into the room
Does the patient look ill or in pain?
Is the patient cachectic or overweight?
What is the patient's ethnic/cultural background?
Does the patient have a normal gait?
Is the patient short of breath on walking into the examination room?
Does the patient require help to get out of a chair?

Fig. 11.1 Points to consider when the patient walks into the room.

The most important component of the examination is inspection.

EXAMINING A SYSTEM

When examining any system, try to answer the following three questions.

1. What is the pathology?

This process requires interpretation of the physical signs so that a deviation from normality can be recognized and the individual signs integrated.

2. What is the aetiology of the pathological process?

It is not enough to identify that a patient has a pleural effusion. An attempt should be made to find the underlying cause (e.g. lymphadenopathy and hepatomegaly suggesting malignancy or green sputum and coarse crackles above the effusion, consistent with pneumonia).

3. What is the severity and functional impact on the patient?

For example, severe aortic stenosis is an indication for valve replacement. This may be assessed clinically by measuring the pulse pressure and noting the presence of a slow-rising pulse.

PHYSICAL SIGNS

Many physical signs are subtle, or simply represent a variation in the normal. It is essential that the doctor has an appreciation of the wide range of normality so that any signs can be placed into context. There are no short cuts. To gain this ability requires practice.

Cardiovascular examination

Objectives

By the end of this chapter you should:

- Be able to perform a cardiovascular examination.
- Be able to recall the features of common valve pathologies.
- Recognize the clinical features of myocardial infarction and heart failure.

EXAMINATION ROUTINE

As with all examinations, the cardiovascular examination follows the format of inspection, palpation, percussion and auscultation. Many cardiovascular pathologies produce multiple signs which, when integrated, allow assessment of the severity and aetiology of the lesion. It is particularly important to perform the examination in a systematic manner so that physical signs can be put into context with each other.

Examination technique

The most important preparation for exams is done by taking histories and examining patients, not reading books. For physical examination stations, a few basic pointers can reap large rewards:

- Adequate inspection is neglected at your peril! If omitted, simple diagnoses can easily be overlooked. It also helps to anticipate subsequently elicited physical signs and to put them in perspective.
- It is essential to know in advance what you are looking for, or it will not be found! Prepare a mental checklist. The whole process need only take a few seconds and is essential exam routine. You know what you're looking for, but you have to demonstrate this to the examiner(s) to get the marks.
- Be careful when describing the arterial pulse, apex beat or heart murmurs, as certain terms often imply specific diagnoses (e.g. tapping apex beat equals mitral stenosis).

- Resist the temptation to auscultate until the rest of the examination routine is completed. It is much easier to auscultate when the sounds can be put into context with the rest of the examination. You should already know what you expect to hear.

Patient exposure and position

Ensure adequate lighting. The patient should be comfortable and seated at 45° to the horizontal, and stripped to the waist. Female patients should remove their bras (it is difficult auscultating through clothing!). It is courteous to provide a blanket as the patient may wish to remain covered until examination of the praecordium.

General inspection

In all systems, the specific examination should be preceded by an inspection of the patient. Vital clues are often found.

General features

Note any obvious features on general observation, including:

- Age, sex, general health.
- Body habitus (obese or cachectic).
- Breathlessness (observe the effort required to climb onto the couch).
- Position in bed (is the patient comfortable, or do they seem to need to sit up or forward?).

Listen for any clicks of metallic prosthetic heart valves. Inspection can then be performed from the head downwards.

Eyes

A brief inspection will reveal abnormalities such as arcus senilis (only significant in young adults, suggesting hypercholesterolaemia) or xanthelasma (suggestive of hyperlipidaemia). The patient may be obviously jaundiced.

Face

Look for evidence of cyanosis, particularly around the lips or under the tongue, or plethora (e.g. due to superior vena cava obstruction, polycythaemia). The patient may have a typical facies for a pathological process (for example malar flush indicating mitral stenosis); these 'classical' presentations are usually rare, but occur frequently in exams as they allow you to demonstrate your underlying knowledge and understanding.

Always inspect the mouth (for example, for dental hygiene in infective endocarditis).

Neck

Note any visible pulsations.

Praecordium

Observe the shape of the chest for:

- Any obvious deformity (e.g. pectus excavatum in Marfan's syndrome).
- Visible collateral veins (e.g. in superior vena cava obstruction).
- Visible apex beat (e.g. in left ventricular hypertrophy).
- Presence of any scars (Fig. 12.1).

Also look in the brachial and femoral regions for scars from cardiac catheterization.

Ankles

Briefly look for swelling suggestive of peripheral oedema. This can be confirmed by later examination.

Specific examination

Specific examination can now be performed.

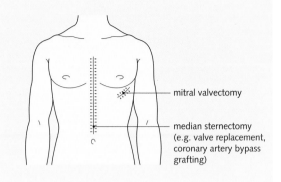

Fig. 12.1 Position of scars related to cardiovascular pathology on the praecordium.

Hands

Careful examination of the hands often reveals clues of underlying cardiovascular disease. Feel the temperature and look for peripheral cyanosis. Note the shape of the hands (e.g. arachnodactyly in Marfan's syndrome).

Clubbing

The cardinal feature of clubbing is loss of angle of the nailfold. Other features include increased convex curvature in both a longitudinal and transverse plane, increased fluctuation of the nailbed and swelling of the terminal phalanx (Fig. 12.2). Remember to look at both hands. Clubbing can rarely be detected in the toes. Hold the nail up to the plane of your eyes to facilitate detection. Although a non-specific sign, its presence should alert the physician to underlying disease (Fig. 12.3). You must be able to tell your examiners the causes of clubbing.

New-onset clubbing is highly significant. If detected, ask the patient whether he or she has noticed any recent change in the shape of the nails.

Splinter haemorrhages

These are linear red or black streaks under the finger or toenails (Fig. 12.4) and are a feature of a vasculitic process, but the most common cause is mild trauma, and a small number (e.g. about five) is normal.

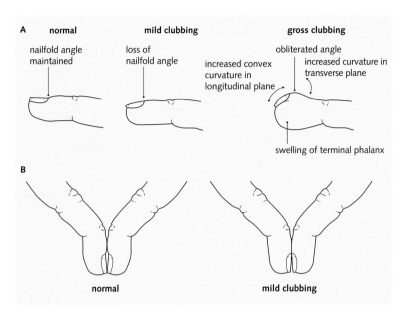

Fig. 12.2 A, Features of clubbing. The cardinal sign is loss of angle of the nailfold. **B,** If the patient is asked to place his index fingers 'back to back', the diamond-shaped gap normally present is obliterated in early clubbing.

Causes of clubbing	
System	**Disease associations**
cardiovascular	infective endocarditis* (late sign); congenital cyanotic heart disease; atrial myxoma (very rare)
respiratory	carcinoma of the bronchus*; fibrosing alveolitis*; chronic suppurative lung disease* (empyema, bronchiectasis, pulmonary abscess, cystic fibrosis)
abdominal	Crohn's disease (unusual); cirrhosis
familial	most common cause*

Fig. 12.3 Causes of clubbing. Asterisks indicate the more common causes.

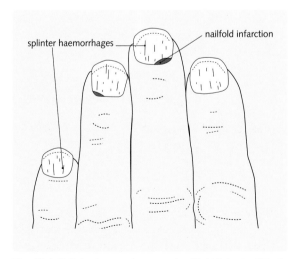

Fig. 12.4 Splinter haemorrhages and nailfold infarcts. If both lesions appear together, infective endocarditis or a vasculitic process is highly likely.

Nailfold infarcts

Nailfold infarcts (see Fig. 12.4) are also a feature of vasculitis, but more specific than splinter haemorrhages. They are often associated with other features of infective endocarditis (see Infective endocarditis, below).

Capillary return

Apart from noting the temperature of the skin, peripheral perfusion may be assessed by capillary return. Light digital pressure to the end of the nail will produce blanching. The speed of capillary return can be visualized. Poor peripheral perfusion can be easily detected. You should press on the nailbed for 5 seconds and the return should take less than 2 seconds (though up to 5 seconds can be normal). Visible pulsation may be seen in aortic regurgitation.

Nicotine staining

Cigarette staining of the fingertips and nails may counter information given in the history.

Other signs of infective endocarditis

Look for other stigmata of infective endocarditis (rare), for example:

- Osler's nodes (tender nodules on the finger pulps).
- Janeway lesions (see Infective endocarditis, below).

Xanthomas

Lipid deposition may occur in tendons, skin or soft tissues in some hyperlipidaemic states, for example:

- Tendon xanthomas (especially of Achilles tendon, extensor tendons of hands) – type II hyperlipidaemia.
- Palmar xanthomas (skin creases of palms and soles) – type III hyperlipidaemia.

Arterial pulse

The arterial pulse may be palpated at various sites (Fig. 12.5). The pulse has various characteristics, which should be defined in each patient.

Presence and symmetry

Compare the radial pulsations synchronously (e.g. large vessel vasculitis, aortic dissection). Obstruction may delay the pulse. Check for radio-femoral delay, especially in hypertension (e.g. due to coarctation of the aorta). Assess the presence of each pulse, especially if there is embolism or peripheral artery disease.

Rate

Normal pulse rate is 60–90/minute. Count for at least 15 seconds at the radial pulse. This may also provide an opportunity for additional visual inspection of the patient. The radial pulse rate may differ from the number of ventricular contractions per minute (e.g. apico-radial deficit in fast atrial fibrillation). Consider the rate within the clinical context (e.g. tachycardia in the presence of fever, hypovolaemia; bradycardia in hypothermia, hypothyroidism).

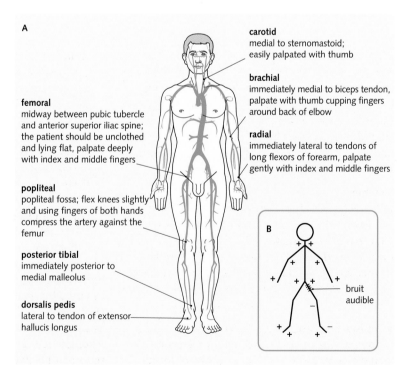

A

carotid
medial to sternomastoid; easily palpated with thumb

brachial
immediately medial to biceps tendon, palpate with thumb cupping fingers around back of elbow

femoral
midway between pubic tubercle and anterior superior iliac spine; the patient should be unclothed and lying flat, palpate deeply with index and middle fingers

radial
immediately lateral to tendons of long flexors of forearm, palpate gently with index and middle fingers

popliteal
popliteal fossa; flex knees slightly and using fingers of both hands compress the artery against the femur

posterior tibial
immediately posterior to medial malleolus

B

bruit audible

dorsalis pedis
lateral to tendon of extensor hallucis longus

Fig. 12.5 A, Locations of the arterial pulses. **B,** Typical notation in the hospital records. (+, pulse present; −, pulse not palpable; bruit audible).

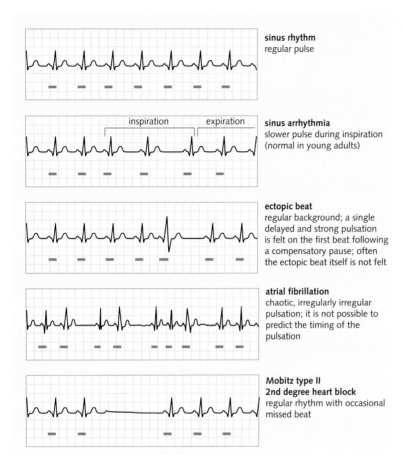

Fig. 12.6 Common arrhythmias. In each recording, the upper trace is the ECG and the lower trace illustrates a bar corresponding to each palpable pulsation. It is often possible to elucidate the underlying rhythm disturbance.

sinus rhythm
regular pulse

inspiration   expiration

sinus arrhythmia
slower pulse during inspiration (normal in young adults)

ectopic beat
regular background; a single delayed and strong pulsation is felt on the first beat following a compensatory pause; often the ectopic beat itself is not felt

atrial fibrillation
chaotic, irregularly irregular pulsation; it is not possible to predict the timing of the pulsation

Mobitz type II
2nd degree heart block
regular rhythm with occasional missed beat

Rhythm

The normal pulse rhythm is regular sinus rhythm. Any irregularity should be characterized and, if possible, confirmed by an electrocardiogram (ECG). Many arrhythmias are characteristic (Fig. 12.6). You should state whether the pulse is regular or irregular and, if irregular, is it regularly irregular or irregularly irregular?

Volume

Pulse volume reflects stroke volume. This is best assessed at the carotid or brachial pulse. There is a wide range of 'normal'. It is important to practice on as many different patients as possible so that you can recognize when abnormal signs are present.

Some of the abnormalities of pulse volume include:

- Pulsus paradoxus. This is detectable in cardiac tamponade or severe asthma. A detectable increase in pulse volume is observed during expiration. Pulsus paradoxus reflects an exaggeration of the normal physiological changes in intrathoracic pressure and the influence of the diaphragm and interventricular septal changes during the respiratory cycle.
- Pulsus alternans. This is a sign of severe left ventricular failure. Alternate pulses are felt as strong or weak due to the presence of bigeminy.
- Coarctation of the aorta. As well as radio-femoral delay (described above), the pulse volume of the femoral pulse is usually noticeably reduced.

Character

With increasing distance from the aorta, particularly in sclerotic vessels, the waveform becomes distorted (Fig. 12.7), so volume and character should be assessed using either the carotid or brachial pulses. A picture of the waveform often correlates with the severity of valvular disorders. This sign requires

Description	Waveform	Associations
normal		–
slow rising		aortic stenosis
collapsing (water hammer)		aortic regurgitation persistent ductus arteriosus

Fig. 12.7 Typical arterial waveforms palpated at the carotid pulse in different conditions. The waveform often provides important information about the severity of the underlying pathology.

considerable practice to elicit, so that the range of normality can be appreciated, and abnormalities put into context. Try to become used to the pulse character of the common pathologies (see Fig. 12.7).

Blood pressure

Measurement of blood pressure is straightforward, yet it is often performed inadequately, with obvious implications for patient management.

> The blood pressure is a vital sign, so should be elicited with great care and given the respect that it deserves! It should be measured to the nearest 2 mmHg.

As with other components of the cardiovascular examination, it is important to adopt a systematic approach, for example as follows:

- Allow the patient to relax – white coat hypertension is a real phenomenon.
- Make sure the cuff is placed centrally over the brachial artery.

- Use a large cuff in obese subjects with an arm circumference over 30 cm – a common cause of overestimating blood pressure.
- Deflate the cuff at a steady controlled rate – ideally no more than 2 mmHg per heartbeat.
- On deflating the cuff, note the point of the first audible sounds (Korotkov I), the point at which sounds become muffled (Korotkov IV), and the point of disappearance of sounds (Korotkov V, which is usually 5–10 mmHg lower than phase IV, but occasionally 0 mmHg). Most observers use phase V as a record of diastolic pressure as this produces less interobserver error.

Occasionally, additional blood pressure readings are indicated:

- Postural – for example if the patient is on antihypertensive medication or has dizziness, or to assess hypovolaemia.
- Comparison of right and left arms – for example for suspected aortic dissection, aortic coarctation.
- Repeated measurements – before diagnosing hypertension, always take readings on a variety of occasions.
- Comparison of the ankle:brachial ratio (called the ABPI) – for example for aortic coarctation, peripheral vascular disease.

All these can be difficult to get as electronic measuring devices are taking over. You need to have a working knowledge of a manual sphygmomanometer and know where one can be found on the ward. Measuring blood pressure is a common question in OSCE style examinations; it should be easy marks.

Neck

Arterial pulse
Follow the usual pattern of:

- Inspection – look at the pulsation (e.g. signs of collapsing pulse).
- Palpation – palpate the carotid artery specifically for volume and character.
- Auscultation – auscultate for the presence of a carotid bruit.

Venous pulse
Assessment of the jugular venous waveform is of fundamental importance to the cardiovascular examination (Fig. 12.8). There are no valves between the right atrium and internal jugular vein, which is

Fig. 12.8 Characteristics of the jugular venous waveform and arterial pulse in the neck.

Characteristics of the jugular venous waveform and arterial pulse in the neck	
Jugular venous pulse	**Carotid pulse**
most prominent deflection is inwards	most prominent deflection is outwards
in sinus rhythm, two deflections for each beat	only one deflection for each beat
height of the wave changes with posture	height is independent of posture
temporary elevation of wave following pressure over the right costal margin (hepatojugular reflux)	constant
can usually be eliminated by light digital pressure over the clavicles	still present after light pressure
not palpable (in absence of tricuspid regurgitation)	palpable

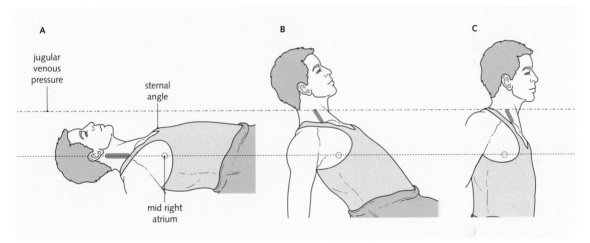

Fig. 12.9 The jugular venous pressure (JVP) is recorded as the vertical height of the visible waveform above the sternal angle. The pressure is fixed, but the anatomical position of the waveform varies according to posture. The jugular venous pressure is usually 2–5 cm.

readily distensible. Therefore it can act as a manometer reflecting the filling pressure of the right heart. Examine the height of the wave. The jugular venous pressure (JVP) is measured as the vertical height of the column of blood in the internal jugular vein above the sternal angle (Fig. 12.9).

Assessment of the JVP should not be neglected. It is the most direct assessment of the filling pressure of the right heart.

The pulsation should be distinguished from arterial pulsation (Fig. 12.10). If it is not visible, the pressure may be:

- Too low (e.g. in hypovolaemia) – lie the patient flat, consider testing the hepatojugular reflux (in an examination, say you would test it by pressing over the liver; it is unnecessary to test in this situation as it is uncomfortable at best).
- Too high (e.g. in right ventricular infarction, volume overload) – sit the patient upright.

Note the character of the waveform. A basic appreciation of the normal jugular venous waveform is needed (see Fig. 12.10). Assessment of the

103

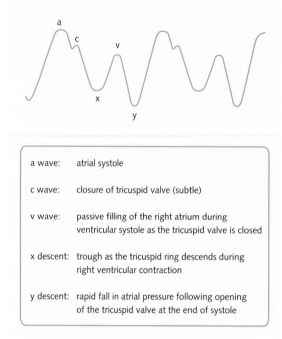

a wave:	atrial systole
c wave:	closure of tricuspid valve (subtle)
v wave:	passive filling of the right atrium during ventricular systole as the tricuspid valve is closed
x descent:	trough as the tricuspid ring descends during right ventricular contraction
y descent:	rapid fall in atrial pressure following opening of the tricuspid valve at the end of systole

Fig. 12.10 The components of the jugular venous waveform. In practice, the c wave is not usually visible. It is only by understanding the normal waveform that pathognomonic signs, such as cannon waves and giant v waves, can be recognized.

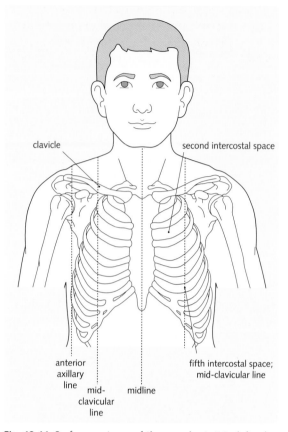

Fig. 12.11 Surface anatomy of the apex beat. It is defined in relation to the intercostal space and imaginary vertical lines related to the clavicle and axilla. The normal apex beat is within the fifth intercostal space in the mid-clavicular line.

waveform requires considerable skill and experience. Unfortunately there are no short cuts. Assessment may give clues to pathologies such as tricuspid regurgitation (giant v waves), atrial fibrillation (no atrial systole, so only one component) and complete heart block (cannon waves at a rate of 40–45 beats per minute).

Praecordium

Although this is the meat of the cardiovascular examination, clues to underlying pathology should have been derived from the peripheral examination. Follow a strict examination routine. Remember that palpation should precede auscultation.

Apex beat

The apex beat is the most downward and outward position where the cardiac impulse is palpable at 90° to the examiner's finger. Define the following.

Position The normal apex beat is within the fifth intercostal space in the mid-clavicular line (Fig.

12.11). A displaced apex beat usually implies volume overload of the left ventricle.

Character Check the character of the beat. For example:

- Forceful, 'sustained', 'heaving' – left ventricular hypertrophy.
- Tapping – mitral stenosis.
- Thrusting – volume overload.

Palpate the rest of the praecordium

Note the presence of:

- Heaves – use either the palm or the medial aspect of the hand; right ventricular hypertrophy may cause a left parasternal lifting sensation.

- Thrills – palpable murmurs (especially in aortic stenosis) feel like a fly trapped in one's hands.
- Palpable heart sounds – for example first heart sound in mitral stenosis felt as a 'tap' at the apex.

Auscultation

An understanding of the cardiac cycle is essential when interpreting findings (Fig. 12.12). To begin with, time what you hear to the patient's pulse, as this helps in determining what you are listening to (Fig. 12.13). You will not pass or fail solely on what you hear. It is only part of the examination.

Consider the most useful regions for auscultation (Fig. 12.14).

Approach in a systematic manner. In each region, listen for:

- The first heart sound (immediately precedes systole and therefore the peripheral pulse).
- The second heart sound (after the pulse is felt).
- Murmurs during systole.
- 'The absence of silence', usually a murmur, during diastole.
- Any extra sounds (e.g. clicks, snaps).

If a murmur is heard, characterize its features as follows:

- Volume (Fig. 12.15).
- Onset – for example presystolic, early systolic.
- Pattern – for example crescendo-decrescendo pattern of aortic stenosis.
- Termination – compare the early systolic murmur of aortic stenosis and systolic murmur of mitral regurgitation.
- Pitch – for example low-pitched murmur of mitral stenosis.
- Location – where it is heard, and where it is most audible.
- Radiation – for example mitral regurgitation murmur radiating to the axilla.

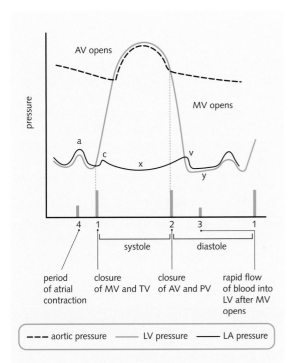

Fig. 12.12 The cardiac cycle. Systole starts at the point of closure of the mitral valve (MV) – first heart sound – when pressure in the left ventricle (LV) exceeds that of the left atrium (LA). There is a period of isovolumetric contraction before the pressure in the LV exceeds that in the aorta, at which point the aortic valve (AV) opens and blood starts to flow into the aorta. Following the onset of relaxation of the LV, the aortic pressure exceeds that in the LV and the aortic valve closes – second heart sound. The ventricle continues to relax until the pressure falls below that in the filled LA, and the MV opens allowing blood to flow rapidly into the LV. Atrial contraction precedes ventricular contraction causing a presystolic accentuation of flow into the LV. PV, pulmonary valve; TV, tricuspid valve.

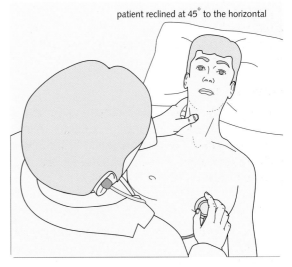

Fig. 12.13 It is essential to listen to the cardiac sounds while timing their point in the cardiac cycle by palpating the carotid pulse at the same time.

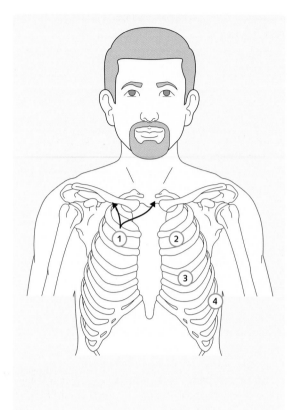

Fig. 12.14 Positions on the chest to auscultate for cardiac sounds. 1. Second right intercostal space (aortic area) – mitral murmurs are very rarely audible here; if a murmur is audible, trace it towards the neck. 2. Second left intercostal space (pulmonary area) – aortic regurgitation may be louder here. 3. Fourth left intercostal space (tricuspid area) – especially for tricuspid regurgitation, but mitral regurgitation and aortic stenosis are often also audible here; aortic regurgitation may be loudest here. 4. Apex (mitral area) – listen specifically for mitral stenosis with the bell of the stethoscope; if a murmur is audible, trace it towards the axilla.

Murmurs

The way to discern murmurs is to listen to lots of them! Reflect upon each murmur you listen to. Why did you get the diagnosis correct (or not)? Did you miss, or misinterpret, clinical signs? This will help you identify areas for improvement in your examination routine. The sign of a good student or clinician is someone who already has a diagnosis in mind before auscultating.

Diastolic murmurs are often difficult to hear. Listen specifically for the 'absence of silence' in diastole. Certain manoeuvres (see below) need to be performed to augment the sounds.

Grade	Intensity
I	just audible under optimal listening conditions
II	quiet
III	moderately loud
IV	loud and associated with a thrill
V	very loud
VI	audible without the aid of a stethoscope

Fig. 12.15 Grading of the intensity of a cardiac murmur. Don't ever be tempted to say a murmur is Grade I or II. You aren't that good . . . yet!

To elicit aortic regurgitation, sit the patient forward and listen over the right second intercostal space and left sternal edge (LSE) in fixed expiration using the diaphragm of the stethoscope. To elicit mitral stenosis, roll the patient into a left lateral position (1/4 turn to the left) and listen over the apex in fixed expiration using the bell of the stethoscope. Mild exertion may accentuate the murmur.

The different heart sounds are detailed in Figure 12.16.

Figure 12.17 is an illustration of a revision card detailing the order in which to conduct a complete cardiovascular system examination and what you are looking to find at each section. Try making your own as a revision tool.

AORTIC STENOSIS

Diagnose the pathology

The features of aortic stenosis are shown in Figure 12.18. The murmur needs to be distinguished from the ejection systolic murmurs associated with:

- Aortic sclerosis – this is common, the patient is usually elderly and there are no haemodynamic effects.
- Bicuspid aortic valve – this is a common cause of an ejection systolic murmur in an asymptomatic young person.
- Subvalvular aortic stenosis – there is no ejection click.

Significance of different cardiac sounds		
Audible heart sounds	**Timing**	**Cause**
first heart sound	immediately presystolic	closure of mitral and tricuspid valves
second heart sound	end of systole	closure of aortic and pulmonary valves
third heart sound	early diastole	corresponds to period of rapid ventricular filling; normal in young fit people; associated with impaired LV function (especially raised end-diastolic pressure) (low-pitched, best heard at apex)
fourth heart sound	immediately presystolic	atrial systole; associated with non-compliant LV (e.g. hypertension); best heard with bell at apex
ejection click	early systole	opening of stenotic aortic valve
opening snap	early diastole	opening of abnormal tricuspid valve and especially mitral valve in mitral stenosis (well-defined short, high-pitched sound best heard with diaphragm at left sternal edge)
pericardial rub	systole and diastole	inflamed pericardium; coarse grating sound (like walking on fresh snow); accentuated by leaning patient forwards; may be very localized

Fig. 12.16 Significance of different cardiac sounds. LV, left ventricle.

- Pulmonary stenosis – this is rare and the murmur is loudest in the left second intercostal space.
- Atrial septal defect – here there is a pulmonary flow murmur, fixed splitting of the second heart sound and an associated tricuspid flow murmur.

Assess severity

Physical features suggestive of significant aortic stenosis include (though the symptoms are the most significant):

- Slow-rising pulse.
- Narrow pulse pressure.
- Displaced apex beat (suggestive of decompensation, unless there is associated aortic regurgitation (mixed aortic valve disease)).

Consider aetiology

The most common causes include:

- Degenerative disease – most common cause.
- Congenital anomaly.
- Rheumatic disease – now less common; look for associated valve pathology.

AORTIC REGURGITATION

Diagnose the pathology

Aortic regurgitation is associated with a diastolic murmur, but there are other clues that indicate the presence of aortic regurgitation (Fig. 12.19).

The murmur of aortic regurgitation is often subtle. A specific attempt should always be made to listen for it. Sit the patient forward and listen with the diaphragm of the stethoscope in fixed expiration. Listen for the 'absence of silence' in diastole.

Features on systemic examination include (describe what you have found and avoid the eponymous names; they are for information only):

- Pulse – collapsing, high volume (Corrigan's pulse).
- Nails – visible capillary pulsation (Quinckes sign).
- Neck – head nods with each systole (De Mussets sign); vigorous arterial pulsation in neck (Corrigan's sign).
- Blood pressure – wide pulse pressure.
- Apex – displaced (due to volume overload – bad sign), thrusting.
- Peripheral pulse – diastolic murmur over lightly compressed femoral artery (several eponymous signs depending upon what is heard).

Cardiovascular examination	
General inspection:	build, obvious pain/distress, features of Down's/Turner's/Marfan's
Hands:	anaemia, cyanosis (poor peripheral perfusion), capillary refill, xanthomata – clubbing – loss of angle, 'boggy' nail, $\uparrow$longitudinal curvature, 'drumstick' – signs of endocarditis: splinter haemorrhage, Osler's nodes, Janeway lesions
Radial pulse:	rate (55–80 at rest), rhythm radio-radial, radio-femoral delay (co-arc^n) – collapsing pulse (aortic regurg^n), bounding pulse ($\uparrow$HR, CO_2 retention, fever, LVF)
Brachial pulse:	character, volume – slow rising (aortic stenosis), collapsing/waterhammer (aortic regurgitation), – biphasic (aortic stenosis+regurgitation). Small vol (shock, $\uparrow$HR; mitral stenosis)
Eyes:	anaemia, jaundice, arcus, xanthomata, conjunctival haemorrhage
Face:	central cyanosis ($\downarrow SaO_2 \rightarrow$blue mucous membranes) – malar flush (mitral stenosis)
Neck:	carotid pulse – separately – JVP – internal jugular vein – just medial to clavicular head of sternocleidomastoid • height of column above sternal angle. >4 cm if: RV failure, $\uparrow$vol, SVC obstruction a = atrial contraction, c = tricuspid valve bulging into atrium, x = atrial relaxation, v = $\uparrow$pressure if atrial filling in ventricular systole, y = tricuspid opening in diastole. • waveform – double impulse No a wave in AF – cannon wave (giant a wave atrial contraction + closed tricuspid) – complete heart block, VT – systolic wave (c and v in ventricular systole) – tricuspid regurgitation
Pericardium:	inspect chest deformity, scars, pacemakers, visible pulsations – palpate (i) apex-position (normal 5th ICS, midclavicular), character (tapping/thrusting), LV heave, thrill (MR) (ii) Ⓛ sternal edge–parasternal heave (RVH), thrill of VSD (iii) aortic area–thrill of aortic stenosis, 2nd RICS (iv) trachea - mediastinal shift – auscultate–apex: heart sounds, added sounds, murmurs. Time a carotids – Ⓛ sternal edge, aortic – pulmonary areas, Ⓛ axilla. Carotid bruits – Ⓛ lateral position – mitral stenosis (low pitched diastolic) – sitting forward, breath held in expiration – aortic regurgitation (Ⓛ lower sternal edge)
Lung bases:	pulmonary oedema
Sacral oedema:	ankle oedema
Peripheral:	pulses, femoral bruits
Hepatomegaly, ascites	
BP	
Fundi	– silver wiring, a/v nipping, hard and soft exudates

Fig. 12.17 Revision card for cardiovascular examination.

Assess severity

Features suggestive of severe disease include evidence of cardiac dilatation and signs of left heart failure.

Consider aetiology

Clues to the underlying pathology may be found, as follows:

- Look at posture and arthropathy – ankylosing spondylitis.
- Look at eyes – Argyll–Robinson pupils associated with syphilitic aortitis.
- Look for high-arched palate, hypermobility, arachnodactyly – Marfan's syndrome (the height of the patient is frequently missed when they are lying down!).
- Check for other valve lesions – rheumatic fever, infective endocarditis.

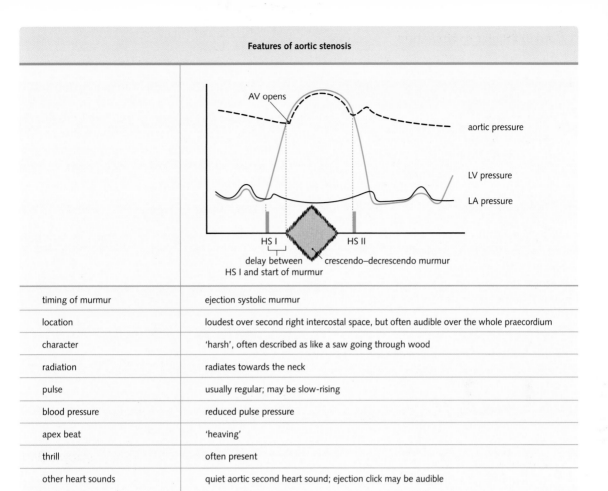

timing of murmur	ejection systolic murmur
location	loudest over second right intercostal space, but often audible over the whole praecordium
character	'harsh', often described as like a saw going through wood
radiation	radiates towards the neck
pulse	usually regular; may be slow-rising
blood pressure	reduced pulse pressure
apex beat	'heaving'
thrill	often present
other heart sounds	quiet aortic second heart sound; ejection click may be audible

Fig. 12.18 Features of aortic stenosis. AV, aortic valve; HS, heart sound; LA, left atrium; LV, left ventricle.

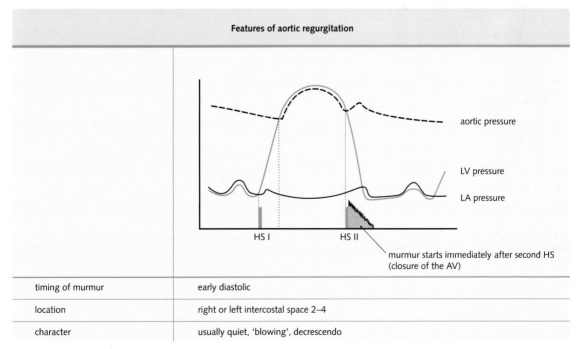

timing of murmur	early diastolic
location	right or left intercostal space 2–4
character	usually quiet, 'blowing', decrescendo

Fig. 12.19 Features of aortic regurgitation.

MITRAL STENOSIS

Diagnose the pathology

A particularly high index of suspicion should be raised in a patient with atrial fibrillation. Look for the classical mitral facies (cyanotic discoloration of the cheeks) and a tapping apex beat. The murmur is often very soft and an attempt should always be made to specifically elicit it – lean the patient on the left hand side, and listen with the bell of the stethoscope in fixed expiration. It may be necessary to accentuate the murmur by exercise. The features of the murmur are illustrated in Figure 12.20.

Assess severity

Look for features of left atrial overload and consequent left heart failure, for example:

- Cyanosis.
- Pulmonary oedema.
- Hypotension (reduced cardiac output).

Look for features of right heart failure (raised JVP, peripheral oedema, left parasternal heave due to right ventricular hypertrophy). Atrial fibrillation occurs later in the natural history of the disease. The opening snap or onset of the murmur is closer to the second heart sound in severe disease as left atrial pressure is raised.

Consider aetiology

Rheumatic heart disease is by far the most common cause.

MITRAL REGURGITATION

Diagnose the pathology

Mitral regurgitation is commonly heard, and produces a pansystolic murmur. It may be due to a number of different disease processes. Clues may be obtained before auscultation by the presence of atrial fibrillation (common, but less than mitral

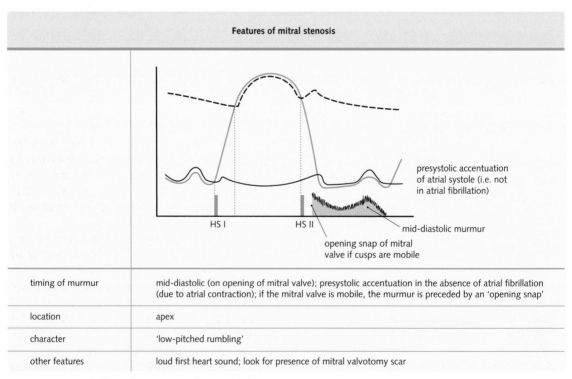

Features of mitral stenosis

presystolic accentuation of atrial systole (i.e. not in atrial fibrillation)

HS I   HS II   mid-diastolic murmur

opening snap of mitral valve if cusps are mobile

timing of murmur	mid-diastolic (on opening of mitral valve); presystolic accentuation in the absence of atrial fibrillation (due to atrial contraction); if the mitral valve is mobile, the murmur is preceded by an 'opening snap'
location	apex
character	'low-pitched rumbling'
other features	loud first heart sound; look for presence of mitral valvotomy scar

Fig. 12.20 Features of mitral stenosis. HS, heart sound.

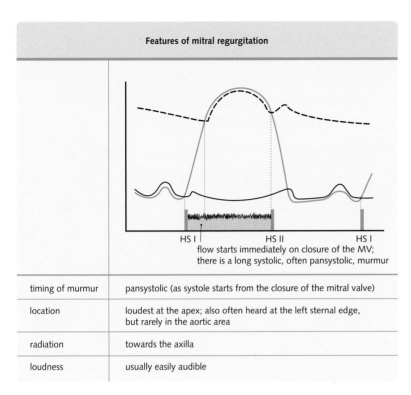

Features of mitral regurgitation

HS I   HS II   HS I

flow starts immediately on closure of the MV;
there is a long systolic, often pansystolic, murmur

timing of murmur	pansystolic (as systole starts from the closure of the mitral valve)
location	loudest at the apex; also often heard at the left sternal edge, but rarely in the aortic area
radiation	towards the axilla
loudness	usually easily audible

Fig. 12.21 Features of mitral regurgitation. HS, heart sound; MV, mitral valve.

stenosis), a displaced thrusting apex beat and, occasionally, a systolic thrill. Features of the murmur are illustrated in Figure 12.21.

Assess severity

It is often difficult to assess the severity clinically, but a third heart sound and a mid-diastolic mitral flow murmur may be detected in severe disease. The severity is usually assessed by the symptoms and haemodynamic consequences.

Consider aetiology

Mitral regurgitation may occur in any process that causes left ventricular dilatation and consequent stretching of the mitral valve annulus, as well as valvular pathology. The most common causes are:

- Ischaemic heart disease – by far the most common cause
- Rheumatic heart disease – relatively common in elderly patients, but now a rare disease in the UK; often associated with mitral stenosis
- Infective endocarditis – look for peripheral stigmata
- Papillary muscle rupture – especially after myocardial infarction

- Mitral valve prolapse – common and due to ballooning of the posterior leaflet into the left atrium; it is associated with a mid-systolic click and late systolic murmur.

TRICUSPID REGURGITATION

Diagnose the pathology

Tricuspid regurgitation can often be recognized from the end of the bed by the observant examiner. It is usually due to primary left heart disease and secondary right ventricular pressure overload, so the signs may be due to a mixture of pathologies.

The jugular venous waveform shows characteristic giant V waves, and may cause oscillation of the ear lobes if the venous pressure is high enough. The patient often has signs of right heart failure with peripheral oedema and, occasionally, ascites. If tricuspid regurgitation is suspected, the liver should be palpated for pulsatile hepatomegaly. In addition, the underlying cause should be sought.

The murmur resembles that of mitral regurgitation in many respects (Fig. 12.22).

Fig. 12.22 Features of tricuspid regurgitation. HS, heart sound.

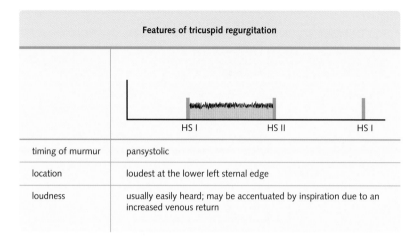

	Features of tricuspid regurgitation
timing of murmur	pansystolic
location	loudest at the lower left sternal edge
loudness	usually easily heard; may be accentuated by inspiration due to an increased venous return

Assess severity

Look for features of right-sided heart failure.

Consider aetiology

Tricuspid regurgitation usually results from right ventricular overload. The more common conditions predisposing to this are:

- Mitral valve disease.
- Cor pulmonale.
- Right ventricular myocardial infarction.

Primary tricuspid regurgitation may occur in:

- Rheumatic fever – rarely an isolated valve lesion.
- Infective endocarditis – especially in drug addicts (look for needle marks).
- Carcinoid syndrome – look for hepatomegaly, flushing, signs of pulmonary stenosis.

INFECTIVE ENDOCARDITIS

Infective endocarditis may affect any of the heart valves. It often presents in a non-specific or insidious manner and should be suspected in all patients with newly diagnosed valvular pathology or those with pre-existing valvular disease who develop pyrexia and malaise or a change in murmur.

Assess valvular pathology and severity

Define the valvular lesions and the haemodynamic consequences by undertaking a thorough cardiovascular examination.

Look for peripheral stigmata suggesting endocarditis

A complete systemic examination is essential as manifestations may arise from many systems. Features on systemic examination include the following.

Pyrexia

Fever is almost universal in infective endocarditis. However, it is often low grade and a single temperature reading may be normal. It is very important to follow the progression of fever through the course of the illness as a marker of successful therapy.

Nails (do not forget the toes!)

There are multiple stigmata of infective endocarditis in the hands. Many of these arise from an associated vasculitis. The following signs may be detected in the nails:

- Splinter haemorrhages. These are a non-specific finding, but common. They are suggestive of a vasculitic process. Although the presence is not specific for vasculitis, the occurrence of new splinter haemorrhages developing in the context of a new murmur and fever is highly suggestive.
- Nailfold infarcts. These are also suggestive of a vasculitic process. They are more specific, but less common than splinter haemorrhages.
- Clubbing. A late sign and hopefully not present!

Hands

Vasculitic signs may also be detected in the hands (both are exceedingly rare):

Signs of right and left heart failure	
Right heart failure	**Left heart failure**
raised jugular venous pressure	third heart sound
peripheral oedema	displaced apex beat (if volume overload)
ascites	pulmonary oedema
hepatomegaly	tachycardia
left parasternal heave	cyanosis
cyanosis	cool, sweaty, pale skin (low output state)
tricuspid regurgitation	mitral regurgitation (due to volume overload of left ventricle)

Fig. 12.23 Features of right and left heart failure.

- Osler's nodes. Do not forget the four Ps (painful, purple, papules on the pulps) of the fingers – this is rare, but traditionally enjoyed by examiners!
- Janeway lesions. These are rare transient macular patches on the palms.

Eyes

Fundoscopy is essential (especially in the exam setting). Look for Roth's spots, which are characteristic flame-shaped haemorrhages on the retina with white centres.

Splenomegaly

Splenomegaly is usually barely palpable, if at all. There is little correlation with the duration or severity of the disease.

Haematuria

An immune complex nephritis is a common feature. Remember that urinalysis is part of the examination. Haematuria usually clears with successful antibiotic therapy. Occasionally confusion may arise as the long-term antibiotic therapy may precipitate interstitial nephritis (e.g. penicillins) or be nephrotoxic (e.g. gentamicin).

Neurological signs

Endocarditis may present with neurological signs due to septic emboli in the brain. The elderly may present non-specifically with confusion.

Consider aetiology

Risk factors for infection may be elicited from the examination:

- Right-sided valve lesions – especially in intravenous drug abusers (*Staphylococcus aureus* and fungal disease more likely).
- Underlying valve lesion – usually present; *Streptococcus viridans* is the most common agent; the most common predisposing valvular lesions are mitral or aortic valve disease, ventricular septal defect (VSD), patent ductus arteriosus (PDA) and coarctation of the aorta.
- Prosthetic valve.
- Dental hygiene – often the mouth is the primary portal of entry.

HEART FAILURE

A diagnosis simply of 'heart failure' is inadequate. An attempt must be made to assess its severity, functional impact and aetiology.

Establish the diagnosis and differentiate the features of right and left heart failure

Clues from the systemic examination are shown in Figure 12.23. High-output cardiac failure may also occur (Fig. 12.24). Not all of the features may be present and they depend upon the severity and underlying cause. Often the two conditions coexist.

Features of different types of heart failure on systemic examination	
Cause of heart failure	Features
impaired myocardial contractility ischaemic heart disease	most common cause; may present as right, left or biventricular failure; look for features to suggest other vascular disease (e.g. carotid bruits, signs of hypertension, etc.)
cardiomyopathy	look for systemic disease (e.g. amyloid, etc.)
myocarditis	look for signs of systemic infection; tachycardia is often a prominent feature; listen for associated pericardial rub
arrhythmia	common; often exacerbates underlying heart disease; may be able to detect arrhythmia from the pulse; aim to distinguish a primary arrhythmia from one resulting from poor myocardial perfusion
volume overload aortic regurgitation mitral regurgitation tricuspid regurgitation	look for signs of an underlying valvular defect; for left-sided lesions, identify a displaced apex beat
pressure overload hypertension aortic stenosis pulmonary embolus	slow-rising pulse; narrow pulse pressure; sustained apex beat right-sided signs associated with hypotension
impaired ventricular filling mitral stenosis cardiac tamponade restrictive cardiomyopathy	right heart failure; pulsus paradoxus; note jugular waveform

Fig. 12.24 Features of different types of heart failure on systemic examination.

Assessment of severity is usually dependent upon the history and the functional limitation imposed on the patient.

Consider aetiology

The principal causes are:

- Impaired myocardial contractility.
- Arrhythmia.
- Volume overload.
- Pressure overload.
- Impaired filling.

Features of these conditions are illustrated in Figure 12.24.

MYOCARDIAL INFARCTION

Often, there are no specific physical signs following a myocardial infarction (MI), but the physical examination can reveal complications and guide management, both in the acute setting and in the ensuing period.

Inspection

A quick visual inspection may reveal signs of pain or discomfort, necessitating better analgesic control. Assess peripheral perfusion – a cold, sweaty, cyanosed, pale patient suggests shock.

Look specifically for complications

Always check for the presence of complications such as left or right heart failure and the presence of arrhythmias.

Left heart failure

Look for features of left heart failure, such as:

- Signs of poor peripheral perfusion (impaired cardiac output).
- Low-volume pulse.
- Inspiratory crepitations at the lung bases.
- Gallop rhythm with third heart sound.
- Dyspnoea.

These signs may indicate that the patient will not tolerate β-blockade or may benefit from diuretics and angiotensin-converting enzyme (ACE) inhibitors.

Right heart failure

It is very important to recognize the patient with a right ventricular infarction. A disproportionately raised JVP in association with very poor peripheral perfusion, hypotension and ECG signs are characteristic. Fluid balance in these patients is critical and demands central monitoring to assess left atrial filling pressure.

In the presence of shock and signs of right heart failure, always consider the presence of a posterior infarct, which may be subtle on the ECG.

Arrhythmia

Check the pulse carefully. Usually there is a mild tachycardia. The presence of a bradycardia suggests an inferior wall MI. The pulse may give clues to an underlying rhythm disturbance, for example:

- Heart block – especially following anterior wall MI.
- Atrial fibrillation – common.
- Ventricular extrasystoles.

These should be confirmed as the patient will have continuous ECG recording.

Subsequent examination

Once patients are on the ward, they should be examined at least daily. Particular features to assess include:

- Pulse rate (β-blocker, primary cardiac rhythm disturbance).
- Blood pressure (e.g. cardiogenic shock, primary hypertension, new therapy with β-blocker or ACE inhibitor).
- Signs of heart failure.
- Murmurs – especially that of mitral regurgitation due to papillary muscle rupture and the long systolic murmur of ventricular septal defect.
- Pericardial rub.
- Psychological rehabilitation – often, the greatest morbidity on discharge from hospital is psychological; this should be recognized by eliciting the patient's concerns, and treated early by appropriate education and reassurance.
- Signs of deep vein thrombosis – especially if there has been prolonged bed rest.

Objectives

By the end of this chapter you should:

- Be able to perform a respiratory examination.
- Be able to elicit the features of consolidation, pleural effusion and fibrosis.
- Know how to manage a tension pneumothorax.

EXAMINATION ROUTINE

Patient exposure and position

The patient should be exposed to the waist in a similar fashion to exposure for the cardiovascular examination and seated comfortably on a couch, inclined at 45° to the horizontal in a well-lit room.

As with ANY examination, remember: inspection, palpation, percussion, auscultation.

General inspection

This should be performed systematically. As with other systems, think about which features you are looking for from the history before inspection. It is convenient to look first at the patient generally, and then to inspect specific features from the head downwards. This process need only take a few seconds. In the exam situation, make sure your examiner sees you observing/inspecting the patient.

General features

Note the following features:

- Age and sex of the patient.
- General health and body habitus.

- Comfort at rest; for example how easily can the patient climb onto the examination couch?
- Respiratory rate. It is imperative that you have counted the respiratory rate. Don't make it obvious you are observing the patient's breathing as this may influence the rate, especially in anxious patients.
- General environment. For example, in a hospital, look on the bedside cabinet for sputum pots, nebulizers, inhalers, peak flow charts.

Head

Look for evidence of:

- Central cyanosis (blue tongue).
- Plethora (e.g. due to secondary polycythaemia due to chronic hypoxia).
- Cigarette staining of hair and smell of cigarettes.

Chest

Note the presence of:

- Scars (Fig. 13.1).
- Breathing pattern (e.g. shallow or pursed lip and use of accessory muscles – sternomastoid, intercostals, abdominal, etc.).
- Chest wall deformity (e.g. pectus excavatus).
- Asymmetry (e.g. due to previous tuberculosis causing upper lobe fibrosis; kyphoscoliosis causing constrictive problems).
- Obvious tracheal deviation.

Specific examination

A more detailed systematic examination can now begin.

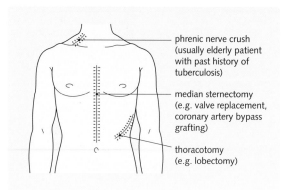

Fig. 13.1 Scars related to the respiratory system.

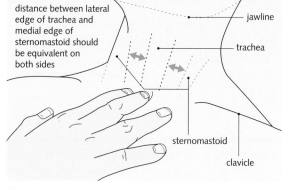

Fig. 13.2 Examination technique for assessing tracheal position. Place two fingers on either side of the trachea and judge the distance between the fingers and the sternomastoid tendons.

Hands

In particular, look for the presence of:

- Clubbing.
- Peripheral cyanosis – if present, check for central cyanosis.
- Tremor – to demonstrate, ask the patient to hold out their hands with the wrists cocked back and to close their eyes, then look for a tremor. The tremor of carbon dioxide retention is classically flapping in nature (a tremor due to β agonist use is more of a fine oscillation). If a patient can see their hands, they can more easily control any tremor.
- Cigarette staining of the nails and fingers.
- Painful swelling of the wrists (and ankles) – this is suggestive of hypertrophic pulmonary osteoarthropathy (HPOA) due to squamous cell cancer of the lung.

Blood pressure and arterial pulse

A quick assessment of the pulse rate and blood pressure is useful. A hyperdynamic circulation may occur in carbon dioxide retention.

Head

An assessment should include specific inspection of:

- Eyes – to reveal, for example, Horner's syndrome (due to carcinoma of the bronchus; rare, but commoner in exams), jaundice, conjunctival pallor.
- Mouth – for example the tongue and mucous membranes may be cyanosed.

Neck

Lymphadenopathy

Examine the cervical and supraclavicular lymph nodes. It may be easier to defer this part of the examination until the patient is sitting forward so that palpation can be performed from the rear.

Tracheal position

Remember that this part of the examination is slightly uncomfortable (Fig. 13.2). The trachea may be shifted by pathology outside the chest (e.g. thyroid enlargement). Mediastinal position can also be assessed by determining the position of the apex beat.

Chest examination

The tracheal position is often neglected by students, but this sign is of fundamental importance as it may indicate the presence of a focal chest expanding (e.g. tension pneumothorax, massive pleural effusion) or constricting process (fibrosis, lung collapse).

If there is unilateral chest pathology, the side of the chest with reduced expansion always indicates the side of the pathology. These factors are often overlooked by students (and doctors!) but they may make the difference between a random assortment of signs and a unifying diagnosis.

Chest

When examining the chest, remember that there are anterior, lateral and posterior aspects to examine. The routine for chest examination is straightforward, but the interpretation of signs requires experience as very often they reflect only a 'qualitative' difference from the normal state. An appreciation of the range of normality demands practice! It is invaluable to have a more experienced observer who can provide constructive criticism of your approach and interpretation of signs with you while you are examining the patient.

Very few chest pathologies are manifest as a single abnormal sign. Rather, a constellation of signs needs to be interpreted in context and then integrated. The competent student doctor will be constantly analysing the elicited signs and refining a differential diagnosis, anticipating how the next sign may modify the assessment. This highlights the importance of adopting an active approach to the examination. Inspiration rarely comes at the end of the examination!

It is a matter of preference whether the whole process is performed on the front of the chest and then the back, or whether each component of the examination is completed for the whole chest in turn, but the former is more comfortable for the patient and is preferred.

Inspection

If tracheal shift is noted, look more closely for asymmetry, especially scalloping below the clavicles suggestive of loss of volume of the upper lobe.

Palpation

There are three main components to this section:

1. Assessment of mediastinal position.
2. Assessment of chest expansion.
3. Tactile vocal fremitus (TVF). (Vocal resonance will also be discussed in this section for ease of explanation.)

Assessment of mediastinal position

If the trachea is deviated, the position of the lower mediastinum can be assessed by determining the position of the apex beat.

Chest expansion

Assess the degree and symmetry of chest expansion. Expansion should be assessed in the infraclavicular,

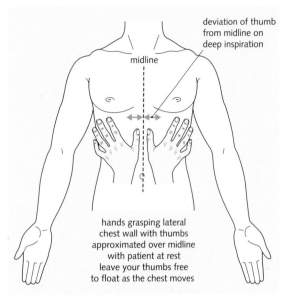

Fig. 13.3 Examination of chest expansion. The primary objective is to assess symmetry. Assess expansion in the upper, lower and middle regions of the chest. Place hands around the lateral chest wall and approximate the thumbs in the midline (but not resting on the chest). Ask the patient to take a deep breath and observe the displacement of the thumbs from the midline.

costal margin, and lower rib region posteriorly (Fig. 13.3).

Tactile vocal fremitus

Place the medial aspect of the hand on the chest wall and ask the patient to make a resonant sound (e.g. say 'ninety nine'). The sound waves are transmitted as low-frequency vibrations, which are palpable. Different regions of the chest should be examined systematically, in particular comparing symmetry of the two sides. Make an attempt to define abnormalities in the upper, middle and lower lobes. This requires an awareness of the surface markings of these lobes.

Vocal resonance

This is included here as it is complementary to TVF (only one of the two need be performed in an exam). Place the diaphragm of the stethoscope on the chest wall and ask the patient to say 'ninety nine'. As with TVF, the lung tissue and airways alter the sound that can be auscultated. When comparing one side to the other, resonance is:

- Increased in consolidation or lung collapse with a patent airway.
- Decreased in pleural effusion, pneumothorax or collapse with an obstructed bronchus.

An appreciation of what is normal, and therefore which side is not, requires practice! Asking the patient to whisper 'one, one, one' over an area of increased resonance may be auscultated more clearly than the opposite side; this is known as *whispering pectoriloquy*.

Tactile vocal fremitus vs. vocal resonance

These techniques potentially offer the same information. In an exam, therefore, you would be well advised to perform only one. In practice, it is easier to detect decreased or absent tactile vocal fremitus and increased vocal resonance from the normal state. Hence, the two signs complement each other. You should make every effort to acquire both skills.

Percussion

Percussion is performed by placing the middle (or occasionally index) finger of the non-dominant hand onto the chest wall and pressing firmly while allowing the digit to conform to the shape of the chest wall. This is then tapped with the index finger of the dominant hand. This action should come from the wrist, rather than being a hammering action, and be just heavy enough to detect reso-nance. The percussing finger should be rapidly with-drawn after striking. Resonance is felt, just as much as it is heard. Practice on yourself with the aid of a senior colleague. Percussion should be performed systematically as for TVF. Findings may include:

- Hyper-resonance of the percussion note. This is often difficult to elicit, but is present if there is more air in the chest cavity (e.g. pneumothorax, emphysema).
- A dull percussion note. This occurs if the lung tissue is replaced by solid material (e.g. consolidation) or if solid material is present between the chest wall and the lung (e.g. pleural effusion, pleural thickening). When this is classical, the note is said to be stony dull.

Auscultation

Breath sounds are produced by turbulent airflow, and transmitted through the airways, lung paren-chyma, pleurae and chest wall. Changes in any of these structures can alter the sounds heard.

Auscultation is usually performed with the bell of the stethoscope as the patient breathes fairly deeply. Once again, it is important to compare find-ings on the two sides.

The aim of auscultation is to define:

- The quality of the breath sounds (Fig. 13.4).
- The presence of added sounds – as with cardiac murmurs, characterize the quality, volume and timing.
- Vocal resonance (see above).

Added sounds should be characterized systemati-cally. Note the presence of the following.

Description of quality of breath sounds		
Breath sounds	**Features**	**Examples**
vesicular breathing	progressively louder during inspiration, merging into expiratory phase with rapid fading in intensity	normal pattern
bronchial breathing	laryngeal sounds transmitted efficiently to chest wall if lung substance becomes uniform and more solid—blowing quality; pause between inspiratory and expiratory phases; expiratory phase as long as inspiratory phase	consolidation; collapse with patent bronchus; fibrosis
diminished volume	impaired conduction due to increased pleural/chest wall thickness or increased air acting as poor conductor	pleural effusion; pleural thickening; obesity; emphysema; pneumothorax

Fig. 13.4 Description of the quality of breath sounds.

Crackles (crepitations) These may be caused by secretions in the larger airways (e.g. in bronchitis, pneumonia, bronchiectasis). They are usually present throughout inspiration and may clear on coughing. Alternatively, reopening of occluded small airways during the later part of inspiration occurs in parenchymal disease (e.g. fibrosis, interstitial oedema) where they tend to have a finer quality. Coughing will then have no effect.

Wheezes (rhonchi) These have a musical quality and usually result from narrowing of the bronchi due to oedema, spasm, tumour or secretions. They occur in:

- Asthma – polyphonic, mainly expiratory, diffuse.
- Bronchitis – may clear on coughing, may have an inspiratory component.
- Fixed obstruction – monophonic (e.g. due to bronchial carcinoma).

Pleural rub This sound is due to friction between the visceral and parietal pleurae in inflammatory conditions (pleurisy). It is often described as the sound of walking on crisp, newly fallen snow.

Other tests

Patients may have their own peak flow meter with them or one may be available. Ask them to perform this, watch and assess their technique. Ask the patient what their peak flow is normally, compare this to today's reading and then against what their predicted peak flow would be. Many wards and examination settings will have spirometry available. Get the patient to do this as well. With both of these tests, always get the patient to have three attempts so that the best can be recorded.

Summary

At the end of the respiratory examination, it is important to take time to reflect on your findings. Remember that few lung pathologies have a single pathognomonic feature and that abnormalities are often a matter of qualitative judgement of deviation from normality rather than the simple presence or absence of a sign. The key to a successful examination is to be attentive and open to differential diagnoses from the start, so that physical signs can be anticipated rather than taking you by surprise. Figure

13.5 shows a revision card for the respiratory system. Try creating one for yourself – a good revision exercise!

PLEURAL EFFUSION

Diagnose the pathology

The physical signs can be anticipated from a basic appreciation of the pathology (Fig. 13.6).

Assess severity

Analyse the significance of physical signs:

- How high up the chest can percussion be demonstrated to be stony dull?
- Define the level of decreased breath sounds and reduced TVF.
- How short of breath is the patient? (This may be altered by coexisting pathology and speed of onset.)
- Is the trachea shifted away from the effusion?

Consider aetiology

Look for clues in the systemic examination:

- Palpable supraclavicular and/or axillary lymphadenopathy – malignancy.
- Hepatomegaly – secondary malignancy.
- Peripheral oedema – hypoalbuminaemia, heart failure.
- Third heart sound – left heart failure.

Consider the more common causes of pleural effusion (Fig. 13.7).

PNEUMOTHORAX

Diagnose the pathology

A small pneumothorax can be hard to identify, but index of suspicion should be high in tall, thin, young adults with sudden-onset pleuritic chest pain and dyspnoea. The basic clinical signs can be elucidated from basic principles (Fig. 13.8).

Respiratory examination

Undress to waist, sit on edge of bed. Look around–O_2, nebulizer, sputum pot, inhalers, spacer, etc.

General appearance — cachexia, respiratory distress, cyanosis/pallor/plethoric etc. Sputum, temperature chart

Hands — clubbing, peripheral cyanosis, nicotine stains, asterixis (CO_2 retention trap)

Pulse, RR
Face — central cyanosis, pallor, nasal flaring, cervical lymphadenopathy, jugular veins

Inspection of chest — RR, pattern of breathing, chest wall deformities, scars, chest wall movements

Palpation of chest — position of trachea, cricosternal distance, supraclavicular lymph nodes, RV heave

Chest expansion — assess upper and lower chest movement
Vocal fremitus — palpate chest wall as patient repeats '99'

Percussion — compare both sides, including clavicles, supraclavicular, axillae
hyper-resonant/resonant or normal/dull/stony dull

Auscultation — breathe in and out thro' mouth. intensity–nature (vesicular/bronchial), air entry
– bronchial breath sounds–high–pitched blowing, pause between insp and exp, consolidation/collapse
– wheeze (rhonchi–high/low pitch, insp/exp, monophonic/polyphonic effect of coughing
– crackles (crepitations) -loud coase/fine and high pitched. Early insp–diffuse airways destruction (eg. COPD)
Late insp–diffuse fibrosis, pulmonary oedema, bronchiectasis, consolidation
– pleural rub–friction betwn pleura →localized creating, e.g. pleurisy, pulmonary infarction, pneumonia
– vocal resonance–auscultate as patient says '99'. Assess volume and clarity, whispering pectoriloquy.

Relevant CVS examination — e.g. heart sounds, JV, hepatomegaly, ankle oedema

	Chest movement	Mediastinum	Percussion	Breath sounds	Visual resonance	Added sounds
Consolidation	↓ affected side	–	Dull	Bronchial	↑	Fine crackles
Collapse	↓ affected side	shift → lesion	Dull	↓ Vesicular	↓	–
Localized fibrosis	↓ affected side	shift → lesion	Dull	Bronchial	↑	Coarse crackles
CFA	↓ both sides	–	Normal	Vesicular	↑	Fine crackles
Pleural effusion	↓ affected side	shift away	Stony dull	↓ Vesicular	↓	–
Pneumothorax	↓ affected side	shift away	Hyper-resonant	↓ Vesicular	↓	–
Asthma	↓ both sides	–	Normal	Prolonged exp	Normal	Exp polyphonic wheeze
COPD	↓ both sides	–	Normal	Prolonged exp	Normal	Exp wheeze, crackles

Overinflation — high shoulders, ↑ant–post. chest diameter, using accessory muscles of resp^n,
limited chest expansion and absent cardiac and/or hepatic dullness on percussion

Diffuse airways obstruction — wheeze, muffled cough, overinflation, costal margin moves in on inspiration, prolonged, expiration, early
insp crackles, ↑forced expiration time

Fig. 13.5 Revision card for respiratory examination.

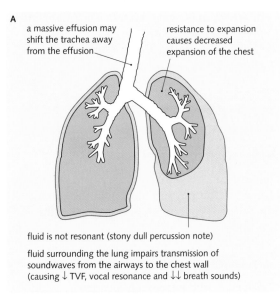

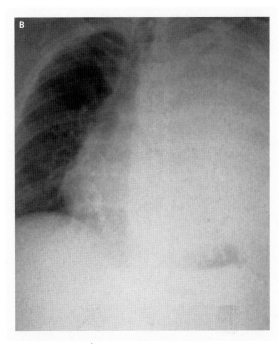

Fig. 13.6 A, Physical signs of a pleural effusion. **B,** Radiograph of a massive left pleural effusion. Note the tracheal deviation to the right. TVF, tactile vocal fremitus.

Common causes of a pleural effusion	
Transudates	**Exudates**
fluid which has passed through a membrane	fluid containing proteins and white cells
left heart failure*	infection* (pneumonia, tuberculosis, empyema, etc.)
hypoalbuminuric states* (e.g. nephrotic syndrome, liver failure)	malignancy* (primary bronchial or metastatic)
fluid overload in renal failure	pulmonary infarction*
	subphrenic abscess
hypothyroidism	pancreatitis
Meigs's syndrome (right pleural effusion associated with ovarian fibroma)	collagen vascular disease (e.g. rheumatoid arthritis)
	haemothorax

Fig. 13.7 Common causes of a pleural effusion. The more common causes are marked with an asterisk.

A tension pneumothorax is a medical emergency and should be identified and treated before requesting a radiograph. Treatment is immediate decompression with a 14G needle placed in the second intercostal space in the mid-clavicular line.

Assess severity

Most pneumothoraces do not cause haemodynamic compromise, but it is important to recognize a tension pneumothorax.

Consider aetiology

Most spontaneous pneumothoraces occur in previously fit young adults and are idiopathic. Look for

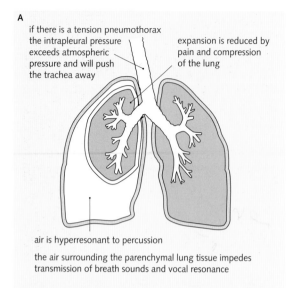

A

if there is a tension pneumothorax the intrapleural pressure exceeds atmospheric pressure and will push the trachea away

expansion is reduced by pain and compression of the lung

air is hyperresonant to percussion

the air surrounding the parenchymal lung tissue impedes transmission of breath sounds and vocal resonance

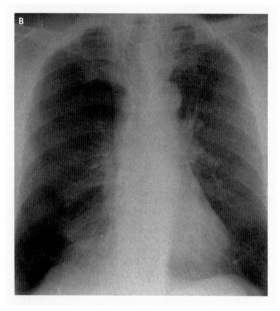

B

Fig. 13.8 A, Physical signs of pneumothorax. **B,** Radiograph of a right pneumothorax.

underlying lung disease that may have precipitated the event. Consider the three major groups of precipitating pathologies, which are:

1. Underlying medical disease (e.g. asthma or emphysema with bullae, carcinoma, tuberculosis).
2. Iatrogenic (e.g. after central venous line insertion, especially subclavian line, intubated patient with positive pressure ventilation, after pleural aspiration or biopsy).
3. Trauma (e.g. fractured ribs, ?surgical emphysema).

LUNG COLLAPSE

Diagnose the pathology

Lung collapse may occur with a patent or occluded bronchus. The physical signs will differ and may depend upon severity and underlying cause (Figs 13.9 and 13.10).

Consider aetiology

Consider the more common causes of collapse.

Extrinsic compression

The most common cause is lymph node compression due to tumour or tuberculosis.

Intrinsic obstruction

Intraluminal obstruction may occlude the airway and cause distal collapse. The more common causes include:

- Tumours – look for clubbing.
- Retained secretions – postoperative, debilitated patients.
- Inhaled foreign body – usually apparent from history, especially right lower lobe – a medical emergency!
- Bronchial cast or plug – for example due to aspergillosis, blood clot.

CONSOLIDATION

Diagnose the pathology

Consolidation implies replacement of air in the acini with fluid or solid material. The lung parenchyma is heavy and stiff, but transmits sound waves to the chest wall more efficiently. In addition, the lower airways often collapse during expiration, but may open explosively during inspiration when negative intrathoracic pressure is generated, producing crepitations (Fig. 13.11).

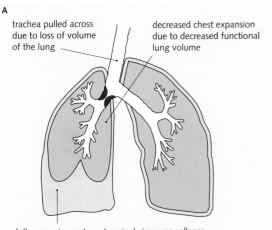

trilogy trachea pulled across due to loss of volume of the lung

decreased chest expansion due to decreased functional lung volume

dull percussion note as terminal airspaces collapse

If the upper airway is occluded, breath sounds and vocal resonance cannot be transmitted efficiently to the chest wall; if the airway is patent, the denser lung parenchyma transmits sounds more efficiently and bronchial breathing with whispering pectoriloquy may be present.

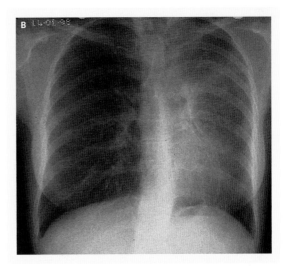

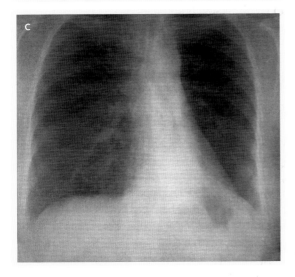

Fig. 13.9 A, Radiographic features and physical signs of lung collapse with and without patent upper airways. **B,** Radiograph of left upper lobe collapse: loss of lung volume is indicated by deviation of the trachea to the left, mediastinal shift to the left and loss of volume of the left lung; in addition, note the classic veil-like opacification over the left lung. **C,** Radiograph of left lower lobe collapse: note the raised left hemidiaphragm, mediastinal shift to the left, loss of volume and hypertranslucency of the left lung, and depressed left hilum indicating loss of lung volume, in association with a change in density behind the heart shadow.

Physical signs of lung collapse with and without patent upper airways		
Sign	**Patent upper airway**	**Obstructed bronchus**
expansion	always reduced on side of lesion	always reduced on side of lesion
trachea	deviated to side of collapse, especially upper lobe collapse	deviated to side of collapse, especially upper lobe collapse
percussion	dull	dull
TVF	usually normal or increased	decreased or absent
vocal resonance	whispering pectoriloquy	—
breath sounds	increased; bronchial breathing	absent or decreased

Fig. 13.10 Physical signs of lung collapse with and without patent upper airways. TVF, tactile vocal fremitus.

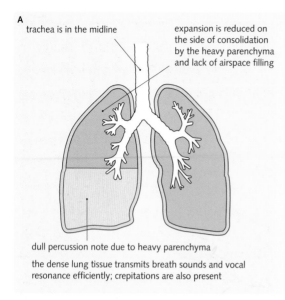

A trachea is in the midline

expansion is reduced on the side of consolidation by the heavy parenchyma and lack of airspace filling

dull percussion note due to heavy parenchyma

the dense lung tissue transmits breath sounds and vocal resonance efficiently; crepitations are also present

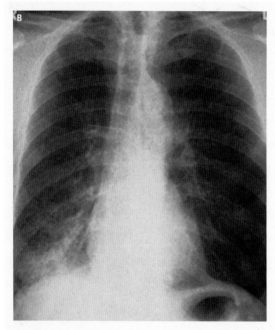

Fig. 13.11 A, Physical signs of consolidation. **B,** Radiograph of right lower lobe pneumonia.

Consider aetiology

The important causes of consolidation include:

- Pneumonia – most common cause, especially if in classical lobar distribution.
- Pulmonary oedema – may be cardiogenic or non-cardiogenic.

- Pulmonary haemorrhage – for example due to pulmonary vasculitis.
- Aspiration.
- Neoplasms – for example alveolar cell carcinoma.

Look for systemic clues of the aetiology:

- Fever, green sputum in sputum pot – pneumonia.
- Third heart sound, mitral murmurs, peripheral oedema – cardiogenic.
- Nailfold infarcts, livedo reticularis, splinter haemorrhages – vasculitis.
- Clubbing – underlying primary bronchial carcinoma.

Differential diagnosis

Clinically, the main differential diagnosis is between pneumonia and pulmonary oedema. This is usually apparent from the clinical setting. It is more likely to be pneumonia if the consolidation is in a lobar distribution or unilateral (especially right sided). In addition, a pleural effusion, raised JVP and peripheral oedema are more common in pulmonary oedema, but not diagnostic.

LUNG FIBROSIS

Pulmonary fibrosis results in lungs that are rigid and resistant to expansion. The fibrotic disease often shrinks the lung, resulting in a constrictive process. The thicker parenchyma will, however, transmit sound waves more efficiently to the chest wall. As with other parenchymal processes, the small airways may open explosively in inspiration causing crepitations. The disease may be unilateral (e.g. tuberculosis) or bilateral, such as in cryptogenic fibrosing alveolitis.

Diagnose the pathology

Distinguish between unilateral and bilateral disease. The tracheal position is most useful, as often there is coexisting pathology (Fig. 13.12).

Consider aetiology

The more common causes of pulmonary fibrosis are illustrated in Figure 13.13.

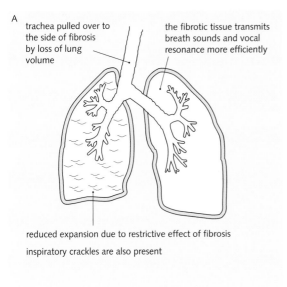

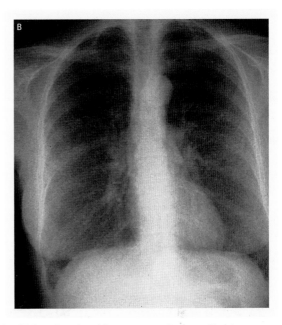

Fig. 13.12 A, Physical signs of pulmonary fibrosis. **B,** Radiograph of bilateral mid and lower zone pulmonary fibrosis.

Causes of pulmonary fibrosis	
Cause	**Examples and signs**
infection*	tuberculosis (typically upper lobe)
collagen disorder	rheumatoid lung (usually basal); scleroderma
extrinsic allergic alveolitis*	especially upper lobes
sarcoidosis*	look for erythema nodosum and other stigmata
radiation	look for radiation burns
drugs	busulfan, bleomycin; etc.
cryptogenic fibrosing alveolitis	rare; begins in lower lobes; look for clubbing
asbestosis	—
ankylosing spondylitis	upper lobes; rigid back; peripheral arthritis; aortic regurgitation; etc.

Fig. 13.13 Causes of pulmonary fibrosis. The more common causes are marked with an asterisk.

INTEGRATING PHYSICAL SIGNS TO DIAGNOSE PATHOLOGY

Figure 13.14 illustrates how the different physical signs elicited in the respiratory system can be integrated so that the basic underlying pathological cause can be identified. Remember that this is only part of the examination process. It is important to assess the severity of the pathology and its underlying cause so that specific therapy can be offered.

ASTHMA

A diagnosis of asthma is usually apparent from the history. The aim of the examination is to assess severity, look for complications and to consider precipitating factors.

Examination findings in the basic lung pathologies					
Lung pathology	Tracheal position	Percussion note	TVF/vocal resonance	Volume of breath sounds	Added sounds
pneumothorax	normal (deviated away in tension pneumothorax)	hyperresonant (often subtle)	decreased (or absent)	decreased or absent	—
consolidation	normal	dull	increased	increased (bronchial breathing)	inspiratory crackles
fibrosis	pulled towards affected side	slightly dull	increased	increased	inspiratory crackles
pleural effusion	normal (deviated away if massive)	stony dull	reduced or absent	decreased or absent	often crackles immediately above effusion
lobar collapse (patent airway)	pulled towards affected side	dull	increased	increased (bronchial breathing)	—
lobar collapse (occluded bronchus)	pulled towards affected side	dull	decreased	decreased	—

Fig. 13.14 Examination findings for the basic lung pathologies. Note that this refers to unilateral lesions. The expansion is always reduced on the side of the lung pathology. TVF, tactile vocal fremitus.

Fig. 13.15 Assessment of severity of an acute asthma attack. In the hospital setting, arterial blood gases (ABG) form part of the routine assessment. Pulsus paradoxus, if easy to elicit, is useful, but often it is difficult – it is a waste of time making an inaccurate 'best guess'.

Assessment of severity of an acute asthma attack			
Feature	Mild	Moderate	Severe
pulse rate (/min)	<100	100–110	>110
respiratory rate (/min)	<20	20–30	>30
peak flow rate	>75% predicted	50–75% predicted	<50% predicted
arterial blood gases	PaO_2 high/normal; $PaCO_2$ low	PaO_2 normal; $PaCO_2$ low or low normal (<5 kPa)	PaO_2 <8 kPa; $PaCO_2$ high normal or high (>5 kPa)

Assess severity

This is assessed by talking to the patient. If the patient cannot finish a sentence, they are having a severe attack and urgent help should be summoned. Use objective reproducible measures of severity and classify the attack as mild, moderate or severe. Remember that not all of the features need to be present in a severe attack (Fig. 13.15). The essential parameters to assess are pulse rate, respiratory rate, peak flow rate and (in hospitals) arterial blood gas estimate.

Other features of a serious or life-threatening attack include:

- Difficulty speaking.
- Bradycardia or hypotension.
- Exhaustion.
- Silent chest.
- Cyanosis.

Look for complications

Examine for the presence of a pneumothorax. This is the main reason for radiography in hospitalized patients.

Consider aetiology

Look for signs of a chest infection, which is common. In older patients with new-onset asthma, look for nasal polyps, which are also associated with aspirin sensitivity. Also inspect for features of atopy, especially in younger patients (e.g. eczema, dry skin, thinning of lateral half of eyebrows from rubbing).

Asthma is potentially fatal. A rapid and objective assessment is essential.

LUNG CANCER

Lung cancer is the most common fatal malignancy in men and its incidence in women is on the increase. It can present in many ways and show many features on examination.

Inspection

Look for clues, such as:

- Cachexia (common).
- Cigarette staining in hair.
- Scar from lobectomy.
- Radiotherapy burn on chest wall.

Hands

There are often signs in the hands, as follows:

- Clubbing – may predate clinical diagnosis.
- Clues to smoking history – nicotine staining on nails.
- Hypertrophic pulmonary osteoarthropathy (HPOA) – pain and swelling of the wrists, especially with small cell carcinoma.

Face

Horner's syndrome (small pupil (miosis), partial ptosis, enophthalmos, anhidrosis due to invasion of the sympathetic ganglion T1 by direct spread) may be a feature of upper lobe disease. This syndrome is frequently missed by final year medical students during finals!

Neck

Palpate for a supraclavicular lymph node. Look for features of superior vena cava obstruction (swollen face and neck, plethora, dilated veins over trunk).

Chest

Look for features of:

- Pleural effusion (common).
- Loss of lung volume due to lobar or lung collapse.

Evidence of spread

Direct spread

Examine specifically for other features of direct spread:

- Pancoast's tumour – apical tumour invading the lower brachial plexus (especially C8, T1, T2) causing sensory loss, wasting and weakness of the small muscles of the hand.
- Phrenic nerve – diaphragmatic palsy.
- Pericardium – effusion (look for features to suggest tamponade).

Metastatic spread

Examine for features of metastatic spread, for example:

- Hepatomegaly.
- Focal neurological signs due to cerebral metastases.
- Localized bony tenderness.

Abdominal examination

Objectives

By the end of this chapter you should:

- Be able to perform an abdominal examination.
- Be able to assess organomegaly of the abdominal organs.
- Recall the common causes of organomegaly of the abdominal organs.

EXAMINATION ROUTINE

A thorough abdominal examination is fundamental whether a surgical or medical condition is suspected, but the emphasis clearly changes according to the presenting complaint.

Patient exposure and position

Ensure good lighting. Patients should be undressed so that a view of the whole abdomen (from nipple to knees) can be obtained. Provide a blanket for warmth and modesty. Lie the patient flat on the couch with a single pillow behind the head (though this may not always be possible if, for example, the patient has orthopnoea or musculoskeletal abnormalities), with arms by their side. If patients are unable to fully relax their abdomen, ask them to flex their hips to 45° and knees to 90° (Fig. 14.1).

Abdominal examination

The key to successful examination is to have a relaxed patient. You will be able to elicit signs more easily if the patient is comfortable.

As with all systems, if asked to examine the abdominal system, always start by looking at the hands and proceed methodically from there. Beware missing or misinterpreting clinical signs; for example jaundice is easily overlooked and is harder to detect in artificial lighting. The recognition and interpretation of signs become much easier once the process of examination becomes second nature; this requires much practice.

General inspection

Observe the general appearance of the patient. Time spent at this stage is invaluable. Take at least 10 seconds, making a mental checklist, for example:

- Is the patient comfortable or distressed at rest?
- Is there any obvious pain?
- Is there any cachexia, pallor, jaundice, abnormal skin pigmentation, distension, etc.?

A rapid but systematic survey of the patient should ensue.

Hands

Careful examination of the hands is fundamental to the abdominal examination and may yield vital clues of underlying abdominal disease.

Note the presence of:

- Metabolic flap (asterixis). This may indicate hepatic encephalopathy (or carbon dioxide retention and uraemia).
- Signs of chronic liver disease. Inspect and palpate both hands for the presence of Dupuytren's contracture, palmar erythema, leuconychia (white nails), spider naevi and clubbing.
- Anaemia. If the patient is profoundly anaemic, palmar skin creases may be pale. Koilonychia (spoon-shaped nails) suggests iron-deficiency anaemia.

Eyes

Inspect the sclerae for jaundice, and the lower eyelid for anaemia.

Fig. 14.1 Ideal position for examination of the abdomen.

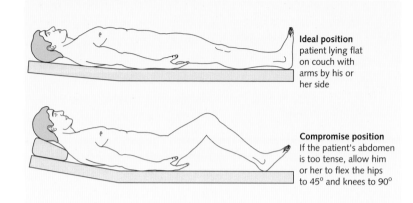

Ideal position
patient lying flat on couch with arms by his or her side

Compromise position
If the patient's abdomen is too tense, allow him or her to flex the hips to 45° and knees to 90°

Face

Note abnormal pigmentation around the lips, or angular stomatitis, which occurs in many medical conditions, especially iron-deficiency anaemia, malabsorption and oral infections.

Oral cavity

Inspect the oral cavity and tongue for:

- Ulceration (e.g. due to inflammatory bowel disease, chemotherapy).
- Inflammation.
- Oral candidiasis (e.g. due to antibiotic therapy, immunodeficiency, diabetes mellitus).
- Halitosis (e.g. due to infection, poor hygiene, hepatic fetor, uraemia, diabetes mellitus).
- Pigmentation (e.g. Addison's disease).

Chest wall

Note the presence of gynaecomastia and spider naevi. The presence of more than five spider naevi is considered to be suggestive of liver disease. These characteristically blanche if the central arteriole is pressed; they refill from the centre outwards.

Supraclavicular lymphadenopathy

Pay particular attention to the left side and look for Virchow's node (Fig. 14.2). If present, this is Troisier's sign, associated with malignancies of the upper third of the gastrointestinal tract

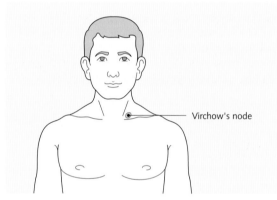

Virchow's node

Fig. 14.2 Virchow's node is a palpable left supraclavicular lymph node. Troisier's sign refers to the presence of a palpable left supraclavicular node in association with gastric carcinoma. This node is easiest to palpate from behind.

Exposure of the abdomen

Following general inspection, the abdomen should be exposed from the nipples to the symphysis pubis. Follow the usual routine of inspection, palpation, percussion, auscultation.

Inspection

Stand at the end of the bed and inspect the abdomen for:

- Symmetry (e.g. massive splenomegaly produces a bulge on the left side).
- Abnormal pulsation (e.g. due to abdominal aortic aneurysm).
- Shape (e.g. distension).

Remember the five Fs from Chapter 5.

Return to right-hand side of the abdomen and actively inspect for the presence of:

- Scars (Fig. 14.3).
- Sinuses (e.g. due to retained suture material).
- Fistulas (e.g. due to Crohn's disease).
- Visible peristalsis (e.g. due to intestinal obstruction).
- Distended veins.
- Flank haemorrhages or apparent bruising (Grey–Turners sign, e.g. due to haemorrhagic pancreatitis or trauma).

Ask patients whether they are aware of any abnormal lumps or areas of tenderness. This may give a clue to the area of pathology. Ask patients to cough, observing pain (peritoneal irritation) and also the hernial orifices.

Palpation

Before palpating, ask the patient where the pain is. Warn the patient that you are going to lay your hand on the abdomen. Lay your hand on the point furthest from the pain and work towards it. The three stages to abdominal palpation are:

1. Light palpation.
2. Deep palpation.
3. Specific palpation of the intra-abdominal organs.

Light palpation

Commence palpation at a site remote from the area of pain. All areas of the abdomen must be palpated systematically. Picture the abdomen in nine regions (Fig. 14.4). (Some people refer to abdominal quadrants as shown in Fig. 14.5.) This helps you to adopt a systematic approach to the examination and when presenting your findings.

Light palpation is performed to elicit any tenderness or guarding. Lie the hands and fingers flat upon the abdomen and press very gently. It is often useful to kneel down beside the bed. It is essential to be as gentle as possible:

- To gain the patient's confidence.
- To prevent voluntary guarding (tensing of the abdominal wall musculature as light pressure is applied), which will mask pathological signs.

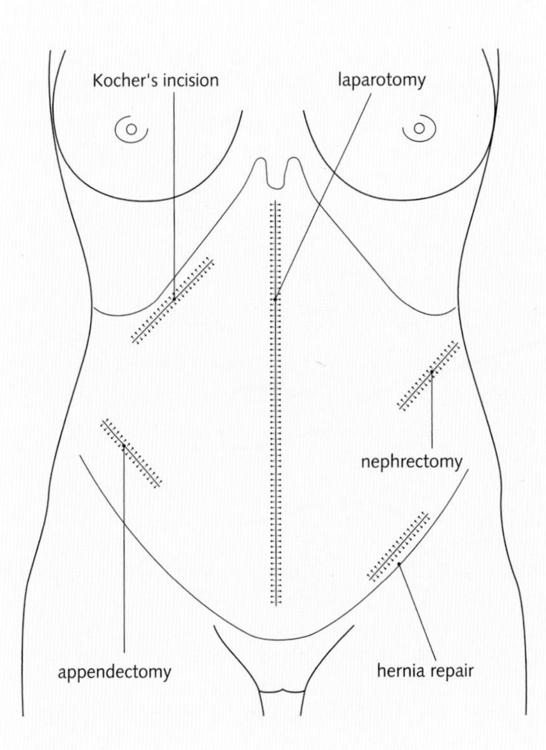

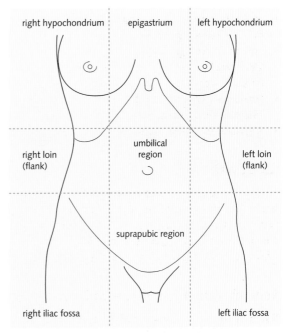

Fig. 14.3 Surgical scars on the abdominal wall.

Fig. 14.4 The regions of the abdomen. It is helpful to be aware of these regions when palpating or presenting physical findings as this helps in the differential diagnosis.

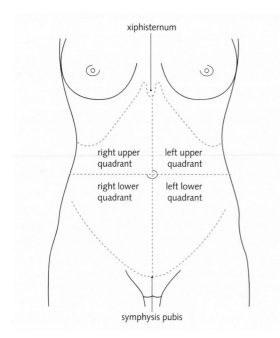

Fig. 14.5 The four quadrants of the abdomen.

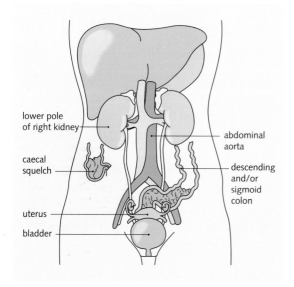

Fig. 14.6 Normal structures that may be palpable on deep palpation of the abdomen.

Deep palpation

Warn the patient that you will be pressing more firmly and feel for any obvious masses (Fig. 14.6) or tenderness in the nine regions. If a mass is identified, determine its characteristics systematically.

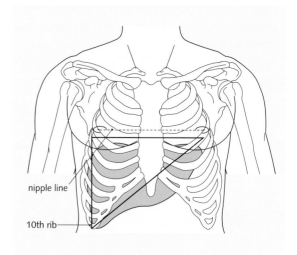

Fig. 14.7 Surface anatomy of the liver.

Specific palpation of the intra-abdominal organs

Liver Always start in the right iliac fossa when examining the liver or the spleen, as both expand towards this region. Place your hand flat with fingers pointing towards the patient's head (or alternatively to the left flank), and palpate deeply while asking the patient to breathe in and out deeply. Keep your hands still while the patient is breathing in, as the liver edge moves downwards on inspiration. If nothing is felt, repeat the process with the hand slightly higher up the abdomen, advancing a few centimetres at a time until the costal margin is reached.

The liver may be palpated in normal subjects, especially if they are thin or if there is chest hyperinflation (Fig. 14.7). If the liver edge is palpable, describe:

- The size of the liver (express as finger breadths below the costal margin).
- Its contour (regular or irregular).
- Its texture (smooth, nodular).
- Any tenderness (see Hepatomegaly, p. 138).

After this, percuss out the superior and inferior borders of the liver, always from resonant to dull. This will tell you whether there is hepatomegaly or whether there is displacment of a normal sized liver.

Spleen The spleen is examined by a similar process as for the liver. Start in the right iliac fossa with fingers pointing towards the left costal margin and ask the patient to breathe in and out while advancing towards the left costal margin. If there is no obvious splenomegaly, ask the patient to roll onto the right side, place your left hand around the lower left costal margin and lift forwards as the patient inspires, while palpating with your right hand; this can be performed after percussing for shifting dullness (see below), therefore necessitating the patient to roll onto their side only once to perform both manoeuvres (Fig. 14.8).

A normal spleen is not palpable.

Kidneys   The kidneys are examined bimanually by the technique of ballottement. The left kidney is felt by placing your left hand in the left loin below the 12th rib, lateral to the erector spinae muscles and above the iliac crest, with the right hand placed anteriorly just above the anterior superior iliac spine. During inspiration, the left hand is then lifted gently upwards towards the right hand (Fig. 14.9).

The kidney may be palpable in thin normal individuals. The right kidney is examined with the right hand posteriorly and the left hand anteriorly.

In normal individuals, the right kidney lies lower than the left (due to downward displacement by the liver) and is more likely to be palpable (Fig. 14.10).

Abdominal aorta   Palpate specifically for an abdominal aortic aneurysm (AAA). This is performed by placing the palmar surfaces of both hands laterally and with the fingertips positioned in the midline a few centimetres below the xiphisternum (Fig. 14.11).

- An AAA is both pulsatile and expansile (fingertips will be pushed outwards).
- A non-aneurysmal abdominal aorta is only pulsatile (fingertips pushed upwards, but not outwards) (Fig. 14.12).

Percussion

Percuss over the whole abdomen and particularly over masses. This is also a sensitive method for eliciting peritonitis. Specifically percuss for ascites by

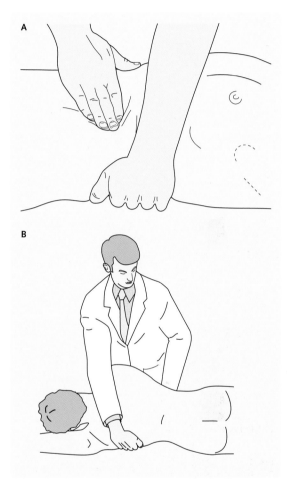

Fig. 14.8 Palpation of the spleen. (**A**) Initial examination. (**B**) If the examination is difficult, the patient should be asked to roll onto the right side; push the spleen forwards with your left hand and palpate with your right hand.

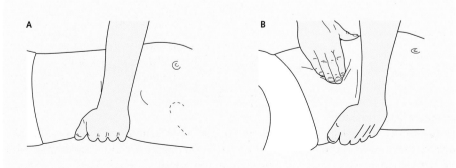

Fig. 14.9 Palpation of the kidneys.

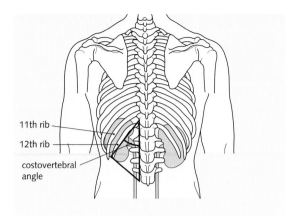

Fig. 14.10 Surface anatomy of the kidneys.

11th rib

12th rib

costovertebral angle

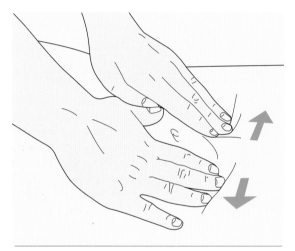

Fig. 14.11 Palpation for the abdominal aorta.

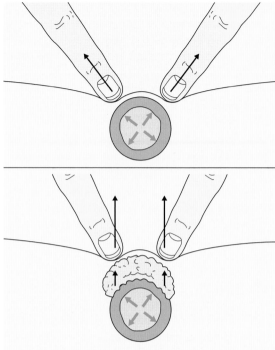

Fig. 14.12 Distinction between aortic pulsation and movement of an overlying structure. True pulsatility is indicated by outward displacement of the palpating hands.

A complete abdominal examination includes assessment of:

- The hernial orifices – see Chapter 16.
- The external genitalia – see Chapter 16.
- A rectal examination.

Digital rectal examination

The digital rectal examination (DRE or PR) is usually performed with the patient in the left lateral position, with both hips and knees fully flexed (Fig. 14.15). It is essential to explain the procedure to the patient and to be gentle! It is usually possible to palpate lesions up to 6–8 cm from the anal verge. Before performing a digital examination:

- Inspect the anus, its margins and surrounding skin.
- Look for skin tags, excoriation, prolapsed or thrombosed haemorrhoids, fistulas, fissures, abscesses or ulceration due to an anal carcinoma.
- Ask the patient to bear down or strain. This may reveal the presence of a rectal prolapse or, occasionally, a polyp.

testing for shifting dullness. Percuss from the midline to the flank. If ascites is present, the initially resonant note will become dull. Note the point of transition on the skin, ask the patient to roll away from that side, wait a few seconds, and percuss over that area. If ascites is present, the initially dull note will become resonant (Figs 14.13 and 14.14).

Auscultation

Listen specifically for bowel sounds. The presence or absence of bowel sounds is important. Listen for 30 seconds before concluding that bowel sounds are absent. Much mythology has been generated about the quality of these sounds and their diagnostic significance, but they should be interpreted with caution. Listen specifically for bruits over the aorta and renal arteries.

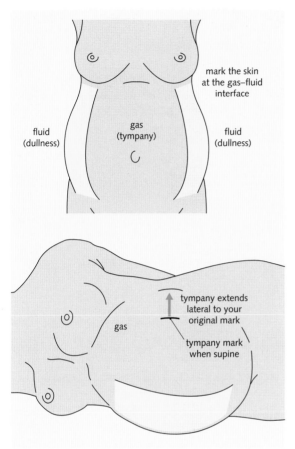

fluid
(dullness)

gas
(tympany)

fluid
(dullness)

mark the skin
at the gas–fluid
interface

tympany extends
lateral to your
original mark

gas

tympany mark
when supine

Fig. 14.13 Shifting dullness is a key sign of ascites.

While performing a digital rectal examination, the sphincter tone should be assessed and any tenderness elicited.

Structures palpable during a normal digital rectal examination

Palpate anteriorly, laterally and posteriorly. Note the following:

- Posteriorly – the tip of the coccyx and sacrum are palpable.
- Laterally – the ischial spines and ischiorectal fossa.
- Anteriorly in males – the prostate (smooth lateral lobes separated by the median sulcus); a prostatic carcinoma may be differentiated from benign prostatic hypertrophy by the loss of the median sulcus and possibly the presence of a palpable hard, craggy, irregular mass; remember to palpate the superior aspect of the prostate.
- Anteriorly in females – the cervix through the vaginal wall and, occasionally, the body of the uterus.

The normal rectum may contain some faeces. Always look at the glove after examination for blood or mucus. Melaena stool has the appearance of sticky tar and an offensive characteristic smell. A rectal carcinoma may be palpable as a shelf-like lesion associated with blood on the glove.

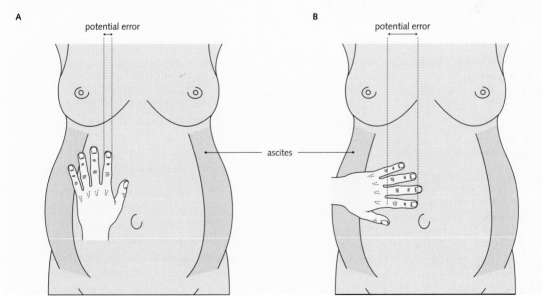

A

potential error

B

potential error

ascites

Fig. 14.14 Correct orientation of your hands is important when percussing the abdomen for the presence of ascites. (**A**) Correct positioning of the hand. (**B**) Incorrect positioning of the hand.

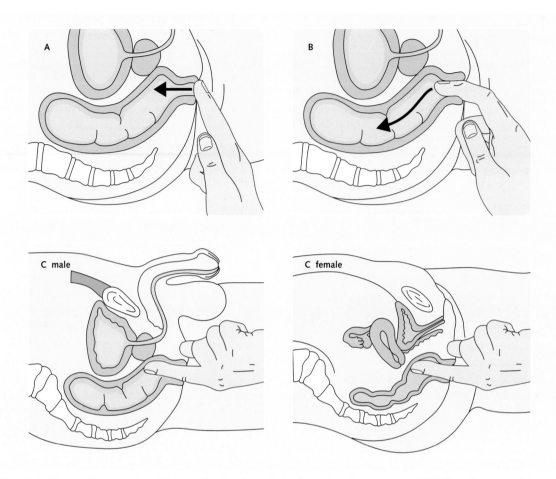

Fig. 14.15 The rectal examination. (**A**) Insert the tip of your index finger into the anal canal. (**B**) Follow the curve of the sacrum. (**C**) Sweep the finger around the pelvis, noting any irregularities, masses or tenderness. Examine the glove on withdrawal of your finger.

Always wipe the patient after examination and offer further tissues.

HEPATOMEGALY

Identify the mass as the liver

The liver is palpable in the right upper quadrant. Hepatomegaly is not usually confused with other organomegaly, but the liver should be distinguished from an enlarged right kidney. The features of hepatomegaly include:

- Palpable below the right costal margin (in gross hepatomegaly, it may extend to the left costal margin).

- Downward movement on inspiration.
- Dullness to percussion.

It is impossible to palpate above the upper margin of the liver.

Define the characteristics of the liver

It is often possible to palpate a liver edge 1–2 cm below the right costal margin. If the liver edge is palpable, it is important to confirm that there is true hepatomegaly rather than a low diaphragm (e.g. due to chronic obstructive pulmonary disease).

The size of the liver may be confirmed by percussion in a sagittal plane recording the 'height' or span of the liver. A normal liver is less than 15 cm.

Record:

- Size of the liver. Once true hepatomegaly is confirmed, trace out the edge of the liver to define its margins. An enlarged right lobe (Riedel's lobe) is a normal finding. In the notes, it is helpful to record accurately the size of the liver in the mid-clavicular line, midline and, if appropriate, the left mid-clavicular line (Fig. 14.16).
- Consistency of the liver (e.g. hard, firm).
- Definition of the liver edge (e.g. smooth, knobbly).
- Tenderness (e.g. engorged liver in right heart failure).
- Pulsatility (e.g. as in tricuspid stenosis).

Consider aetiology

It is important to look for other features on systemic examination if hepatomegaly is found, as they may give clues to the underlying pathology (Fig. 14.17). In particular, look for:

- Signs of chronic liver disease (e.g. due to alcoholic cirrhosis).
- Splenomegaly (e.g. due to portal hypertension, lymphoma).
- Generalized lymphadenopathy (e.g. due to lymphoma, carcinoma).

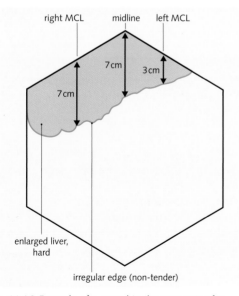

Fig. 14.16 Example of a record in the case notes for hepatomegaly. MCL, mid-clavicular line.

- Jugular venous wave (e.g. due to right heart failure, tricuspid regurgitation).
- Features of underlying malignancy.

Assess severity

Look for features of hepatic decompensation, for example:

- Features of chronic liver disease (e.g. testicular atrophy, loss of axillary hair, gynaecomastia, spider naevi, leuconychia). These features suggest that the underlying disease process is chronic.
- Signs of portal hypertension. Always specifically check for ascites and splenomegaly. Look for a 'caput medusae' (dilated collateral veins radiating from the umbilicus).
- Signs of hepatic encephalopathy. Check the patient's mental state (especially for level of consciousness and constructional apraxia; Fig. 14.18). Specifically check for a metabolic flap (asterixis) and fetor hepaticus.
- Jaundice.

SPLENOMEGALY

Identify the mass as the spleen

A mass in the left upper quadrant is usually a spleen. It is not normal to be able to palpate the spleen. It must be enlarged 2- to 3-fold before it becomes palpable. It must be distinguished from the left kidney.

The characteristics of the spleen on physical examination include:

- Presence in the left upper quadrant.
- Upper edge not palpable.
- Expansion towards the right lower quadrant.
- On inspiration, movement towards the right lower quadrant.
- A notch that may be palpable.
- Dullness to percussion. The dullness extends above the costal margin.
- Not ballottable.

Assess spleen size

In a manner analogous to assessing the degree of hepatomegaly, it is important to measure the descent of the spleen from the left costal margin (Fig. 14.19).

Fig. 14.17 Causes of hepatomegaly. Asterisks indicate the most common causes.

Causes of hepatomegaly	
Causes	**Features on examination**
cirrhosis*	features of chronic liver disease; features of portal hypertension; hard irregular, knobbly liver common
alcoholic	common; look for evidence of alcoholic toxicity in other systems (e.g. neuropathy)
primary biliary	usually middle-aged female; pruritus common (look for excoriation); xanthelasma
haemochromatosis	skin pigmentation; gonadal atrophy; more common in men
α-1-antitrypsin deficiency	signs of chronic obstructive pulmonary disease
secondary carcinoma*	hard irregular knobbly liver edge; systemic features of malignancy (e.g. cachexia, etc.); lymphadenopathy; signs of primary (e.g. palpable breast lump, etc.)
congestive cardiac failure*	raised jugular venous pressure; peripheral oedema prominent; third heart sound; look for features of tricuspid regurgitation
infections (hepatitis A, B, C—rarely; glandular fever; cytomegalovirus; leptospirosis; hydatid; amoebic)	features are usually apparent from the history, but look for generalized lymphadenopathy
lymphoproliferative disorder (lymphoma; leukaemia; polycythaemia)	splenomegaly; generalized lymphadenopathy; anaemia or plethora; petechiae; etc.
miscellaneous amyloid polycystic fatty liver	splenomegaly; waxy skin; chronic disease; palpable kidneys; signs of uraemia

Consider aetiology

The more common causes of splenomegaly are illustrated in Figure 14.20. In particular, note the presence of:

- Hepatomegaly – portal hypertension, lymphoproliferative disorder.
- Generalized lymphadenopathy – lymphoproliferative disorder.
- Size of spleen – massive splenomegaly is usually due to chronic malaria, myelofibrosis or chronic myeloid leukaemia (CML); a barely palpable spleen has a much wider differential diagnosis.

Figure 14.21 is an outline of the approach to the abdominal examination.

ANAEMIA

The causes of anaemia are widespread. Remember the section in Chapter 5. The diagnosis is largely made from detailed laboratory testing and imaging investigations. However, it is essential that the investigation is focused, and this relies upon a systematic assessment during the history and examination.

Anaemia is usually detected clinically if the haemoglobin concentration is less than 10 g/dL. The signs of anaemia include:

- Conjunctival or mucosal pallor.
- Loss of colour in the palmar skin creases.

The more common causes of anaemia are:

Figure provided by doctor	Attempt at copying by encephalopathic patient
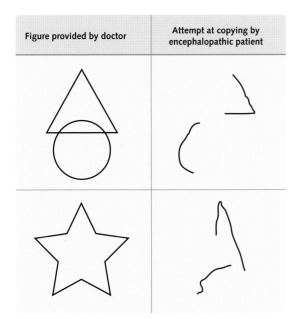	

Fig. 14.18 Hepatic encephalopathy is associated with constructional apraxia. Ask the patient to copy a simple figure such as a five-pointed star or simple overlapping geometric shapes.

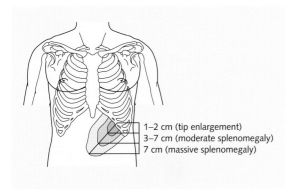

1–2 cm (tip enlargement)
3–7 cm (moderate splenomegaly)
7 cm (massive splenomegaly)

Fig. 14.19 Different degrees of splenomegaly.

- Iron-deficiency anaemia.
- Folate deficiency.
- Vitamin B_{12} deficiency.
- Haemolytic anaemia.

However, these diagnoses are insufficient, as an underlying cause still needs to be identified in order to provide suitable treatment and prognostic information.

Attempt to define the cause of anaemia

A thorough systematic examination is essential. Clues may be found by considering the following.

General inspection

Inspect the face carefully. Look at the general health of the patient (e.g. cachexia suggests chronic disease), and note specifically any obvious disorders (e.g. rheumatoid arthritis). Some clues to iron deficiency and megaloblastic anaemia are illustrated in Figure 14.22.

Note racial origin, for example:

- Mediterranean – thalassaemia.
- Afro-Caribbean – sickle cell anaemia.
- Northern European – hereditary spherocytosis.

Look for causes of blood loss

A systematic survey for a potential source of blood loss should be performed, for example:

- Abdominal scars.
- Gastrointestinal (GI) bleeding – look for abdominal masses; a rectal examination is mandatory.
- Genitourinary source – look for a palpable bladder or kidneys, perform urinalysis.

Causes of splenomegaly		
Large (past umbilicus)	**Moderate (to umbilicus)**	**Mild (just palpable)**
Myelofibrosis	Chronic lymphatic leukaemia	Portal hypertension
Chronic myeloid leukaemia	Lymphoma	Lymphoma
Malaria	Portal hypertension	Rheumatoid arthritis (Felty's syndrome)
		Chronic lymphatic leukaemia

Fig. 14.20 Causes of splenomegaly.

Gastrointestinal examination	
General	lie patient flat with arms at side wasting, scars, liver flap (asterixis), cock wrists – Hands–clubbing, leuconychia, spider naevi, palmar erythema, Dupuytren's – Head–jaundice, anaemia, purpura – Mouth–telangiectasia, pigmentation, ulceration, tongue, hepatic fetor (sweet smell) – Supraclavicular node. (Virchow's Ⓛ: Troisier's sign.) – Chest wall-gynaecomastia, spider naevi, bruising/purpura, muscle wasting
Observe abdomen	– breathe in deeply, cough – areas of fullness, masses, ascites – visible pulsation (aneurysm) – scars, striae – peristalsis – distended veins, direction of flow – hernias – everted umbilicus (ascites)
Palpate abdomen	– ask if tender, watch patient's face, kneel down – gentle palpation in each quadrant, for masses, tenderness – deep palpation – palpate liver from Ⓡiliac fossa, with inspiration, size, border smoothness, tenderness – palpate spleen from Ⓡiliac fossa with inspiration (confused with kidney–cannot get above spleen, dull to percussion, moves with resp, can't ballot). Turn onto Ⓡ feel under Ⓛcostal margin – palpate kidneys–bimanually, ballot – palpate for aortic aneurysm
Percussion	– liver spleen – shifting dullness – centre of abdomen to flank, mark point, roll patient to side, back towards umbilicus – fluid thrill – examiner's hand on midline, flick one side, detect on other
Auscultate	– bowel sounds – bruits aorta, hepatic, renal
Hernial orifices	– cough
Genitalia	
PR	
Urinalysis	

Fig. 14.21 A summary of the abdominal examination.

Features on general inspection in an anaemic patient	
Cause	Features
iron deficiency*	koilonychia (spoon-shaped nails); painless glossitis; angular stomatitis
hereditary haemorrhagic telangiectasia	visible telangiectasia on face and mouth
Peutz–Jegher's syndrome	pigmented macules around the lips and mouth
megaloblastic anaemia*	mild jaundice (lemon-yellow tinge) due to ineffective erythropoiesis; beefy red swollen tongue; angular stomatitis
pernicious anaemia	usually middle-aged or elderly female; look for features of other autoimmune disease (e.g. vitiligo)

Fig. 14.22 Features on general inspection of a patient with anaemia. Asterisks indicate the more common causes.

Look for features of a chronic disease

Look for features of chronic disease such as:

- Infections (e.g. tuberculosis, osteomyelitis, infective endocarditis).
- Connective tissue disease.
- Crohn's disease.
- Malignancy.

Exclude pregnancy

Pregnancy may be associated with folate and iron deficiency.

Perform a thorough abdominal examination

Pathology of the GI tract may cause anaemia, for example:

- GI bleeding, malabsorption (e.g. due to coeliac disease) – iron-deficiency anaemia.
- Gastrectomy, blind loop syndrome, Crohn's disease – anaemia due to folate or vitamin B_{12} deficiency.

In addition, an intra-abdominal malignancy may be detected.

Organomegaly is associated with different types of anaemia:

- Liver disease is associated with macrocytic anaemia.
- Splenomegaly may be responsible for haemolytic anaemia.
- A large uterus may be due to pregnancy or be a cause for blood loss.
- Polycystic kidneys may be a cause of chronic renal failure and consequent anaemia.

Look for signs of haemolysis

Splenomegaly or mild jaundice may indicate that the underlying cause is haemolysis.

Assess severity of anaemia

Try to make an assessment of the functional consequences of the anaemia. It is often hard to correlate the degree of pallor with the haemoglobin level. The functional impact of anaemia depends upon the underlying condition, the age and fitness of the patient, and the speed of onset. Look for signs of decompensation, for example:

- Hypotension – rapid blood loss may result in hypovolaemia and hypotension; postural blood pressure is the most sensitive indicator.
- Tachycardia – this develops early as a means of increasing oxygen delivery to the peripheral tissues in anaemia.
- Dyspnoea – note the exercise tolerance of the patient (e.g. short of breath at rest or on climbing onto the examination couch).
- Heart failure – especially in the elderly.

ACUTE GASTROINTESTINAL BLEED

Assess the functional impact

This is a medical emergency. Do not struggle on your own trying to sort this out. It is a team effort, which requires senior support and involvement. If you come across a patient with a significant GI bleed as a student, get help. Page 183 has a section on the approach to the unwell patient. The initial examination should follow the standard routine of Airway, Breathing and Circulation (ABC). When a problem is identified, it must be treated before moving on. It is important to recognize which patients are in danger of exsanguination (see Rockall score in Ch. 5). The processes of assessment and emergency treatment should run in parallel.

Determine the site of bleeding

It is often clear that the bleeding is from the upper or lower GI tract. The vast majority of patients presenting with an acute GI bleed have a lesion at the level of the duodenum or above. Features to suggest an upper GI tract source of bleeding include:

- Haematemesis. Exclude haemoptysis or epistaxis with swallowing and subsequent vomiting of blood.
- Frank blood or 'coffee ground' material in the nasogastric (NG) aspirate.
- Absence of a bilious NG aspirate.
- Melaena. It is essential to perform a rectal examination. A melaena stool indicates bleeding proximal to the right colon and bleeding of usually more than 500 mL in the

previous 24 hours. Note that the presence of blood per rectum does not always indicate a lower GI bleed, as a very brisk upper GI bleed can result in apparently fresh blood per rectum.

Features of a lower GI bleed include the passage of bright red blood per rectum (haematochezia). This is not pathognomonic (see above), but if the bleeding is from the upper GI tract, the patient will invariably be profoundly hypovolaemic. Common causes of a lower GI bleed are haemorrhoids and diverticular disease.

Perform a detailed abdominal examination

A systematic abdominal system examination may provide further clues to the cause, for example:

- Abdominal masses (tumours, diverticular masses, abdominal aortic aneurysm (AAA)).
- Signs of liver disease. Gastrointestinal bleeding is common in liver failure, particularly of an alcoholic aetiology.
- Surgical scars. May indicate previous peptic ulcer disease, for example.
- Rectal examination findings. Note stool colour. This may indicate the location of the bleeding point as well as the speed of bleeding.

Acute GI bleeding

The examination is aimed at assessing the degree of hypovolaemia, rate of blood loss, source of bleeding and urgency of resuscitation. The initial haemoglobin estimate may be misleading, so the requirement for blood transfusion relies upon thorough clinical assessment. It is useful to remember the vital signs associated with a diagnosis of shock: respiratory rate >25; heart rate of >100 bpm; systolic BP <100 mmHg; O_2 saturation <90%. The presence of any of these factors should prompt you to seek senior support.

ACUTE ABDOMINAL PAIN

Acute abdominal pain is one of the most common causes of presentation to a casualty department. The differential diagnosis is vast, ranging from trivial conditions to life-threatening surgical emergencies. Abdominal pain is also a common 'functional', or 'medically unexplained' symptom. It is important to adopt a systematic approach to the examination. Consider the differential diagnosis throughout the examination (see Fig. 5.2, p. 34) so that further management strategies can be instituted efficiently. The main aims of the examination are:

- To establish the cause of the pain.
- To assess whether the patient would benefit from admission to hospital.
- To assess whether the patient requires surgical intervention.

General inspection

Before specifically examining the abdomen, look at the patient as a whole. It is helpful to ask the following questions.

Does the patient look unwell?

Patients with acute peritonitis usually look obviously unwell. They are disinterested in their surroundings and lie still so as not to aggravate the pain. Patients with renal colic may also appear distressed, but tend to be restless and in obvious pain. Conversely, patients who are laughing, smiling or eating are most unlikely to have any significant acute surgical disease.

How old is the patient?

Different diseases are more common in different age groups. For example:

- Acute diverticulitis or AAA are much more common with increasing age.
- Acute appendicitis is commonest in young children and adolescents, but occurs in all age groups.
- Ectopic pregnancy is only going to occur in women of childbearing age.

What sex is the patient?

In women, the differential diagnosis needs to be broadened to include gynaecological conditions (see Chs 5 and 15). The other causes of intra-abdominal pain can occur in either sex, but some have a tendency to be more common in one sex. For example:

- Gallstones are more common in women.
- Peptic ulceration is more common in men.

Don't forget the possibility of a gynaecological (e.g. pelvic inflammation) or obstetric (e.g. ectopic pregnancy) cause.

Specific inspection

Perform a more specific inspection starting at the head and working down to the feet, noting the following.

General appearance

Cachexia may be due to chronic illness or malignancy.

Jaundice

The presence of jaundice should alert the doctor to:

- Gallstones – obstruction of the common bile duct, cholecystitis, pancreatitis.
- Chronic liver disease – associated with gastritis, acute alcoholic hepatitis and pancreatitis as well as oesophageal varices.

Conjunctival pallor

In the context of acute abdominal pain, it may be hard to assess skin colouration as the patient often appears pale, grey and sweaty. However, conjunctival pallor may suggest the presence of a chronic bleeding lesion, for example:

- Peptic ulcer.
- Colonic tumour with subsequent obstruction or intussusception.

Stigmata of chronic liver disease

See explanation of jaundice above.

Fever

The presence of fever suggests that an active inflammatory process is present.

Left supraclavicular lymphadenopathy

This is suggestive of intra-abdominal malignancy.

Check vital signs

It is essential to check the vital signs as a baseline and to determine the urgency of therapy. Check the following.

Oral temperature

Even in the presence of peritonitis and active infection, the patient may not have a fever, especially if shocked. However, the presence of pyrexia indicates that organic pathology is almost invariably present.

Pulse rate

It is unusual to have a tachycardia in the absence of active pathology. However, a very anxious patient may have tachycardia. The sequential recording of pulse rate is an accurate indicator of systemic disturbance if there is a progressive tachycardia. Equally, a completely normal pulse in the presence of severe acute abdominal pathology is unusual. However, beware reliance on the pulse if the patient is on β-blockers.

Blood pressure

Check supine blood pressure. If the patient is able to cooperate, it is useful to check postural blood pressure. If the patient has shock or hypovolaemia, there will be a drop in blood pressure on standing.

Assess fluid status

If the jugular venous waveform is easily visible, the jugular venous pressure (JVP) provides a useful marker of fluid status.

Abdominal examination

Inspection

Inspect the abdomen, noting:

- Visible peristalsis – suggestive of obstruction.
- Abdominal distension – may be due to obstruction.
- Rigidity – a tense, board-like abdomen occurs in the presence of peritonitis.
- Any skin discoloration – pancreatitis may be associated with a bluish discoloration in the loins due to extravasation of bloodstained pancreatic juice into the retroperitoneum.
- Obvious hernias.
- Abdominal scars – their presence raises the possibility of obstruction due to adhesions.
- Obvious organomegaly – for example massive polycystic kidneys may cause bulging in the flank. Bleeding into a cyst may be the cause of the pain.

Note the site of the pain

Ask the patient to show you exactly where the pain is on the abdomen. The location of the pain is the key to the underlying cause (see Fig. 5.2, p. 34).

Examine the abdomen in detail

Perform a detailed abdominal examination as described at the start of this chapter. In particular, note:

• Presence or absence of signs of peritonitis.
• Presence of any abdominal masses.
• Location of the tenderness.

Signs of peritonitis

The presence of unexplained peritonitis is an indication for surgical intervention. The features of peritonitis are:

• Signs of shock (tachycardia, hypotension, which becomes progressive on serial observation; Fig. 14.23).
• Tenderness.
• Guarding (a sign of severe tenderness).
• Rebound tenderness. This is a useful discriminatory sign as many anxious patients have involuntary guarding upon palpation, but do not expect tenderness on withdrawal of the palpating hand. The tenderness may be distant to the site of palpation. Watch the patient's face for signs of rebound tenderness.

• Localized pain distant to the site of palpation.
• Absent bowel sounds.

Presence of an abdominal mass

Examination of palpable liver, spleen and kidney is discussed on pages 134–135. In the context of acute abdominal pain, a mass in the right or left iliac fossa is most relevant.

Mass in the right iliac fossa

The most common causes are an appendix mass or carcinoma of the caecum. The differential diagnosis is shown in Figure 14.24.

Mass in the left iliac fossa

The same diseases as shown in Figure 14.24 may also cause a left iliac fossa mass, except for carcinoma of the caecum, Crohn's disease and tuberculosis. Diverticulitis is common and may cause a mass. Carcinoma of the colon usually presents with weight loss or a change in bowel habit, but occasionally presents as a left iliac fossa mass, especially if it is causing obstruction.

Location of the pain

Generalized

Generalized abdominal pain is likely to be due to generalized peritonitis. The history is central to the diagnosis.

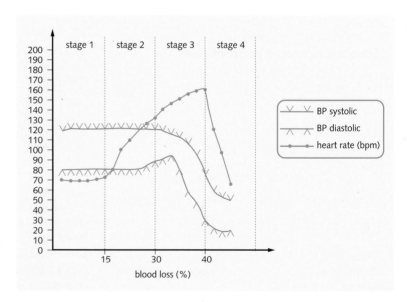

Fig. 14.23 This figure shows the four stages of shock. Note how much blood you have to lose before your blood pressure drops.

Causes of a right iliac fossa mass	
Causes	**Features**
appendix mass	preceding history of central abdominal pain moving to the right iliac fossa; anorexia; tender mass; persistent fever and tachycardia; tender per rectum (PR)
carcinoma of the caecum	firm distinct mass; often non-mobile; usually non-tender; patient does not look acutely unwell
tuberculosis	more common in patients from the Indian subcontinent or Africa
Crohn's disease	patient may appear hypovolaemic owing to diarrhoea; oral aphthous ulcers; skin tags; mass usually mobile and of rubbery consistency; tender mass
psoas abscess	ill-defined mass; lumbar tenderness
iliac lymph nodes	
iliac artery aneurysm	

Fig. 14.24 Causes of a right iliac fossa mass.

Epigastric

The most common causes of epigastric pain are:

- Peptic ulcer.
- A biliary cause (biliary colic or cholecystitis).
- Pancreatitis.

Peptic ulcer usually produces no signs unless perforation has occurred, though pyloric stenosis may result in visible peristalsis.

Biliary pain is more usually in the right upper quadrant. Often there are no abdominal signs, though there is commonly tenderness in the right upper quadrant upon inspiration.

Pancreatitis often produces surprisingly few abdominal signs for the degree of shock.

The conditions that cause abdominal pain, shock and a soft abdomen are:

- Pancreatitis.
- Bowel infarction.
- Dissection of an AAA.
- Referred pain from a myocardial infarction.

Loin pain

The main causes of loin pain are:

- Renal colic.
- Pyelonephritis.
- Musculoskeletal pain.

Right iliac fossa pain

The most important cause of right iliac fossa pain is acute appendicitis, although this is relatively rare. However, the differential diagnosis is wide. Other causes include:

- Gastroenteritis.
- Mesenteric adenitis.
- Ruptured ovarian cyst.
- Acute salpingitis.
- Perforated peptic ulcer.
- Acute cholecystitis.
- Crohn's disease.
- Acute diverticulitis (rarely on the right).
- Renal colic.
- Ectopic pregnancy.

Medical conditions

It is important to remember that medical conditions may present with acute abdominal pain, so a detailed systemic examination is essential. In particular, examine:

- The cardiovascular system. Inferior myocardial infarction or angina occasionally present with predominantly upper abdominal pain.
- The respiratory system. Pneumonia (especially with lower lobar disease) may cause right or left upper quadrant pain. Look for signs of consolidation.
- Diabetic ketoacidosis.

CHRONIC RENAL FAILURE

Patients with chronic renal failure often have specific problems, which should always be considered during an assessment. For dialysis patients, consider the following points.

Assess dialysis access

Haemodialysis

Note the presence of the arteriovenous fistula (AVF), its site (e.g. radial, brachial) and the palpable thrill. Look for access sites used and other possible access sites; include Goretex graft, shunts or central venous catheters. Look specifically for signs of infection (especially with venous catheters).

Peritoneal dialysis

Look at the exit site of the peritoneal dialysis catheter for evidence of tunnel infection or exit site infection. If the patient has reported abdominal symptoms, it is important to inspect the dialysis fluid for turbidity, blood or cloudiness suggestive of peritonitis.

Assess fluid balance

Fluid overload is a common problem in anuric patients.

Clinical assessment of the kidneys

In chronic renal disease, measuring the patient's weight at every visit is the most sensitive guide to changes in fluid balance on a day-to-day basis. Dialysis patients will have an ideal dry weight and most of them will be able to tell you what this is; you can then assess their fluid status with reference to the dry weight. Do not forget to ascertain the original disease causing the renal failure!

Do not forget to check the urinalysis via dipstick and also by microscopy, if available, as this provides an estimate of the degree of proteinuria and/or haematuria.

It is worth remembering that transplanted kidneys are normally placed in the right iliac fossa; these occur frequently in exams.

Look for signs of fluid overload, such as:

- Raised JVP.
- Peripheral oedema – if the patient has a normal plasma albumin, peripheral oedema usually indicates a fluid overload of approximately 3 kg.
- Uncontrolled hypertension.
- Pulmonary oedema – often a combination of fluid overload and cardiac failure.

Some patients develop symptoms such as light-headedness, fainting or malaise towards the end of dialysis. Look for signs of dehydration, such as:

- Dry mucous membranes.
- Postural hypotension.

Check lying and standing blood pressure

Most dialysis patients have hypertension, and pristine blood pressure control is central to reducing long-term morbidity from cardiac disease. Postural blood pressure assessment is useful for determining fluid balance. Erythropoietin therapy may exacerbate hypertension.

Look for activity of underlying disease

It is important to remember the underlying cause of renal failure as this may cause special problems in management. For example:

- The diabetic patient has a particularly high risk of cardiovascular disease.
- A patient with a vasculitic illness may develop recurrent vasculitis with extrarenal complications (e.g. pulmonary haemorrhage).
- Amyloidosis may progress causing systemic complications.

Perform a full systemic examination

Renal failure is a multisystem disease, and a full assessment is essential.

PALPABLE KIDNEYS

It is unusual to be able to palpate normal kidneys in any but very lean patients. If a kidney is palpable, it is necessary to:

- Identify the mass as a kidney.
- Consider the underlying cause.

Identify the mass as a kidney

The right kidney is often palpable in a thin subject; the left is palpable less often. Features of an enlarged palpable kidney include the following:

- Location in the loin (paracolic gutter).
- It is usually only possible to palpate the lower pole.
- Downward movement on inspiration (the spleen tends to move towards the right iliac fossa).
- Usually resonant to percussion (see Hepatomegaly, p. 134, Splenomegaly, p. 135 and Fig. 14.6; it is overlaid by the colon).
- It can be balloted (almost a pathognomonic sign).
- It may be possible to 'get above' the mass (compare directly with the liver or spleen).

A palpable kidney is rarely confused with any other organ, but it should be clearly differentiated from the spleen on the left and the liver on the right (Fig. 14.25).

Define the characteristics of the kidney

Once the organ has been identified as kidney, define the size, consistency and shape. Listen over the organ for a bruit, which may be present if there is renal artery stenosis or a tumour.

Consider aetiology

Bilateral palpable kidneys

The most common causes of bilaterally palpable kidneys are:

- Polycystic kidneys (the most common cause). The kidneys may be massive. Most other causes

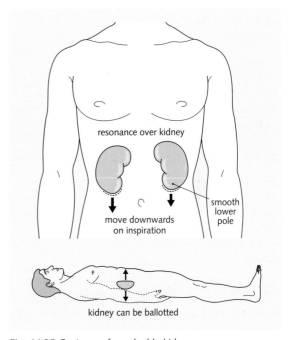

resonance over kidney

move downwards
on inspiration

smooth
lower
pole

kidney can be balloted

Fig. 14.25 Features of a palpable kidney.

Causes of enlarged kidneys, single or bilateral	
Causes	**Features**
polycystic kidneys	usually bilateral; large cysts may be individually detected; check blood pressure; note uraemic complications; may detect hepatomegaly
hydronephrosis	unilateral or bilateral; may be tender if acute; check prostate and palpable bladder

malignancy
   Wilms' tumour (child)
   Renal cell carcinoma (hypernephroma)
	look for systemic features of malignancy

miscellaneous
   pyonephrosis
   single cyst
   amyloid
	tender
	usually incidental finding
	rare; look for underlying disease

Fig. 14.26 Causes of single or bilaterally enlarged kidneys.

result in smooth hemiovoid masses, but occasionally individual cysts may be palpable. Look for an associated polycystic liver, signs of uraemia or an arteriovenous fistula.

- Bilateral hydronephrosis.
- Amyloidosis. The patient may have a typical facies, hepatosplenomegaly, peripheral neuropathy or obvious underlying inflammatory disease (e.g. rheumatoid arthritis, chronic osteomyelitis).

Other causes are shown in Figure 14.26.

Unilateral palpable kidney

The most common causes are similar but, in order of frequency, include:

1. Polycystic kidneys.
2. Renal cell carcinoma.
3. Hydronephrosis.
4. Hypertrophy of a single functioning kidney (kidney only just palpable).

Obstetric and gynaecological examination

Objectives

By the end of this chapter you should:

- Understand the principles behind examination in obstetrics and gynaecology.
- Be able to perform a basic obstetric abdominal examination.
- Be able to perform a basic pelvic examination.

This chapter is not intended to be a stand-alone guide to examination of obstetric and gynaecological patients. More comprehensive information can be obtained from Panay et al 2004 (upon which this chapter is based); see Further reading. The following is a guide to the basic examinations that could be expected of a clinical medical student.

OBSTETRIC EXAMINATION

It can be helpful to think of obstetric examination as a specialized form of abdominal examination. As such, it has some very specific exceptions; for example the patient should not be examined lying flat due to the risk of postural supine hypotensive syndrome (weight of abdominal contents reduces systemic and placental blood flow leading to maternal and fetal hypotension). General examination should always be performed, looking for general well-being, cardiovascular problems (including blood pressure (pre-eclampsia) and anaemia, with reference to the normal ranges in pregnancy) and respiratory symptoms. No obstetric exam is complete without urinalysis.

Communication skills

Due to the intimate and invasive nature of examination in obstetrics and gynaecology, good communication skills are vital. A significant proportion of complaints against doctors are made because of suboptimal communication between patient and doctor. You should familiarize yourself with the procedure for each examination, offer to explain the examination to the patient, arrange a chaperone and talk through each step as you are performing it. This also allows supervising doctors (and examiners!) to appreciate that you understand what you are doing and looking for.

Inspection

Stigmata of pregnancy should be noted during the general examination, including striae gravidarum (stretch marks) and skin pigmentation – the linea nigra, nipples and scars can all be affected. The appropriately exposed abdomen should be inspected closely for surgical scars, the degree of distension and any abdominal masses.

Palpation

Before palpation, enquire about abdominal pain or tenderness. Any areas of tenderness should be examined, exrcising more caution toward the end of the examination if possible. The patient should always be examined from the patient's right hand side.

Fundal height

Assessment of uterine size can be made by assessment of the fundal height. By placing the dominant (examining) hand below the umbilicus, running the hand superiorly until the upper border of the uterus is felt and running a tape measure from the examining hand to the symphysis pubis, an estimation of gestation can be obtained, where number of centimetres is equal to number of weeks. After 24 weeks, this number ±3 weeks is usually accurate. It is also possible to equate the fundal height to certain levels, where the uterus is palpable at 12 weeks, the umbilicus at 16 weeks and the xiphisternum at 36 weeks.

Fetal poles

A fetal pole is the head or breech of a fetus. In a single pregnancy, both poles should be palpable. In a multiple pregnancy, the number of poles palpable is one less than would be expected; a twin pregnancy should have three poles and triplets should have five poles. In order to palpate a fetal pole, begin by dividing the abdomen into thirds, with the left hand on the left side of the abdomen and the right hand on the right side of the abdomen, and palpate the lower third of the abdomen. Ballot any pole gently between your examining hands, flexing at the meta-carpophalangeal joint and keeping the fingers straight. Move your hands up to the mid-abdominal section and repeat this process. If fetal poles are present in this third, the left and right of the middle third should be examined separately, with both examining hands balloting on the left side of the middle third and then separately on the right side. To examine the upper third, reverse hand positions so that the right hand is on the left side of the abdomen and vice versa; again ballot between the examining hands (Fig 15.1).

There are several things that should be assessed in relation to the poles.

Fetal presentation

The pole that presents into the pelvis should be determined. If this is the head, this is a cephalic presentation. This can be further divided into vertex, brow and face depending on the degree of neck flexion. If the buttocks are present in the lower third of the abdomen, this is a breech presentation.

A

B

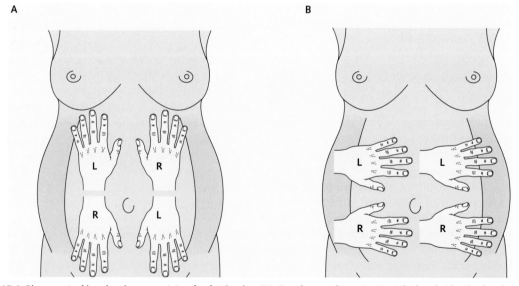

Fig. 15.1 Placement of hands when examining for fetal poles. (**A**) Standing at the patient's right hand side, the hands should be placed as shown in the upper and lower thirds. Any fetal poles can be felt between the hands. (**B**) Again, standing to the patient's right hand side, the hands should be placed as shown for examination of fetal poles in the centre third of the abdomen. This will allow assessment of a longitudinal lie.

Fig. 15.2 Fetal lie. The axis between the poles and its relation to the uterine axes (reproduced from Panay et al 2004, Fig. 46.2).

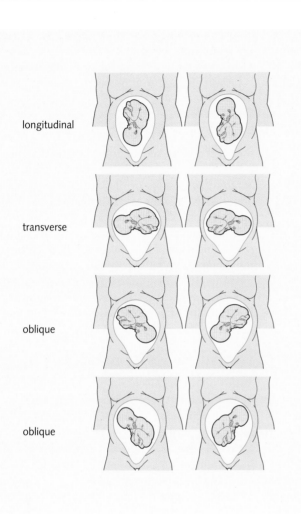

longitudinal

transverse

oblique

oblique

Fetal lie

This is the relation between the two poles of the fetus and the axes of the uterus, and is either longitudinal, oblique or transverse (Fig. 15.2).

Engagement

This is the relation of the fetal head and the pelvic brim. Engagement occurs when the widest part of the fetal head has passed the pelvic brim. This can be further divided into the number of fifths of fetal head present below the pelvic brim.

Liquor volume

This is difficult to assess and not as reliable as ultrasound assessment. An assessment of an increased or decreased volume can be made. If the uterus is small for dates and the fetal parts are easily palpable, this would indicate reduced volume. If the uterus is large

for dates and the fetal parts are not easily palpable, this would indicate increased volume.

Fetal heart

It is vital to assess the fetal heart rate, as this is a sensitive indicator of fetal distress. The most reliable way to assess this is to use a Doppler ultrasound probe over the anterior shoulder of the fetus. The heart rate should be between 110 and 150 bpm. For more accurate assessment of the fetal heart rate, a cardiotocograph (CTG) should be employed.

GYNAECOLOGICAL EXAMINATION

Internal pelvic examination is potentially the most intimate and invasive examination that will be performed by most doctors. It raises specific questions regarding consent and chaperoning. For all intimate examinations, a female chaperone must be present. Many medical schools insist on written consent from patients regarding pelvic examination under anaesthesia; you should be aware of any restrictions or rules regarding examinations from your medical school or NHS trust.

> Pelvic examination can be a daunting prospect. Most medical schools will have pelvic examination models available for practice; become comfortable with the procedure before progressing to anaesthetized patients in theatre or fully awake ones in the clinic.

Positioning

As with all examinations, once introductions are made, the procedure is explained and the patient has consented, the key thing is positioning the patient. With pelvic examination, the patient should be allowed to undress in absolute privacy. Often, leaving the patient in the examining room until she is ready is necessary if space does not permit an adequateley screened area. Ensure that you cannot be disturbed during the examination. With the patient still covered by a sheet, ask her to keep her ankles together while bending her knees as much as is comfortable. Then ask the patient to let her knees

'flop' to the side. There may well be a wall next to the examining couch; this should not cause any problems as long as the patient is comfortable. This is known as the dorsal position.

Inspection

As with any examination, inspection is key and is the most overlooked section by many students. The external genitalia should be inspected for normal anatomy, any signs of discharge (physiological or abnormal), inflammation, ulceration, swellings or atrophy. Inspection should include the clitoris and urethral meatus as well as the vagina. Particular note should be made of previous scars (from episiotomy, for example) and the presence of any prolapse. If the purpose of the examination is investigation of a urogynaecological complaint, the presence of prolapse should be reassessed with the patient bearing down.

Speculum examination

At this point, the pelvic examination can be tailored depending upon the patient's symptoms. It is normal practice to inspect the vagina and cervix with the use of a speculum before bimanual palpation if swabs are to be taken, as there is at least a theoretical risk of introducing pathogens and external flora with the gloved examining finger.

Speculae broadly come in two types (Fig. 15.3): the Sims' speculum is used for assessing the anterior and posterior vaginal walls; the patient must be repositioned in the left lateral, or Sims' position, and the speculum used to retract the anterior or

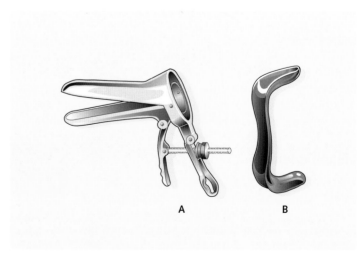

A   **B**

Fig. 15.3 A, Cusco's and **B,** Sims' specula (reproduced from Panay et al 2004, Fig. 46.8).

posterior walls. The Sims' is used extensively in urogynaecological assessment: for more information please see Panay et al 2004.

Introducing the speculum

The more common speculum is the Cusco's, which is a bivalve speculum introduced with the patient in the dorsal position, as above. To introduce the speculum, the thumbscrew or ratchet at the top should be inspected to ensure it will lock the speculum open once introduced. With the speculum in the dominant hand, the labia should be parted gently with the non-dominant hand. The suitably lubricated speculum should be rotated so that the ratchet at the top faces the patient's right thigh. This is because the anteroposterior dimension of the vaginal vestibule and introitus is greatest, while the lateral dimension of the tip of the speculum is greatest, facilitating an easier introduction. Once the speculum has been introduced and advanced, it should be gently rotated so that the ratchet is now pointing superiorly. The valves of the speculum can be opened and retained with the use of the ratchet or thumbscrew.

Inspection

With the valves of the speculum open, the vaginal vault and walls can be inspected along with the cervix. Signs of inflammation, bleeding, discharge, ulceration and masses should be looked for and the specifics noted, such as where the discharge is present, or how deep, wide and extensive ulcers are.

Swabs

In the presence of any suspicious findings or history, it may be desirable to perform some swabs, or a cervical smear. Swabs are either high vaginal, endocervical or low vaginal swabs taken from the posterior fornix, endocervix or within the introitus, respectively. Swabs should be handed to you in sterile fashion, introduced to the area to be sampled, gently rotated and placed back into the supplied sterile container, which usually contains a transport medium.

Separate swabs are used for chlamydia screening; these are sent for polymerase chain reaction (PCR) analysis. They are taken by placing the tip of the swab in the endocervix and gently rotating, and then sent in a separate container for analysis.

To perform a cervical smear, a cytology brush is used. This is introduced with the middle bristles in the endocervix, the brush is then turned clockwise five times, removed and placed in a cytology pot for analysis (Fig. 15.4).

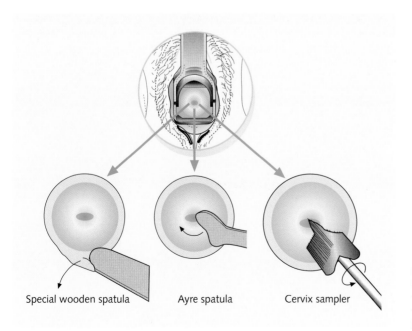

Special wooden spatula   Ayre spatula   Cervix sampler

Fig. 15.4 Placement of cytology brush during cervical smear (altered from Panay et al 2004, Fig. 47.6, to include only the cervix sampler brush).

Removal

Once the examination is complete, the speculum can be removed – remember to unlock the blades first! The speculum should then close by itself and can be withdrawn. Asking the patient to take a deep breath and removing the speculum during expiration can help keep the patient relaxed.

Bimanual examination

Once the speculum is removed, it is possible to perform a bimanual pelvic examination. This is usually conducted with the index and middle fingers of the dominant hand. In women who have not given birth (nulligravid) and in postmenopausal women, it may be possible to pass only the index finger. Palpation of the vaginal walls may reveal cysts, scars or tumours; the fornices should also be palpated for thickening and scarring.

Cervix

The cervix should be palpable and feel rubbery (like touching the end of your nose). In obstetrics, the cervix is graded according to its favourability for induction of labour (the Bishops' score) and is used to assess progress in the first stage of labour.

Moving the cervix from side to side may cause pain (this is known as cervical excitement).

Uterus

With the finger(s) in the posterior fornix, the non-dominant hand palpates the abdomen beneath the umbilicus; a normally positioned (anteverted) uterus will be palpable between the examining hands. The size, position and mobility of the uterus should be noted. If the uterus is retroverted, it may be felt as a swelling in the posterior fornix.

Adnexae

The ovary and its adjacent fallopian tube are referred to as an adnexa. With the finger(s) of the dominant hand in a lateral fornix, the non-dominant hand should be used to palpate the ipsilateral quadrant of the abdomen. If the hand is placed parallel to and below the inguinal ligament, the adnexa may be palpated between the examining hands. It is normal for the adnexa to be impalpable in all but the slimmest premenopausal women. If any masses are palpated, then the shape, size, consistency, mobility and fixation should always be noted. Great care must be taken if an ectopic pregancy is suspected as rupture of the fallopian tube can be precipitated.

SUMMARY

There are a great many variations of the above examinations and different factors that must be taken into account in certain situations, for example assessment of the cervix in labour. Interested readers are referred to Panay et al 2004.

Further reading

Panay N, Dutta R, Ryan A et al 2004 *Crash course: obstetrics and gynaecology*. St Louis: Mosby

Surgical examination

Objectives

By the end of this chapter you should:

- Be able to examine lumps and bumps and describe them accurately.
- Be able to examine the peripheral vascular system.

Surgery is a rich hunting ground for the Objective Structured Clinical Examination (OSCE) question setter! There are lots of different examinations that you need to be proficient at. Hopefully your course handbooks or study guides will detail which lumps and bumps you should be able to examine. Be aware: not all surgical units will have a full range of surgical specialties so the onus is on you to find a willing middle-grade doctor who will help you perfect your examination routine on patients. This chapter details how to examine some surgical pathologies. It should be used in conjunction with hands-on bedside teaching.

Practice makes perfect. In an exam you should look as if you've done this a thousand times before. Practice with and on your friends. Also, don't forget to wash your hands before and after examining each patient!

LUMPS IN THE GROIN

This is one method that will take you through how to examine a lump in the groin that covers all the salient special tests you will be expected to perform.

First, expose the patient's abdomen. The standard teaching is nipples to knees but umbilicus to knees will normally suffice.

- Look at the patient lying down.
- Can you see any obvious swelling? Where is it, is it superior or inferior to the inguinal ligament or is it in the scrotum? What colour is it; for example is it erythematous?
- Are there any scars from previous surgery?
- If nothing can be seen, then ask the patient to stand. Can you see anything now?

Remember, we have two of most things, allowing easy comparison. Don't forget to check for contralateral pathology in the heat of the moment.

Next, define the swelling. Lumps have three dimensions so measure the length, breadth and depth. There is likely to be a measuring tape located at the bedside, so use it. Now you have the size, define the shape of it.

- Is the lump fluctuant and can you transilluminate it? When you feel it, does it extend beyond the obvious skin markings?
- Test the lump for a cough impulse. Place your hand over the swelling and ask the patient to cough. Does the swelling get worse?

Ask the patient to reduce the swelling themselves. Put pressure over the deep inguinal ring (half way between the pubic tubercle and the anterior superior iliac spine) and ask the patient to cough. If a swelling appears medial to the pressure, then it is a direct

157

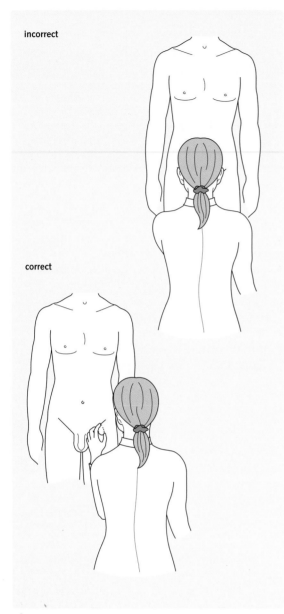

incorrect

correct

Fig. 16.1 Examination of a hernia. If examining a swelling, stand at the patient's side. Try to imagine what it will look like you're doing if you don't!

Differential diagnoses of lumps in the groin	
hernias	inguinal/femoral
vascular	saphena varix/femoral aneurysm
lymph nodes	lymphadenopathy (infection/neoplasm/lymphoma)
muscles	psoas abscess
testicular	ectopic testes
skin/subcutaneous	lipoma/sebaceous cyst

Fig. 16.2 Differential diagnoses of lumps in the groin.

Differential diagnosis of groin swellings

Above the inguinal ligament

- Sebaceous cyst, lipoma.
- Direct/indirect inguinal hernia.
- Incompletely descended testis.

Below the inguinal ligament

- Sebaceous cyst, lipoma, lymph nodes.
- Femoral hernia.
- Saphena varix (dilation of the saphenous vein at the confluence with the femoral vein).
- Femoral aneurysm (expansile pulsation, bruit, not compressible with no cough impulse).
- Incompletely descended testis.
- Psoas abscess (rare).

Alternatively, think of the mass in terms of structure as in Figure 16.2.

SCROTAL SWELLINGS

You are unlikely to be able to practice this on your friends! However, a sock, two kiwi fruits and a banana simulate a passable male genitalia scrotum for practice purposes!

Examination

Observe the swelling from the anterior and posterior aspects of the scrotum. Define its size and shape and note the skin colour.

Gently palpate the swelling. This is best achieved by rolling the testes between the thumb and finger. Find and feel the epididymis and feel the spermatic cord.

hernia. Release the pressure and ask the patient to cough again; if the hernia appears now, it is an indirect hernia (Fig. 16.1).

Next check the other side for any similar defects.

Ask the examiner if you may now proceed to examine the abdomen and scrotum. You will probably be told you've done enough, so cover the patient up and wash your hands!

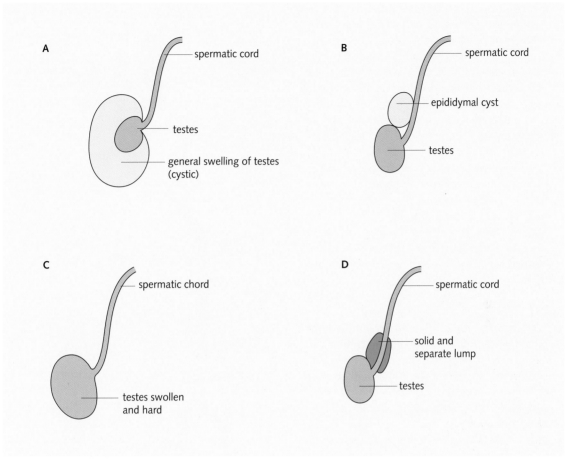

Fig. 16.3 Differential diagnoses of scrotal swellings.

Try to assess the swelling. What is its size, shape, fluctuance and does it transilluminate? There ought to be a light at the bedside of these patients. Can the upper edge of the mass be felt (i.e. can you get above it) and is it separate from the testes?

Your differential diagnosis should run along these lines:

- Cystic, testicular and not usually tender: hydrocele (Fig. 16.3A).
- Cystic, separate and not usually tender: epididymal cyst (Fig. 16.3B).
- Solid and testicular: tumour, orchitis (very tender), granuloma, gumma (Fig. 16.3C).
- Solid, separate and usually tender: chronic epididymitis (Fig. 16.3D).

PERIPHERAL VASCULAR DISEASE

Arteriopaths are challenging patients to examine. They will have multiple medical problems all of which will need to be investigated. This is an approach to examing their limbs.

Expose both legs completely, while preserving the patient's dignity; this includes taking their socks off. If the patient has a dressing on, this should be taken down and the underlying wound inspected. You should have a nurse along to help you if this needs to be done (this should not be necessary in an exam).

Inspect both legs and feet for:

- The colour of the skin. Is it white, red or black? Each is associated with differing degrees of vascular insufficiency.
- Trophic changes; for example is the skin smooth and shiny, is there loss of hair (note where this occurs) or wasting of subcutaneous tissue? Careful note should be made of any ulcers. If there is an ulcer present, this should be examined. Please see page 162 for how to do this.
- Look specifically at the pressure points in the limb for ulcers. Pay special attention to the lateral aspect of the foot, the head of the first metatarsal, the heel and malleoli.
- Finally, inspect the tips of the toes and between the toes. Patients are often immobile and will be unable to care for their feet. A small lesion here can rapidly progress.

Palpate both legs:

- Feel both legs for a difference in temperature. Note the level of any temperature change.
- Count the capillary refill time in both feet or stumps (to do this, press on the nailbed for 5 seconds, release and count how long it takes to turn pink. The capillary refill should be less than 2 seconds).
- Feel for all the pulses in the legs (femoral, popliteal, dorsalis pedis and posterior tibial). Classify the pulse as normal, diminished or absent. Figure 16.4 shows the arterial tree of the lower limb (this is much loved by examiners).

Auscultate and listen along the major vessels for arterial bruits.

Special tests

Elevate the leg to 15° and look for venous guttering. Keep elevating the leg until it becomes white (ischaemic) and note the angle. This is known as Buerger's angle and is normally >90°. From this elevated position, lower the leg over the side of the bed and look for reactive hyperaemia (this is Buerger's test).

There will often be a Doppler ultrasound probe around (especially in examinations). This is used to measure the ankle/brachial pressure index. Take a blood pressure in the ankle with the Doppler probe and a brachial blood pressure with the Doppler probe. Divide the ankle pressure by the brachial pressure to give a ratio (Fig. 16.5).

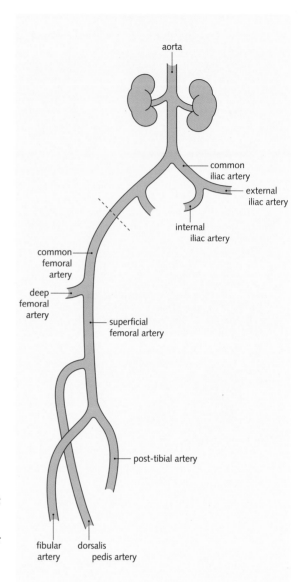

Fig. 16.4 The arterial tree of the lower limb.

ABPI	Significance
1–1.2	Normal
0.8	Claudication pain
0.4	Rest pain
<0.4	Ulceration and gangrene

Fig. 16.5 Ankle/brachial pressure index (ABPI).

You should also examine the patient's cardiovascular system for other signs of arterial disease or an aortic aneurysm.

VARICOSE VEINS

Varicose veins are a common problem. They are often found in patients in outpatient clinics and day surgery wards. They are often good examination topics as well. When asked to examine a patient's varicose veins, follow the usual pattern.

Inspection

Keep the patient decent and expose both legs with the patient standing up. Observe the patient's legs and the distribution of the varicose veins. Note the nutritional state of the patient's legs (especially the area superior to the medial malleolus). You should also look for any eczematous changes, pigmentation and varicose ulcers.

Palpation

Palpate and compare both legs. Is there a difference in temperature? Is the patient tender over the medial aspect of the lower leg? You should specifically palpate the ankle for dermatoliposclerosis. Is there ankle oedema? These signs give an indication of the chronicity of the problem.

Specific tests

There are some special tests that you should perform when examining varicose veins.

Feel the saphenofemoral junction (4 cm inferior and lateral to the pubic tubercule – N.B. remember Fig. 16.2). Is there a swelling here? If so, it is likely to be a saphena varix. With your fingers on the saphenofemoral junction, ask the patient to cough. If you feel an impulse, it is suggestive of venous incompetence. Last, perform the percussion test. Tap the top of the vein and feel below for an impulse – if one is present, it is suggestive of superficial venous incompetence.

You should now ask the patient to lie down and elevate the leg to empty the veins. Place two fingers on the saphenofemoral junction and ask the patient to stand. As you are doing this, carefully observe the leg. If the veins rapidly fill, then the lower leg perforators are incompetent. Now release your fingers. If there is rapid filling of the leg, then there is an incompetent saphenofemoral valve.

The school of perfection says that you should auscultate over the veins for bruits in case there is an arteriovenous malformation.

To conclude your examination, you should examine the abdomen for any masses. If you are suspicious of an obstruction, a per rectum examination should be undertaken. It is rare to do this in practice.

EXAMINATION OF AN ULCER

An ulcer is a break in the continuity of an epithelial surface. Ulcers are associated with a number of conditions, some of which have been discussed above. At some point in your career, someone will present to you with an ulcer and ask your opinion about it. It helps to have a system to help you describe it.

Inspection

Take off all of the dressing and gently clean off any topical applications. Expose the ulcer completely. Inspect the ulcer:

- Note its position, size and shape.
- Note the base (colour, penetration of underlying structures, e.g. tendon/bone).
- Note any discharge from the ulcer (blood, pus or serous fluid).
- Measure the depth of the ulcer in mm.
- Inspect the edge (Fig. 16.6).
- Feel the surrounding tissues for any tenderness or temperature changes.
- Note the nutrition of surrounding tissue and check for regional lymph nodes.
- You should also make an assessment of the neurovascular supply to the area, e.g. sensation and muscle power.

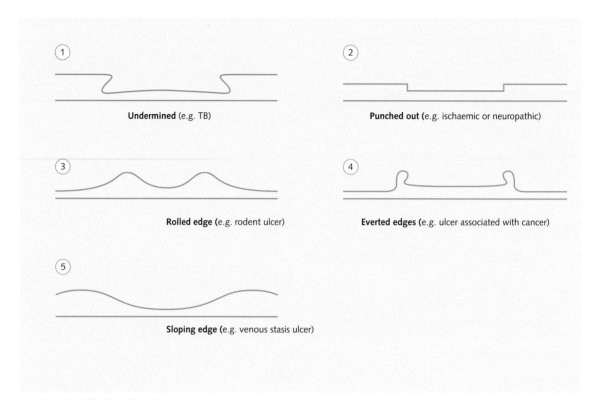

Fig. 16.6 Classification of an ulcer.

Neurological examination

Objectives

By the end of this chapter you should:

- Be able to examine all major aspects of the neurological system including the cranial nerves and upper and lower limbs.
- Recall the key dermatomes.
- Understand the principles behind defining the level of a neurological lesion with specific reference to upper and lower motor neuron injury.

EXAMINATION ROUTINE

The neurological examination is often considered to be the most difficult part of the examination by students and is commonly omitted or performed badly. In a full routine assessment, it is essential to perform at least a basic neurological assessment. This need not take more than a few minutes, but provides essential information.

Probably more than in any other system, it is vital to be systematic, objective and methodical, and to be aware of the pathological significance of any elicited sign, and how each sign may change. It is essential to record exactly what has been assessed, rather than putting down meaningless phrases (e.g. 'CNS – NAD' (nothing abnormal detected)). Interpretation of neurological conditions often relies on changes in neurological signs with time, highlighting the need to ensure accuracy when writing up the medical record.

Cranial nerves

It is important to understand the basic anatomy and function of the individual cranial nerves (CNs) when interpreting physical findings. For each cranial nerve, function, anatomy, examination routine and interpretation of the physical signs will be considered.

Olfactory nerve (cranial nerve I)

Function
The olfactory nerve is a sensory nerve conveying the sense of smell.

Anatomy
Nerve fibres pass from sensory receptors in the nasal cavity through the cribriform plate to the olfactory bulb, where they synapse, and then pass towards the anterior perforated substance.

Examination
(This is rarely performed.) Ask patients whether they have noticed any change in their sense of smell. Test smell in each nostril separately using a sniff test. Use common, easily recognizable, non-irritant substances (e.g. vanilla, orange, coffee).

Interpretation
It is relatively unusual to detect lesions of the olfactory nerve on physical examination. Formal testing is rarely needed unless a lesion of the anterior cranial fossa is suspected. If a lesion is detected, note the following:

- Anosmia (no sense of smell) is usually due to nasal rather than neurological disease.
- The olfactory nerve is vulnerable as it passes through the cribriform plate, especially if there is a head injury. Also consider frontal lobe tumours and meningism (infective or neoplastic).

Optic nerve (cranial nerve II)
Function
The optic nerve is a sensory nerve conveying the sense of vision from the retina.

Anatomy
The optic nerve leaves the eye via the optic foramen, partially decussates at the optic chiasm and synapses

at the lateral geniculate nucleus. Secondary fibres pass to the occipital cortex via the optic radiation (see 'Interpretation' below).

Examination

Visual acuity   Assess distant and near vision formally using Snellen and Jaeger charts, allowing patients to wear their spectacles, or crudely at the bedside (e.g. count fingers from 2 metres or read newsprint) for each eye (see Ch. 22).

Visual fields   Test by confrontation. Sit opposite the patient so that you are approximately 1 metre apart at the same level. Both of you then cover one eye, and you bring a test object (traditionally a white hat pin but, in the clinic, other objects may need to be

substituted, for example a pen top or your own fingers) into the field of vision from each quadrant, midway between yourself and the patient. The patient states when he or she first sees the object, and you can then compare the patient's visual field directly with yours.

Pupillary reflexes   These are discussed under Oculomotor nerve (see p. 165).

Fundoscopy   See Chapter 22, page 221 for detail.

Interpretation

Visual field defects should be correlated with the anatomical site of the lesion. It is helpful to understand the visual pathway (Fig. 17.1).

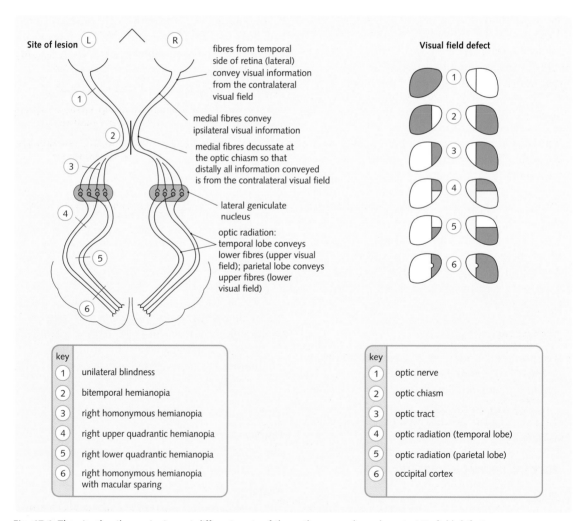

Fig. 17.1 The visual pathway. Lesions at different parts of the pathway produce characteristic field defects.

Oculomotor, trochlear and abducens nerves (cranial nerves III, IV, VI)

Function

The oculomotor, trochlear and abducens nerves are considered together as they supply the extraocular muscles (Fig. 17.2). The oculomotor nerve also supplies the levator palpebrae superioris, which opens the upper eyelid. In addition, it also has parasympathetic fibres supplying the sphincter pupillae (constrict the pupil) and the ciliary muscle of the lens.

Anatomy

The oculomotor nucleus lies in the midbrain. It passes close to the posterior communicating artery before entering the lateral wall of the cavernous sinus on the way to the orbit.

The trochlear nucleus lies lower in the midbrain. Fibres pass dorsally, decussate, pass around the midbrain and enter the lateral wall of the cavernous sinus.

The abducens nerve originates close to the facial nerve in the pons, emerges in the cerebellopontine angle and has a very long intracranial course, passing over the petrous temporal bone on the way to the cavernous sinus.

Examination

Inspection   Look at the eyelids for ptosis and symmetry.

Pupils   Look at pupil size and symmetry. Test the pupillary reflex by shining light on the pupil from the side, looking at both the direct and consensual response.

Ocular movements   Observe the patient following a target up, down, to either side, and for convergence. Note diplopia or nystagmus.

Interpretation

When interpreting physical signs, note the following:

- Ptosis (Fig. 17.3).
- Abnormal pupillary reflexes. The afferent limb is from the optic nerve (CN II) and the efferent pathway is via the oculomotor nerve (CN III).

Nerve supply and movement produced by the extraocular muscles		
Nerve	Muscle	Movement
oculomotor	medial rectus	adduction
	inferior rectus	inferior movement (especially when eye abducted)
	superior rectus	superior movement
	inferior oblique	superior movement (especially when eye adducted)
trochlear	superior oblique	inferior movement (especially when eye adducted)
abducens	lateral rectus	abduction

Fig. 17.2 Nerve supply and movement produced by the extraocular muscles.

Causes of ptosis	
Cause	Examples
third nerve palsy complete ptosis, associated with widely dilated pupil, and eye paralysed with outward and downward deviation	posterior communicating artery aneurysm 'coning' of the temporal lobe mononeuritis multiplex (e.g. due to diabetes mellitus, vasculitis) midbrain lesion
Horner's syndrome loss of sympathetic supply to eye, partial ptosis, pupillary constriction, enophthalmos and decreased sweating on affected side	brain lesion (e.g. CVA lateral medullary syndrome); cervical cord lesion (e.g. syringomyelia) T1 root lesion (e.g. apical lung cancer, cervical rib) sympathetic chain lesion (e.g. neoplasia)
neuromuscular disease	myasthenia gravis botulism
myogenic	senile degenerative changes dystrophia myotonica

Fig. 17.3 Causes of ptosis. CVA, cerebrovascular accident.

An intact consensual reflex with an absent direct reflex implies a lesion of the IIIrd nerve. Conversely, pupil constriction only when light is shone into the opposite eye implies a sensory deficit (CN II).

- Holmes–Adie pupil. This is a common normal finding in women. The pupil is large with an absent light reflex and delayed accommodation reflex, which is sustained. It is often associated with absent ankle reflexes.
- Nystagmus. This may be due to visual disturbances or lesions of the labyrinth, cerebellum, brainstem or central vestibular connections. See Nystagmus, below.
- VIth nerve palsy (loss of eye abduction). This is often a false localizing sign. The VIth nerve has a very long intracranial course and is vulnerable to compression as it passes over the petrous temporal bone. Any pathology causing raised intracranial pressure may result in a VIth nerve palsy.
- VIth nerve lesions may be due to a lesion in the pons or cerebellopontine angle. This often occurs in association with a VIIth (or VIIIth) nerve palsy and may result in contralateral pyramidal tract signs.
- IVth nerve palsy. This rarely occurs in isolation. Orbital trauma often damages the tendon, causing muscular weakness.

Nystagmus

Nystagmus often causes anxiety in the exam candidate; however it is relatively simple to describe. It is caused by posterior fossa disease or ear pathology.

First, which eye is the nystagmus most obvious in? Second, is the nystagmus greater when looking to the affected side (e.g. present in the right eye and worse when looking right)? This is the most common situation. Nystagmus is caused by:

- Contralateral vestibular lesion (associated with vertigo and deafness).
- Multiple sclerosis.
- Middle ear surgery.
- Ménière's disease.
- Ipsilateral cerebellar lesion (associated with other cerebellar signs).
- Neoplasia.
- Cerebrovascular accident (CVA).
- Ipsilateral brainstem lesion. These can be infective, vascular (the most likely), neoplastic or demyelinating in origin.

If the patient has vertical nystagmus, this implies a central lesion. If it is down gaze, consider pathologies around the foramen magnum. If it is up gaze, the lesion will be around the superior colliculus.

Sixth nerve palsy is commonly a false localizing sign and results from raised intracranial pressure. Its presence warrants further investigation.

Trigeminal nerve (cranial nerve V)

Function

The trigeminal nerve conveys sensory and motor nerve fibres. The main functions are:

- Sensory – somatic sensation to the face.
- Motor – muscles of mastication (masseters, temporalis, pterygoids).

Anatomy

The trigeminal nerve is split into the following three divisions (Fig. 17.4):

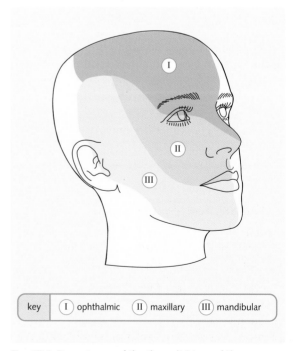

| key | I ophthalmic | II maxillary | III mandibular |

Fig. 17.4 Dermatomes of the three divisions of the trigeminal nerve.

1. Ophthalmic.
2. Maxillary.
3. Mandibular.

The trigeminal ganglion lies near the pons. The nerve fibres pass near the medial lemniscus to the thalamus.

Examination

Sensory   Test modalities of sensation over the three distributions of the nerve (e.g. forehead, cheeks, chin) bilaterally.

Corneal reflex   Lightly touch the cornea with cotton wool, approaching from the side (unseen), and observe a brisk contraction of the orbicularis oris (blinking).

Motor   Inspect for wasting of the temporalis and masseter. Test jaw opening against resistance (unilateral pterygoid weakness will cause the jaw to deviate to the side of the weakness).

Jaw jerk   This is often difficult to interpret.

Interpretation

When examining the Vth nerve, note the following points:

- An absent corneal reflex may be the first sign of ophthalmic herpes. Lesions of the Vth and VIIth nerves can be distinguished by comparing the contralateral responses. The afferent limb is from the Vth nerve and the efferent limb is provided by the facial nerve, and is usually bilateral.
- Central lesions are often associated with other localizing signs (e.g. first division in association with the IIIrd, IVth and VIth nerves in the cavernous sinus; cerebellopontine angle lesions).
- Sensory lesions are much more common than motor lesions.

Facial nerve (cranial nerve VII)

Function

The facial nerve is primarily a motor nerve, but conveys fibres of three different modalities:

1. Motor – to muscles of facial expression.
2. Parasympathetic – to lacrimal, submaxillary and sublingual glands.
3. Sensory – taste for the anterior two-thirds of the tongue and an insignificant part of the external ear.

Anatomy

The motor nucleus lies in the pons close to the VIth nerve. It emerges in the cerebellopontine angle and enters the internal auditory meatus with the VIIIth nerve, giving off the nerve to the stapedius and chorda tympani (taste) before emerging through the stylomastoid foramen and passing peripherally through the parotid gland, giving off various branches.

Examination

Inspection   Inspect the face for:

- Asymmetry (e.g. loss of nasolabial fold, drooping and dribbling from corner of mouth, weak smile).
- Facial expression.
- Involuntary movements.

Muscle strength   Examine the individual muscles. The lower face may be assessed by smiling, whistling, pursing the lips; the upper face by closure of the eyes, elevation of the eyebrows, frowning.

Taste   Taste is rarely formally tested.

Interpretation

By far the most important component of the facial nerve is motor. Upper motor neuron (UMN) lesions often result in relative preservation of movements of the upper face due to crossed innervation (Fig. 17.5). In addition, emotional expression may be preserved. Lower motor neuron (LMN) VIIth nerve lesions do not spare the muscles around the eyes. The site of the lesion can often be localized, for example to the:

- Pons – for example due to a CVA; associated with VIth nerve lesion and contralateral pyramidal tract signs.
- Cerebellopontine angle – for example due to an acoustic neuroma; associated with lesions of the VIth, VIIth and VIIIth nerves as well as cerebellar ataxia.
- Facial canal – for example due to Bell's palsy, herpes zoster; associated with loss of taste and hyperacusis as well as muscles of facial expression.
- Parotid gland – for example due to sarcoidosis, parotid tumour; individual facial muscles may be affected.

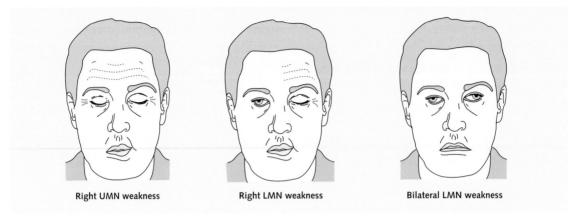

Right UMN weakness   Right LMN weakness   Bilateral LMN weakness

Fig. 17.5 Facial weakness. Patients are asked to close their eyes and purse their lips. Note the failed eye closure in lower motor neuron (LMN) lesions, the preserved forehead wrinkles in upper motor neuron (UMN) lesions and the nasolabial fold with drooping mouth in both UMN and LMN lesions.

Vestibulocochlear nerve (cranial nerve VIII)

Function

The vestibulocochlear nerve is a sensory nerve. It has two primary functions:

1. Auditory – sense of hearing.
2. Labyrinthine – sense of balance.

Anatomy

Auditory Sensory fibres from the cochlea enter the cerebellopontine angle in association with the facial nerve, synapse in the lower pons and ascend in the lateral lemnisci.

Vestibular Fibres pass with the auditory division, but synapse in the vestibular nucleus in the medulla, from which there are connections to the cerebellum, extraocular muscles and higher centres.

Examination

Auditory Crude assessment can be made for each ear (e.g. 'can hear whispered voice'). If a defect is found, conductive or sensory deficits may be identified using tuning fork tests (Fig. 17.6), as follows:

- Weber's test – apply the base of the tuning fork to the middle of the forehead and ask the patient whether he or she hears the sound in the midline or to one side – the test is abnormal if the sound is lateralized.
- Rinne's test – using a 512 Hz tuning fork (>300 Hz so as not to stimulate pacinian

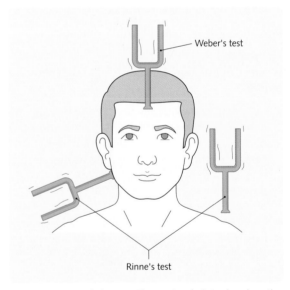

Fig. 17.6 Tuning fork tests. The tuning fork is placed on the vertex of the head in Weber's test. In Rinne's test, a tuning fork is struck and placed close to the external ear or rested on the mastoid process.

corpuscles), compare subjective loudness when it is presented close to the external auditory meatus and when the base is applied to the mastoid – a positive test (normal) occurs if the former appears louder.

Vestibular This function is not routinely tested. Positional nystagmus can be observed by holding the patient's head over the end of the examination couch and then fully extending and turning it with the eyes open. This test is called Hallpike's manoeuvre (Fig. 17.7).

Interpretation

If hearing loss is identified, try to identify the underlying cause. Note that:

- Tuning fork tests are useful for identifying the hearing deficit as primarily conductive or sensory in origin (Fig. 17.8).
- Deafness is commonly conductive (e.g. due to otitis media, ear wax).
- Sensory deafness can be further defined by formal audiometry to ascertain the frequency of loss.

Vertigo is most commonly peripheral. Central lesions (e.g. cerebellar lesions) are not associated with deafness or tinnitus, but are often associated with pronounced ataxia between the episodes of vertigo and persistent nystagmus.

Glossopharyngeal nerve (cranial nerve IX)

Function

The glossopharyngeal nerve has three functions:

1. Sensory – taste for the posterior two-thirds of the tongue, most of the oropharynx and soft palate.
2. Parasympathetic.
3. Motor – to the stylopharyngeus.

Anatomy

The motor and sensory nuclei lie in the medulla. The nerve leaves the skull with the Xth and XIth nerves at the jugular foramen.

Examination

Gag reflex Ask the patient to say 'aah' with his or her mouth open and observe palatal movement. Touch the posterior wall of the oropharynx with an orange stick. This elicits constriction and elevation. If there is no response, ask the patient whether he or she felt the stimulus. The presence or absence of the gag reflex does not correlate with whether a patient has a safe swallow reflex after a CVA.

The gag reflex is unpleasant, and should only be performed if a lesion of the IXth or Xth cranial nerves is suspected.

Interpretation

The afferent limb is via the IXth nerve, the efferent from the Xth.

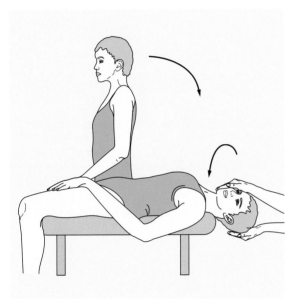

Fig. 17.7 Hallpike's manoeuvre for testing positional nystagmus.

Assessment of tuning fork tests			
Condition	Rinne's (left ear)	Rinne's (right ear)	Weber's
normal hearing	positive	positive	heard in midline
conductive deficit in right ear	positive	negative	heard on right side
sensory deficit in right ear	positive	positive	heard on left side

Fig. 17.8 Assessment of tuning fork tests.

Vagus nerve (cranial nerve X)

Function

The vagus nerve supplies innervation to the viscera in the thorax and foregut as well as having smaller motor and sensory functions. The main functions are:

- Parasympathetic – visceral innervation to the heart, lungs, foregut.
- Motor – to the larynx, soft palate, pharynx.
- Sensory – for dura mater of the posterior cranial fossa, small part of the external ear.

Anatomy

The nucleus lies in the medulla.

Examination

Speech   Listen for dysphonia (altered voice production) or a bovine cough (like a cow!) associated with recurrent laryngeal nerve palsy.

Soft palate   Observe the uvula. In a unilateral lesion, it will droop away from the lesion.

Gag reflex   See above for the glossopharyngeal nerve.

Interpretation

In the presence of dysphonia, the vocal cords should be examined.

Accessory (spinal accessory) nerve (cranial nerve XI)

Function

The spinal accessory nerve is a motor nerve supplying the sternomastoid and trapezius muscles.

Anatomy

The anterior horn cells of the cervical cord innervate these muscles, but fibres pass up to the medulla before descending again through the jugular foramen.

Examination

Test the bulk and power of the sternomastoid. Ask the patient to:

- Force the chin downwards against the resistance of your hand (bilateral).
- Turn the chin to one side against resistance (unilateral weakness affects turning to the opposite side).

The power of the trapezius can be tested by asking patients to shrug their shoulders against resistance.

Hypoglossal nerve (cranial nerve XII)

Function

The hypoglossal nerve is a motor nerve supplying innervation to the muscles of the tongue.

Anatomy

The nucleus is in the medulla.

Examination

Inspection   Look at the tongue for wasting and fasciculation.

Protrusion   Ask the patient to protrude the tongue. If there is a unilateral lesion, the tongue will deviate towards the side of the lesion.

MOTOR SYSTEM

When examining the motor system, the aims are to:

- Identify any lesions
- Ascertain whether the lesion is an UMN or LMN lesion
- Locate the anatomical site of the lesion
- Consider the differential diagnosis of lesions at that site.

The fundamental distinction is between UMN (above the anterior horn cells) and LMN lesions (Fig. 17.9).

The examination should follow a strict routine of:

- Inspection.
- Palpation.
- Assessment of muscular tone.
- Assessment of power.
- Assessment of tendon reflexes.
- Assessment of coordination.
- Assessment of gait (this may occur as the patient walks into the room, but should always be tested in an exam).

Inspection

Inspection should begin as the patient enters the examination room. Note posture, gait, coordination, abnormal movements, etc. The patient should be fully exposed on the examination couch so that individual muscle groups can be observed. Inspect specifically for:

Features of UMN and LMN lesions	
UMN	**LMN**
no muscle wasting (but there may be disuse atrophy)	wasting
increased tone ('clasp-knife')	flaccid
weakness of characteristic distribution	marked weakness
hyperreflexia	depressed or absent reflex
abnormal plantar response	normal plantar response
no fasciculation	fasciculation

Fig. 17.9 Features of upper motor neuron (UMN) and lower motor neuron (LMN) lesions.

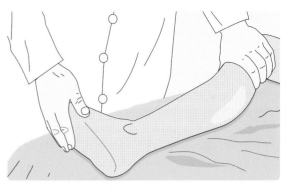

Fig. 17.10 Technique for eliciting ankle clonus. Bend the patient's knee slightly, supporting it with one hand. Grasp the forefoot with the other hand and gently rotate before suddenly dorsiflexing the foot. Clonus is made up of regular oscillations of the foot. Sustained (>4 beats) clonus indicates an upper motor neuron lesion. One or two beats is normal.

- Wasting – note symmetry; look specifically for distribution (e.g. proximal wasting).
- Fasciculation (spontaneous contraction of small groups of muscle fibres) – usually implies a LMN lesion (e.g. motor neuron disease).
- Tremors – note whether coarse and whether a resting tremor or an intention tremor.

Palpation

Palpate the muscle groups, specifically noting:

- Muscle bulk.
- Tenderness (e.g. myositis).

Assessment of muscular tone

Normally, there is a limited resistance through the range of movement.

Assessing tone

When assessing muscular tone, it is essential that the patient is properly relaxed and lying in a neutral position. The first thing most patients do when asked to relax is to tense, so try using other phrases such as 'Just let yourself go all floppy'. During examination of the limbs, it is important to ascertain whether increased tone is voluntary or not. Remember to support the patient's limb while performing movements of the joints. If you move two joints simultaneously (flexing the wrist and elbow, for example), it is much more difficult for the patient to voluntarily resist.

If increased tone is suspected, attempt to elicit clonus (Fig. 17.10). Assessment of tone is subjective and requires experience, but specifically consider:

- Hypertonia (increased tone) – for example 'clasp-knife' (high resistance to initial movement and then sudden release; characteristic of UMN lesions), 'lead-pipe rigidity' (resistance through the range of movement; in Parkinson's disease, this in combination with the tremor produces 'cogwheel rigidity').
- Hypotonia (decreased tone) – for example due to LMN and cerebellar lesions.

Assessment of power

Individual muscle groups should be tested to assess power. When testing, patients should be at a slight mechanical advantage so that if they have normal power, they can just overcome the resistance of the examiner. Muscle strength can be classified according to the Medical Research Council (MRC) grade (Fig. 17.11).

It is usually sufficient to test:

- Movements of the neck.
- Shoulder abduction and adduction.
- Movements of the elbows, wrists and hands.
- Movements of the hips, knees and ankles.

When detecting a weakness, it should be categorized as either UMN or LMN. If classified as LMN, the physical signs should be integrated to identify the lesion anatomically (Fig. 17.12).

When testing each muscle group, you always need to consider:

- The myotome.
- The peripheral nerve supplying the muscles.

The major movements in the upper and lower limb are illustrated in Figures 17.13 and 17.14. It is essential to be systematic if a weakness is identified, so that the pattern of involvement can be recognized

MRC classification of muscle power	
Grade	MRC grade of muscle strength
0	no movement
1	flicker of movement visible
2	movement possible with gravity eliminated
3	movement possible against gravity, but not resistance
4	movement possible against resistance, but weakened (often subdivided to 4−, 4, 4+)
5	normal power

Fig. 17.11 Medical Research Council (MRC) classification of muscle power.

as corresponding to a nerve root, peripheral nerve or an individual muscle group (Fig. 17.15). This can be correlated with the other features of the examination of the peripheral nervous system.

Specific tests for the more common peripheral neuropathies are considered later.

Assessment of tendon reflexes

The reflexes should be elicited using a tendon hammer. Compare the relative responses, both against normality and with each side. Make a note of the response – whether it appears normal, brisk or reduced. If no response is obtained, try methods to reinforce the reflex. For example, ask patients to clench their teeth or hook their fingers around each other and try to separate their hands without disentangling their fingers (Fig. 17.16). The nerve roots of the more commonly elicited reflexes are listed in Figure 17.17 and the reflex arc is shown in Figure 17.18.

Disruption of the reflex may be due to a lesion at the level of the:

- Peripheral nerves (peripheral neuropathy). Typically the reflex is depressed early in the course of the pathology.
- Spinal cord.

Fig. 17.12 It is usually possible to localize a lower motor neuron (LMN) lesion as originating in the spinal cord, nerve root, peripheral nerve, neuromuscular junction or muscle.

Features of LMN lesions originating in the spinal cord, nerve root, peripheral nerve, neuromuscular junction and muscle		
Location of lesion	Examples	Features
anterior horn cells	motor neuron disease; polio	usually symmetrical; no myotome/nerve root distribution; often distal initially; no sensory involvement
nerve root (radiculopathy)	nerve root compression	distribution of affected muscles according to myotome; may have associated dermatomal sensory loss
peripheral nerve (neuropathy)	carpal tunnel	weakness according to nerve supply of affected nerve; usually associated sensory loss; early loss of reflexes
neuromuscular junction	myasthenia gravis	loss of power fluctuating in severity; not in distribution of peripheral nerve or myotome
muscle (myopathy)	n.a.	often has characteristic distribution of a particular disease; reflexes may be preserved early in disease; no sensory loss

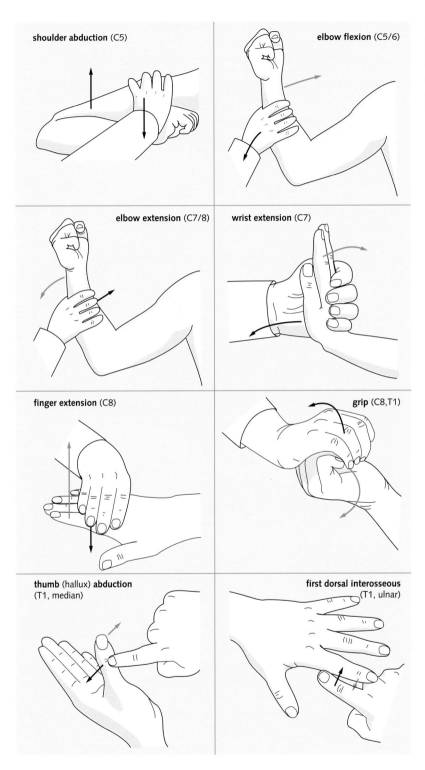

shoulder abduction (C5)

elbow flexion (C5/6)

elbow extension (C7/8)

wrist extension (C7)

finger extension (C8)

grip (C8,T1)

thumb (hallux) abduction
(T1, median)

first dorsal interosseous
(T1, ulnar)

Fig. 17.13 The major muscle groups tested in an assessment of power in the upper limb. Patients should be at a slight mechanical advantage so that they can just overcome the resistance offered by the examiner.

Fig. 17.14 Testing muscle groups of the lower limb.

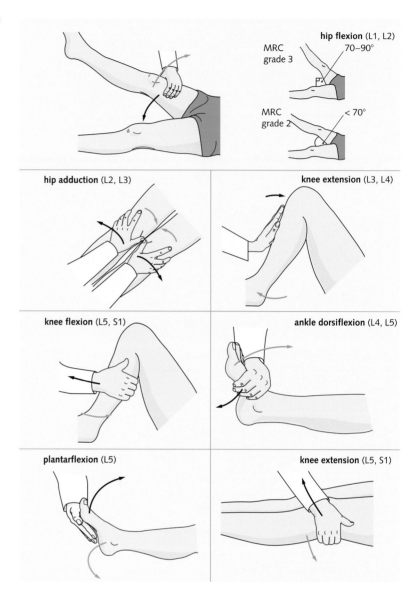

- Neuromuscular junction.
- Muscle (myopathy). The reflex is usually retained until late in the natural history of the disease.

Assessment of coordination

Coordination in the upper limb is tested by the finger–nose test. Hold a finger at arm's length from the patient and then ask them to rapidly touch the tip of their nose, and then the tip of your finger with their index finger. The smoothness and accuracy of the movements should be interpreted and put into context with any muscle weakness.

The lower limbs may be assessed by the heel–shin test. Ask the patient to place the right heel over (not in contact with) the left shin and to slide it down and up the shin, and then to repeat the test using the left heel.

If there is an intention tremor or dysmetria (irregular error in the distance and force of limb movements), cerebellar pathologies can be investigated by looking for dysdiadochokinesis. Ask patients to rapidly slap one palm with the other hand, alternating between the palm and back of the hand.

Nerve roots, peripheral nerves, and muscles responsible for each movement			
Movement	Myotome	Peripheral nerve	Main muscle groups
shoulder abduction	C5,6	axillary	deltoid
shoulder adduction	C5,6,7,8	lateral pectoral, thoracodorsal	pectoralis major, latissimus dorsi
elbow flexion	C5,6	musculocutaneous	biceps
elbow extension	C6,7,8	radial	triceps
wrist extension	C7,8	radial	long extensors
wrist flexion	C8	ulnar and median	long flexors
pronation	C6,7	median	pronator teres, pronator quadratus
supination	C6,7	musculocutaneous, radial	biceps, supinator
finger abduction	T1	ulnar	dorsal interossei
finger adduction	T1	ulnar	palmar interossei
opposition of thumb	T1	median	opponens pollicis
extension of thumb at the interphalangeal joint	T1	radial	extensor pollicis longus
hip flexion	L4,5	inferior gluteal	glutei
hip extension	L2,3	femoral	iliopsoas
knee extension	L5, S1	sciatic	hamstrings
knee flexion	L2,3,4	femoral	quadriceps
inversion of ankle	L4,5	peroneal	tibialis anterior, long extensors, peroneus, extensor digitorum brevis
eversion of ankle	S1,2	tibial	gastrocnemius, tibialis posterior
dorsiflexion of foot	L4	peroneal, tibial	tibialis anterior, tibialis posterior
plantar flexion of foot	S1	peroneal	peronei, long extensors, extensor digitorum brevis
extension of great toe	L4,5, S1	deep peroneal	extensor hallucis longus

Fig. 17.15 Nerve roots, peripheral nerves and muscles responsible for each movement. By integrating the pattern of weakness, it should be possible to recognize a pattern corresponding to these anatomical subdivisions.

Assessment of gait

No neurological assessment is complete without an assessment of the patient's gait.

At a basic level, gait should be assessed as the patient walks into the examination room. Certain gaits are characteristic of certain pathologies. For example:

- Spastic gait – the extensor muscles are stiff and the foot is plantar flexed so patients have a stiff gait and avoid catching their toes on the ground by circumducting the leg at the hips.
- Ataxia – the gait is wide based and there is also marked clumsiness when patients are asked to walk in a straight line, placing their heel immediately in front of the toe of the opposite foot (e.g. in cerebellar ataxia).
- High-stepping gait – in patients with foot drop, the foot is lifted high off the ground and then slapped back down.

175

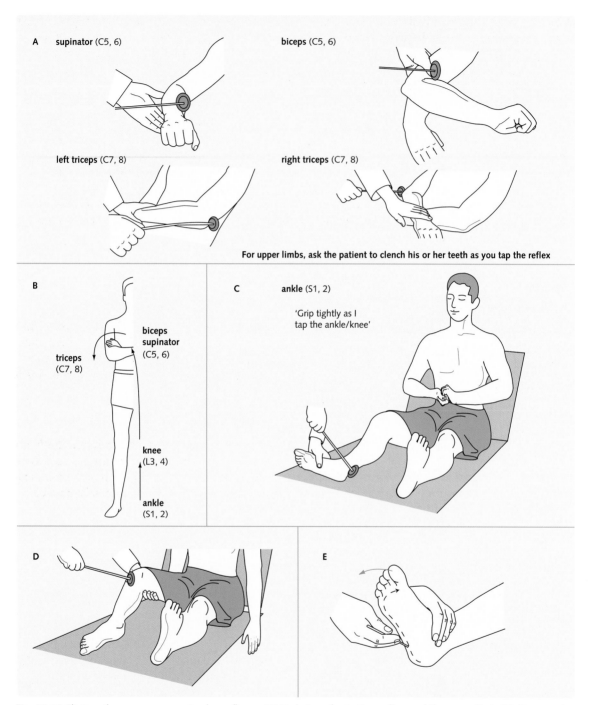

Fig. 17.16 Eliciting the more common tendon reflexes. (**A**) Technique for testing reflexes of the upper limb. (**B**) Nerve roots tested during reflexes. (**C**) Reinforcement of the ankle reflex. (**D**) The knee reflex. (**E**) Plantar reflexes.

- Parkinsonism – the gait is slow and shuffling and there is no associated arm swinging; patients often find it difficult to stop and turn around.
- Waddling gait – for example due to proximal myopathy.

SENSORY SYSTEM

Patients are usually aware of numbness, paraesthesiae or altered sensation indicating a sensory pathology (all are common, and are worrying presenting symptoms), but examination of the sensory system forms part of a routine assessment. Attempt to identify the modality of sensory loss and its distribution (e.g. correlating with a dermatome, peripheral nerve). A knowledge of the dermatomes is essential (Fig. 17.19).

Useful dermatomes to remember are: C5, deltoid; C6, thumb; C7, middle finger; C8, little finger; T4, nipple; T8, xiphisternum; T10, umbilicus; T12, symphysis pubis; L4, medial leg; L5, between great and second toe; S1, lateral border of foot.

Nerve roots supplying the major reflexes	
Reflex	Nerve root
biceps	C5*, C6
brachioradialis	C6*, C7
triceps	C6, C7*,C8
knee	L2, 3*, 4*
ankle	S1, S2
anal	S2, 3, 4

Fig. 17.17 Nerve roots supplying the major reflexes. Asterisks indicate the major nerve root supplying each reflex.

Light touch

Dab (do not stroke) the skin lightly with a small wisp of cotton wool. If there is decreased sensation, this should be mapped out. Start from the area of decreased sensation and move outwards as this is more sensitive.

Pin prick (pain)

Use a disposable pin or needle gently and check that the patient can identify the stimulus as sharp. Temperature sensation is also conveyed in the spinothalamic tracts, so is not usually routinely assessed.

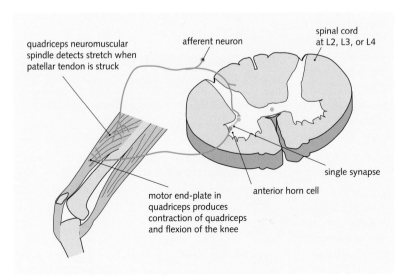

Fig. 17.18 Neurological pathway for the knee jerk.

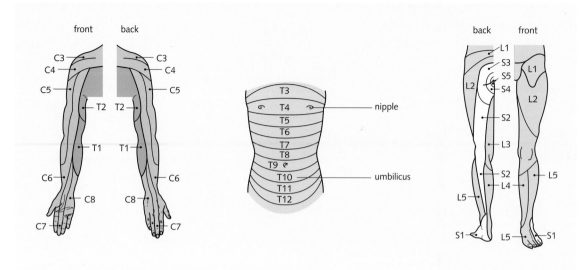

Fig. 17.19 Distribution of the dermatomes on the surface of the body.

Vibration

Place the base of a vibrating 128 Hz tuning fork on the distal phalanx of the great toe. Patients should be aware of a buzzing sensation. Ask patients to close their eyes and to indicate when they think that the tuning fork has stopped vibrating. Usually vibration sense is impaired peripherally. If absent, move proximally (i.e. from the lateral malleolus to the upper tibia and then the iliac crest).

Loss of vibration sense is one of the earlier physical signs indicative of peripheral neuropathy in diabetes mellitus.

Joint position sense

Start distally. Move the great toe passively, either extending or flexing, and observe whether the patient can identify the direction of movement. Be careful not to contact the second toe with your hand as this may give additional sensation to the patient. When you hold the toe, hold it by the side; this prevents the patient feeling the increase in pressure when moving the toe up or down. If there is impairment, test the ankle and knee.

Romberg's test

Romberg's test may be positive if there is impaired position sense. Patients are asked to stand upright with their eyes closed. Marked swaying when patients close their eyes is 'Rombergism'; that is, a positive test (Fig. 17.20).

HIGHER FUNCTIONS

There are many sophisticated tests of mental function. A crude screening test is the mini-mental test. This may reveal the presence of dementia and is an essential part of any examination of an elderly person. Furthermore, a change in mental functioning may be identified over a period of time.

Other tests should include:

- State of consciousness.
- Emotional state.
- Speech (e.g. to reveal receptive or expressive dysphasia).

If a cortical lesion is suspected, attempt to localize it (Fig. 17.21).

PATTERNS OF NEUROLOGICAL DAMAGE

It is helpful when assessing neurological lesions to be aware of common patterns of deficits so that the anatomical site of the lesion can be determined (e.g. myopathy, peripheral neuropathy, nerve root, cortical). Some of the more important lesions are described below.

Myopathy

Weakness and wasting of muscles occurs in a distribution that is characteristic of the particular type of

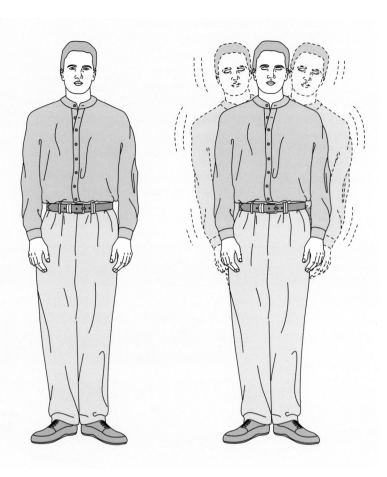

Fig. 17.20 Romberg's test. A positive test indicates impaired position sense. Patients can maintain a good posture when their eyes are open, but sway when asked to close their eyes.

Typical features of localized cortical lesions	
Lobe affected	**Features**
frontal lobe	predominant mood and behavioural changes; disinhibition; motor cortex may be involved; pout reflex
temporal lobe	dominant: sensory dysphasia, alexia (agraphia); non-dominant: may cause visuospatial deficits
parietal lobe (dominant)	aphasia; dysgraphia; dyslexia; right–left disorientation
parietal lobe (non-dominant)	neglect of contralateral sensation; altered body image; dressing apraxia; constructional apraxia

Fig. 17.21 Typical features of localized cortical lesions.

myopathy (e.g. facioscapulohumeral muscular dystrophy, Duchenne muscular dystrophy). Reflexes are preserved until the disease is advanced.

Peripheral nerves

Radial nerve

Damage to the radial nerve may cause wrist drop. Test sensation over the first dorsal interosseous muscle. Test extension of the interphalangeal joint of the thumb (e.g. compression neuropathy after sleeping with arm over a chair).

Median nerve

Damage to the median nerve (e.g. carpal tunnel syndrome) characteristically produces:

- Sensory loss over the palmar aspect of the lateral/radial three-and-a-half fingers.
- Wasting of the thenar eminence.
- Weakness of opposition, flexion and abduction of the thumb.

Ulnar nerve

Ulnar nerve lesions (e.g. due to trauma at the elbow) result in:

- Sensory loss over the little finger and medial half of the ring finger.
- Wasting of the hypothenar eminence.
- Weakness of finger abduction and adduction.

Fibula/common peroneal nerve

The common peroneal nerve is sometimes damaged during a fracture to the head of the fibula. This may result in:

- Impaired sensation of the lateral calf.
- Foot drop.
- Weakness of dorsiflexion and eversion of the foot.

Peripheral neuropathy

Generalized neuropathy is usually in a 'stocking and glove' distribution (i.e. hands and lower forearms, and feet and lower legs). Tendon reflexes are lost early in the course of the disease.

Cerebellum

The main features of cerebellar pathology are (note the mnemonic DANISH):

- Dysdiadochokinesis.
- Ataxia of gait (wide based).
- Nystagmus.
- Intention tremor.
- Staccato speech (dysarthria).
- Hypotonia.
- Dysmetria.

Extrapyramidal system

There is a wide range of extrapyramidal syndromes. They tend to be characterized by:

- Decreased movement (e.g. bradykinesia in Parkinson's disease).
- Involuntary movements (e.g. tardive dyskinesia with drug therapy).
- Rigidity.

SUMMARY OF NEUROLOGICAL EXAMINATION

In summary, when carrying out a neurological examination:

- It is of paramount importance to be systematic. Most of the signs represent a qualitative change from normality. This can only be recognized with practice.
- Identifying a lesion is the easy part. A given lesion may be due to multiple pathologies at various locations in the nervous system. The next stage is to prepare a list of all the elicited signs and consider whether a single pathological lesion could account for them. If so, identify its anatomical site.
- Remember that lesions produce characteristic patterns of physical signs. For example, if a weakness is identified, consider whether the pattern fits into an UMN or LMN pattern. If it is LMN, look for features that identify the site of the lesion (i.e. at the muscle itself, the neuromuscular junction, a peripheral nerve, a nerve root or anterior horn cells). If the lesion is UMN in pattern, there are usually other localizing signs that allow identification of the anatomical point of pathology.
- Once the anatomical location of the lesion has been identified, prepare a differential diagnosis of possible pathologies that could produce a lesion at that site. Consider collateral

information from the history or systemic examination to narrow down the differential diagnosis.

Usually a diagnosis of stroke is apparent from the history, but the examination is crucial to confirm the presence of a focal neurological deficit, to document baseline function objectively and to consider aetiological factors.

Assess level of consciousness

See Unconscious patient, below.

Define the neurological deficit

A full neurological assessment is essential to ascertain the degree of damage. It is often possible to identify the vascular territory affected: these are often documented as *posterior*, *anterior* or *lacunar* syndromes (Fig. 17.22). Occasionally, the lesion produces a more subtle lesion such as impaired cognition. The anatomical site of damage may offer prognostic information.

Consider aetiology

The systemic examination may offer clues to the underlying cause.

Pulse

Arrhythmias, especially atrial fibrillation, predispose to emboli. Consider the possibility of a recent myo- cardial infarction, which may have caused a watershed infarct (due to reduced cerebral perfusion) or been complicated by transient arrhythmias.

Blood pressure

Hypertension is one of the major risk factors for stroke. In the acute setting, it should be interpreted with caution and treated even more cautiously, as an abrupt drop in blood pressure may cause further ischaemia as vascular autoregulation will be impaired.

Eyes

Argyll Robertson pupils may suggest syphilis (now a rare cause of stroke in the UK) or diabetes mellitus.

Fundoscopy

Look for evidence of hypertensive and diabetic retinopathy.

Face

The facial appearance may suggest an underlying pathology, for example plethora (polycythaemia), mitral facies.

Neck

Listen carefully for a carotid bruit as a source of embolus. Have a low threshold for considering a Doppler scan of the carotid vessels, as a patient with over 70% stenosis of the internal carotid artery may benefit from subsequent endarterectomy. Although detection of carotid bruits is not particularly sensitive, it is highly specific for the probability of significant stenosis.

Classification	Symptoms
total anterior circulation syndrome (TACS)	new higher cerebral dysfunction (dysphasia, dyscalculia or visuospatial disorder); homonymous visual field defect and ipsilateral motor and/or sensory deficit of at least two of the face, arm and leg
partial anterior circulation syndrome (PACS)	patients with two of the three components making up the TACS
posterior circulation syndrome (POCS)	any of the following: ipsilateral cranial nerve palsy with contralateral motor and/or sensory deficit; bilateral motor and/or sensory deficit; disorder of conjugate eye movement; cerebellar dysfunction without ipsilateral long-tract deficit (i.e. not ataxic hemiparesis); isolated homonymous visual field defect
lacunar infarct syndrome (LACS)	a pure motor stroke, pure sensory stroke, sensorimotor stroke or ataxic hemiparesis; two of three areas (face, arm, leg) must be involved with involvement of the whole of any affected limb

Fig 17.22 Stroke classification by vascular area.

Heart

Listen for any murmurs. Mitral stenosis (especially in association with atrial fibrillation) is a potent risk factor for left atrial thrombus and subsequent embolus. Consider the possibility of endocarditis (especially if there is a murmur). If there is fixed splitting of the second heart sound in association with a flow murmur, look for sources of paradoxical embolus from the venous circulation.

EPILEPTIFORM SEIZURE

The diagnosis of epilepsy usually relies upon an objective eyewitness account of at least two seizures – one seizure is not epilepsy. The doctor rarely sees a fit in an individual patient.

Acute seizure

The priority is to ensure that patients do not harm themselves, and then to assess and, if necessary, provide specific therapy for the seizure. One should:

- Protect the airway.
- Ensure that the patient will not harm him or herself.
- Observe the nature of the seizure activity (e.g. tonic–clonic, focal, absence attack).
- Check the pulse rate and, if possible, blood pressure.
- Check blood sugar (BM), obtain blood for electrolytes (toxins).
- Undertake measures for status epilepticus if the seizure is prolonged.

Postictal

Usually the patient is seen in the postictal period, having been brought to hospital after collapsing. The history is central to the diagnosis (see Ch. 6 Presenting problems), and very often the examination is unremarkable. The examination may be useful when considering the differential diagnosis or for assessing the aetiology and functional consequences of the seizure (Fig. 17.23). Remember, any neurological symptoms may be observed after a seizure but they should be monitored and seen to resolve.

Perform detailed neurological assessment

In particular, focus on:

- Level of consciousness. Check the Glasgow Coma Score (Fig. 17.24) until a normal level of consciousness is achieved, as seizure may be the presentation of head injury or intracranial bleed.
- Paralysis. A postictal focal weakness (Todd's paralysis) may be present for 24 hours following a seizure, but consider the presence of an acute CVA or intracranial space-occupying lesion if neurological signs are present.
- Eyes. Check pupillary responses and fundoscopy to exclude papilloedema.

Perform systemic examination to identify precipitating cause

It is particularly important to perform a detailed systemic examination to try to identify an underlying cause for the epileptic seizure so that specific therapy may be offered. Consider:

Assessment of a patient during a seizure to reveal conditions other than typical epilepsy	
Cause of collapse	Discriminatory features
syncope	usually apparent from prodromal history; check pulse rate (e.g. tachyarrhythmia, Stokes–Adams attack); check postural blood pressure; pallor during episode
narcolepsy	no convulsions; patient rousable
hysteria	often many atypical features; usually only occur when there is an audience; no urinary incontinence/tongue biting

Fig. 17.23 Assessment of a patient during a seizure may reveal the presence of a condition other than typical epilepsy.

Glasgow Coma Score	
Eyes (E)	
open spontaneously (with blinking)	4
open to command of speech	3
open in response to pain (applied to limbs or sternum)	2
not opening	1
Motor function (M)	
obeys commands	6
localizes to pain	5
withdraws from pain	4
flexor response to pain (decorticate)	3
extensor response to pain (decerebrate)	2
no response to pain	1
Vocalization (V)	
appropriate speech	5
confused speech	4
inappropriate words	3
groans only	2
no speech	1

Fig. 17.24 The Glasgow Coma Score.

- Alcohol withdrawal – signs of chronic liver disease, smell of alcohol on breath, unkempt condition, common.
- Trauma.
- Pyrexia – infants.
- Encephalitis – fever, level of consciousness, focal neurological signs, herpetic ulcer.
- Malignancy – look for signs of primary bronchial, breast, colonic or kidney tumour.
- Degenerative brain disease (e.g. dementia).

UNCONSCIOUS PATIENT

Assessment of the unconscious patient provides a great challenge for the diagnostic and management skills of the doctor, and may be something you encounter on your first shift as an F1 doctor or intern, so you need to have a system. The first priority is to ensure that the patient is stable before performing a detailed assessment. This all needs to be practised until it is second nature. You can't learn the skills from a book as they are practical procedures, but you can learn the framework. You are likely to be required to demonstrate you are competent at the management of the unconscious patient (e.g. an Objective Structured Clinical Examination (OSCE) question). The trust resuscitation officer is a great source of experience and will be able to help you become proficient at this.

Adopt a SAFE approach

- S – Shout for help.
- A – Approach with care.
- F – Free the patient from danger.
- E – Evaluate the patient's ABC (see below).

The SAFE approach is designed to ensure that you do not become a casualty yourself. Whilst approaching a patient in A&E may not present any danger, stopping to help at a road traffic accident might. If the patient is in harm's way, either remove the harm or remove the patient from the harm. Only when both you and the patient are safe can you start to assess the patient.

Keep it simple. You are only aiming for two things – to ensure that:

- Air goes in and out.
- Blood goes round and round.

The best way to achieve this is with the ABC principles:

- *Airway.* Check that it is patent; this includes looking in the mouth to remove any obstructions, for example false teeth or debris. Can the patient maintain their own airway? If they cannot, you will need to help with this (e.g. triple airway manoeuvre with or without a Guedel Airway).
- *Breathing.* Now that the airway is patent, is the patient breathing? If not, you need to breathe for them, either mouth-to-mouth or with a bag and mask, depending on where you are. The patient needs to be on 100% oxygen in this situation. Look at the chest: does it rise and fall equally? If it doesn't, do they need a chest drain?
- *Circulation.* Does the patient have a pulse? If they've got a pulse, do they have a blood pressure? If there is no pulse, cardiopulmonary resuscitation (CPR) needs to be started and an electrocardiogram attached. The guidelines for the management of a cardiac arrest are under constant revision, so look up the Resuscitation Council Web site at www.resus.org.uk for the latest algorithm. If the patient is shocked (e.g. remember the section on gastrointestinal bleeding), then they need a fluid challenge. In the adult, this is 500–1000 mL of Hartmann's solution (depending on the patient's size); in children, it is 20 mL/kg of a colloid of your

choice. The patient needs two large intravenous cannulae. If you haven't got access after two attempts, stop wasting time trying. In adults, do a venous cut down; in children under age 6, place an intraosseous needle. When you get blood or marrow, do a BM and cross match.

Remember you do not move from A to B until you have dealt with A, etc. Resuscitation is a team effort. This is why you must shout for help right at the beginning. Practise the practical things on mannikins so you know what you are trying to do. When you are presented with a major trauma or a cardiac arrest, it is frightening. You are allowed to be scared – I still am. Afterwards, sit down and have a cup of tea with everyone involved and share your feelings and appraise what happened. There are always things you have done well and things to remember for next time.

Assess level of consciousness

The Glasgow Coma Score (GCS) is widely used to provide a simple reproducible objective assessment of conscious level. This is based on the best responses obtained (see Fig. 17.24). If you don't have time to assess the GCS, use the AVPU scoring system: is the patient *A* alert, responding to *V* voice, responding to *P* pain or *U* unresponsive?

> The Glasgow Coma Score is particularly useful if repeated observations are made so that a rapid and unambiguous diagnosis of a deteriorating level of consciousness can be made.

Perform full neurological examination

Look for any localizing signs. Pay particular attention to:

- Pupil size and response to light (Fig. 17.25) (including symmetry) – some drugs (e.g. opiates) constrict pupils; other drugs (e.g. phenothiazines, amfetamines) dilate pupils; 'coning' results in a single fixed dilated pupil; pontine lesion results in pinpoint pupils; brain death results in fixed dilated pupils.
- Fundi (especially for papilloedema).
- Gag and corneal reflexes.
- Motor responses.
- Tendon reflexes.
- Abnormal tone.

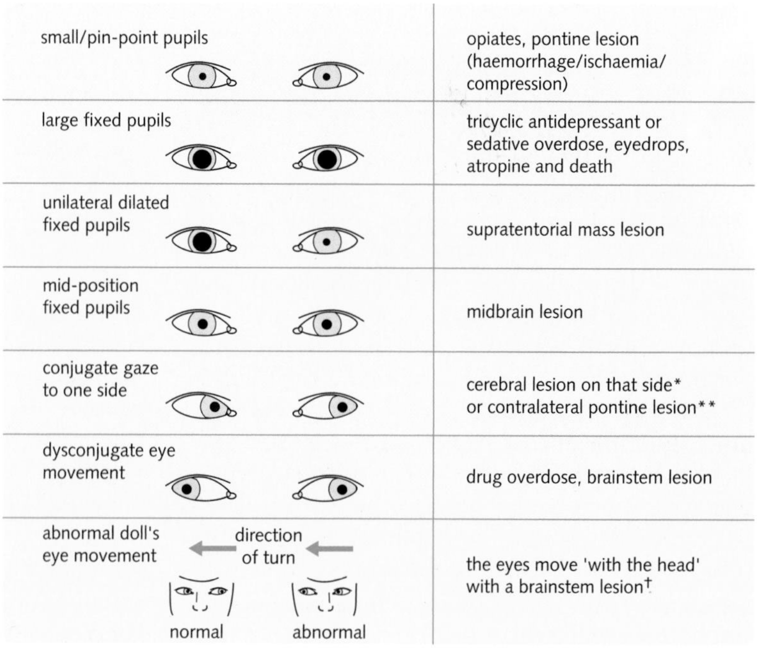

small/pin-point pupils	opiates, pontine lesion (haemorrhage/ischaemia/compression)
large fixed pupils	tricyclic antidepressant or sedative overdose, eyedrops, atropine and death
unilateral dilated fixed pupils	supratentorial mass lesion
mid-position fixed pupils	midbrain lesion
conjugate gaze to one side	cerebral lesion on that side* or contralateral pontine lesion**
dysconjugate eye movement	drug overdose, brainstem lesion
abnormal doll's eye movement	the eyes move 'with the head' with a brainstem lesion†

Fig. 17.25 Examination of the pupils in an unconscious patient. Note the size and reaction to light. In addition, note the position of gaze fixation. (*Looking towards the lesion; **looking away from the lesion; †normally, if the head is held and turned quickly from side to side, the eyes swivel in the opposite direction to the head.)

Perform detailed systemic examination

The six traumatic things that kill you quickly and can be treated are best remembered as ATOM FC:

- **A** Airways obstruction.
- **T** Tension pneumothorax.
- **O** Open pneumothorax.
- **M** Massive haemothorax.
- **F** Flailed chest.
- **C** Cardiac tamponade.

Coma may be due to metabolic, infective, cardiovascular pathology, etc. In particular, note:

- Evidence of trauma or head injury (look for otorrhoea or cerebrospinal fluid leaking from the nose).
- Temperature (e.g. hypothyroidism, infection).
- Evidence of needle marks (indicating diabetes mellitus or drug addiction).
- Jaundice or other features of chronic liver disease.
- Breath (e.g. revealing hepatic fetor, alcohol, diabetic ketoacidosis).
- Respiratory pattern (e.g. Cheyne–Stokes respiration, Kussmaul's respirations of diabetes mellitus or uraemia).
- Cyanosis (e.g. due to hypoxia).

Blood glucose

Exclude hypoglycaemia or hyperglycaemia early in the assessment of an unconscious patient by performing a BM stick analysis. This is often performed by paramedics. Always repeat it: if nothing else, the trend may tell you about the patient's metabolic state and, in an acute situation, the information may have been incorrectly recorded.

MULTIPLE SCLEROSIS

Multiple sclerosis causes neurological lesions that are disseminated in time and place. It can be diagnosed with reasonable confidence from the history of relapses and remissions and thorough examination, if it follows a typical course. Areas of demyelination may occur anywhere in the central nervous system, but certain patterns are more common.

Sites of involvement

The more common sites of involvement are the:

- Optic nerve (optic neuritis).
- Brainstem.
- Cerebellum.
- Cervical cord.
- Periventricular region.

Eyes

Certain patterns of disease should be distinguished:

- Retrobulbar neuritis – relative afferent pupillary defect, central scotoma.
- Optic neuritis – also note swelling of the optic disc acutely, and temporal pallor following recovery.
- Optic atrophy – this is a common finding in long-standing multiple sclerosis, but should be distinguished from other pathologies (Fig. 17.26).
- Nystagmus – often jerking or ataxic; pronounced in late disease.
- Internuclear ophthalmoplegia – diplopia on lateral gaze due to failure of the adducting eye to cross the midline and demyelination of the medial longitudinal bundle.

Brainstem

Multiple brainstem or cerebellar signs may be present, especially nystagmus, diplopia, intention tremor and scanning speech.

Differential diagnosis of optic atrophy	
Site of lesion	Causes
retina	central retinal occlusion; toxic (e.g. quinine, methylated spirits)
optic nerve	optic and retrobulbar neuritis, (e.g. multiple sclerosis); chronic glaucoma; any cause of papilloedema; toxin (e.g. alcohol, tobacco, ethambutol); tumour (e.g. meningioma, optic glioma)
optic chiasm	pituitary tumour; craniopharyngioma; meningioma

Fig. 17.26 Differential diagnosis of optic atrophy.

Spinal cord

Spastic paraparesis is common in late disease. Other signs can include bladder dysfunction, decreased limb sensation and loss of posture sensibility, which is often marked.

Mental

Both euphoria and depression or irritability are common.

Multiple sclerosis

This is another disease with limited treatment options and no cure. It is therefore vital that your communication with the patient is excellent. The medical interview is often a source of therapy for the patient, referred to by Balint as the 'drug doctor'. Remember that the dose and frequency are determined by the patient and doctor, not the doctor alone. Always attempt to make a functional assessment of how multiple sclerosis affects daily activities and underlying concerns and beliefs, as these are ultimately what determines a patient's quality of life.

Objectives

By the end of this chapter you should:

- Be able to examine a diabetic patient with specific reference to the diabetic foot, diabetic ketoacidosis and hyperosmolar non-ketotic coma.
- Be able to perform an examination of thyroid status.
- Be able to recognize features of endocrine disease in general examinations (such as cardiovascular/neurological system examinations).

DIABETES MELLITUS

Inspection

Note weight and height. Calculate the patient's body mass index (BMI); diabetic control may improve with a normal BMI.

Macrovascular complications

Ischaemic heart disease and cerebrovascular disease

Diabetics have a much increased risk of stroke and ischaemic heart disease. It is important to modify any reversible risk factors. Assess these other risk factors, especially:

- Obesity – obesity is not only a risk factor for macrovascular disease, but predisposes to poor glycaemic control.
- Blood pressure control.
- Smoking – nicotine staining of fingers and hair.
- Xanthoma, arcus senilis – hyperlipidaemia.

Peripheral vascular disease

Examine the peripheral pulses (femoral, popliteal, dorsalis pedis and posterior tibial arteries). Consider measuring the ratio of ankle : brachial artery pressure. Listen for bruits, especially over the abdomen and carotid and femoral arteries, which indicate turbulent flow and are suggestive of stenosis.

Look for evidence of ulcers, especially between the toes. Ulcers may be painless and large without patients knowing that they have got them.

Microvascular complications

Eyes

Examination of the eyes is important in diabetic patients as they are at risk of many diseases. Diabetic patients should have a detailed fundoscopic assessment with dilated pupils (ideally retinal photography as well) at least once a year. Note the following:

- Cataracts – these are more common in diabetics.
- Evidence of background retinopathy ('dot and blot' haemorrhages, exudates).
- Proliferative retinopathy (neovascularization, vitreous haemorrhage).
- Macular oedema.

Assess eye movements (especially in long-standing diabetes mellitus) for a third nerve palsy due to mononeuritis multiplex.

Check visual acuity, which is often reduced due to maculopathy.

Peripheral neuropathy

Test for evidence of a 'stocking and glove' sensory loss. Loss of vibration and joint position sense are often the first modalities to be affected, especially in the legs. A defined peripheral nerve (e.g. median nerve) lesion or mononeuritis multiplex (simply meaning multiple named nerves are affected) may also be present.

Diabetic nephropathy

Nephropathy is particularly common in the presence of retinopathy and neuropathy. It is also a marker for an increased risk of ischaemic heart disease. Look for proteinuria. At every assessment, a urinalysis should be performed.

Urinalysis is an essential component of examination of any diabetic patient!

Feet

Diabetic foot

This is an important examination, and is a common component of finals. The foot can show the earliest signs of micro- and macrovascular disease. The key is to observe closely, looking at the patient's footwear, between the toes, on the sole and in the gaiter area for ulcers and infection. Fine touch (usually with a 10 g monofilament), vibration and joint position sense are tested before finishing with palpation of the lower limb pulses. Pin-prick sensation should only be assessed with caution due to the risk of introducing infection through breaks in the skin.

The feet of diabetic patients should be examined at three-month intervals. Look for the presence of ulcers and the state of the nails. Test the temperature, pulses and shape of the foot to determine vascular and neuropathic causes. The presence of abnormally sited callosities on the sole may indicate uneven weight distribution.

Assess diabetic control

Different patients use different methods for recording glucose control.

Urinalysis

Look for glycosuria, but be aware of the different renal thresholds for glycosuria. If a patient consistently has negative glycosuria, it may be worthwhile checking a blood HbA1c level.

BM sticks

Look at the patient's own record of BM stick assessments. This is most informative in type I diabetes mellitus. You may wish to ask patients to perform

an estimate in front of you so that you can assess their technique.

Evidence of skin infections (e.g. boils, abscesses, candidiasis) may suggest chronically poor control.

Assess injection sites

Look for complications such as scarring, abscesses and lipodystrophy, which result from not rotating injection sites frequently enough. Note any amyotrophy (painful wasting of a muscle group, e.g. quadriceps).

Diabetic coma

A young patient may first present with diabetes in diabetic ketoacidosis. Any treated patient may have hypoglycaemia complicating therapy.

If there is any doubt about the cause of a coma in a diabetic patient and the glucose level is not known, treat with intravenous glucose first.

Diabetic ketoacidosis (DKA)

This typically occurs in a younger person. In particular:

- Note the respiratory pattern – Kussmaul's respiration, also known as 'air hunger', suggests acidosis.
- Note the mental state – may be alert, but usually confused or stuporose.
- Smell the breath – ketones may be detectable (a sweet, fruity smell, like pear-drops).
- Consider fluid status – the patient is invariably dehydrated with dry skin and mucous membranes, decreased (or undetectable) jugular venous pressure (JVP) and hypotension.
- Consider the underlying cause.
- Note that fever is often absent before therapy, even in the presence of infection, but sources of infection should be carefully sought.
- Do arterial blood gases. Patients can be extremely acidotic and this needs to be monitored.

Most centres now have DKA protocols. Know where to find yours and have a look at it before you start working there.

Hyperosmolar non-ketotic coma

Hyperosmolar non-ketotic coma usually presents in a similar fashion as diabetic ketoacidosis, but in elderly and middle-aged subjects who are usually type II diabetics. Many of the clinical signs are the same, with the exception of ketonuria, as are the basic management strategies, as follows:

- Rehydrate.
- Optimize acid–base balance.
- Replace insulin deficiency.
- Look for and treat the underlying cause.
- Provide general supportive care.

The management differs only in the fine tuning.

HYPERTHYROIDISM

Like hypothyroidism (see below), many systems may show signs of hyperthyroidism, and although marked hyperthyroidism due to Graves' disease is unmistakable, the signs are often non-specific, and hyperthyroidism occurs in the differential diagnosis of many symptoms and signs.

Inspection

General inspection often provides clues that are easily overlooked (Fig. 18.1). In particular, note:

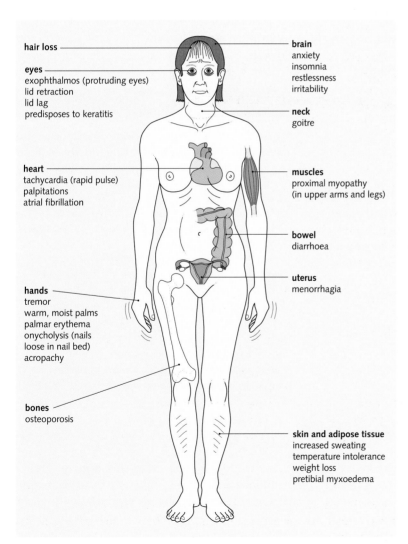

hair loss

eyes
exophthalmos (protruding eyes)
lid retraction
lid lag
predisposes to keratitis

heart
tachycardia (rapid pulse)
palpitations
atrial fibrillation

hands
tremor
warm, moist palms
palmar erythema
onycholysis (nails
loose in nail bed)
acropachy

bones
osteoporosis

brain
anxiety
insomnia
restlessness
irritability

neck
goitre

muscles
proximal myopathy
(in upper arms and legs)

bowel
diarrhoea

uterus
menorrhagia

skin and adipose tissue
increased sweating
temperature intolerance
weight loss
pretibial myxoedema

Fig. 18.1 Symptoms and signs of hyperthyroidism.

- General demeanour – patients are typically agitated, restless, irritable and have poor concentration.
- Facies, for example proptosis/exophthalmos.
- Goitre – associated with Graves' disease, toxic multinodular goitre.

Cardiovascular system

Cardiovascular abnormalities are common in thyroid disease and may provide a sensitive measure of assessing thyroid status in a treated patient. Look for the presence of:

- Tachycardia – very common.
- Atrial fibrillation – although this is a common arrhythmia, its presence should always raise a suspicion of hyperthyroidism – this is important as hyperthyroidism is an easily treatable cause of this arrhythmia.
- Warm vasodilated peripheries with a bounding arterial pulse.
- Hypertension – occasionally a feature.
- Ischaemic heart disease.
- Presence of a goitre.
- Mitral valve disease (especially mitral stenosis).

Neurological system

A brief examination often reveals discriminatory signs, for example:

- Fine resting tremor.
- Agitated, restless, hyperactive, shaky, irritable mental state. The patient may have a frank psychosis.
- Proximal myopathy, which may be profound.

Features to suggest Graves' disease

Eye signs

Graves' disease is particularly associated with eye signs, and this can be used to differentiate it from other causes of hyperthyroidism. Note the presence of:

- Lid lag – the upper eyelid does not keep pace with the eyeball as it traces a finger moving downwards from above.
- Exophthalmos.
- Chemosis.

- Ophthalmoplegia – extraocular muscles become swollen and develop secondary fibrotic changes.
- Proptosis – most common cause of unilateral and bilateral proptosis.

Pretibial myxoedema, thyroid acropachy and features to suggest general autoimmune predisposition

Deposition of mucopolysaccharides in the subcutaneous tissues of the legs produces a non-tender infiltration on the front of the shins, known as pretibial myxoedema. Occasionally, it can present acutely.

Thyroid acropachy is a syndrome resembling clubbing with new bone formation in the fingers. It is classically associated with exophthalmos and pretibial myxoedema.

Thyroid disease is associated with other autoimmune processes. In particular, look for alopecia and vitiligo.

HYPOTHYROIDISM

Hypothyroidism often develops insidiously and non-specifically, especially in the elderly. A keen level of awareness is important, and the diagnosis is often made by an observant doctor who has never seen the patient previously.

It is very easy to overlook a diagnosis of hypothyroidism. Always maintain a high index of suspicion, especially if the patient is seen on a regular basis, as it is easy to miss. Non-specifically unwell patients, especially those labelled TATT (tired all the time) are more likely to have a diagnosis of hypothyroidism. Slow-relaxing reflexes are a particularly useful sign for confirming a clinical suspicion of hypothyroidism, though they are rare.

Inspection

It is easy to overlook hypothyroidism, so always consider the diagnosis, especially for patients presenting with chronic fatigue, dementia, slow thought or non-specific difficulty coping with previously simple tasks (Fig. 18.2). Note the:

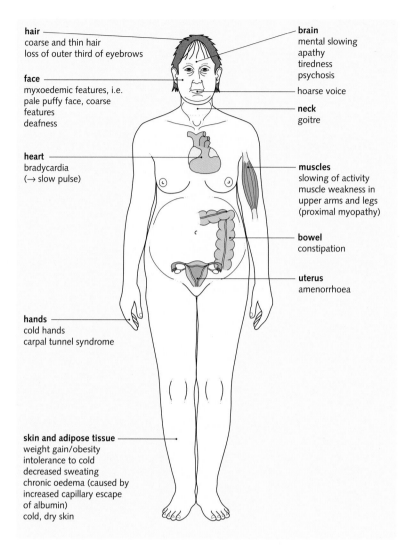

Fig. 18.2 Symptoms and signs of hypothyroidism.

hair
coarse and thin hair
loss of outer third of eyebrows

face
myxoedemic features, i.e.
pale puffy face, coarse
features
deafness

heart
bradycardia
(→ slow pulse)

hands
cold hands
carpal tunnel syndrome

skin and adipose tissue
weight gain/obesity
intolerance to cold
decreased sweating
chronic oedema (caused by
increased capillary escape
of albumin)
cold, dry skin

brain
mental slowing
apathy
tiredness
psychosis

hoarse voice

neck
goitre

muscles
slowing of activity
muscle weakness in
upper arms and legs
(proximal myopathy)

bowel
constipation

uterus
amenorrhoea

- Facies.
- Body shape – obesity, which is usually mild.
- Presence of a goitre – for example if due to Hashimoto's thyroiditis, iodine deficiency.
- Non-pitting oedema.
- Dry, scaly skin.

Cardiovascular system

Discriminatory features may include:

- Cold peripheries.
- Bradycardia – a useful sign, especially if it is out of context with the patient's condition.
- Hypothermia.

- Pericardial effusion (i.e. difficult to locate apex beat, quiet heart sounds), heart failure.

Neurological system

It is common to find neurological signs, for example:

- Slow-relaxing deep tendon reflexes (classically at the ankle joint).
- Proximal myopathy.
- Carpal tunnel syndrome.
- Deep, hoarse voice.
- Mental slowing, which may present with dementia, stupor or even coma.

Reticuloendothelial examination

By the end of this chapter you should:

- Be able to examine the major lymph node groups.
- Be able to accurately describe abnormal lymph nodes.
- Have a methodological approach to examination of the skin.

Examination of lymph node groups is important in many disease states. It is important to know the anatomical drainage pattern of the major organs as regional lymphadenopathy may be the first manifestation of local disease.

Normally, lymph nodes are not palpable, except in some thin people. If a single lymph node is found to be enlarged, it is important to adopt a systematic approach.

EXAMINATION ROUTINE

How to examine any lump

You need to expose the area in question, and then:

Inspect

Look at the shape of the lump and its position. Is there any associated colour change or change in the skin overlying the lump? Now, remember to ask if the lump is painful, before:

Palpation

Ask yourself if there is any difference in temperature between the lump and the surrounding tissue. Feel and measure the shape, size (it is a three-dimensional structure, so length, breadth and depth) and surface.

- Determine the edge of the lump. Is it well or poorly demarcated? What is the consistency of the lump?
- Is it pulsatile? If so, is it expansile or is it a transmitted pulse?
- Is the lump compressible or even reducible (e.g. a form of hernia)? Is there a cough impulse present?

- Is there a fluid thrill present in the lump or is it fluctuant?
- Now try moving the lump. What is it fixed to? Is it in the skin or is it fixed to muscle?

Auscultate

You should listen over the lump for both bowel sounds and bruits. And lastly, try to transilluminate the lump to see if it is fluid filled. You may need to switch off the light to get a clear idea of this.

Now that you've finished with the lump, examine the surrounding tissue. Is there any change in power or sensation?

Examine all lymph node groups

Examine all the lymph node groups systematically to define the anatomical distribution of the enlarged lymph nodes (Fig. 19.1). It is important to examine the liver and spleen as well, as these are also reticuloendothelial organs and may be enlarged in the presence of generalized lymphadenopathy.

Define the characteristics of the enlarged lymph node(s)

Define the texture, size, mobility and fixation to superficial and other tissues of the enlarged lymph node(s) (as with the examination of any lump). Certain characteristics are suggestive of different disease processes. For example:

- Rubbery texture is suggestive of lymphoma.
- Hard, matted, fixed lymph nodes suggest malignancy.
- Tender lymph nodes suggest infection or other inflammatory states.

Fig. 19.1 Position of the major lymph node groups and lymphoid organs.

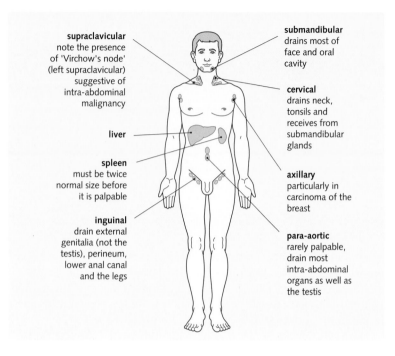

supraclavicular
note the presence of 'Virchow's node' (left supraclavicular) suggestive of intra-abdominal malignancy

submandibular
drains most of face and oral cavity

cervical
drains neck, tonsils and receives from submandibular glands

liver

spleen
must be twice normal size before it is palpable

axillary
particularly in carcinoma of the breast

inguinal
drain external genitalia (not the testis), perineum, lower anal canal and the legs

para-aortic
rarely palpable, drain most intra-abdominal organs as well as the testis

Explore the region drained by the enlarged lymph node group

If a single lymph node group is enlarged, try to identify the cause. The broad causes of lymphadenopathy are:

- Infection.
- Metastatic tumour.
- Lymphoproliferative disorder.
- Sarcoidosis.

In cases of cellulitis or other bacterial infection, it is sometimes possible to see lymphangitis. It is visible as thin red streaks following the line of the lymphatics in the skin.

Perform a systemic examination

Consider the pathological cause by performing a systemic examination, concentrating on inflammatory or malignant conditions draining into that lymph node group. Consider the more common causes of localized and systemic lymphadenopathy (Fig. 19.2).

Usually the cause of lymphadenopathy is apparent from a detailed history and systemic examination, but if further investigation is needed, the most useful tests include:

- Full blood count and film – for infection, leukaemia.
- Erythrocyte sedimentation rate (ESR) and C-reactive protein (CRP) – for malignancy, systemic inflammatory disease, systemic lupus erythematosus (SLE) – these are both markers of an acute inflammatory response, and a differential rise in ESR or CRP is occasionally discriminatory.
- Biochemical profile, especially liver function.
- Chest radiography – for example for sarcoid, malignancy, chest infection.
- Viral screens, autoantibody profiles, blood culture.
- Lymph node biopsy – often provides diagnostic information if sinister pathology is suspected.

If a single lymph node is enlarged, explore the region drained by that lymph node in detail.

Regional lymphadenopathy is common, but usually transient. Persistent lymphadenopathy always warrants investigation.

Differential diagnosis of localized and generalized lymphadenopathy		
	Lymphadenopathy	Features and examples
localized	infection (bacterial, viral, fungal)	pharyngitis, dental (cervical); lymphogranuloma venereum (inguinal)
	lymphoma (Hodgkin's; non-Hodgkin's)	can present anywhere, but cervical group is the most common
	malignancy	Virchow's node (left supraclavicular lymphadenopathy due to intra-abdominal or thoracic disease); breast cancer (axillary or supraclavicular)
generalized	infection	infectious mononucleosis; syphilis; tuberculosis; toxoplasmosis; HIV
	malignancy	lymphoma; leukaemia (especially CLL); carcinoma (unusual)
	autoimmune disease	SLE, rheumatoid arthritis, sarcoidosis, other connective tissue diseases
	drugs	
	hyperthyroidism	

Fig. 19.2 Differential diagnosis of localized and generalized lymphadenopathy. CLL, chronic lymphocytic leukaemia; SLE, systemic lupus erythematosus.

SKIN EXAMINATION

A definition of a dermatologist: someone who tells you in Latin what you just told them in English!

The skin is the single largest organ in the body and yet is often overlooked. Rashes are common things to be asked about and you need to be able to describe the lesion, even if you cannot make a precise diagnosis, so you can communicate with others about it. Different institutions have differing expectations about which terms they expect students to be familiar with (Fig. 19.3). During your dermatology attachment(s) you will probably be provided with a glossary or list of what is expected. Try to assimilate some of it!

When examining a rash, you will need to expose the patient in a well-lit room. The patient may well be shy about their lesion and feel it is unsightly, so be sensitive. Have a chaperone present. First, consider the distribution of the rash, so take a step back and look. Is it only on areas exposed to sunlight; is it dermatomal in distribution (e.g. herpes zoster); is it related to jewellery or buttons? Is the rash generalized or localized, bilateral or unilateral? Are there any areas that are spared?

Next, what is the morphology of the lesion? You need your dictionary to help with this. Try and determine the shape, size, colour and the margins of any lesions you can see. If you can't remember the special phrase, just describe exactly what you see in plain English.

Some dermatological terms	
alopecia	absence of hair where it normally grows
bulla	a circumscribed elevation of skin greater than 5 mm, containing fluid
crusting	scale composed of either dried fluid or blood
erythema	a flushing of the skin due to the dilation of the capillaries
excoriation	an erosion or ulcer secondary to scratching
lichenification	thickening of the epidermis of the skin with exaggeration of the normal creases
macule	small flat area of altered skin colour or texture
nodule	a solid mass in the skin greater than 5 mm
papule	a small solid elevation of skin less than 5 mm
petechia	pinhead sized flat collection of blood in the skin
plaque	elevation area of skin greater than 20 mm, without depth
purpura	a large flat or raised collection of blood in the skin
pustule	visible accumulation of pus in the skin
ulcer	a loss in the continuity of an epithelially lined surface
vesicle	circumscribed elevation of skin less than 5 mm, containing fluid

Fig. 19.3 Some dermatological terms.

Breast examination

Objectives

By the end of this chapter you should:

- Understand the importance and sensitive nature of breast examination.
- Be able to perform a breast examination.
- Be able to accurately describe any examination findings.

EXAMINATION ROUTINE

Breast examination forms an integral part of a full medical clerking. However, a more detailed breast examination is always necessary if the presenting symptoms:

- Are specific breast symptoms (e.g. lump, pain, nipple discharge, change in appearance).
- Arouse suspicion of disseminated malignancy with an undiagnosed primary (e.g. presentation with pleural effusion, hepatomegaly, bony tenderness).
- Include fever of unknown cause.

As with any system, a methodical approach is needed. Anticipate what information might be obtained from each part of the examination and how each elicited sign is placed into context with the presenting illness. Breast examination findings are more easily appreciated in postmenopausal breast tissue, since, in younger women, the density of breast tissue may make detection of abnormalities difficult.

Patient exposure and position

Clearly great sensitivity is essential. Very often the patient feels uncomfortable or embarrassed, especially if the doctor is male. Remember that many patients will be terrified, not only of the examination, but of the potential underlying diagnosis. It is not uncommon for women to 'ignore' a breast mass for several months or even years.

Explain clearly why the examination is being performed and the useful information that is likely to be gained. Ensure complete privacy. It is clearly unacceptable for a secretary or another doctor to burst through the door or screens revealing a semi-clad patient to the waiting room or ward! The room should be warm and a blanket should always be provided so that the patient can remain covered until the examination is performed.

Explain clearly what the examination will entail before asking the patient to undress. Ask the patient to remove all clothing (including bra) from the waist upwards, and to sit on the side of the examination couch. You must have a female chaperone with you at all times.

The examination follows the usual sequence of:

- Inspection.
- Palpation.
- Systemic examination.

Inspection

Remember to look at the whole patient. Make a mental note of:

- Age. Breast carcinoma is more common in older women, but can occur in any age group from the third decade onwards. Fibroadenoma is more common in premenopausal women. Abscess is much more common in women of childbearing age.
- Sex. Men also get breast disease!
- General health.

Breast substance

Ask patients to sit still, facing you with arms by their side and then with arms raised above the head – a mass tethered to the skin may then become apparent, and the undersurface of the breasts can be seen (Fig. 20.1). Note:

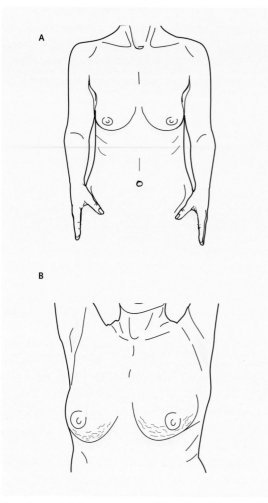

- Symmetry. It is not uncommon for one breast to be slightly larger than the other, but underlying masses, infections or nipple disease can also cause asymmetry of size, shape or the nipples.
- Any obvious mass.
- Skin discoloration. Infections and occasionally malignancy may cause a red discoloration.
- Skin puckering or 'peau d'orange' (Fig. 20.2). Peau d'orange is caused by infiltration of the skin lymphatic system and, as the name implies, its appearance resembles the peel of an orange.

Nipples

Inspect the nipples carefully, especially noting the following:

- Symmetry.
- Retraction or deviation. If one or both nipples are retracted, ask the patient if this is a new phenomenon. This is an ominous sign. As well as carcinoma, it is associated with chronic abscess and fat necrosis.
- Discharge. If present, note whether it appears milky (galactorrhoea), bloody or pustular.
- Skin colour. Particularly note the presence of an eczematous rash suggestive of Paget's disease of the nipple.

Palpation

The normal consistency of a breast varies considerably. It is recommended to start the examination with palpation of the normal breast. Palpate with

Fig. 20.1 A, Ask the patient to sit facing you with hands by their sides. **B,** Then ask the patient to lift her arms in the air. Skin tethering may become apparent, and abnormalities on the undersurface of the breast will become visible.

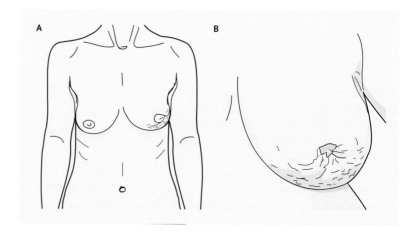

Fig. 20.2 Inspection often reveals obvious asymmetry. (**A**) A retracted nipple and skin tethering on the left breast. (**B**) Peau d'orange in association with a retracted nipple.

the palmar surface of the fingers. It is helpful to divide the breast into quadrants (Fig. 20.3) and the axillary tail of Spence. Palpate each quadrant in turn, noting:

- Consistency and texture of the breast.
- Tenderness.
- Presence of any mass.

Examine the breast with symptoms. Before palpating, ask patients to point to the area of tenderness or to any lump that they may have felt. Once again, examine each quadrant in turn. If a mass is felt, it should be systematically examined, as described below.

Breast mass

If any lump or mass is identified on systemic examination, the essential features should be described. This is well illustrated with a breast mass. The following characteristics should be noted.

Position

It is usual to describe a breast mass in relation to the quadrant it is located in. A breast carcinoma is more common in the upper outer quadrant. Remember to palpate the axillary tail as this also contains breast tissue.

Size

Describe the size of the mass in three dimensions. Ideally, measure the mass objectively with a tape

measure or calipers to decrease interobserver error – much more helpful than using terms such as 'grape-sized'. This is essential when assessing the progression or regression of a lesion (e.g. judging the response of breast carcinoma to chemotherapy).

Consistency

Note the consistency of any mass. In practice, it is easiest to use terms such as:

- 'Craggy' – literally like a rock.
- 'Hard' – like pressing on your forehead.
- 'Rubbery' – like pressing on the tip of your nose.
- 'Soft' – like pressing on your lips.

Relation to the skin

Note the presence of tethering or fixation to the skin. Fixation suggests an infiltrating carcinoma; tethering occurs in carcinoma, abscess or fat necrosis.

Relation to underlying tissue

Note the mobility of any lump. A mass fixed to deeper tissue is much more likely to be a carcinoma.

Fibroadenomas are typically described as highly mobile, and may be difficult to palpate. Trapping a fibroadenoma between finger and thumb can be tricky, and has been likened to chasing a mouse, hence the term 'breast mouse'.

Tenderness

Breast carcinoma is rarely tender on presentation. A tender mass is much more likely to be an abscess, cyst or fat necrosis.

Skin discoloration

Note any change in the appearance of the skin overlying the mass. Erythema is common in association with infections. Paget's disease of the nipple presents with an eczematous rash; a carcinoma occasionally has a red or blue hue.

Temperature

Inflammatory lesions often produce palpable warmth.

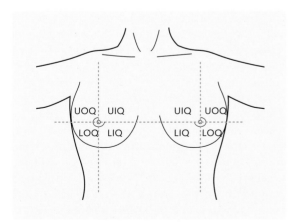

Fig. 20.3 Quadrants of the breast. Upper outer quadrant (UOQ); upper inner quadrant (UIQ); lower outer quadrant (LOQ); lower inner quadrant (LIQ).

Fig. 20.4 Example of a recording of the presence of a breast mass in the medical notes. A simple diagram provides unambiguous objective information.

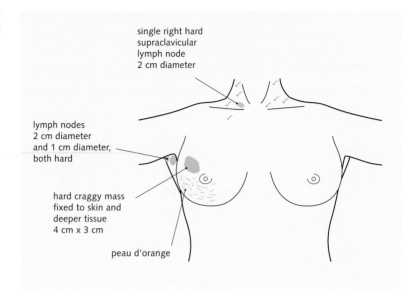

single right hard supraclavicular lymph node 2 cm diameter

lymph nodes 2 cm diameter and 1 cm diameter, both hard

hard craggy mass fixed to skin and deeper tissue 4 cm x 3 cm

peau d'orange

Associated lesions

It is essential to examine the rest of the breast for the presence of a second mass. It is not unusual for breast carcinoma to present with bilateral disease. Fibroadenosis often presents with multiple lumps.

If breast carcinoma is suspected, a full systemic examination is essential. Look for evidence of metastatic spread. In particular, check for the presence of:

- Hepatomegaly – liver metastases result in a grim prognosis; note the presence of jaundice.
- Pleural effusion – the lung is a common site of metastatic spread.
- Bony tenderness – bony metastases are the most common site of secondary breast. malignancy after axillary lymph nodes, and the axial skeleton is often involved; disease may be present for several years.
- Ascites – peritoneal deposits often present with ascites.

Lymphatic drainage

Palpation of the axillary lymph nodes forms part of the routine breast examination (see below).

An example of a recording of the presence of a breast mass in the medical notes is illustrated in Figure 20.4.

Axillary examination

Breast examination is incomplete without examination of the axilla, as these are the natural sites of

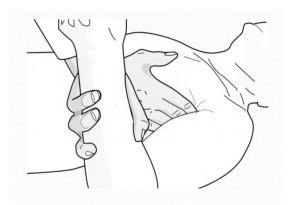

Fig. 20.5 Examination of the axilla for lymphadenopathy. It is important to support the patient's arm to relax the muscles of the axillary folds.

lymphatic drainage of breast tissue. Inspect the axilla for any obvious lumps. The axillary tail of Spence is composed of breast tissue and runs from the upper outer quadrant to the axilla; it is important not to forget to palpate this area separately.

Axillary lymphadenopathy can only be palpated if the muscles forming the walls of the axilla are relaxed. Stand on the patient's right-hand side, support the patient's right arm with your right hand and encourage the patient to allow you to take the weight of the arm (Fig. 20.5). Palpate the axilla with your left hand flat against the lateral chest wall, reaching high up into the axilla with your fingertips, sweeping around all of the walls. The left axilla is examined in a similar manner, but with the opposite hand.

The key to successful examination of the axilla is to ensure that the arm is totally relaxed and supported by your hand.

If any mass is detected in the breast or axilla, the supraclavicular and cervical lymph nodes should also be examined.

Finally, note the size of the patient's arms. Lymph-oedema can result from tumour invasion of the lymphatics, axillary dissection or radiotherapy.

BREAST MASS

Presentation with a breast mass is common. Clearly, the most important diagnosis to exclude is carcinoma of the breast. The most common causes are listed below.

Carcinoma of the breast

In the presence of a breast mass, features to suggest carcinoma include:

- A hard, craggy mass.
- Fixation of the mass to skin or deeper tissues.
- Nipple deviation or retraction.
- Skin changes (peau d'orange, tethering, ulceration).
- Axillary lymphadenopathy.
- Signs of systemic disease.

A 64-year-old female has a probability of invasive carcinoma of the breast of 0.35% in the next year; the finding of a mass raises this to 0.73%, Presence of the features of malignancy raises this further to 8.8%. Clearly, examination findings must be interpreted accurately.

Fibroadenoma

A fibroadenoma is typically a well-demarcated, highly mobile, non-tender mass in a young or middle-aged woman.

Fibroadenosis (cystic hyperplasia, fibrocystic disease)

The features of fibroadenosis are highly variable and depend upon the degree of cystic change, fibrosis and masses. Often there is a diffuse change in texture in the breasts and there are multiple masses, which are poorly defined. Mild tenderness is common.

Abscess

Abscesses usually present in lactating women and are easily distinguished by the presence of:

- Tenderness.
- Erythema.
- Poor definition.
- Axillary lymphadenopathy.
- Systemic features of infection (e.g. fever, tachycardia).

Fat necrosis

Fat necrosis is much more common in large breasts following a history of trauma. The lump is usually hard and may be tethered to the skin. It may be distinguishable from a breast carcinoma by the lack of axillary lymphadenopathy or peau d'orange.

BREAST PAIN

The more common causes of breast pain are:

- Fibroadenosis.
- Premenstrual tension.
- Mastitis.
- Abscess.

The cause is usually apparent from the history and physical examination. Breast carcinoma rarely presents with pain.

GYNAECOMASTIA

Gynaecomastia results from an increase in breast tissue. It may be unilateral or bilateral, and is confirmed by palpation. The most common causes are:

- Puberty. Normal finding (very common) including in boys.
- Old age.

- Liver failure (look for stigmata of chronic liver disease).
- Carcinoma of the lung.
- Testicular tumours.
- Adrenal tumours.
- Drugs (e.g. spironolactone, cimetidine, digoxin).
- Testicular feminization (very rare).
- Pituitary tumours.

INVESTIGATIONS

The cause of many breast lesions is often apparent from a careful history and physical examination. However, in many circumstances it is important to exclude breast carcinoma. The most reliable procedure for such exclusion is to perform an excision biopsy, but it is clearly desirable for this to be avoided for benign lesions.

Mammography is performed as a screening procedure in the UK for women over 50 years of age. In the presence of a lump, mammography provides a useful adjunct by detecting areas of calcification, which are indicative of an underlying carcinoma. In addition, a second suspicious area may also be revealed, which should also be investigated clinically.

Ultrasound can be used to identify the composition (i.e. solid versus cystic) of a mass. Cystic lesions can be aspirated as a diagnostic or therapeutic procedure. Ultrasound is often requested to localize masses before surgery. In addition, ultrasound can be used to identify masses for fine needle aspiration (FNA). Fine needle aspiration may provide adequate cytological information to increase or decrease the clinical index of suspicion of malignancy, and so determine the urgency of treatment.

Within every region, women should have access to triple assessment clinics via their GP or primary care physician. At the clinic, the woman will be able to have a clinical examination of her breast by an experienced clinician. This will be complemented with FNA of the lump (with reporting facilities) and mammography with experienced reporting available. This ensures a one-stop visit where the woman can have full investigation of any lump and get the results that day. There must also be counselling services on site if bad news is going to be given.

One last thought: men also get breast cancer, although it is rare. It also carries a worse prognosis, as often the lump will already be adherent to the chest wall at presentation.

Locomotor examination

Objectives

By the end of this chapter you should:

- Be able to perform general and directed regional examinations (e.g. of the knee) of the locomotor system.
- Be able to assess gait and relate it to further examination findings.
- Appreciate the functional importance of any findings (including the psychological impact of some findings).
- Be able to recognize patterns of symptoms such as those that can be found in rheumatoid or osteoarthritis.

EXAMINATION ROUTINE

Locomotor assessment is fundamental to even the shortest medical clerking, as information on the patient's functional ability is assessed and integrated with other physical signs. Clearly, if a patient describes decreased mobility, weakness or joint pain, a more detailed general assessment is necessary. Equally, if a focal abnormality is detected, this should be fully examined.

This chapter is not intended to provide a detailed description of examination of every joint and muscle group, but illustrates a methodical and functional approach to examining the system. Your own medical school may have its own format for examination of the locomotor system – be familiar with the specific outcomes and approach expected of you.

Patient exposure and position

When examining muscle groups or mobility, it is important to ensure that patients are properly exposed. They should be provided with a blanket and asked to strip to their underwear in a warm, well-lit room. The initial formal assessment is usually performed on the examination couch. The approach to examining a joint should follow the scheme of 'Look, Feel, Move'.

The most important part of the locomotor examination is inspection.

Inspection

General inspection

Remember that the examination begins as soon as the patient walks into the examination room! It is often possible to form an impression of functional ability by observing how easily the patient gets out of the chair, walks to the examination room and climbs onto the examination couch. Note the following.

Always try to relate pathological signs to a functional disability such as difficulty in performing routine daily activities (e.g. getting out of bed, writing, walking, picking up a knife and fork).

Age and sex

Different disease processes are more likely to occur at different ages, for example:

- Osteoarthritis and polymyalgia rheumatica are more common in the elderly.
- Osteoporosis is primarily a disease of postmenopausal women.
- Ankylosing spondylitis usually presents in young men.
- Many inflammatory arthritides first present in young women.

Racial origin

Many forms of arthritis or diseases presenting with impaired mobility have a strong genetic predisposition and are more common in certain racial groups. Others have a predominantly environmental cause that varies geographically, for example:

- Systemic lupus erythematosus (SLE) is much more common in Afro-Caribbean women.
- Paget's disease is more common in Caucasians.
- Multiple sclerosis is more common in patients from temperate climates.

General health and appearance

Note whether the patient appears well or is cachexic. Obesity predisposes to osteoarthritis of the back and weight-bearing joints. Cachexia may indicate chronic systemic disease or carcinoma. Note whether the patient appears to be in pain at rest or on walking. An obvious focal weakness may be apparent. These visual clues may contrast with the information obtained from the history or when the patient knows that a formal examination is being performed.

Facial appearance

Note the appearance of facial asymmetry or obvious weakness. In addition, note the general appearance (e.g. myopathic facies of muscular dystrophy with unlined expressionless facies and wasting of the facial muscles).

How easily does the patient get out of a chair?

Proximal muscle weakness (e.g. due to polymyalgia rheumatica, steroid myopathy, hyperthyroidism) will have a profound effect on getting up from a seated position. The significance of a patient wincing with pain when they rise from the chair is unclear.

Does the patient require any aids for walking?

Observe whether the patient walks unaided into the examination room or walks with the aid of a walking stick or zimmer frame, or holding onto another person for support. The patient will hold a walking stick in the hand opposite the weakest leg.

If the patient has a zimmer frame, observe how he or she uses it. It is not unusual for nervous but mobile patients to become reliant upon a frame and, on observation, they will be seen to carry the frame in front of themselves, rather than rely upon it for support.

Gait

If the patient walks unaided, make a quick assessment of gait. A more formal assessment should be made later in the detailed examination.

How easily does the patient climb onto the examination couch?

A certain amount of agility is needed for elderly patients to climb onto the examination couch. It may be instructive to observe the patient during this process.

Detailed inspection

Once the patient is properly exposed and positioned, a detailed inspection should be performed. This process forms the key to a successful examination.

A detailed examination of the locomotor system is time-consuming and tiring for the patient. With the benefit of the history, a focused examination is possible after detailed inspection. The more important features to note are outlined below.

Skin rash

Skin rashes may suggest an underlying inflammatory condition:

- Psoriasis (there are a number of psoriatic arthropathies).
- Infection.
- Malignancy.
- Pain (erythema ab igne from a hot-water bottle).

Muscle bulk

Note the general muscle bulk and the distribution of any atrophy, for example:

- Disuse atrophy in a hemiplegic limb.
- Atrophy of the quadriceps in the presence of a proximal myopathy.
- Atrophy of the distal muscles of the lower limbs (Charcot–Marie–Tooth disease).
- Old polio.

Deformity or swelling

Observe any obvious deformity that may be the result of bone, joint or muscle disease, for example:

- Scoliosis.
- Varus or valgus deformity. This is common in the knees in osteoarthritis.
- Rigid back with loss of lumbar lordosis and fixed posture in ankylosing spondylitis.
- Obvious joint swelling. The distribution of swollen (or inflamed) joints should be mapped out, as the distribution is often characteristic of the underlying disease and variations can be correlated with changes in disease activity (Fig. 21.1).

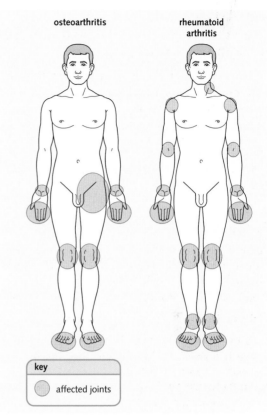

Fig. 21.1 Typical patterns of joint involvement in two different forms of polyarthritis. In rheumatoid arthritis, there is usually a symmetrical polyarthropathy. Osteoarthritis is less likely to be symmetrical, but often is symmetrical in its widespread form.

Gait

A formal assessment of the gait should be part of the routine examination, especially in an elderly patient. Apart from highlighting a possible aetiological factor for impaired mobility, it provides a direct functional assessment and an impression of the problems with daily living.

Ask the patient to walk in a straight line, turn around and then walk back to you. Balance and ataxia may be additionally tested by asking the patient to walk 'heel-to-toe'. Characteristic gaits may be noted as follows.

Ataxic gait

An ataxic gait is characteristically wide based. The arms are often held out wide to aid balance. Marked clumsiness is obvious on walking heel-to-toe. The patient often staggers to the left or right. This is associated with cerebellar pathology.

Spastic gait

If the patient has spastic paraplegia, the gait is stiff and described as a 'scissor gait'. The appearance may resemble someone wading through water. If the patient has a hemiplegia, the affected leg is extended and the leg is swung around the hip joint.

Sensory ataxia

Peripheral neuropathy with sensory ataxia results in a high-stepping gait with the appearance that the feet are being 'thrown'. The feet tend to be 'slapped' on the floor and the patient walks on a wide base. Romberg's test is positive (see Ch. 17). Patients often appear to be concentrating hard on where their feet are being placed.

High-stepping gait

In the presence of foot drop, the affected foot is lifted high off the ground to avoid scraping the toes on the floor. Such patients are unable to stand on their heels.

Parkinsonian gait

The patient has a characteristic stooped appearance. The gait is shuffling and hesitant with short steps and a lack of associated arm movement. The arms do not usually swing during walking. The gait is described as 'festinant' – having the appearance that the patient is always chasing his or her own sense of gravity. The patient often has great difficulty when asked to stop and turn around and there are usually other features of Parkinson's disease.

Osteogenic gait

Patients with legs of unequal length may walk normally in shoes with appropriate shoelifts, but the abnormality should be obvious when they walk barefoot.

Waddling gait

A proximal myopathy is associated with a waddling gait. The patient walks on a wide base with the trunk moving from side to side on each step and the pelvis drooping as the leg leaves the ground.

Observe whether the patient has any pain on walking and, if so, which movements appear to provoke the pain.

REGIONAL EXAMINATION

The history from the patient will point you to the region of the body that needs to be examined. Specific regional examination can now begin. As with other systems of the body, it is important to follow a strict routine. When examining any region of the body or joint, use the following routine.

Inspection

Note the bones, alignment, joint swelling, redness, deformity, local swelling and the presence of any scars.

Palpation

Palpate the area concerned, paying particular attention to:

- Skin temperature. In particular, note any areas of increased warmth. Compare the two sides.
- Tenderness. Map out any areas of tenderness and try to relate these to the affected structure.
- Deformity. Note the bony contours and any varous, valgus or fixed-flexion deformity.
- Soft tissues. Note the presence of any abnormal swellings (e.g. fluid in the joint, cysts, bursae, tumours). Palpate the muscles for bulk, etc.

Assessment of joint movement

Assess the range of active and passive movement, muscular power and whether movement is accompanied by pain. It is important to have an appreciation of the range of normal movement around each joint. It is often helpful to test the unaffected side first so that any deviation can be more easily appreciated.

Sensation

Test the sensory modalities (see Ch. 17). The light touch, pin prick and vibration tests are the most discriminatory.

Function

The most useful part of the examination is to assess the function of the relevant body part. Try to relate your assessment to daily activities, for example:

- In the lower limb, test the ability to walk, jump, hop or (possibly) run.
- In the upper limb, test the ability to comb the hair and touch the lumbar spine (an indirect test of ability to attend to personal hygiene).
- In the hand, test the ability to hold a pen and grip strength.

GENERAL EXAMINATION

It is very important to perform a systemic examination. The localized joint symptoms may form part of a systemic disease or be referred from another site. Furthermore, it is only possible to place the impairment in context when the patient is considered as a whole.

Hands

Most systemic examinations start with examination of the hands, as stigmata of systemic disease are often manifest in the hands. This is also true for a patient presenting with impaired mobility. Furthermore, many patients specifically complain of symptoms directly related to their hands (e.g. paraesthesiae, weakness, joint pain). A methodical approach to examination of the hands is therefore useful.

Inspection

Remember to look at the face of the patient first for any clues to the underlying disease or treatment (e.g. scleroderma, Cushingoid facies). Look at both hands. Ask the patient to place his or her hands flat on the table, palmar surface downwards. Look at:

- The general shape of the hands and note any deformity (e.g. ulnar deviation in rheumatoid arthritis).
- The colour of the skin (e.g. pigmentation, icterus, erythema, rash).
- The nails. Look for signs of psoriasis (e.g. pitting, onycholysis), clubbing, splinter haemorrhages, nailfold infarcts.
- Soft tissue. Note any swellings (e.g. Heberden's nodes in osteoarthritis on the proximal and distal interphalangeal joints, gouty tophi).
- Joints. Look for any swelling or redness suggestive of an active arthritis.

Ask the patient to turn his or her hands over so that the palmar aspect can be inspected. Repeat the same process of inspection. In particular, note the presence of palmar erythema, and pay close attention to the muscle bulk of the thenar eminence (atrophy in carpal tunnel syndrome) and hypothenar eminence (atrophy in T1 root lesion or in the presence of severe rheumatoid arthritis). Note the presence of any scars (e.g. carpal tunnel decompression). The hands of patients with rheumatoid or psoriatic arthritis may contain a huge number of clinical signs, making them an effective and discriminative case for examinations. Make a list of every sign you can find and learn them well.

Palpation

Palpate over the joints of the hand and wrist gently, noting any tenderness of the joints. Record the distribution of any tender joints (Fig. 21.2). Many forms of polyarthritis have a characteristic distribution of joint involvement. In addition, palpate for the presence of any swellings or palmar thickening (e.g. Dupuytren's contracture, trigger finger).

Movement of the joints

Assess the movement of the joints of the hand. Try to relate this to functional activity. First, test passive movement and then active movement. If the primary problem is joint disease, useful tests of function might include:

- Grip strength.
- Pincer grip (ask the patient to pick up a pen and write his or her name).

Assess whether movement is limited by deformity, pain or muscular weakness. Some of the signs of rheumatoid arthritis are illustrated in Figure 21.3.

Neurological assessment

Symptoms in the hand may result from a nerve lesion. Test sensation on the middle finger (median nerve), little finger (ulnar nerve) and the anatomical snuff box (radial nerve). The most common pathologies are:

- Peripheral nerve/radial nerve injury – most commonly at the spiral groove of the humerus due to fracture of the proximal humeral shaft, and leading to wrist drop and a weak grip. The ulnar nerve is most commonly injured at the elbow and this leads to claw hand with wasting of the muscles between the metacarpals. The medial nerve is also commonly injured at the wrist (as in carpal tunnel syndrome) and this leads to wasting of the thenar eminence.
- Nerve root or brachial plexus lesion (e.g. Pancoast's tumour).
- Sensory neuropathy.

When examining motor and sensory function, consider the implication of each elicited physical sign and whether the underlying problem is likely to be nerve root, peripheral nerve, muscular or joint pathology.

Motor function tests

The tests of motor function are illustrated in Figure 21.4. If the problem is unilateral, it is very helpful to directly compare strength in the two hands, testing the normal hand first.

Fig. 21.2 Patterns of joint involvement in some systemic polyarthropathies. (**A**) Osteoarthritis. Joint swelling of the first carpometacarpal joints is characteristic. Joint involvement is usually symmetrical in the hands and typically affects the distal interphalangeal joints. (**B**) Rheumatoid arthritis. Active synovitis is detected by joint warmth, swelling, redness and tenderness. During a flare, arthritis is usually symmetrical, affecting the wrists and metacarpophalangeal and proximal interphalangeal joints. The distal interphalangeal joints are usually spared. (**C**) Systemic lupus erythematosus (SLE). The distribution of joint involvement is similar to that of rheumatoid arthritis in the hands, but the signs are usually less marked and pain is often out of proportion to the signs of synovitis. (**D**) Gout. Initial attacks of gout usually present in the lower limb, especially the first metatarsophalangeal joint. However, it may present in the hands. It is often monoarticular. The distal interphalangeal joints are more prone to attacks.

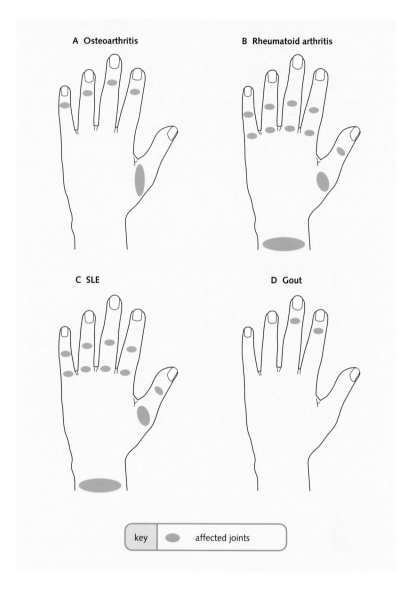

Sensory function tests

Test sensory function in the hand, including the different modalities (pin prick, light touch, vibration and joint position). The distribution of sensory loss is illustrated for the three peripheral nerves as well as the dermatomes in Figure 21.5. Remember that the areas of skin providing sensory fibres for the peripheral nerves or nerve roots often overlap, so it is important to test in the more discriminating areas.

Clues from the systemic examination

It should be apparent from the history and assessment of the hands whether the lesion is articular, vascular, neurological, etc. There are often clues to be obtained from a systemic survey. For example:

- Look at the elbows for rheumatoid nodules or a psoriatic rash.
- Note the presence of gouty tophi on the ear lobes.
- If the patient has an arthritis, it is essential to examine each joint so that the distribution of joint involvement can be mapped out.

Shoulder

The shoulder is a difficult joint to examine as it moves in so many different ways! Expose the patient.

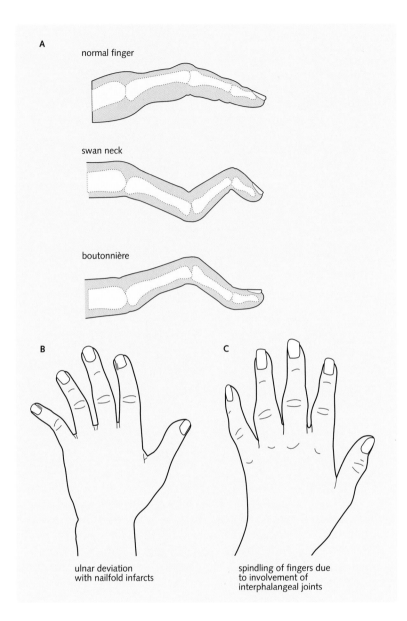

A

normal finger

swan neck

boutonnière

B

C

ulnar deviation
with nailfold infarcts

spindling of fingers due
to involvement of
interphalangeal joints

Fig. 21.3 Signs of rheumatoid arthritis in the hands. (**A**) Deformities of the fingers result from tendon rupture and joint laxity. Characteristic patterns include swan neck and boutonnière deformities. The thumb may develop a Z deformity. (**B**) Ulnar deviation results from subluxation at the metacarpophalangeal joints. Nailfold infarcts are one of the manifestations of vasculitis. (**C**) Spindling of the fingers is an early sign and is due to involvement and swelling of the interphalangeal joints.

Ask them to strip to the waist and sit on the end of the bed.

Look

Inspect the patient for any signs of skin erythema or scars. Note any asymmetry between the two shoulders. Look for evidence of wasting of the deltoid muscles and effusions or joint swelling. Observe the position of the shoulders.

Feel

Palpate over all the joints that encompass the shoulder: the sternoclavicular joints, the acromioclavicular joints and the glenohumeral joints. Pay attention to any joint line tenderness or swelling.

Move

First, ask the patient to do the moving (active movement). Ask them to move their shoulder through

Fig. 21.4 Movements of (**A**) the fingers and (**B**) the thumb. Finger abduction and adduction (e.g. 'Grip a piece of paper between your fingers') rely on the ulnar nerve. Thumb extension relies on the radial nerve; opposition, flexion and abduction rely on the median nerve.

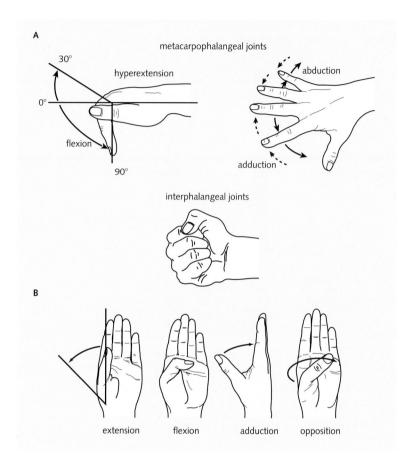

abduction, flexion, extension and internal and external rotation (Fig. 21.6). A good way to do this is to ask them to place their hand behind their neck and slide their hand down between their shoulders.

Second, you should move the shoulder through a full range of movement (passive movement) and see how far you can move the shoulder before the patient experiences pain. You should have a hand over the joint, feeling for crepitus or restrictions to movement.

Then test the power of the patient's shoulder against resistance.

Examining the glenohumeral joint

Immobilize the scapula by placing a hand on it to restrain it. Ask the patient to abduct their arm (which must start down by their side). If they cannot initiate this movement, there is probably a rotator cuff tear. If abduction is restricted and further passive movement causes more pain, this is classed as 'impingement pain', and the patient is likely to have painful arc syndrome.

The rotator cuff comprises the supraspinatus, subscapularis and infraspinatus tendons. An incomplete tear leads to painful arc syndrome (pain on movement from 45–140°). A complete tear results in the patient being unable to initiate abduction. If you abduct the patient's arm to greater than 45°, then the patient should be able to continue the abduction to 180°.

Elbow

Look

Expose the patient to the waist and inspect both elbows from behind, with the arms in extension. Look for any obvious deformity, or for an effusion filling out the hollow at the head of the radius.

Feel

Palpate the bony contours of the elbow. Palpate over the epicondyles and radial head. Feel for any signs of inflammation or bursae.

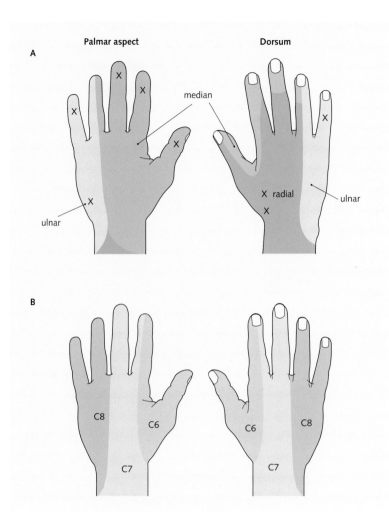

Fig. 21.5 A, Dermatomes corresponding to the peripheral nerve supply of the hand. The crosses indicate the more useful places for assessing sensation in a quick examination. **B,** Dermatomes corresponding to the nerve root supply.

Move

The elbow should move freely from 0 to 150° (normal people also get a little extension). While doing this, palpate over the elbow joint and the head of the radius, feeling for crepitus.

Now fix the elbow at the patient's side and flex the arm to 90°. Now test pronation and supination. With the elbow fully extended, test the integrity of the collateral ligaments.

To test for tennis elbow (lateral epicondylitis), ask the patient to fully extend their elbow and then to squeeze your hand. Then ask them to slightly flex their elbow and squeeze your hand. If the patient has tennis elbow, the second manoeuvre should be less painful. Figure 21.7 shows how to test for golfer's elbow (medial epicondylitis). They will also be tender over the respective epicondyles.

Back

Back pain is a very common presentation, both to GPs and hospital doctors. It is necessary to develop a systematic approach when examining patients with back pain so that potentially serious disease can be recognized early and investigated, while appropriate advice can be offered to the vast majority with less serious but nonetheless disabling problems.

211

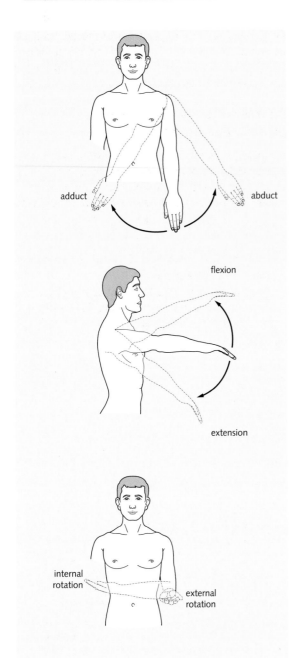

adduct   abduct

flexion

extension

internal rotation

external rotation

Fig. 21.6 Shoulder movements.

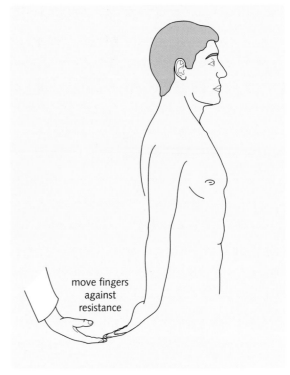

move fingers against resistance

Fig. 21.7 Golfer's elbow. If moving the fingers against resistance causes pain, then the patient has golfer's elbow.

Back pain

In the context of back pain, neurological examination of the legs is important. In acute presentations, this should include sensation of the limbs and perianal area and digital rectal examination. In situations of spinal trauma, a degree of spinal shock may be present (the spinal cord effectively shuts down after trauma, similar to a neuropraxia). This necessitates testing basic reflexes (such as the bulbo-cavernosus reflex), as only when basic reflexes are present can the degree of neurological deficit be interpreted accurately. Spinal cord compression warrants urgent investigation, usually in the form of magnetic resonance imaging: immediate discussion with a spinal surgeon is advised.

Inspection

A brief survey of the patient often provides invaluable clues. In particular, note:

- General health (e.g. cachexia suggestive of underlying malignancy).
- Posture and deformity (e.g. kyphosis, scoliosis, loss of lumbar lordosis).
- Scars (e.g. previous surgery to the back).
- Pain. Note if the patient appears to be in pain and is lying very still for fear of provoking worse pain, comfortable at rest and whether or not they are pain free on moving around.

Examination of the back

Palpate the back for local tenderness suggestive of an inflammatory process. Record the site of elicited tenderness. This may indicate the underlying disease (Fig. 21.8).

Examine the movements involving the vertebrae (flexion, extension, lateral flexion, rotation) and record both the range of movement and any pain elicited (Fig. 21.9). If ankylosing spondylitis is sus-

pected, the sacroiliac joints must be assessed, as these are often the first site of inflammation. Lateral compression of the pelvis may elicit pain in the presence of sacroiliitis.

Peripheral joints and systemic examination for arthritis

A full survey of other joints may reveal a more widespread arthropathy such as ankylosing spondylitis, psoriatic arthropathy. Note the distribution of joint involvement and the presence of active synovitis.

In addition, there may also be non-articular clues to the presence of a systemic arthritis, for example:

- Psoriatic arthropathy is associated with a classic rash and nail changes.
- Ankylosing spondylitis is associated with decreased chest expansion, upper lobe fibrosis, iritis and aortic regurgitation.

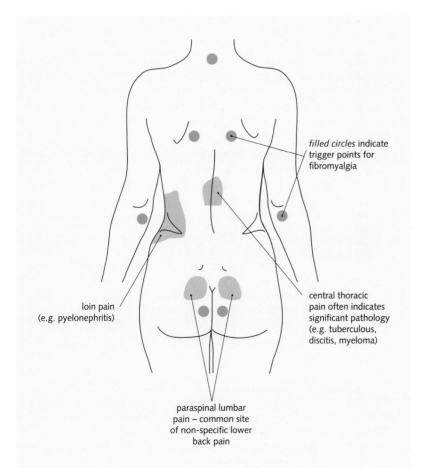

filled circles indicate trigger points for fibromyalgia

central thoracic pain often indicates significant pathology (e.g. tuberculous, discitis, myeloma)

loin pain (e.g. pyelonephritis)

paraspinal lumbar pain – common site of non-specific lower back pain

Fig. 21.8 The site of back tenderness is important as this may correspond to the aetiology.

213

Fig. 21.9 Assessing movements of the (**A**) lumbar and (**B**) thoracic spine. Flexion of the lumbar spine can be objectively measured. Ask patients to touch their toes, keeping their knees straight. Mark the spine at the lumbosacral junction 5 cm and 10 cm above this point. The distance between the upper two points should move approximately 5 cm on full flexion. This movement is impaired in ankylosing spondylitis.

Test for evidence of nerve root entrapment

Acute or chronic back pain may be due to a prolapsed intervertebral disc which may be associated with sciatic nerve root entrapment. The presence of nerve root entrapment and its localization may be elicited by testing:

- Straight leg raising.
- Femoral stretch test (Fig. 21.10).

Neurological examination of the legs

It is important to exclude nerve root pressure or spinal cord compression. A description of the neurological assessment of the legs is given in Chapter 17.

Detailed systemic examination

A detailed systemic examination should always be performed in a patient presenting with new-onset or progressive back pain. The back pain may be a manifestation of a systemic disease (e.g. metastatic carcinoma) or be referred from a source in the abdomen or pelvis, for example:

- Chronic pancreatitis.
- Carcinoma of the pancreas.
- Posterior duodenal ulcer.
- Aortic aneurysm.
- Retroperitoneal fibrosis (rare, associated with certain β-blockers, methyldopa, LSD and amfetamines).

Hips

The hip is a common site of osteoarthritis in the elderly, but pathology sometimes begins in infancy or childhood. It is important to recognize disease early in its natural history so that appropriate treatment can be instituted.

The initial part of the examination is performed with the patient lying flat and properly exposed. Later, posture and gait can be formally assessed.

Position

Start by ensuring that the pelvis is set square so that leg length and deformity can be assessed accurately (Fig. 21.11). Attempt to position the line joining the anterior superior iliac spines perpendicular to the legs. If this is not possible, there is a fixed abduction or adduction deformity.

Inspection

Note the presence of any scars (e.g. previous hip replacement), abnormal bony or soft tissue contours and any abnormalities of the skin (e.g. erythema, sinuses).

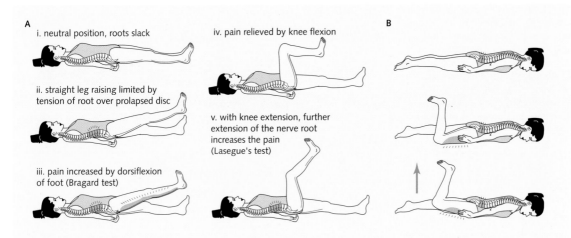

Fig. 21.10 Stretch tests. (**A**) Straight leg raising. Record the angle (normally 80–90°) through which each leg can be raised. (**B**) Femoral stretch test.

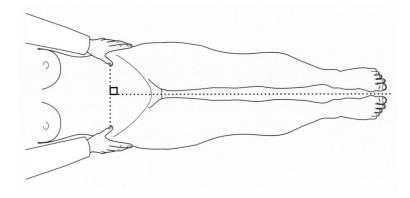

Fig. 21.11 Setting the pelvis square. Ask the patient to lie flat upon the examination couch. Palpate the anterior superior iliac spines. Move the pelvis so that they are square to the lower limbs.

Palpation

Palpate for any tenderness or warmth.

Measurement of leg length

If the pelvis is square, it is easy to estimate relative leg length by inspection. However, if there is any doubt, the leg can be measured from the anterior superior iliac spine to the medial malleolus (Fig. 21.12). An apparent discrepancy can be excluded by measuring the distance from the xiphisternum to the medial malleolus.

Examination for fixed deformity

Long-standing arthritis commonly results in a contracture of the joint capsule or muscles and subsequent fixed-flexion deformity. Often, patients compensate by increasing their lumbar lordosis (Fig. 21.13). This may be assessed by placing a hand behind the lumbar spine to detect a lordosis and then asking patients to fully flex their good leg. Push the good leg further into flexion to obliterate the lordosis and observe the angle of fixed flexion deformity in the affected hip. This is known as Thomas' test.

Fig. 21.12 Assessing relative leg length. (**A**) Measure from the anterior superior iliac spine to just below the medial malleolus on each side. (**B**) If the pelvis is tilted, the leg length may appear discrepant. (**C**) Apparent shortening may be detected measuring the distance from the xiphisternum to the medial malleolus.

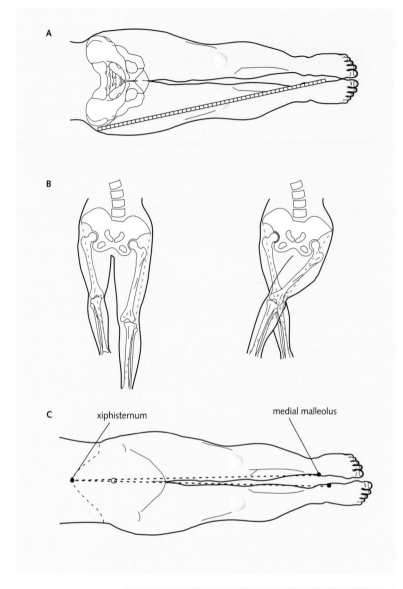

Fig. 21.13 Examination for fixed flexion deformity of the hip. A fixed deformity may be hidden by increasing the lumbar lordosis. This should be eliminated.

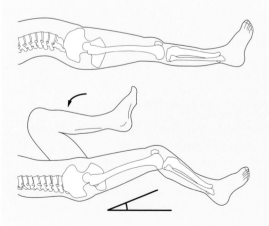

Assessment of movements about the hip

It is important to assess movement about the hip and to eliminate movement of the pelvis that may compensate for deficiencies. Assess the range of movement for passive and active movements. The normal range of movement is illustrated in Figure 21.14.

Active movement should also be tested to assess power using the Medical Research Council (MRC) grading of strength (see Fig. 17.11, p. 172).

Gait and posture

Ask the patient to stand up. The Trendelenburg test should be performed to assess the postural stability of the hip joint (particularly the gluteal muscles). Normally, if one leg is lifted off the ground, the abductors will stabilize the leg and the pelvis will tilt up on the side of the lifted leg. If the abductors are ineffective, the body weight is too much for the adductors and the hip will tilt downwards (Fig. 21.15).

The causes of a positive Trendelenburg test are:

- Paralysis of the abductor muscles (e.g. polio).
- Absence of stability (e.g. unimpacted or malunited fracture of the femoral neck). Finally assess the gait.

Systemic survey

Remember to perform a systemic survey for other causes of hip symptoms.

Normal range of movement at the hip joint	
Movement	Range (degrees)
flexion	0–120
extension	0 (extension occurs by rotating the pelvis)
abduction	0–30
adduction	0–30
lateral rotation	40
medial rotation	40

Fig. 21.14 Normal range of movement at the hip joint.

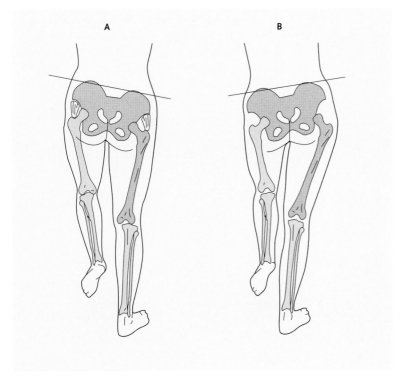

Fig. 21.15 Trendelenburg test. (**A**) Normally the hip abductors will tilt the pelvis upwards when the leg is lifted off the ground. (**B**) If the abductors cannot sustain the weight, the pelvis will droop.

Knee

Inspection

With the patient properly exposed and supine on a couch, inspect the knee, thigh and lower leg, noting:

- Muscle bulk and evidence of wasting.
- Bony deformity (e.g. genu varum, genu valgum; Fig. 21.16).
- Evidence of soft tissue swelling or effusion.

Palpation

Palpate the bony contours of the knee joint, noting any areas of tenderness or warmth. Specifically

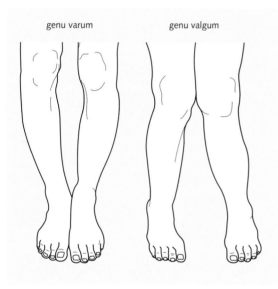

genu varum   genu valgum

Fig. 21.16 Genu varum and genu valgum.

examine for an effusion. A small effusion may only be detectable by massaging fluid into the suprapatellar pouch and observing accumulation in the medial compartment by pressure over the superior and lateral aspects of the joint. A larger effusion is detectable by the patellar tap.

Movements

Assess for the presence of a fixed-flexion deformity, the range of passive movement and strength of active movement.

Tests of stability

The four major ligaments should be tested in turn, as follows:

- Medial and lateral ligaments (Fig. 21.17A). Support the knee in a position of 15–20° of flexion and ask patients to relax their muscles. Apply an abduction and adduction force in turn to test the integrity of the medial and lateral ligaments, respectively.
- Anterior and posterior cruciate ligaments (Fig. 21.17B). The anterior cruciate ligament prevents anterior displacement of the tibia on the femur. The posterior cruciate prevents posterior displacement. Again, with the knee in 15–20° of flexion (with your knee underneath the patient's on the couch, if necessary) place your non-dominant hand on the femur above the patella, and place your dominant hand with the fingers in the popliteal fossa and circling the lateral aspect of the tibia, with your thumb placed over the tibial tuberosity and into the joint line. Apply force on the tibia anteriorly and press posteriorly on the femur: laxity

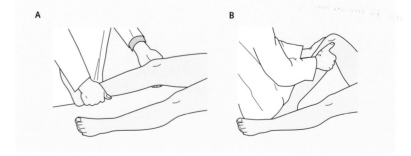

A   B

Fig. 21.17 Testing for stability of the knee joint. (**A**) Medial and lateral collateral ligaments. (**B**) Cruciate ligaments.

indicates rupture of the anterior cruciate ligament with high positive predictive value. With the knee flexed at 90°, observe for posterior sag which indicates rupture of the posterior cruciate ligament.

Alternatively, flex the knee and fix the foot firmly on the couch by sitting lightly on it. Clasp the knee joint with both hands, holding your fingers behind the joint and thumbs laterally so that the tips are resting on each femoral condyle. Alternately push and pull the tibia to assess anteroposterior stability.

Further reading

Solomon DH, Simel DL, Bates DW et al 2001 The rational clinical examination. Does this patient have a torn meniscus or ligament of the knee? Value of the physical examination. *Journal of the American Medical Association* **286**(13): 1610–1620

Ophthalmic examination

Objectives

By the end of this chapter you should:

- Appreciate the different modalities of testing the eye, including fundoscopy, visual fields and acuity.
- Be able to perform these tests confidently.
- Be able to recognize retinal changes associated with diabetic and hypertensive retinopathy.

Ophthalmic examination is essential in any detailed assessment of a patient. Not only may the cause of visual symptoms be determined, but the retina is the only place where the small blood vessels of the body can be directly visualized, providing clues to a host of systemic diseases.

EXAMINATION ROUTINE

Inspection

Before fundoscopy, look at the eyes for:

- Red eye (e.g. conjunctivitis, iritis, acute glaucoma, scleritis).
- Pupil size, symmetry and irregularity.
- Pupil reflexes.
- Arcus senilis (significant in adults <40 years old).
- Squint.
- Ptosis.

Visual acuity

Test near and distant vision. Visual acuity should be tested in any complete physical examination. Test each eye individually, with patients wearing their own spectacles or contact lenses to correct any refractive error. If these are unavailable, a pinhole in a piece of card used in place of a lens can correct most refractive errors. This assessment need take only a few seconds, and can be adapted to different circumstances. For example:

- Read a newspaper headline from the other side of the room.
- Count fingers from the end of the bed.
- Identify light from dark, perceive hand movements.

If time permits, perform a formal assessment with a Snellen chart (Fig. 22.1). The patient should be placed 6 metres from a standard chart (or 3 m from a mini Snellen chart) and asked to read the letters. The last line that can be clearly distinguished by the patient should be recorded for each eye. For example, if the patient can read the line with '12' written beside it, but not the next line, acuity in that eye is recorded as 6/12. Accepted normal vision is 6/6. The numerator refers to the distance from the chart that the patient is seated and the denominator is the last line that can be read.

Near vision can be formally tested with special books with text of a defined font and pitch size.

Eye movements and nystagmus

These are discussed in Chapter 17. Squint may be assessed by the cover test. If the eye fixating an object is covered, the squinting eye will move to take up fixation.

Fundoscopy

Examine in a darkened room to maximize pupil size. Ophthalmologists and diabetologists will dilate the pupil if there is no contraindication. The patient should look straight ahead and focus on the far wall.

Fig. 22.1 Use of Snellen chart to test visual acuity.

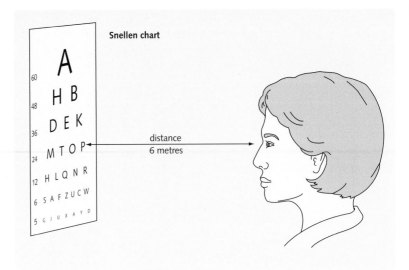

Snellen chart

distance 6 metres

Use the same eye to look into the patient's eye; that is, your right eye to the patient's right eye. Placing your hand on the patient's head will prevent you from colliding with him or her during the examination. First set the ophthalmoscope so that you can see through it in an undistorted manner (especially if you normally wear glasses). Familiarize yourself with the way a fundoscope works so that, if asked in an exam, you can at least demonstrate you have lifted one up before!

Practice fundoscopy on as many normal people as possible. It will become easier to recognize any pathological signs.

Red reflex

Start by shining the light from about 30 cm on the pupil to look for a red reflex. Loss of red reflex is usually due to vitreous haemorrhage or a dense cataract. Bring the ophthalmoscope closer to the eye and focus on the retina, looking systematically at the following.

Anterior chamber

Using a +10 lens, focus on the iris and examine for rubeosis, hypopyon, etc. Then by decreasing the power of the ophthalmoscope, focus through the

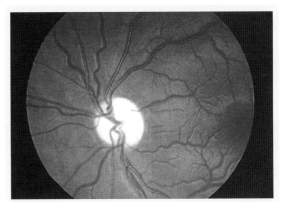

Fig. 22.2 Optic atrophy. (Courtesy of Myron Yanoff.)

anterior chamber, lens and posterior chamber. Small cataracts appear black and well demarcated using this technique.

Optic disc

Note the size of the disc and colour. The most important pathologies to note are:

- Optic atrophy (Fig. 22.2) – may be due to multiple sclerosis, compression of the optic nerve (e.g. by pituitary tumour, aneurysm).
- Papilloedema (Fig. 22.3) – may be due to accelerated hypertension, a space-occupying lesion, hydrocephalus, benign intracranial hypertension (especially in obese women), cavernous sinus thrombosis, central retinal vein thrombosis.

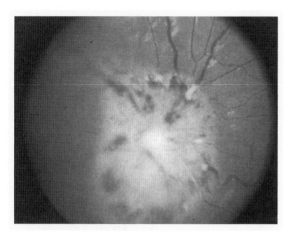

Fig. 22.3 Papilloedema (swelling of the optic disc) caused by acute lymphoblastic leukaemia. (Courtesy of Myron Yanoff.)

- Glaucoma – pathological cupping of the disc due to gradual loss of nerve fibres and supporting glial cells, resulting in a pale disc with an enlarged cup.
- Myelinated nerve fibres.

Retina

Note the retinal vessels. Trace the vessels away from the optic disc towards each quadrant of the retina in turn (you can also find the disc by tracing a vessel back to its origin). The veins are darker and appear wider than the arteries. The main features to observe about the retinal vessels are:

- Engorgement of the veins, which implies slow flow (e.g. retinal venous occlusion, polycythaemia).
- Attenuation of the arterioles (e.g. due to retinal artery occlusion, widespread retinal atrophy).
- Arteriovenous (AV) nipping in hypertension.

Note the retinal background in each quadrant. In particular, look for:

- Haemorrhages. The most common haemorrhages are flame-shaped haemorrhages, which are superficial and occur in severe hypertension. 'Dot haemorrhages' are not true bleeding areas but represent microaneurysms, which are prone to rupture, forming 'blot' haemorrhages
- Hard exudates (true retinal exudates). These are usually small, sharply defined and intensely white

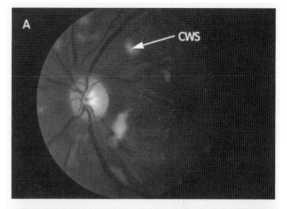

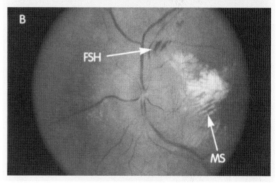

C	Features of hypertensive retinopathy
Grade I	silver wiring of arterioles
Grade II	AV nipping
Grade III	soft exudates (due to small infarcts), flame-shaped haemorrhages
Grade IV	papilloedema

Fig. 22.4 A, B, Grade III hypertension. CWS, cotton wool spot; FSH, flame-shaped haemorrhage; MS, macular star. **C,** Features of hypertensive retinopathy. Grades III and IV indicate accelerated hypertension. (Courtesy of Myron Yanoff.)

- Soft exudates (areas of infarction). These have a fluffy appearance resembling cotton wool. They usually have an ill-defined edge.
- Neovascularization. New blood vessel formation is an important sign of diabetic retinopathy. These new blood vessels are fragile and appear as a tuft of delicate vessels on the surface of the retina.
- Photocoagulation scars.
- Pigmentation (retinitis pigmentosa).

Some of the more important fundal abnormalities are illustrated in Figures 22.4 and 22.5.

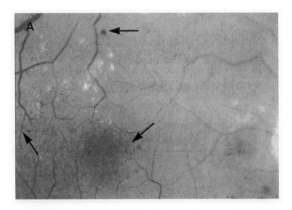

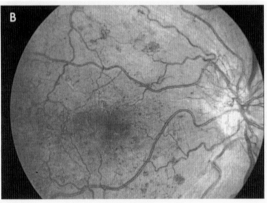

C	Features of diabetic retinopathy

background diabetic retinopathy
  dot haemorrhages (micro-aneurysms)
  blot haemorrhages (discrete bleed)
  hard exudates

proliferative retinopathy
  neovascularization
  (laser coagulation scars)

Fig. 22.5 Diabetic retinopathy. (**A**) Background changes. *Arrows* indicate haemorrhages. (**B**) Neovascularization at the optic disc. (**C**) Features of diabetic retinopathy. (A, B courtesy of Myron Yanoff.)

Fundoscopy

Diabetic patients should have detailed fundoscopy at least once per year, often as part of a comprehensive annual review, as fundal changes are sensitive markers of end-organ damage – not only in diabetes mellitus but also in hypertension. Assessment is usually performed by retinal photograph. Take any opportunity, in hospital or in the community, to look at such photographs. Only by seeing many of these will you begin to appreciate the range of normal and the subtleties of some diagnoses. These photographs also appear frequently in exams!

Macula

Ask the patient to look briefly, but directly, at the light of the ophthalmoscope. Macular disease is common in the elderly.

Visual fields

The examination routine for testing by confrontation is discussed in Chapter 17 (p. 164). More formal visual field testing can be performed by assessing the visual threshold in different regions, but this relies upon special equipment and skilled interpretation.

Writing medical notes

23

Objectives

By the end of this chapter you should:

- Understand the importance of accurate record keeping.
- Be able to record systematically a medical clerking.
- Be able to accurately record findings and instructions from a ward round.

GENERAL POINTS

When writing up the medical notes, it is extremely important to adopt a systematic and objective style. Your notes form a permanent record of your impression of the patient at that moment in time. It should be possible for another healthcare professional to read your notes and to understand them and your conclusions. Although the following points appear to be obvious, it is alarming how often simple good clinical practice is ignored. Always ensure that:

- Each piece of paper has the patient's name at its head – notes have an uncanny knack of falling apart!
- Each entry is dated and, ideally, a time recorded.
- Each entry is followed by a legible signature, your name, grade (printed), specialty and your bleep number. Scrawled initials are inadequate. Someone else has to read your notes and be able to identify who has written them.
- Your handwriting is legible. This point sounds ludicrously obvious, but unfortunately anyone who has looked through a set of notes will testify that it is often forgotten.
- Each statement is objective. It is no longer acceptable to write value judgements in the notes if they cannot be justified. Avoid the use of frivolous and offensive phrases such as the commonly written NFN (Normal For Norfolk) or NLM (Nice Looking Mum). These are not acceptable. You could end up in a world of trouble in court if a lawyer got hold of these.

Your handwriting must be clear and legible. It is potentially dangerous to write in hieroglyphics!

- You write in the notes every time you see the patient. You do not have to write an essay – a short statement of the patient's progress is an important record. Even noting a lack of any change since your last consultation is important.
- You avoid the use of abbreviations as much as possible. If they have to be used, use only accepted terms such as MI (myocardial infarction) or LVF (left ventricular failure).
- You use diagrams where appropriate, as they are often much more descriptive than long paragraphs.
- You keep your record as concise as possible.

Remember that patients (or their lawyers) may gain access to the notes. Avoid the use of statements that you cannot justify.

Often the history given by the patient is recounted in an unconventional manner. When writing your notes, it is usual practice to record the history in the order described in Part I of this book. This helps to structure your, thoughts and also those of anyone who subsequently reads the notes. An example is given later in the chapter. Clearly the history is a dynamic process, and there are infinite variations and exceptions to this general aim. Remember that patients come to us with a story and we effectively translate this into symptoms, signs and diagnoses. This is where the art of medicine comes into play.

Finally, it is important to make a note of what information has been given to the patient. Poor communication between the doctor and patient is responsible for the majority of instances of patient dissatisfaction, complaints and litigation. It is helpful, not only to yourself but also for other doctors, to be aware of exactly what information the patient has when starting the next consultation – beware of information getting 'lost in translation'.

STRUCTURING YOUR THOUGHTS

When you first start to take medical histories, it is a struggle to remember the traditional order of questions to ask and the normal examination routine. However, your job has only just begun! Remember that the aim of a clerking is to:

- Identify any problems, as well as the patient's ideas, concerns and expectations.
- Formulate a differential diagnosis of the problems.
- Consider a plan of initial investigations to elucidate the underlying cause, severity and prognosis of each problem.
- Initiate treatment and advice for the patient.

Once your examination has been completed, it is helpful to go through a routine and ask yourself the following questions.

In an examination setting, it is important to spend time reflecting on the significant points of the history and examination. Always allow enough time to gather your thoughts before presenting your findings.

What is the patient's presenting complaint?

Never forget the initial reason for the patient seeking medical attention. Although you may have identified more significant medical problems during your history and examination, you must show patients that you are addressing their primary concerns. Patients may have a very different list of priorities to yours, so always seek to explain what your concerns are and why.

What problems have I identified?

Problems can take many forms and be physical, social or psychological. Before trying to dissect the differential diagnosis, it may be helpful to write down a list of each problem. A problem may be:

- A proven diagnosis (e.g. diabetes).
- A pathological state (e.g. renal failure).
- An abnormal symptom (e.g. coughing up blood – haemoptysis).
- An abnormal sign (e.g. pulmonary consolidation, tachypnoea, pitting oedema).
- An abnormal investigation result (e.g. raised plasma creatinine concentration, increased carbon monoxide transfer coefficient, a raised C-reactive protein, haematuria and red cell casts in the urine).
- A past medical history (e.g. pneumonia).
- A social or psychological problem (e.g. unemployed, homeless, intravenous drug abuser).
- A significant risk factor for illness (e.g. smoking).

It is only by identifying these individual problems and considering them as a group that their relative importance and relationship to each other can be systematically assessed.

What action is needed for each problem?

Try to prioritize the importance of each problem and decide upon the urgency of treatment and investigation of each.

What is the differential diagnosis of each problem?

By assessing each problem in turn, it may become apparent that a unifying diagnosis could explain a number of features. Equally, some diagnoses may be excluded by the presence of certain features. When you see a patient, always remember that unusual presentations of common problems are still more frequently encountered than the common presentations of rare diseases.

What is the most likely underlying diagnosis?

Once a differential diagnosis has been formulated, it is important to make a mental note of the order of probability of each diagnosis. This is important, as immediate treatment and investigations are aimed at the diagnoses at the top of your list and those that require urgent therapy. Furthermore, it has been shown that making a commitment to a specific diagnosis increases your diagnostic accuracy.

What investigations should be requested?

In the examination setting, as in real life, it is important to consider which investigations are appropriate and their urgency. When requesting investigations, it is essential to consider how the result of a particular test is going to help refine your management of the patient. It is no longer acceptable to adopt the mentality of 'I always request chest radiography, an electrocardiogram, full blood count and biochemical profile for all medical patients!' Consider each problem in turn and assess which investigations are appropriate. It is by this systematic approach that potentially important and useful investigations are not omitted.

Start by considering the simple and non-invasive tests, and then consider the value of more discriminatory but invasive investigations. Weigh up the potential usefulness of the result against the potential morbidity and cost (and mortality) of each test. For example:

- Many simple blood tests are cheap and non-invasive so a low threshold is needed.
- A positron emission tomograph (PET scan) is expensive, but in limited situations offers very specific information that cannot be reliably obtained from other sources.
- Coronary angiography is the gold standard for diagnosing structural causes of angina, but does have a defined mortality.

Ideally you should have an idea about the sensitivity (ability to detect something when it's there), specificity (ability to recognize when something isn't there) and the predictive values of the test in the population of the patients you are involved with.

PRESENTING YOUR FINDINGS

It is important to get as much practice as possible in presenting your findings to your colleagues. In this respect, the easiest part of a consultation is taking a history and performing an examination. However, most students struggle when trying to collate the vast amount of information obtained from a clerking and reformat it in a presentable and digestible form.

Remember how dull it can be listening to inexperienced students presenting their findings. It is essential to be as concise as possible so that important information is not lost among masses of irrelevant detail. Further, examination formats such as structured long cases, mini-CEXs and OSCEs, which are increasingly common, require students to give focused case presentations.

When presenting a clerking, put yourself in the audience's position and consider what information you would like, or need, to hear.

Start with a pithy introductory phrase describing the patient and his or her reason for presenting for medical attention. This is important as it provides a frame for the subsequent information. Your audience will already have started the process of differential diagnosis and be anticipating certain specific facts.

Describe the detailed history, *using the patient's words where possible*, and avoid making value judgements at this stage. You will have asked the patient many specific questions. Your listeners are interested in a limited number of these. Using your judgement, introduce important negative responses, but do not overburden your audience with irrelevant detail. For example, if your patient presents with exertional chest pain, your audience needs to know about risk factors for ischaemic heart disease and other cardiac symptoms, even if there are none, but introducing information about myopia or the patient's stool habit will only cloud the important issues. The rest of the history should be presented in a similarly edited form. Unrelated symptoms described by the patient can be mentioned at the end of the history.

When presenting your examination findings:

- Start by describing the vital signs.
- Then concentrate on the system primarily affected. Discuss your findings in this system at length.
- For the other systems of the body, describe any positive findings and any important negative signs that may affect interpretation of the main findings.

It is important to sound confident. It is very difficult when you first begin, but you must commit yourself. As stated above, good problem representation has been shown in itself to aid the process of clinical reasoning. Many students start by making excuses for having difficulty in eliciting a sign. This does not impress examiners and is not helpful – in the real world, your interpretation of physical signs affects your management of the patient.

At the end of your examination findings, provide:

- A short summary encompassing the main features of the history and examination.
- Your interpretation and analysis of the problems, with a brief differential diagnosis and a note of features in favour of, or against, each differential diagnosis.

SAMPLE CLERKING

A sample medical clerking is shown in Figure 23.1 and illustrates some of the points discussed at the start of this chapter.

WARD ROUND ENTRIES

In contrast to clerking patients, ward round entries are, in your early career at least, often about making sure that the correct information as elicited by another health professional is well documented. There are a huge number of different ways in which ward rounds are conducted. Some doctors often dictate entries to be typed and included in the notes later; some departments write entries on stickers which are then stuck in the notes and signed (making sure the signature passes from the sticker onto the notes page itself). By far the most conventional approach is that found on most medical wards, whereby the entire team is present on the round, the interview is conducted by the most senior and the documentation is made by the most junior member.

Each entry should contain a date and time, Ward round (or W/R) followed by the surname of the most senior person present, the vital signs and any relevant symptoms or findings (e.g. 'CT scan shows diverticular change in sigmoid colon, discussed with patient.').

The patient's concerns should be included, together with the information given regarding each one. Examination findings can be included by diagram as in the sample clerking. An impression may be included; this can be a diagnosis or differential, followed by a plan. If there are no changes, the entry may be brief and the plan may be 'continue'; this is especially true of patients undergoing elective procedures. Every entry should be followed by your signature, name, grade and bleep/DECT number. A sample ward round entry is shown in Figure 23.2.

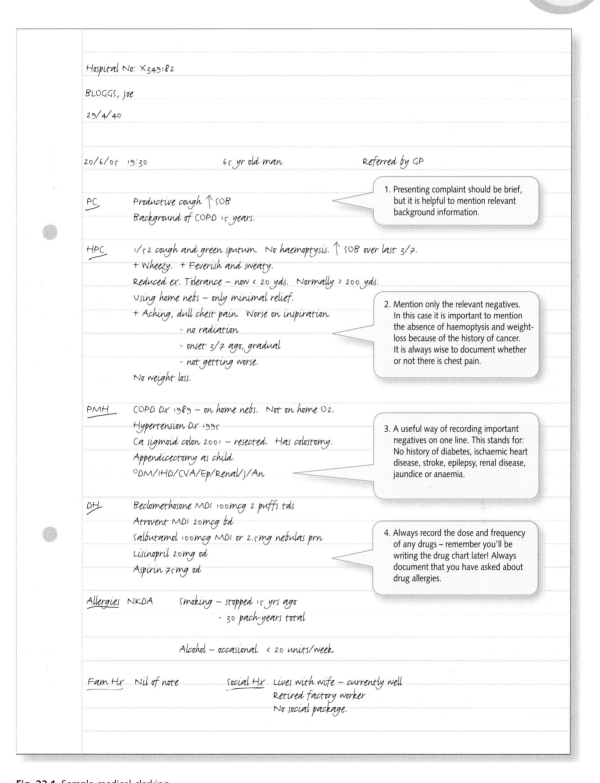

Hospital No: X349182

BLOGGS, Joe

29/4/40

20/6/05 19:30   65 yr old man   Referred by GP

PC   Productive cough ↑SOB
        Background of COPD 15 years.

> 1. Presenting complaint should be brief, but it is helpful to mention relevant background information.

HPC   1/52 cough and green sputum. No haemoptysis. ↑SOB over last 3/7.
        + Wheezy. + Feverish and sweaty.
        Reduced ex. Tolerance – now < 20 yds. Normally > 200 yds.
        Using home nebs – only minimal relief.
        + Aching, dull chest pain. Worse on inspiration.
                - no radiation
                - onset 3/7 ago, gradual
                - not getting worse.
        No weight loss.

> 2. Mention only the relevant negatives. In this case it is important to mention the absence of haemoptysis and weight-loss because of the history of cancer. It is always wise to document whether or not there is chest pain.

PMH   COPD Dx 1989 – on home nebs. Not on home O2.
        Hypertension Dx 1995
        Ca sigmoid colon 2001 – resected. Has colostomy.
        Appendicectomy as child.
        °DM/IHD/CVA/Ep/Renal/J/An

> 3. A useful way of recording important negatives on one line. This stands for: No history of diabetes, ischaemic heart disease, stroke, epilepsy, renal disease, jaundice or anaemia.

DH   Beclomethasone MDI 100mcg 2 puffs tds
        Atrovent MDI 20mcg bd
        Salbutamol 100mcg MDI or 2.5mg nebulas prn
        Lisinopril 20mg od
        Aspirin 75mg od

> 4. Always record the dose and frequency of any drugs – remember you'll be writing the drug chart later! Always document that you have asked about drug allergies.

Allergies NKDA   Smoking – stopped 15 yrs ago
                              - 30 pack-years total

                   Alcohol – occasional. < 20 units/week.

Fam Hx Nil of note   Social Hx Lives with wife – currently well
                                    Retired factory worker
                                    No social package.

Fig. 23.1 Sample medical clerking.

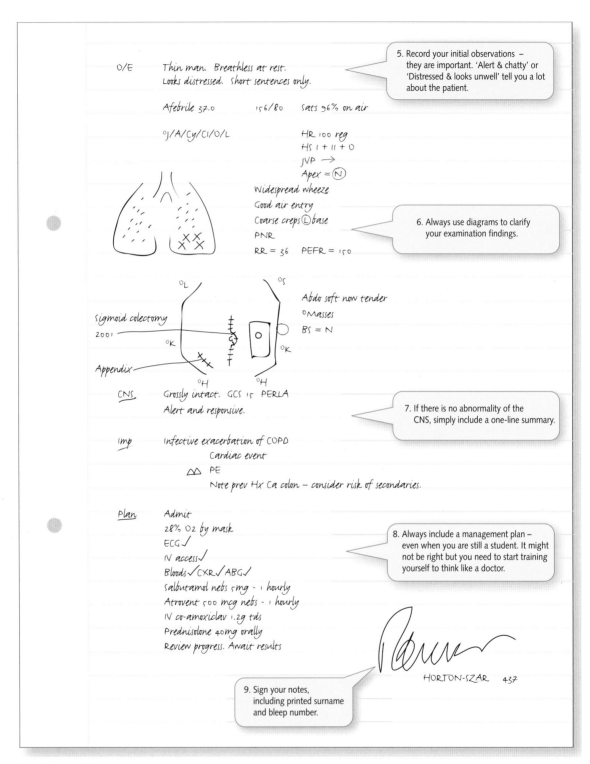

O/E   Thin man. Breathless at rest.
      Looks distressed. Short sentences only.

5. Record your initial observations – they are important. 'Alert & chatty' or 'Distressed & looks unwell' tell you a lot about the patient.

Afebrile 37.0   156/80   Sats 96% on air

°J/A/Cy/Cl/O/L   HR 100 reg
                                 HS I + II + 0
                                 JVP →
                                 Apex = (N)
                         Widespread wheeze
                         Good air entry
                         Coarse creps (L) base

6. Always use diagrams to clarify your examination findings.

                         PNR
                         RR = 36   PEFR = 150

Sigmoid colectomy
2001

Appendix

                                 Abdo soft now tender
                                 °Masses
                                 BS = N

CNS   Grossly intact. GCS 15 PERLA
      Alert and responsive.

7. If there is no abnormality of the CNS, simply include a one-line summary.

Imp   Infective exacerbation of COPD
            Cardiac event
         △△ PE
            Note prev Hx Ca colon – consider risk of secondaries.

Plan Admit
      28% O2 by mask
      ECG ✓
      IV access ✓
      Bloods ✓ CXR ✓ ABG ✓

8. Always include a management plan – even when you are still a student. It might not be right but you need to start training yourself to think like a doctor.

      Salbutamol nebs 5mg - 1 hourly
      Atrovent 500 mcg nebs - 1 hourly
      IV co-amoxiclav 1.2g tds
      Prednisolone 40mg orally
      Review progress. Await results

                         HORTON-SZAR   437

9. Sign your notes, including printed surname and bleep number.

Fig. 23.1, cont'd

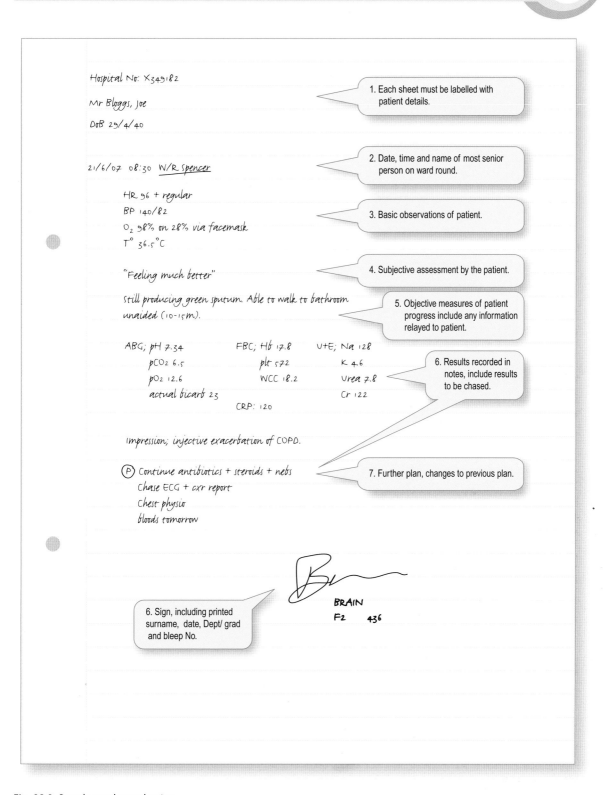

Hospital No: X349182

Mr Bloggs, Joe

DoB 29/4/40

1. Each sheet must be labelled with patient details.

21/6/07 08:30 W/R Spencer

2. Date, time and name of most senior person on ward round.

HR 96 + regular
BP 140/82
O₂ 98% on 28% via facemask
T° 36.5°C

3. Basic observations of patient.

"Feeling much better"

4. Subjective assessment by the patient.

Still producing green sputum. Able to walk to bathroom unaided (10-15m).

5. Objective measures of patient progress include any information relayed to patient.

ABG; pH 7.34   FBC; Hb 17.8   U+E; Na 128
      pCO₂ 6.5   plt 572   K 4.6
      pO₂ 12.6   WCC 18.2   Urea 7.8
      actual bicarb 23   Cr 122
                      CRP: 120

6. Results recorded in notes, include results to be chased.

Impression; injective exacerbation of COPD.

P Continue antibiotics + steroids + nebs
   Chase ECG + cxr report
   Chest physio
   bloods tomorrow

7. Further plan, changes to previous plan.

BRAIN
F2   436

6. Sign, including printed surname, date, Dept/ grad and bleep No.

Fig. 23.2 Sample ward round entry.

Objectives

By the end of this chapter you should:

- Understand which investigations may be requested for conditions affecting the major systems.
- Be able to interpret basic blood test findings.
- Be able to request appropriate basic investigations for the conditions listed in this book.

Once the history and examination have been performed, it is necessary to assess your differential diagnosis and to plan a management strategy for the patient. Part of this process includes arranging further investigations in order to refine the differential diagnosis. The tests may be performed to:

- Exclude serious conditions.
- Confirm the presence of a suspected pathology.
- Obtain a baseline against which further progress may be assessed.
- Assess the severity of the current disease process(es).
- Assess the response to therapy.
- Predict prognosis.

Before requesting any investigations, think (a) why you are doing it, (b) what you expect to find and (c) how the patient will benefit from the result.

Remember that it is easy to request investigations, but the results may require great skill in interpretation. Furthermore, each test has a cost and may potentially have unwanted consequences for the patient. It is essential to anticipate how the results of the investigation may alter your management at the time of the request. If you cannot see how the results of an investigation will alter your management, there is absolutely no point requesting that test. For each test, be prepared to justify:

- Why it is being requested.
- How the result will affect management.(i.e. considering what you will do if the test is positive, negative or equivocal).
- That the potential benefit of the information outweighs the cost and morbidity incurred.
- The urgency of the request.

These attributes are the fundamental requirements of a good house officer/F1, as a great deal of time is spent ordering and organizing investigations requested by yourself and your seniors.

A few general points:

- With many test results, normality is presented as a 'reference range', which (usually) covers 95% of results for patients who do not have the abnormality or disease. Thus 1 in 20 results will be outside the range and will flag up as 'abnormal' (i.e. false positives).
- Information is available about the attributes of many tests (and, indeed, many symptoms and physical signs), usually in terms of their specificity (i.e. ability to rule in the presence of disease – think 'SpIn') and sensitivity (ability to rule out disease – think 'SnOut'); thus their positive or negative predictive value. However, the latter vary with the prevalence of the abnormality or disease in a particular population. A more useful parameter is the 'likelihood ratio' (LR), calculated by combining the specificity and sensitivity. Likelihood ratios can be either positive or negative; LR >1 increases the possiblity of the

disease being present and LR between zero and 1 decreases it.

- If you do enough tests on a patient, you will eventually throw up an abnormal result; after undertaking 14 tests, the chance of an abnormal result rises from 1 in 20 to an astonishing 50%. It has been said that a normal patient is simply someone who has not yet been investigated enough!
- Indiscriminate testing, for lack of thinking it through, to cover your back, or because the consultant demands it, is potentially harmful to the patient, costly to the system and, ultimately, unethical practice.

The lists given below are not intended to be comprehensive, and certainly do not imply that every test should be requested for each system. The tests requested need to be tailored to the clinical scenario (and in 'real life', to take account of individual hospital and national policies and guidelines). For each system, consider:

- Urine.
- Blood.
- Radiological imaging.
- Electrical recording.
- Special investigations.

CARDIOVASCULAR SYSTEM

Cardiovascular investigations

There are often specific protocols for the investigation of important conditions. It is important to become familiar with protocols in your own hospital and the evidence on which they are based. In cardiology:

- The diagnosis of acute myocardial infarction (MI) is made on the basis of two of the following three factors being present: chest pain, electrocardiogram (ECG) changes consistent with MI and an increase in cardiac enzymes. Appropriate investigation involves assessing these factors as an absolute minimum.
- If infective endocarditis is suspected, obtain at least six sets of blood cultures. These should be obtained from three or more separate locations. Endocarditis can then be grouped into 'culture positive' and 'culture negative' types.

Blood tests

The following blood tests may help in the diagnosis of cardiac pathologies:

- Full blood count. Anaemia may be the cause of, or exacerbate, heart failure or angina.
- Erythrocyte sedimentation rate (ESR). Inflammatory conditions (e.g. endocarditis) are associated with a raised ESR. The C-reactive protein (CRP) is an acute-phase protein and is often more sensitive in changing inflammatory states (e.g. monitoring the response of infective endocarditis to antibiotic therapy).
- Cardiac enzymes. Troponin T (or troponin I in some centres) is released by cardiac muscle breakdown. It should be measured 12 hours post chest pain or a change in rhythm.
- Biochemical profile. Exclude electrolyte disturbance as a cause of arrhythmia. Potassium should be maintained above 4 mmol/L to reduce the incidence of post MI arrhythmia.
- Thyroid function tests. Hyperthyroidism is a common cause of atrial fibrillation. Hypothyroidism may present as a pleural effusion or heart failure.
- Blood cultures. Essential if endocarditis is suspected.

Urinalysis

Look for microscopic haematuria if infective endocarditis is suspected.

Imaging

Chest radiography

A chest radiograph is usually requested for a patient who presents with cardiological symptoms. Note the presence of:

- Cardiac enlargement – cardiothoracic ratio should be less than 50% (posteroanterior (PA) film only).
- Signs of increased left atrial filling pressure – upper lobe blood diversion, septal lines, pulmonary oedema, pleural effusion (Fig. 24.1).
- Signs of left atrial enlargement – prominence of atrial appendage (straight left heart border), double contour of right heart border, splaying of the carina.
- Left ventricular aneurysm – post myocardial infarction.

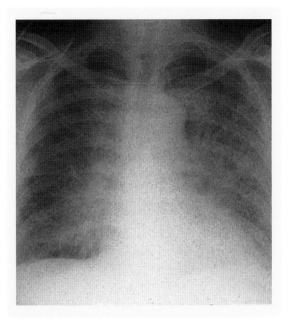

Fig. 24.1 Radiograph of left heart failure. Note the cardiomegaly, bilateral alveolar shadowing in a perihilar distribution and the presence of fluid in the horizontal fissure.

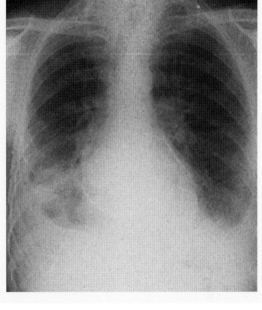

Fig. 24.2 Radiograph of pericardial calcification. Note the associated pleural effusions.

- Abnormal calcification – valvular calcification in rheumatic heart disease, tuberculous pericardial disease (Fig. 24.2).

Echocardiography

Echocardiography may be transthoracic (non-invasive) or transoesophageal, which provides better images of the left atrium and aorta. It is useful for assessing chamber size, valvular pathology, pericardial disease and contractility of the heart. Echocardiography is usually requested to:

- Comply with the guidance of the National Service Framework that now says all cases of heart failure should have an echo to investigate the cause of heart failure.
- Investigate heart murmurs.
- Look for vegetations in suspected infective endocarditis.
- Investigate pericardial effusions and tamponade.
- Assess the severity of cor pulmonale.
- Investigate aortic aneurysms.

Modifications of echocardiography such as stress echo and contrast echo may also be used to investigate angina and septal defects, respectively. The two most common views are illustrated in Figures 24.3 and 24.4.

Nuclear imaging

Nuclear imaging (e.g. thallium scan) can be used to investigate suspected angina in a patient who cannot perform an exercise test.

Electrocardiography (ECG)

12-lead ECG

Electrocardiograms are very widely performed. They can be used to assess rhythm disturbances, ischaemia or infarction, left ventricular hypertrophy and right ventricular hypertrophy. There are many books that deal just with the interpretation of ECG findings; a useful introductory ECG resource can be found at http://library.med.utah.edu/kw/ecg/.

24-hour tape (Holter monitor)

A 24-hour tape is used to investigate paroxysmal cardiac arrhythmias that may be associated with symptoms. Patients should be instructed to indicate when they have palpitations or other symptoms

Fig. 24.3 Diagrammatic representation of an echocardiogram of a normal long-axis view of the left ventricle. The probe is placed in the left parasternal region. This provides a good view for assessing chamber size, left ventricular wall motion and ejection fraction as well as mitral and aortic valve regurgitation using colour Doppler. AV, aortic valve; IVS, interventricular septum; LA, left atrium; LV, left ventricle; MV, mitral valve; RV, right ventricle.

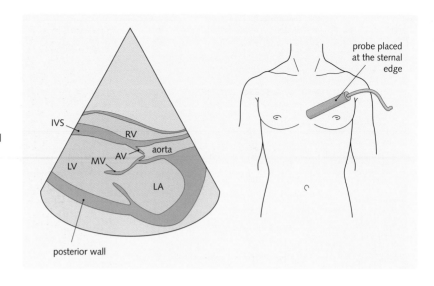

Fig. 24.4 Diagrammatic echocardiogram representing a four-chamber view. The probe is placed at the apex. This view allows assessment of the left ventricular apex and quantification of the severity of aortic and tricuspid valve pathology. LA, left atrium; LV, left ventricle; RA, right atrium; RV, right ventricle.

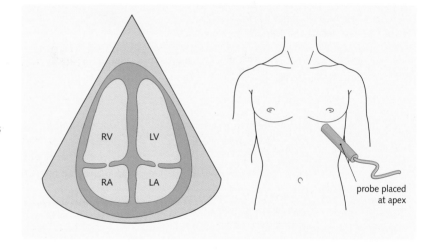

such as lightheadedness, so that their symptoms can be correlated with electrical disturbances.

Event recorder (cardiac memo)

Similar to a 24-hour tape, patients record any symptoms and an ECG is recorded. If palpitations last long enough, it is desirable to record the ECG at the time of the arrhythmia. Symptoms can then be directly related to the electrical activity of the heart.

Exercise ECG

The exercise ECG is used as a screening test in the investigation of ischaemia. Many cardiologists do not refer patients for angiography if the exercise test is normal. In addition, exercise testing provides collateral information on exercise tolerance. Remember that it is a waste of time requesting an exercise test for a wheelchair-bound 90-year-old!

Cardiac catherization

These invasive investigations are used:

- To assess flow and stenoses in the coronary arteries.
- To measure pressures in the heart chambers in the assessment of valve pathology or cor pulmonale.

RESPIRATORY SYSTEM

Blood tests

Patients with respiratory disease often have infections or unexplained dyspnoea. The more common blood tests requested include:

- Full blood count. A raised white cell count suggests infection. A raised eosinophil count may occur in rare conditions (and should be noted in asthmatic patients).
- C-reactive protein. This can be checked as a non-specific marker of inflammation. As the disease state resolves, the level should drop.
- Blood cultures – may be important in the diagnosis of pneumonia.
- Arterial blood gases – essential in the assessment of severe asthma attacks and for providing baseline function for patients with chronic obstructive pulmonary disease (COPD).

Sputum assessment

Sputum culture is part of the routine assessment of a patient with a chest infection. When investigating pneumonia, especially in sick patients, liaison with the laboratory is important for diagnosing some of the less common infections (e.g. *Pneumocystis*, mycobacteria, *Nocardia*).

Sputum cytology is used in the investigation of malignancy and certain pneumonias.

Imaging

Chest radiography

In the context of respiratory disease, the important features to note are:

- Areas of consolidation – for example whether confined to a lobe or widespread.
- Evidence of COPD – paucity of lung vascular markings, hyperinflation, flat diaphragms, narrow mediastinum, bullae.
- Pneumothorax or areas of collapse in asthmatic patients.
- Hilar masses – lung carcinoma may underlie many respiratory disorders (Fig. 24.5); bilateral lymphadenopathy occurs in sarcoidosis.
- Pleural effusions.
- Fibrosis.

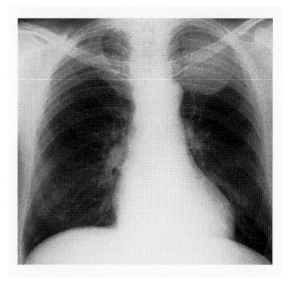

Fig. 24.5 Radiograph of a Pancoast tumour with destruction of the first left rib.

Computed tomography (CT) scan

Occasionally, CT scanning is needed to clarify features on the radiograph, for example:

- Investigation of direct or metastatic spread of suspected lung cancer.
- Investigation of bronchiectasis.
- Investigation of pulmonary fibrosis.

Lung function tests

Assessment of gas exchange and airway function may be performed by the following methods.

Peak expiratory flow rate

This should be considered as part of the routine examination of the asthmatic patient, though not if the patient is in acute respiratory distress!

Arterial blood gases

These are useful for assessing gas exchange in acute pulmonary disease. In addition, they provide a baseline assessment for patients with COPD when they are stable (e.g. three months after any infective exacerbation).

Spirometry

The most simple assessments of forced expiratory volume in 1 second (FEV_1) and forced vital capacity

(FVC) can provide invaluable information in respiratory disease. The ratio of FEV_1/FVC and absolute values will help in the diagnosis of:

- Obstructive disease – low ratio (<70%) and low absolute values.
- Restrictive disease – normal or high ratio, reduced FVC (Fig. 24.6).

In addition, the absolute value of FEV_1 is often used to assess patients with respiratory muscle weakness (e.g. myasthenia gravis, Guillain–Barré syndrome).

Variations can be used to record bronchial reactivity to allergens or potential reversibility with bronchodilators.

The transfer coefficient of carbon monoxide (K_{CO}) provides a measure of the efficiency of gas exchange and permeability of the alveolar membrane:

- Diseases resulting in impaired ventilation or perfusion reduce K_{CO}.
- Diseases such as pulmonary haemorrhage increase K_{CO}.

Bronchoscopy

Bronchoscopy allows direct visualization of the upper airways and abnormal areas can be biopsied. In addition, bronchoalveolar lavage can be performed to collect specimens for culture and cytology and to assess the differential cell types in the alveoli.

ABDOMINAL SYSTEM

Blood tests

Blood tests are required for a wide range of abdominal presentations, including the investigation of anaemia, jaundice, palpable masses, abdominal pain and bowel disturbance.

The more commonly requested investigations include:

- Full blood count. This may reveal anaemia. The mean cell volume (MCV) provides a starting point for further investigation. A raised white count suggests an inflammatory process.
- ESR. Raises the suspicion of inflammatory lesions.
- Biochemical profile. Assesses liver function, renal function and calcium. Electrolyte levels may fluctuate during diarrhoeal illnesses or fluid replacement.
- Vitamin B_{12}, iron studies, red cell folate levels. These are initial investigations in anaemia.
- Amylase (and serum lipase, where available). Normal levels exclude pancreatitis as a cause of abdominal pain.

Urinalysis

Urinalysis is part of the routine assessment during a detailed examination. Note the presence of:

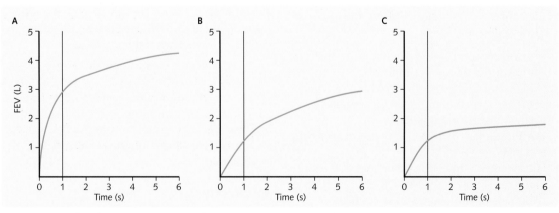

Fig. 24.6 Examples of spirometry. (**A**) Normal patient: $FEV_1/FVC = 3.0/4.0$ (75%). (**B**) COPD: $FEV_1/FVC = 1.5/3.0$ (50%). (**C**) Fibrosis: $FEV_1/FVC = 1.5/1.8$ (83%).

- Glycosuria – diabetes mellitus.
- Haematuria – for example due to glomerulonephritis, renal stone, bladder lesion.
- Proteinuria – chronic kidney disease.
- Ketonuria – diabetic ketoacidosis.
- Nitrites or leucocytes – indicate infection.

Imaging

Chest radiography

An erect chest radiograph is often requested for patients presenting with acute abdominal pain. The important features to note are:

- The presence of free air under the diaphragm – suggests perforated viscus (Fig. 24.7).
- Lower lobe consolidation – pneumonia masquerading as acute abdominal pain.

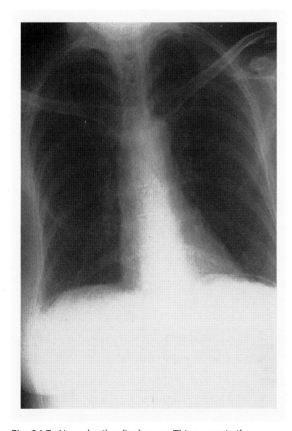

Fig. 24.7 Air under the diaphragm. This suggests the presence of a perforated abdominal viscus.

Abdominal radiography

The main features to note are:

- Distended loops of bowel – suggestive of bowel obstruction. 'Ladder' patterns may show if the large or small bowel are distended. If lines are visible across the whole bowel, these are known as 'valvulae connivente' (small bowel); if 50–70% across the diameter of the bowel, they are 'haustra' (large bowel).
- Double contour to the bowel – perforation with free air in the abdomen (Wriggler's sign); gas in the abdomen acts as a contrast, allowing the wall to be visualized.
- Radio-opaque gallstones – unusual (only 10%).
- Radio-opaque renal calculi – common (i.e. 90%).

Abdominal ultrasound and CT scan

These are performed to localize and identify any abnormal masses, detect free fluid in the abdomen and assess the size and parenchyma of intra-abdominal organs.

Ultrasound is particularly useful in the assessment of jaundice, as biliary tree dilatation and gallstones are readily identified.

Following trauma, CT or ultrasound may be used to diagnose damage to the liver, kidney or spleen.

Barium studies

Barium studies are used to assess abnormalities of the mucosa or motility disorders. The four main studies are:

1. Barium swallow – to assess the oesophagus (e.g. for dysphagia, heartburn).
2. Barium meal – to assess mucosal abnormalities of the stomach and duodenum (e.g. for dyspepsia, iron-deficiency anaemia).
3. Small-bowel meal or enema – to assess the small bowel, in particular the terminal ileum (e.g. for malabsorption, suspected Crohn's disease).
4. Barium enema – to investigate the colon (e.g. for anaemia, change in bowel habit, rectal bleeding).

Endoscopy

Endoscopy is performed to visualize the mucosa of the bowel directly and to biopsy any abnormal area.

Upper gastrointestinal endoscopy will assess as far as the duodenum. A good colonoscopy will reveal the whole of the large bowel.

Endoscopic retrograde cholangiopancreatography (ERCP) can be performed to image the pancreatic or hepatic and bile ducts and to treat any strictures or remove stones. Increasingly, magnetic resonance imaging (MRI) cholangiopancreatography (MRCP) is used, though this has no therapeutic potential.

In the presence of decreased level of consciousness or focal neurological signs, obtain a CT scan of the brain before performing a lumbar puncture.

Stool assessment

Stool assessment may involve:

- Culture – for diarrhoeal illnesses.
- Microscopy – for ova cysts and parasites.
- Faecal fat estimation.

NEUROLOGICAL SYSTEM

Imaging

Computed tomography scans and magnetic resonance imaging (MRI) of the brain or spinal cord are often requested to investigate acute and chronic neurological symptoms. Computed tomography is the modality of choice in the investigation of acute trauma or subarachnoid haemorrhage.

Electroencephalography (EEG)

The EEG relates to the brain in the same way that the ECG relates to the heart. However, correlation between the traces and physiological function is less understood. The main uses of the EEG are in the diagnosis of epilepsy and encephalitis.

Lumbar puncture

Lumbar puncture provides essential information in the assessment of patients with neurological disease, especially those with suspected meningitis. The main indications are:

- Investigation of meningitis.
- Pyrexia of unknown origin.
- Subarachnoid haemorrhage (chronic/subacute when xanthochromia – blood broken down into bilirubin – is present).
- Inflammatory central nervous system (CNS) disease (e.g. multiple sclerosis, vasculitis).

Lumbar puncture is contraindicated in the presence of:

- Suppuration of the skin overlying the spinal canal.
- Raised intracranial pressure.
- Undiagnosed papilloedema.

Electromyography (EMG)

Electromyography is used in the assessment of the peripheral nervous system. It is particularly useful if the patient reports:

- Weakness or wasting.
- Undue fatiguability.
- Sensory impairment or paraesthesiae.

Electromyography will establish a diagnosis of a neuropathy and identify the pathological process as a demyelination or axonal neuropathy. In addition, diagnostic information may be provided for some myopathies.

Evoked potentials

Visual evoked potentials are sometimes used in the assessment of a patient with suspected multiple sclerosis.

LOCOMOTOR SYSTEM

X-rays are the usual investigation in the diagnosis of muscular and joint problems. Figures 24.8 and 24.9 show the changes associated with osteoarthritis in the knee and hip, respectively. These are common in exams.

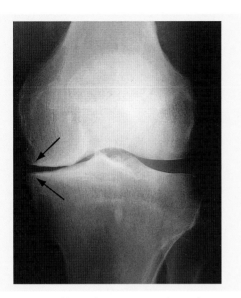

Fig. 24.8 X-ray of knee showing osteoarthritic changes. Note the loss of joint space, bone sclerosis, osteophytes and subchondral cysts.

Further reading

The following books and web site are good starting points to delve into this aspect of evidence-based practice.

Sackett DL, Strauss SE, Richardson WS 2000 *Evidence-based medicine. How to practise and teach EBM.* London: Churchill Livingstone

Summerton N 2007 *Patient-centred diagnosis.* Oxford: Radcliffe Medical Press

http://www.shef.ac.uk/scharr/ir/netting/
The 'Netting the Evidence' web site hosted by ScHARR in Sheffield. A 'portal' to all sorts of resources related to evidence-based practice.

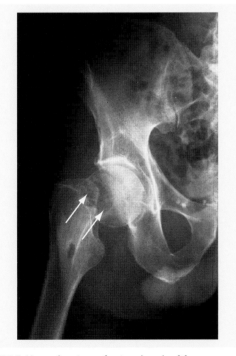

Fig. 24.9 X-ray showing a fractured neck of femur.

SELF-ASSESSMENT

Examination technique 245

Multiple-choice questions (MCQs) 247

Short-answer questions (SAQs) 255

Extended matching questions (EMQs) 257

MCQ answers 273

SAQ answers 283

EMQ answers 287

Examination technique

By the time you read this book you will probably have passed lots of exams at medical school. The aim of this short section is to share with you some of the tips my friends and I found useful.

Study groups are useful; a minimum of three people is really needed and everyone needs to prepare for them. I found that groups worked best for refining physical examination technique. The group first brainstormed all the things we thought we could be asked to examine. I have included the list here:

- Cardiovascular system
- Respiratory system
- Gastrointestinal system
- Cranial nerves
- Eye function
- Upper limbs
- Lower limbs
- Cerebellar screen
- Speech/language
- Higher functions
- Hands
- Elbows
- Shoulders
- Back
- Knees
- Hips
- Neck
- Thyroid screen/goitre
- Describe the lump/mass
- Varicose veins
- Peripheral vascular system
- Skin lesions
- Lumps in groin/scrotum
- Developmental assessment
- Baby check
- Examination of legs
- Examination of ears/nose/throat
- Breast examination
- Cardiopulmonary resuscitation
- Fluid balance charts
- Prescription kardexes
- Warfarin/insulin prescriptions
- Death certificates.

This list is not exhaustive so add to it if you wish. Remember, examiners get more devious all the time!

Then we pooled our resources to try to come up with the best answers. We all brought with us some good ideas, and so will you. We set it up as follows. One person would be the patient, one the medical student and one the examiner.

Once we had all the questions and answers, we wrote the Objective Structured Clinical Examination (OSCE) questions onto slips of paper and placed them all in a bag. The session then went as follows. The 'examiner' put his hand into the bag and pulled out a question. The 'student' then had to examine the 'patient'. The 'examiner' timed and marked the 'student'. At first we were hopeless but we all got there in the end and 'passed'. This has the advantage that it mimics the surprise of walking into an OSCE and it means you do not concentrate on one area to the detriment of the others.

If you want to practise alone, then your old teddy bear makes a good patient. One of my friends made a life-size paper man and took it to her attachments throughout the region so she had something to practise on!

The groups should be sociable and are also a good way of helping share the stress of examinations.

Spin the bottle has always been popular. Try it with a bottle of wine. Get a group of friends together. Everybody should bring some sample questions and put them in a bag. Have some food, drink the wine, then sit in a circle and spin the bottle. Whoever the bottle ends up pointing to has to answer a question pulled from the bag. This worked best for Friday night revision sessions.

When you are revising, allow yourself time off. Say 'I'm not doing anything on Sunday evening. I'll see friends or go to the cinema but I'm not going to feel guilty about not revising'. There comes a point with revision when all you are doing is appeasing your conscience.

The aim of revising at this stage is not to have a detailed knowledge of everything; this is impossible. It is also why, in the postgraduate world, people specialize. The aim is rather for broad strokes. Know a little about as much as you can. Study guides and/or learning outcomes should detail which conditions your examiners feel important. Let that guide your learning.

On the day, there are some points to remember. Read the questions and do what you are asked to do. It sounds simple but every year people do not. If you mess up a question, put it behind you and move on. There will not usually be 'killer stations' that you must pass, so accept that you have failed a question and move on.

There are two key facts that are the absolute bottom line when it comes to exams:

1. The world will still turn.
2. Your family still love you.

Everyone will tell you that the examiners are actually out to pass you by the time you come to finals (after all, if they don't, who is going to be their house officer in two months' time?). It does not matter how many times you are told this, you will not believe it until you have sat the finals, and only then will you understand what people meant.

Multiple-choice questions (MCQs)

In the following, indicate whether each answer is True or False.

1. **In the fetal circulation:**
   a. The placental blood flow is perfusion dependent
   b. Venous blood flow is deflected from the right atrium to the left atrium via the ductus arteriosus
   c. There is no blood flow through the fetal lungs
   d. Systemic venous return is a mixture of oxygenated and deoxygenated blood
   e. At birth, the first breath increases pulmonary arterial pressure

2. **In matters pertaining to contraception:**
   a. Eisenmenger's syndrome is an indication for early female sterilization
   b. Focal migraine is not a contraindication to the combined oral contraceptive
   c. A condom is as efficacious as a diaphragm at preventing the spread of sexually transmitted diseases
   d. The return of normal menstruation may be delayed after progesterone-only contraceptives
   e. Male sterilization is associated with less morbidity than female sterilization

3. **Features associated with aortic stenosis include:**
   a. A systolic murmur which radiates to the neck
   b. Left ventricular hypertrophy on ECG
   c. Right axis deviation on ECG
   d. Fainting on exertion
   e. A collapsing pulse

4. **Regarding acute myocardial infarction (MI):**
   a. A diagnosis can be made solely on elevated troponin levels
   b. ST elevation of >2 mm in two or more chest leads associated with chest pain is diagnostic
   c. ST elevation in II, III and aVR would be consistent with an inferior lesion
   d. The ST segment can be usefully interpreted in people with left bundle branch block
   e. VF is a common arrhythmia following MI

5. **Causes of dyspnoea on exertion include:**
   a. Anaemia
   b. Pulmonary embolism
   c. Pulmonary hypertension
   d. Hypothyroidism
   e. Constrictive pericarditis

6. **A collapsing radial pulse may be associated with:**
   a. Hypertrophic cardiomyopathy
   b. Aortic reflux
   c. AV fistula
   d. Persistent ductus arteriosus
   e. Hypovolaemic states

7. **Risk factors of a pulmonary embolus include:**
   a. Long bone fracture
   b. First trimester abortion
   c. Malignancy
   d. Atrial fibrillation
   e. Thrombocytopenia

8. **In patients with gastrointestinal bleeding:**
   a. A digital rectal examination is mandatory to confirm or deny the presence of melaena
   b. A drop in haemoglobin is a marker of blood loss
   c. One blue venflon is sufficient for effective fluid resuscitation
   d. A normal BP in the presence of tachycardia is not a cause for concern
   e. Those with cirrhosis and oesophageal varices have a good prognosis

9. **Gastrointestinal causes of clubbing include:**
   a. Crohn's disease
   b. Cirrhosis
   c. Rectal cancer
   d. Irritable bowel syndrome
   e. Chronic pancreatitis

10. **During the examination of the abdomen:**
    a. Start at the site of the pain
    b. Palpate from the left iliac fossa (LIF) when examining the liver
    c. Palpate from the right iliac fossa (RIF) when examining the spleen
    d. Look for spider naevi
    e. Ascites can be detected by shifting dullness

11. **Regarding oesophageal cancer:**
    a. There is a higher incidence in Japan compared to the UK
    b. Five year survival is around 65%
    c. Patients present with difficulty swallowing solids more than liquids
    d. Twenty per cent of patients will have lymph node involvement at presentation
    e. Heavy smoking is a risk factor

12. **Patients with peptic ulcers are at an increased risk of:**
    a. Perforation
    b. Haemorrhage
    c. Short bowel syndrome
    d. Pancreatitis
    e. Peripheral vascular disease

13. **In patients with COPD:**
    a. Their activities of daily living are a useful guide in assessing their premorbid state in acute exacerbations
    b. Home oxygen should be used for at least 16 hours a day
    c. Home oxygen is prescribed to help with the symptomatic relief of shortness of breath
    d. An FEV_1 of <1 L when well would be grounds for an admission to ITU for artificial ventilation
    e. An arterial blood gas measurement prior to discharge is a useful prognostic indicator

14. **Physical features of hypercapnia include:**
    a. Peripheral vasodilation
    b. A resting tremor
    c. A low volume pulse
    d. Confusion progressing to drowsiness
    e. A raised alanine aminotransferase

15. **Causes of haemoptysis include:**
    a. Pulmonary oedema
    b. Pulmonary TB
    c. Infective exacerbation of COPD
    d. Oesophageal cancer
    e. Thrombophilia

16. **Regarding aspiration of a foreign body:**
    a. It is treated in babies by placing them head down on your arm and firmly patting them on the back
    b. The Heimlich manoeuvre is now no longer considered safe
    c. It can lead to lung abscess
    d. The foreign body typically goes down the left main bronchus
    e. The sensation of something stuck in the throat warrants laryngoscopy

17. **In the management of asthma:**
    a. Serial peak flows are a guide to treatment
    b. Volumatics are as efficacious as nebulizers
    c. You can be reassured by a normal arterial $PaCO_2$ in someone having an asthma attack
    d. The inability to complete a sentence is a concerning feature
    e. The incidence of asthma is falling in developed nations

18. **Risk factors for the successful completion of suicide include:**
    a. Traumatic means
    b. Young and female
    c. Schizophrenia
    d. Alcoholism
    e. Having children

19. **Common features of a significant depressive episode include:**
    a. A persistently low mood of one week's duration
    b. Anhedonia
    c. Flight of ideas
    d. Increased libido
    e. Early morning wakening

20. **Neurological causes of difficulty in walking include:**
    a. Tourette's syndrome
    b. Epilepsy
    c. Parkinson's disease
    d. Proximal weakness
    e. Cerebellar ataxia

21. **Common causes of headache include:**
    a. Temporomandibular joint dysfunction
    b. Alcohol excess
    c. Increased intracerebral pressure
    d. Migraine
    e. Middle ear infection

22. **When performing a lumbar puncture:**
    a. A CT scan excluding increased intracerebral pressure must be available
    b. Clotting abnormalities should be excluded
    c. The pressure of the cerebrospinal fluid (CSF) should be measured
    d. Oligoclonal bands are normally present in the CSF
    e. Cellulitis of the overlying skin is not a contraindication

23. **In patients with peripheral vascular disease:**
    a. Medical management is the first-line treatment
    b. Patients should be encouraged to exercise within their claudication distance
    c. Ankle : brachial pressure indices (ABPI) should be measured pre and post angioplasty
    d. As a PRHO you should take consent for angioplasty
    e. Diabetes is a risk factor for peripheral vascular disease

24. Definitions:

a. A hernia is the protrusion of a viscus or tissue out of the body cavity which contains it
b. A fistula is a normal communication between two epithelial surfaces
c. An ulcer is a break in the continuity of an epithelial surface
d. A sinus is a blind-ending tract communicating with an epithelially lined surface
e. An aneurysm is the abnormal dilation of a vein

25. In patients presenting with a fractured neck of femur:

a. Clinically they will have a shortened leg
b. The affected leg will be internally rotated
c. A positive family history is likely
d. They are at high risk of a deep venous thrombosis (DVT)
e. Minor trauma is a common cause

26. Diffuse alopecia:

a. Is associated with pernicious anaemia
b. Can be caused by iron deficiency
c. Can be caused by zinc deficiency
d. Can be caused by IV heparin administration
e. Is genetically determined in women

27. Acute tonsillitis:

a. An elevated anti-streptolysin titre suggests bacterial infection
b. Does not occur following tonsillectomy
c. The most common lymph glands affected are the preauricular glands
d. Is associated with a leucocytosis
e. Can cause referred pain to the ear

28. Deep vein thrombosis after hip replacement:

a. Is commoner in patients with an increased body mass index (BMI)
b. Usually leads to pulmonary embolism (PE)
c. Is easy to clinically diagnose
d. Can be prevented by the use of subcutaneous (SC) heparin prophylactically
e. Happens in more than 20% of cases

29. Urinary tract infections (UTIs) in children:

a. Are more common in boys than girls at six years of age
b. Are commoner in boys than girls in the newborn period
c. Can present as febrile convulsions
d. Can present as failure to thrive
e. May be diagnosed on microscopy

30. A small-for-dates baby is liable to suffer from:

a. Significant weight loss in the neonatal period
b. Meconium aspiration
c. Intrauterine hypoxia
d. Hypoglycaemia
e. Hyaline membrane disease

31. Breath-holding attacks:

a. Can be precipitated by frustration
b. Are a risk factor for the development of epilepsy
c. Rarely present in infants over the age of one
d. Can be associated with unconsciousness
e. Can be prevented with phenobarbital

32. The following are contained within the broad ligament:

a. The ovary
b. The ureter
c. The uterine artery
d. The fallopian tube
e. The ovarian artery

33. The management of bronchiolitis in infants includes:

a. A reducing course of hydrocortisone
b. Clarithromicin
c. Intravenous aminophylline
d. Humidified oxygen
e. Tube feeding

34. Stress incontinence of urine:

a. Should be investigated by cystoscopy before surgery
b. Can be a transient difficulty post delivery
c. Is commoner in multiparous women
d. Can be distinguished from urge incontinence by a cystogram
e. Can coexist with detrusor instability

35. Difficulty with swallowing can occur as a result of:

a. A cerebrovascular accident (CVA)
b. Recurrent larygneal nerve trauma
c. Carcinoma of the stomach
d. Dementia
e. Motor neuron disease

36. Absent ankle jerk reflexes can be caused by:

a. Syphilis
b. Diabetes mellitus
c. Gonorrhoea
d. Parkinson's disease
e. Motor neuron disease

37. In alcoholic liver disease:

a. The sensitivity to sedative drugs is increased
b. Prothrombin time (PT) is increased
c. Synthetic liver function is best assessed by clotting studies
d. The pattern of alcohol excess determines the development of liver disease
e. Liver damage is mainly due to dietary insufficiencies

249

38. The normal metabolic response to trauma includes:

a. The release of antidiuretic hormone
b. Potassium loss
c. A negative nitrogen balance
d. Glycolysis
e. A decrease in the lean body mass

39. In Colles' fracture, the distal radial fragment:

a. Is usually impacted
b. Is associated with median nerve damage
c. Is deviated to the radial side
d. Is ventrally angulated on the proximal radius
e. Is usually torn from the intra-articular triangular disc

40. Regarding multiple pregnancies:

a. They are commoner in younger women
b. They are associated with a shorter pregnancy
c. There is an increased risk of postpartum haemorrhage
d. They are usually monozygotic following IVF
e. Twin pregnancies occur in roughly 1:50 pregnancies

41. The following confirm ovulation:

a. In a 28-day cycle, blood progesterone level on day 21
b. In a 28-day cycle, blood oestrogen level on day 14
c. A basal body temperature drop of at least 0.5°C on day 12
d. 'Spinbarkeit' in cervical mucus
e. Histological examination of a premenstrual endometrial biopsy

42. Termination of pregnancy:

a. Must be approved by three different medical professionals
b. Can be performed with prostaglandins
c. Can be legally performed at any point in pregnancy
d. Can be safely performed by suction termination of pregancy before 14 weeks' gestation
e. Can be carried out with mifepristone (RU 486) up to 12 weeks' gestation

43. Following a medial nerve injury, the patient will demonstrate:

a. The inability to abduct the thumb
b. The inability to oppose the thumb and little finger
c. Altered sensation of the forearm
d. Pain in the upper arm
e. Sensory loss over the medial fingers

44. Regarding developmental milestones in childhood:

a. A 10-month-old will demonstrate opposition
b. A 12-month-old will release objects on request
c. A 3-month-old can roll from back to front
d. An 18-month-old can put three words together
e. A 4-month-old can localize sound

45. Successful suicide is:

a. Rare in young men
b. At a similar rate in all developed countries
c. An event which commonly occurs without warning
d. Associated with unemployment
e. Increased in mental illness

46. Schizophrenia:

a. Carries a 1 in 10 risk of lifetime suicide
b. Is strongly associated with violence to others
c. Is a lifelong diagnosis
d. Is influenced by the expressed emotion within the Family
e. If chronic is associated with marked cognitive defects

47. A swelling just below the angle of the mandible in a 25-year-old male may be:

a. Ectopic thyroid tissue
b. A thyroglossal cyst
c. An enlarged lymph node
d. A pharyngeal pouch
e. A carotid body tumour

48. The following are associated with hyperthyroidism:

a. Weight loss
b. Atrial fibrillation
c. Lid lag
d. Onycholysis
e. Feeling tired all the time

49. Dementia:

a. Is invariably progressive
b. If onset occurs before 75, is termed pre-senile
c. The sleep–wake cycle is preserved
d. May be caused by a head injury
e. Can be diagnosed with a mini-mental test score of 8 or less

50. In Parkinson's disease:

a. Urinary incontinence is an early symptom
b. The gait is typically shuffling
c. The onset of symptoms is typically before the age of 50
d. There is typically cog-wheel rigidity
e. Tremor is most noticeable on sustained movement

51. In asthma:

a. Nocturnal cough is not a symptom
b. Young patients suffering an asthma attack are always obviously unwell
c. Peak expiratory flow rate (PEFR) is a useful marker of disease activity
d. A history of atopy is not unusual in child asthmatics
e. A type 4 hypersensitivity reaction is involved

52. Deep vein thrombosis:

a. Is common in the first three days postoperatively
b. D-dimers can be raised for up to 6 weeks after an operation
c. In the lower limb can be assessed using the Wells' score
d. Can be treated with low molecular weight heparin
e. Is uncommon in patients with ovarian cancer

53. In the biliary system:

a. Gallstones are always symptomatic
b. Common bile duct dilatation of more than 10 mm indicates an obstruction
c. Ninety per cent of gallstones are radiolucent
d. Cholecystectomy prevents further gallstone formation
e. Impacted gallstones predispose to infection

54. In the confused patient:

a. Confusion can be caused by a raised serum calcium
b. Confusion due to a urinary tract infection can be treated empirically with trimethoprim
c. Such patients should be nursed in darkened rooms
d. Confusion is rarely caused by respiratory tract infections
e. Sedation is mandatory until the episode passes

55. When assessing the full blood count:

a. Haemoglobin may be slightly reduced in pregnancy
b. Raised white cell count is always due to infection
c. Platelets are raised in inflammatory processes
d. Red cell distribution width can demonstrate a dimorphic blood picture
e. Raised eosinophils can be seen in asthma

56. Renal failure:

a. Can be caused by inadequate fluid replacement postoperatively
b. Can be due to heart failure
c. Always requires dialysis
d. Is uncommon in hospital inpatients
e. Always causes a high serum potassium

57. In patients with pacreatitis:

a. Serum calcium of 1.90 mmol/L is a good prognostic marker in pancreatitis
b. Intravenous morphine should not be given as this causes contraction of the sphincter of Oddi
c. Fluid losses are large
d. Grey–Turner's sign is frequently present
e. They may lean forwards in bed to reduce their pain

58. With regard to autoimmune diseases and autoantibodies:

a. Anti-smooth muscle antibodies are associated with systemic lupus erythematosus
b. The presence of anti-double-stranded DNA antibodies is diagnostic of Sjögren's syndrome
c. Anti-mitochondrial antibodies can be found in primary biliary cirrhosis
d. Lambert Eaton myasthenic syndrome is associated with anti-LKM antibodies
e. Subsets of antinuclear antibodies can distinguish between limited and diffuse scleroderma

59. With regard to viral hepatitis:

a. Hepatitis A is a common cause of liver failure
b. The presence of anti-hepatitis B e antigen indicates that the patient is infectious
c. The presence of anti-hepatitis B s antibodies can be indicative of immunization
d. Hepatitis C is commonly transmitted by sexual contact
e. Hepatitis C rarely leads to chronic infection

60. With regard to ischaemic heart disease (IHD):

a. All patients with ischaemic heart disease should lower their cholesterol
b. Exercise tolerance testing can cause a myocardial infarction
c. The most predictive risk factor for IHD is a positive family history
d. It cannot be treated with heart transplantation
e. Should not be treated with beta blockers due to the risk of arrhythmia

61. Respiratory failure:

a. Is diagnosed by a PaO_2 of <8 kPa at sea level
b. Is classified type 2 if the $PaCO_2$ is consistently >4 kPa
c. Can be treated effectively by non-invasive ventilation
d. Can be caused by opioid analgesia
e. Due to COPD can be treated symptomatically with long-term oxygen therapy

62. With regard to diabetes:

a. Type II diabetics are at increased risk of diabetic ketoacidosis
b. It forms part of the metabolic syndrome
c. It can be treated with transplantation of the pancreas
d. It can cause delayed gastric emptying
e. It can cause autonomic neuropathy which can mask angina

63. With regard to the haemolytic anaemias:

a. Splenectomy is an appropriate treatment for hereditary spherocytosis
b. The normal lifespan of a red blood cell is 120 days
c. Haemolysis of red blood cells across a prosthetic heart valve is a common cause of haemolytic anaemia
d. Plasma haptoglobins are raised in haemolytic anaemia
e. Patients with longstanding haemolytic anaemia are at increased risk of gallstones

64. With regard to *Mycobacterium tuberculosis* infection:

a. Pulmonary TB usually affects the bases of the lungs
b. Cavitation is a result of *Aspergillus* infection
c. TB is the commonest cause of Addison's disease worldwide
d. The primary focus of infection may be known as the Ghon focus in the lung
e. Open TB refers to the absence of acid-fast bacilli in the sputum

65. With regard to septic shock:

a. Septic shock is defined as sepsis with organ dysfunction
b. Urinary catheterization allows surrogate measurement of renal perfusion
c. Confusion in a septic patient indicates cardiac insufficiency leading to poor cerebral perfusion
d. Patients will have an increased systemic vascular resistance due to compensation for third space fluid loss
e. Infected intravascular lines are a common cause of sepsis

66. With regard to calcium metabolism:

a. Paget's disease of bone is a potential cause of hypercalcaemia due to increased renal uptake
b. Hypercalcaemia in patients with malignancy may indicate bony metastases
c. Hypocalcaemia may be caused by phosphate retention in chronic renal failure
d. Chvostek's sign may be present in hypocalcaemia and is carpopedal spasm induced by inflation of a blood pressure cuff around the arm for greater than three minutes
e. Vitamin D deficiency can cause hypercalcaemia

67. With regard to polycystic kidney disease:

a. It is an autosomal recessive disorder
b. Seventy-five per cent of patients will have polycystic liver disease
c. Around 10% will have berry aneurysms of the intracranial arteries
d. Mitral valve prolapse is found in 40% of patients
e. Most patients are hypertensive

68. In nephrotic syndrome:

a. Reduced cholesterol is a feature
b. Urinary protein is greater than 4.5 g in 24 hours
c. Haematuria is common
d. Patients are at increased risk of bleeding due to loss of clotting factors
e. Hypoalbuminaemia is not a feature

69. Hyponatraemia:

a. Is common postoperatively
b. If due to syndrome of inappropriate antidiuresis, will give a serum osmolarity greater than 270 mOsmol/kg
c. Can be treated with furosemide
d. Can cause confusion, cramps and seizures
e. Can be caused by hypothyroidism

70. With regard to colorectal cancer:

a. If it is in the proximal colon, it can present with iron deficiency anaemia and weight loss
b. If it is in the rectum, it can cause tenesmus
c. Carcinoembryonic antigen (CEA) is used as a diagnostic test
d. Ulcerative colitis predisposes to colorectal cancer
e. Duke's grade D is the least invasive and carries the best prognosis

71. With regard to pancreatic cancer:

a. It is predominantly adenocarcinoma
b. It produces symptoms early in its natural history
c. Sixty per cent occurs in the head of the pancreas
d. It can be treated surgically by Whipple's procedure
e. It is a cause of hepatic jaundice

72. Crohn's disease:

a. Is characterized by superficial ulceration of the colon
b. Can cause fistula formation
c. May produce few examination findings
d. Has surgery as a mainstay of treatment
e. Is associated with ankylosing spondylitis

73. Iron deficiency:

    a. Is common in the elderly
    b. Has koilonychia as a feature of disease
    c. Can be caused by chronic gastrointestinal blood loss
    d. Causes a raised mean cell volume
    e. Should be treated with an intravenous iron infusion

74. With regard to bronchiectasis:

    a. It is characterized by fine crackles over the affected area
    b. Clubbing may be present
    c. It can be a sequela of pneumonia
    d. Plethora is rarely a feature
    e. Anaemia of chronic disease may be present

75. With regard to leg ulcers:

    a. Venous ulcers are rarely present in the gaiter area
    b. Arterial ulcers are commonly found on the ball of the foot
    c. Chronic ulcers are at risk of becoming cancerous
    d. Eczematous changes may be present in arterial disease
    e. Venous and arterial disease rarely coexist

Short-answer questions (SAQs)

It is important to only answer the question set, especially in short-answer questions. This is not only easier to mark – examiners are looking for specific points and will not give marks for superfluous information. For example, the first question only asks for features of severe aortic stenosis, so do not waste time discussing the auscultatory signs of aortic stenosis.

Only list three causes in any answer, as only the first three will score marks.

1. What features from the history and examination would suggest the presence of severe aortic stenosis in the presence of an ejection systolic murmur?

2. An elderly patient presents with cough and dyspnoea. What features on physical examination would favour a diagnosis of heart failure?

3. What features on the examination would alert you to a severe attack of asthma?

4. List three of the more common causes of a median nerve palsy at the wrist and describe the physical signs.

5. Describe the features of Parkinson's disease on physical examination.

6. A patient complains of acute lower back pain radiating down the legs. What signs on physical examination would raise the suspicion of an acute disc prolapse of L5–S1?

7. List three causes of massive splenomegaly. Describe the features of a palpable spleen on physical examination.

8. List seven features that may be found on examination of the upper limbs of a patient with longstanding rheumatoid arthritis.

9. List three major causes of papilloedema, and a suggestive feature for each on physical examination.

10. What mode of transmission is illustrated in Figures 1 and 2? Give two examples of each.

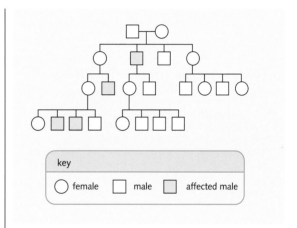

key

○ female □ male ▨ affected male

Fig. 1

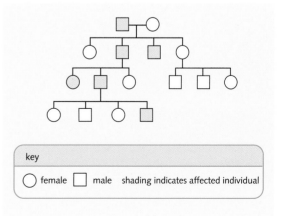

key

○ female □ male shading indicates affected individual

Fig. 2

11. A patient has a breast lump. Describe features on examination that would make you concerned that she has a carcinoma.

12. Give an example of opportunistic infections that may cause disease in the following:
    a. Eyes.
    b. Lungs.
    c. Central nervous system.
    d. Gastrointestinal tract.
    Give two groups of patients who may be susceptible to opportunistic infections.

13. A patient presents with anaemia. Give some examples of food that may provide a good dietary source of iron, folate and vitamin B_{12} to the patient. Give two other examples of causes for each type of anaemia.

14. How may a patient present with hypercalcaemia? Give four causes.

15. A nulliparous woman who is 30 weeks pregnant complains of swollen ankles. What are the three most likely causes? What features on clinical examination would help the differential diagnosis? Which investigations may exclude the more serious causes?

16. List five skin rashes that may be associated with an occult malignancy.

17. Indicate two features in the history and three findings on physical examination that may help to differentiate a hypoglycaemic coma from diabetic ketoacidosis in a diabetic patient.

18. What four questions might you ask a patient with suspected hypothyroidism? Indicate three discriminatory signs that may be present.

19. List four classes of drug that might cause hyperkalaemia and indicate the mechanism.

20. List four risk factors for human immunodeficiency virus (HIV) infection.

Extended matching questions (EMQs)

1. Theme: Limp (musculoskeletal)

a. Chondromalacia patellae
b. Compartment syndrome
c. Congenital dislocation of the hip
d. Fractured neck of femur
e. Gout
f. Osteoarthritis of hip
g. Osteoarthritis of knee
h. Perthe's disease
i. Slipped femoral epiphysis
j. Trochanteric bursitis

Instruction: For each scenario described below, choose the SINGLE most likely diagnosis from the above list of options. Each option may be used once, more than once or not at all.

1 On a GP visit to a nursing home, you are asked to see an 85-year-old woman with dementia and recurrent falls. She can only walk with assistance. On examination, when lying supine, she has a shortened externally rotated left leg. ☐

2 A 65-year-old man with high blood pressure and ischaemic heart disease presents with a hot erythematous knee joint. His regular medications include a thiazide diuretic. ☐

3 A 13-year-old boy presents with niggly pain in the hip which is much worse after being tackled at football. ☐

4 A 25-year-old man presents in A&E with a limp and excruciating pain in the lateral aspect of his lower leg following training for a marathon. ☐

5 A 58-year-old man presents to his GP with pain in his left knee of several months' duration. On examination he has a restricted range of movement in his left hip. ☐

2. Theme: Gastrointestinal bleeding

a. Anal fissure
b. Colonic carcinoma
c. Crohn's disease
d. Diverticular disease
e. Duodenal ulcer
f. Gastric carcinoma
g. Haemorrhoids
h. Laryngeal carcinoma
i. Mallory–Weiss tear
j. Oesophageal varices

Instruction: For each scenario described below, choose the SINGLE most likely diagnosis from the above list of options. Each option may be used once, more than once or not at all.

1 A 60-year-old man presents with altered bowel habit, crampy abdominal pain and blood mixed in with his stools. ☐

2 A 40-year-old woman presents with a long history of indigestion, worse recently, black tarry stools and feeling light-headed. ☐

3 A 32-year-old woman presents to her GP following childbirth, worried about the fresh, painless red blood on the toilet paper. ☐

4 A 22-year-old medical student presents to A&E with haematemesis and chest pain following celebrations on passing his finals. ☐

5 A 17-year-old woman with cystic fibrosis presents to A&E vomiting profuse volumes of fresh blood. ☐

3. Theme: Liver problems

a. Alcoholic liver disease
b. Biliary atresia
c. Gallstones
d. Hepatitis A
e. Hepatitis B
f. Hepatitis C
g. Metastatic liver disease
h. Physiological jaundice
i. Primary biliary cirrhosis
j. Wilson's disease

Instruction: For each scenario described below, choose the SINGLE most likely cause from the above list of options. Each option may be used once, more than once or not at all.

1 An obese 60-year-old man presents to the medical admissions unit following haematemesis with unsteadiness, a flapping tremor and erythematous palms. ☐

2 Two years after major abdominal surgery, a 75-year-old woman presents to her GP with weight loss and jaundice. On examination she has an enlarged knobbly liver. ☐

3 A 4-week-old baby girl is taken to see the GP by a concerned breast-feeding mother. The baby is deeply jaundiced. ☐

4 A 20-year-old soldier returning from the Gulf has abdominal pain, vomiting and pale stools with jaundice. His liver edge is just palpable, smooth and slightly tender. ☐

5 An overweight 45-year-old woman presents with right upper quadrant pain associated with meals. On examination she is slightly tender in the right upper quadrant on deep inspiration. ☐

4. Theme: Cardiac murmurs and added sounds

a. Aortic regurgitation
b. Aortic stenosis
c. Bicuspid aortic valve
d. Mitral regurgitation
e. Mitral stenosis
f. Patent ductus arteriosus
g. Transposition of the great vessels
h. Tricuspid regurgitation
i. Ventricular septal defect

Instruction: For each scenario described below, choose the SINGLE most appropriate cardiac lesion from the above list of options. Each option may be used once, more than once or not at all.

1 A 60-year-old man presents with heart failure. On examination he has a collapsing pulse, an early diastolic murmur and a displaced apex beat. ☐

2 A 56-year-old man gives a history of rheumatic fever. He has flushed cheeks, an irregularly irregular pulse and a mid-diastolic murmur. ☐

3 A 75-year-old woman presents to A&E following a drop attack. Her ECG shows left ventricular hypertrophy. On examination a harsh systolic murmur is heard over both carotid arteries. ☐

4 Three days following his myocardial infarction, Mr Thompson has a sudden deterioration, developing left ventricular failure. He has a new pansystolic murmur radiating from the apex to the axilla. ☐

5 A baby who is cyanotic at birth with a loud long systolic murmur in whom Fallot's tetralogy is suspected. ☐

5. Theme: Finger clubbing

a. Alcoholic liver disease
b. Crohn's disease
c. Emphysema
d. Familial clubbing
e. Fibrosing alveolitis
f. Infective endocarditis
g. Pulmonary abscess
h. Scleroderma
i. Squamous cell carcinoma

Instruction: For each scenario described below, choose the SINGLE most likely cause of finger clubbing from the above list of options. Each option may be used once, more than once or not at all.

1 A 40-year-old man with right-sided chest pain with weight loss of three months' duration and a grumbling pyrexia. ☐

2 A 20-year-old man with intermittent lower abdominal pain on defaecation associated with blood and mucus. ☐

3 A 48-year-old woman presents with dyspnoea. On examination she has bilateral, fine, late inspiratory crackles. ☐

4 An 18-year-old intravenous drug abuser who presents with fever and weight loss and who is short of breath. ☐

5 A 15-year-old boy who presents to his GP surgery with asthma. ☐

6. Theme: The painful joint

a. Ankylosing spondylitis
b. Fibromyalgia
c. Gout
d. Osteoarthritis
e. Pseudogout
f. Psoriatic arthritis
g. Reiter's syndrome
h. Rheumatoid arthritis
i. Septic arthritis
j. Systemic lupus erythematosus

Instruction: For each scenario described below, choose the SINGLE most likely cause of pain from the above list of options. Each option may be used once, more than once or not at all.

1 A 27-year-old man with low back pain and morning stiffness. He had an episode of iritis a year previously. ☐

2 A 40-year-old woman with symmetrically painful fingers and wrists. There is little evidence of synovitis. Her erythrocyte sedimentation rate is raised and she has an erythematous rash over her nose and cheeks. ☐

3 A 37-year-old man with gonococcal urethritis presents with pain in the hip and a limp. ☐

4 An 82-year-old woman presents with a painful right hip and knee, worse on exercise. On examination she has a positive Trendelenburg test and walks with a limp. ☐

5 A 75-year-old man with a painful great toe, one week after starting therapy for mild heart failure. ☐

7. Theme: Trauma

a. Airway obstruction
b. Cardiac tamponade
c. Flail chest
d. Haemothorax
e. Myocardial infarction
f. Open pneumothorax
g. Pulmonary contusion
h. Ruptured diaphragm
i. Ruptured oesophagus
j. Ruptured spleen
k. Tension pneumothorax
l. Traumatic aortic rupture

Instruction: For each scenario described below, choose the SINGLE most likely diagnosis from the above list of options. Each option may be used once, more than once or not at all.

1 An unconscious 30-year-old male with weak pulse who previously complained of left-sided pleuritic chest pain. Examination reveals the left chest to be hyper-resonant to percussion, with decreased breath sounds. ☐

2 A 20-year-old girl is brought in with a stab wound to the right side of her chest. She is in respiratory distress and air can be heard moving through the defect in the chest wall. ☐

3 A 60-year-old man is hit by a lorry while out walking his dog. He has injured his chest and is in obvious respiratory distress. Initial observations show he has a pulse of 120, blood pressure of 120/60 and respiratory rate of 45. Examination reveals paradoxical chest wall movement. ☐

4 A 3-year-old girl who has been eating peanuts and who now has a cough and stridor. ☐

5 A 38-year-old man is brought into A&E after a high-speed road traffic accident. He has chest pain and a blood pressure of 70/50 and a widened mediastinum on his chest X-ray. ☐

8. Theme: Heavy periods (or menstrual problems)

a. Adenomyosis
b. Break-through bleeding
c. Endometrial polyp
d. Fibroids
e. Hypothyroidism
f. Hyperthyroidism
g. Pelvic inflammatory disease
h. Physiological bleeding
i. Salpingitis
j. Thrombocytopenia
k. von Willebrand's disease

Instruction: For each scenario described below, choose the SINGLE most likely diagnosis from the above list of options. Each option may be used once, more than once or not at all.

1 A 34-year-old woman who presents with a 15-month history of worsening painless menorrhagia. The bleeding is so heavy she is having to take time off work. On examination she has a bulky uterus and is anaemic. ☐

2 A 36-year-old woman who complains of feeling tired all the time. She has been gaining weight and losing her hair. She now has heavy irregular periods. ☐

3 A 20-year-old woman presents to her GP four months after stopping the combined oral contraceptive pill. She has developed prolonged and heavy periods. Investigations and blood tests are all normal. ☐

4 A 25-year-old woman complains of lower abdominal pain and menorrhagia. Her smear shows *Chlamydia trachomatis*. ☐

5 A 17-year-old A-level student presents to her GP with fatigue and heavy periods. She is anaemic with a prolonged bleeding time. ☐

9. Theme: The diabetic eye

a. Arc light damage
b. Background retinopathy
c. Central retinal artery occlusion
d. Central retinal vein occlusion
e. Cortical blindness
f. Maculopathy
g. Optic atrophy
h. Pre-proliferative changes
i. Proliferative retinopathy
j. Snowflake cataracts
k. Snow blindness
l. Vitreous haemorrhage

Instruction: For each scenario described below, choose the SINGLE most likely diagnosis from the above list of options. Each option may be used once, more than once or not at all.

1 Fundoscopy in a 75-year-old woman with type 2 diabetes (non-insulin-dependent diabetes mellitus) demonstrates widespread microaneurysms, blot haemorrhages and hard exudates with macular sparing. ☐

2 A 50-year-old man with longstanding type 1 diabetes (insulin-dependent diabetes mellitus) presents to A&E with acute-onset complete loss of vision in the left eye. Fundoscopy reveals loss of the red reflex with grey haze obscuring the retina. ☐

3 Fundoscopy in a 29-year-old man with poorly controlled type 1 diabetes reveals microaneurysms and blot haemorrhages with hard exudates, cotton wool spots and neovascularization. ☐

4 A 68-year-old man presents to his GP with sudden loss of vision in his left eye. Examination reveals that acuity in the left eye is diminished to the perception of light only. The fundus is pale with a cherry red spot at the macula. ☐

5 A 57-year-old man with type 2 diabetes presents with sudden loss of vision in his left eye. Fundoscopy reveals a swollen optic disc with cotton-wool spots and haemorrhages across the retina. ☐

10. Theme: Delirium

a. Concussion
b. Diabetic ketoacidosis (DKA)
c. Encephalitis
d. Hepatic encephalopathy
e. Hypercalcaemia
f. Hypercapnia
g. Hyperosmolar, non-ketotic state
h. Hyponatraemia
i. Opiate analgesia
j. Subarachnoid haemorrhage
k. Subdural haematoma

Instruction: For each scenario described below, choose the SINGLE most likely cause of delirium from the above list of options. Each option may be used once, more than once or not at all.

1 A previously fit and well 13-year-old boy presents to A&E with a 24-hour history of increasing confusion. On examination he is unwell, smells of ketones and is dehydrated. ☐

2 A 70-year-old lifelong smoker who is attending the oncologist presents to the medical admission unit with a 48-hour history of increasing confusion and vomiting. His blood gases are normal but he is dehydrated. ☐

3 A 60-year-old smoker with a 75-pack year history presents with a productive cough and feeling short of breath. He is drowsy but rousable. He has a bounding pulse and a flapping tremor. ☐

4 An 82-year-old lady with a history of dementia and falls presents with increasing confusion and more frequent falls. She is drowsy with a mental test score of 2/10. She has a right-sided weakness and up-going plantars. ☐

5 A previously fit 25-year-old man presents to A&E with a severe headache and increasing confusion. While you are examining him, he starts to fit. While applying the oxygen mask, you notice a cold sore. ☐

11. Theme: Respiratory symptoms

   a. Acute asthma attack
   b. Aspiration pneumonia
   c. Asthma
   d. Bronchopneumonia
   e. Hayfever
   f. Inhaled foreign body
   g. Pleural effusion
   h. Pneumothorax
   i. Right basal pneumonia
   j. Tracheoesophageal fistula

Instruction: For each scenario described below, choose the SINGLE most likely option from the above list. Each option may be used once, more than once or not at all.

1 An 87-year-old man with known COPD and previous 60-cigarette pack year history presents with green sputum, shortness of breath and tachypnoea. Chest X-ray shows patchy changes over both mid-zones but no focal consolidation or effusions. ☐

2 A 12-year-old boy presents to his GP with sneezing and difficulty breathing after playing football in June. He also has eczema. ☐

3 A 24-year-old man presents with acute-onset shortness of breath while playing basketball. On examination you notice that the left side of his chest has reduced expansion with reduced breath sounds. His feet are overhanging the edge of the bed. ☐

4 A 55-year-old lady with recently diagnosed motor neuron disease presents with new cough and sputum over the past four days. She has coarse crackles at her right base and finds it difficult to communicate with you. ☐

5 A 26-year-old GP trainee presents to A&E having self-medicated with a five-day course of amoxicillin for a chest infection. Her sputum has subsided but she is coughing more and is short of breath. On examination she has reduced air entry and dullness over the left base. ☐

12. Theme: Pancreatitis

   a. Alcohol
   b. Endoscopic retrograde cholangiopancreatography (ERCP)
   c. Gallstones
   d. Grey-Turner's sign
   e. Hyperlipidaemia
   f. Pancreatic divisum
   g. Pancreatic necrosectomy
   h. Saponification
   i. Scorpion venom
   j. Trauma

Instruction: For each scenario described below, choose the SINGLE most likely option from the above list. Each option may be used once, more than once or not at all.

1 A 45-year-old care home assistant presents with acute onset of upper abdominal pain. His abdomen is guarded, with bowel sounds. He denies any alcohol consumption in the past four years. His serum amylase is 800 mmol/L; his serum sodium is 118 mmol/L. The lab calls to inform you of the low sodium, and you notice a milky flashback from the cannula in the man's hand.

2 A 23-year-old woman is involved in a traffic collision. Her motorcycle is hit from the side by a careless driver and she is found unconscious, lying on her back, by the ambulance crew. On admission, during the secondary survey, you notice she has bruising over her flanks.

3 A 47-year-old man presents with recurrent pancreatitis. He claims he is abstinent from alcohol. At endoscopic retrograde cholangiopancreatography (ERCP), the sphincter of Oddi cannot be cannulated, and the diagnosis is confirmed by magnetic resonance cholangiopancreatography (MRCP).

4 A 25-year-old man presents to hospital with severe abdominal pain having returned from holiday in Gran Canaria. His serum amylase and lipase are significantly elevated. He denies having had a drink since his return two days ago.

5 A 40-year-old woman is admitted to hospital for treatment of acute pancreatitis. She describes a recent history of sharp right upper quadrant pain that would last for two hours at a time.

13. Theme: Symptoms of heart failure

    a. Ankle swelling
    b. Cough
    c. Hepatomegaly
    d. Orthopnoea
    e. Paroxysmal nocturnal dyspnoea
    f. Pink frothy sputum
    g. Pulmonary fibrosis
    h. Pulmonary oedema
    i. Shortness of breath

Instruction: For each scenario described below, choose the SINGLE most likely option from the above list. Each option may be used once, more than once or not at all.

1 A patient has been on a cardiology ward for three weeks with a diagnosis of pulmonary oedema. He is receiving large doses of furosemide. His systolic blood pressure is 74 mmHg and increases on being placed head down in bed. On auscultation he has fine crackles throughout both lung fields. ☐

2 A patient complains of waking in the middle of the night and racing out of bed to the window, gasping for breath. ☐

3 A patient sleeps on four pillows at night so he can get his breath. ☐

4 A patient with a diagnosis of heart failure complains of this symptom several weeks after starting an angiotensin-converting enzyme (ACE) inhibitor. ☐

5 Congestion due to right heart failure can cause a nutmeg appearance and this symptom. ☐

14. Theme: Knee injuries

    a. Anterior cruciate ligament rupture
    b. Bucket handle tear of meniscus
    c. Dislocation of the knee
    d. Lateral collateral ligament rupture
    e. Medial collateral ligament rupture
    f. Meniscal tear
    g. Patella fracture
    h. Patella ligament rupture
    i. Posterior cruciate ligament rupture
    j. Tibial plateau fracture

Instruction: For each scenario described below, choose the SINGLE most likely option from the above list. Each option may be used once, more than once or not at all.

1 A 13-year-old girl presents with locking of the knee, having sustained previous knee injuries while playing netball. Arthroscopy diagnosis and treats the condition successfully. ☐

2 A 37-year-old man is hit side on by a van while riding his motorbike. You, the orthopaedic senior house officer, are called to see the man after several failed attempts at reducing his patella have been tried. There is obvious deformity of the knee. X-ray shows medial displacement of the femur on the tibia. ☐

3 A 52-year-old lady is involved in a car accident. She is hit by an uninsured driver as she crosses the road to work. Her knee is very painful and she is unable to weight bear. X-ray demonstrates the abnormality which is reported as 'Schatzker grade 3'. ☐

4 A 32-year-old man is brought into A&E having injured his right knee in a strong tackle playing football. The mechanism of injury is unclear. On examination pulses are present, there is no posterior sag but anterior draw is positive. ☐

5 A 30-year-old professional woman is brought into A&E. She fell down some stairs at a conference, landing on her right knee. She is in pain and is unable to straight leg raise. ☐

263

15. Theme: Hernias

a. Congenital diaphragmatic hernia
b. Epigastric hernia
c. Femoral hernia
d. Incarcerated hernia
e. Incisional hernia
f. Inguinal hernia
g. Obturator hernia
h. Rolling hiatus hernia
i. Sliding hiatus hernia
j. Spigelian hernia
k. Umbilical hernia

Instruction: For each scenario described below, choose the SINGLE most likely option from the above list. Each option may be used once, more than once or not at all.

1 A 64-year-old labourer presents to his GP with increasing pain in his groin. He has had a lump for several years which usually disappears on lying. It no longer does so. On examination you can feel a mass in the right groin arising above the inguinal ligament. It is irreducible but bowel sounds are present. ☐

2 A 57-year-old woman is referred to the surgeon who performed her laparoscopic cholecystectomy two years previously. She has developed a bulge, which on examination is in the mid-clavicular line in the right upper quadrant of the abdomen. It disappears on lying down and 'gurgles'. ☐

3 A 92-year-old frail lady is admitted to hospital with severe abdominal pain, colicky in nature, with nausea and vomiting. A nasogastric tube is passed, but on abdominal X-ray, no bowel distension is seen. Close inspection of the film reveals the diagnosis. ☐

4 A 31-year-old man presents to his GP with a mass in his left groin. It disappears on lying, is above the inguinal ligament and reappears on coughing. His brother had the same problem. ☐

5 A 29-year-old woman presents to her GP with an abdominal mass following the birth of her first child. During her pregnancy, she had had an everted umbilicus and this has worsened after giving birth. ☐

16. Theme: Nerves of the upper limb

a. Axillary nerve
b. Long thoracic nerve of Bell
c. Lower brachial plexus
d. Medial cutaneous nerve of forearm
e. Median nerve
f. Musculocutaneous nerve
g. Nerve to coracobrachialis
h. Radial nerve
i. Ulnar nerve
j. Upper brachial plexus

Instruction: For each scenario described below, choose the SINGLE most likely option from the above list. Each option may be used once, more than once or not at all.

1 A 14-year-old boy presents with an inability to grip in his right arm after falling from a tree and catching himself on a low-lying branch. ☐

2 A 30-year-old pregnant woman presents with numbness and reduced grip in her left hand, symptoms which are worse at night. On examination the numbness appears limited to the thumb, index and middle fingers. The thumb is particularly weak in flexion, opposition and abduction. ☐

3 A 25-year-old man presents to A&E with a broken arm following an arm wrestling contest. X-ray demonstrates a fracture of the proximal third of the humerus. He has wrist drop and has a loss of sensation over the webspace between the thumb and index finger. ☐

4 The same gentleman as above, on futher examination, has anaesthesia over the regimental patch and is unable to abduct his arm. ☐

5 A 58-year-old woman presents having had a mastectomy and axillary node clearance. On asking her to press her hands against the wall in front of her, you notice that the scapula on the affected side is more prominent, with the medial border standing proud. ☐

17. Theme: Headache

a. Bacterial meningitis
b. Benign intracranial hypertension
c. Cerebral abscess
d. Cerebral metastasis
e. Cluster headache
f. Giant cell arteritis
g. Subarachnoid haemorrhage
h. Subdural haemorrhage
i. Temporomandibular joint disease
j. Tension headache

Instruction: For each scenario described below, choose the SINGLE most likely option from the above list. Each option may be used once, more than once or not at all.

1 A 22-year-old woman presents to her GP with frontal headaches, worse in the morning. She has no other symptoms. On examination she has a body mass index of 32 with nothing else to note. Computed tomography of the head and lumbar puncture are normal. Her condition settles with dietary advice. ☐

2 A 60-year-old man presents to his GP with severe headaches over the frontal region and shoulders. The examination is unremarkable except for a thick band palpable over the right temple. The erythrocyte sedimentation rate is 97 mm/hour. ☐

3 A 54-year-old woman presents with severe headaches; she is completely incapacitated by the pain which is present over her right eye. She often sees zig-zag lines before the headache and they are so severe she presses over her eye to try and relieve the pain. These attacks last several hours and disappear as quickly as they start. ☐

4 A 27-year-old man complains of headaches to his GP. They are temporal and are relieved by sleeping. He is anxious as his company is merging and he risks losing his job. ☐

5 A 19-year-old medical student presents with headache and photophobia of three hours' duration. She is unable to touch her chest with her chin. There are no signs of raised intracranial pressure. The cerebrospinal fluid shows reduced glucose and raised lymphocyte count. ☐

18. Theme: Diabetes

a. Acanthosis nigricans
b. Diabetes insipidus
c. Diabetic ketoacidosis
d. Fasting serum glucose of >7.1 mmol/L
e. Glycosylated haemoglobin
f. Hyperosmolar non-ketotic coma
g. Pyoderma gangrenosum
h. Random serum glucose >7.1 mmol/L but below 12 mmol/L
i. Type 1 diabetes mellitus
j. Type 2 diabetes mellitus

Instruction: For each scenario described below, choose the SINGLE most likely option from the above list. Each option may be used once, more than once or not at all.

1 A patient presents with polydipsia and polyuria, often drinking 15–20 L of fluid per day. Fasting serum glucose is 5.2 mmol/L; serum osmolarity is reduced. ☐

2 A 56-year-old man presents with a one week history of polyuria and polydipsia; he has been drinking copious amounts of cola to rehydrate himself and feels very thirsty. Serum glucose is 37 mmol/L. ☐

3 A 14-year-old boy presents with a two day history of lethargy, shortness of breath and polydipsia. His GP started amoxicillin for a chest infection three days ago. On examination he is very dehydrated, smells of nail varnish remover and is making deep sighing breaths. ☐

4 A patient is diagnosed with impaired glucose tolerance. ☐

5 This diagnosis is associated with a total lack of insulin production and presents in the young. ☐

19. Theme: Cancer

a. Carcinoid syndrome
b. Cerebral metastases
c. Ectopic adrenocorticotropic hormone syndrome
d. Erythema gyratum repens
e. Hepatic cell carcinoma
f. Hypercalcaemia
g. Mesothelioma
h. Oesophageal carcinoma
i. Small cell lung cancer
j. Squamous cell lung cancer

Instruction: For each scenario described below, choose the SINGLE most likely option from the above list. Each option may be used once, more than once or not at all.

1 A 72-year-old woman with known non-small cell lung cancer presents with a two day history of confusion, abdominal pain and constipation and on examination is dehydrated. ☐

2 A 56-year-old man presents with difficulty swallowing of several months' duration. He has early satiety and nausea with weight loss and a hoarse voice. On examination he has increased keratinization of the soles of his feet. ☐

3 A 48-year-old man presents with abdominal pain, flushing of the face and diarrhoea. On examination he has marked hepatomegaly. On compressing the liver during examination, facial flushing is evident. ☐

4 A 68-year-old lady with known small cell lung cancer presents with obesity and increased hair growth. On examination her face is round and she has elevated blood pressure. Her random blood glucose is 14 mmol/L. ☐

5 A 62-year-old ex-shipyard boiler maker presents with shortness of breath, cough and haemoptysis. His occupation has exposed him to asbestos. ☐

20. Theme: Diarrhoea

a. ACE inhibitor
b. Amoebic dysentery
c. β-Blocker
d. Cancer of the colon
e. Diverticular disease
f. Giardiasis
g. Irritable bowel syndrome
h. Metformin
i. Ulcerative colitis
j. Zollinger–Ellison syndrome

Instruction: For each scenario described below, choose the SINGLE most likely option from the above list. Each option may be used once, more than once or not at all.

1 A 23-year-old man with type A personality presents with a change in bowel habit. He has alternating constipation with pellet-like stool and diarrhoea. He occasionally passes mucus but never any blood. ☐

2 A 55-year-old man with newly diagnosed diabetes complains of diarrhoea after his last consultation with you. ☐

3 A 68-year-old man presents with diarrhoea alternating with constipation and passage of blood per rectum. He has no weight loss or tenesmus. Colonoscopy demonstrates no masses but makes the diagnosis. ☐

4 A 25-year-old woman presents with profuse diarrhoea with passage of blood and mucus. She describes cramping abdominal pains. Examination reveals left iliac fossa tenderness. She has also lost weight. ☐

5 A 45-year-old man presents with diarrhoea in association with acid reflux. Gastroscopy reveals multiple duodenal ulcers, and serum gastrin levels are 10 times the upper limit of normal. ☐

21. Theme: Electrocardiograph findings

a. Anterior myocardial infarction
b. Atrial fibrillation
c. Atrial flutter with 2:1 block
d. Digitalis effect
e. Inferior myocardial infarction
f. Pericarditis
g. Posterior myocardial infarction
h. Sick sinus syndrome
i. Third-degree heart block
j. Ventricular tachycardia

Instruction: For each scenario described below, choose the SINGLE most likely option from the above list. Each option may be used once, more than once or not at all.

1 ST depression in V1–3 and tall R wave in V1 and V2. ☐

2 ST elevation that is present in all leads and is saddle shaped. ☐

3 ST elevation of 3 mm in leads II, III and aVF. ☐

4 Broad complex QRS with a rate of 45 bpm; p waves are present. ☐

5 No p waves, narrow complex QRS rate of 150 bpm with sawtooth baseline. ☐

22. Theme: Blood test results

a. Alkaline phosphatase of 2000 mmol/L
b. High white cell and C-reactive protein
c. Low caeruloplasmin levels
d. Low haemoglobin and raised C-reactive protein
e. Markedly high prothrombin time
f. Normal bilirubin with raised alkaline phosphatase and gamma glutamyltransferase
g. Positive anticardiolipin antibodies
h. Raised haemoglobin and ferritin levels
i. Serum potassium of 6.6 mmol/L
j. Serum sodium of 120 mmol/L; urinary sodium of 25 mmol/L

Instruction: For each scenario described below, choose the SINGLE most likely option from the above list. Each option may be used once, more than once or not at all.

1 A 67-year-old woman is admitted to the ward with a chest infection. She is producing green sputum and has a right basal pneumonia on her chest X-ray. ☐

2 A tanned 40-year-old man presents to his GP with polyuria and polydipsia. He denies any foreign travel in the past two years. ☐

3 A 78-year-old lady fractures her hip. In theatre, she loses 1500 mL of blood and is given fluid, but no blood has been cross-matched. The anaesthetist sends blood samples and a request for cross-matching. ☐

4 A medical emergency requires administration of calcium gluconate, electrocardiogram and an insulin and dextrose infusion. ☐

5 A 46-year-old man is an elective admission for laparoscopic cholecystectomy. ☐

23. Theme: Liver and biliary tree anatomy

a. Bare area of the liver
b. Calot's triangle
c. Caudate lobe
d. Coronary ligament
e. Cystic duct
f. Falciform ligament
g. Hepatoduodenal ligament
h. Porta hepatis
i. Quadrate lobe
j. Rouviere's sulcus

Instruction: For each scenario described below, choose the SINGLE most likely option from the above list. Each option may be used once, more than once or not at all.

1 A fissure lying between the right lobe and caudate process. ☐

2 The anterior border of the epiploic foramen of Winslow. ☐

3 Corresponds to vascular segment 1 of the liver. ☐

4 Lies anterior to the upper pole of the right kidney. ☐

5 Formed by the cystic artery, hepatic duct and cystic duct. ☐

24. Theme: Psychiatry

a. Agoraphobia
b. Anorexia nervosa
c. Depression
d. Drug-induced psychosis
e. Generalized anxiety disorder
f. Lewy body dementia
g. Obsessive compulsive disorder
h. Personality disorder
i. Post-traumatic stress disorder
j. Schizophrenia

Instruction: For each scenario described below, choose the SINGLE most likely option from the above list. Each option may be used once, more than once or not at all.

1 A 72-year-old man complains that the staff in his nursing home have stolen his false teeth. He states that he is sick of seeing the same army regiment marching through the living room: 'Haven't they got anything else better to do?' On examination he has a pill rolling tremor and a mask-like face. ☐

2 A 38-year-old man is dragged to his GP by his wife. He lost his job six months ago and since then has 'given up'. He states that he thinks he is worthless: 'That's why they sacked me, can't blame them really!' He wakes at 5 am each morning but lies in bed all day. He has little to eat and has marked weight loss. ☐

3 A 16-year-old boy is admitted to A&E having been found by his parents covering the windows with tin foil. He states that he is in mortal danger and the tin foil is the only thing that can prevent the 'death rays' from penetrating the windows and killing him. His parents deny any drug use; however you notice a tobacco tin in his pocket. When questioned, he states that he has been smoking cannabis. ☐

4 A 42-year-old man presents to his GP with difficulty sleeping. He has problems falling asleep and wakes early each morning. He has nightmares and flashbacks to a car accident he had several months ago. ☐

5 A 'nervous' 32-year-old woman visits her GP. She complains of shaking, generalized aches and trouble concentrating and has had to take time off work. She feels very uncertain about her future and often has a feeling that something very bad is about to happen. ☐

25. Theme: Heart disease

a. Acute myocardial infarction
b. Aortic stenosis
c. Atrial myxoma
d. Complete heart block
e. Cor pulmonale
f. Dilated cardiomyopathy
g. Fibrinous pericarditis
h. Infective endocarditis
i. Mitral valve stenosis
j. Tricuspid incompetence

Instruction: For each scenario described below, choose the SINGLE most likely option from the above list. Each option may be used once, more than once or not at all.

1 A 58-year-old woman with known asymptomatic aortic stenosis presents with fever and feeling generally unwell. She has just had three fillings performed by her dentist. ☐

2 A 49-year-old man presents to his GP with shortness of breath on walking up hills. He also experiences central chest pain which is relieved by resting, as well as occasional dizzyness. He has a loud systolic murmur. ☐

3 On the second day post myocardial infarction, a 71-year-old man develops pleuritic sounding chest pain. This is relieved by ibuprofen. The ECG demonstrates saddle-shaped ST changes. ☐

4 A 63-year-old man presents to A&E with shortness of breath. He has a loud pansystolic murmur and a displaced apex beat. Chest X-ray demonstrates a straight left heart border and cardiothoracic ratio of >50%. He admits to previous excess alcohol intake. ☐

5 A 44-year-old woman attends her respiratory clinic appointment. She has known COPD and is a previous heavy smoker. Her most recent symptoms are ankle swelling and an exercise tolerance of 15 yards. ☐

26. Theme: Strokes

a. Lacunar infarct
b. Partial anterior circulation infarct (PACI)
c. Partial anterior circulation syndrome (PACS)
d. Partial occipital circulation syndrome (POCS)
e. Pontine stroke
f. Posterior inferior cerebellar artery (PICA) occlusion
g. Total anterior circulation syndrome (TACS)
h. Transient ischaemic attack (TIA)
i. Vertebrobasilar insufficiency

Instruction: For each scenario described below, choose the SINGLE most likely option from the above list. Each option may be used once, more than once or not at all.

1 A 60-year-old hypertensive man with ischaemic heart disease presents to A&E with a 'funny turn'. On examination he has right hemiparesis, right inattention and a right homonymous hemianopia. ☐

2 A 72-year-old diabetic woman with renal disease presents to A&E with a right hemiparesis affecting her right arm and leg; sensation is normal on the right and left. ☐

3 A 55-year-old obese man is admitted to the ward following a stroke. On admission he was found to have loss of pain and temperature sensation on the left side of his body and the right side of his face, nystagmus to the right and a right Horner's syndrome. ☐

4 A 90-year-old woman attends her GP following an episode of weakness that affected her right side – she was unable to stand or use her arm. She is quite distressed, and worries that the symptoms may return. On questioning, the episode seems to have lasted around four hours. ☐

5 A 78-year-old man is admitted with a stroke. Computed tomography demonstrates an infarct in the anterior circulation. On examination he has a right hemiparesis and a right homonymous hemianopia. ☐

27. Theme: Breast lumps

a. Breast abscess
b. Cyst
c. Duct ectasia
d. Fat necrosis
e. Fibroadenoma
f. Fibrocystic change
g. Galactorrhoea
h. Lipoma
i. Mastitis
j. Papilloma

Instruction: For each scenario described below, choose the SINGLE most likely option from the above list. Each option may be used once, more than once or not at all.

1 A 19-year-old woman presents with a highly mobile lump in her breast. Fine needle aspiration shows no malignancy. ☐

2 A 28-year-old woman, four weeks post partum, presents to her GP with a tender lump in the upper outer quadrant of her left breast. On examination it is ill-defined, erythematous and hot to touch. ☐

3 A 38-year-old very anxious woman presents with a hard irregular lump in her breast. On examination it is craggy and well-defined. Fine needle aspiration shows no evidence of malignancy. On further questioning she reveals that she was involved in a car crash six months previously. ☐

4 A 42-year-old woman presents with a yellow-brown discharge from her nipples. On examination her discharge appears to be coming from multiple ducts. Examination is otherwise normal and mammography is unremarkable. ☐

5 A 32-year-old woman presents with a lump in her breast which is painless. On examination it is well demarcated, fluctuant and transilluminable. ☐

28. Theme: Rashes

a. Acanthosis nigricans
b. Atopic eczema
c. Dermatitis herpetiformis
d. Erythema ab igne
e. Herpes zoster
f. Necrobiosis lipoidica
g. Pemphigoid
h. Pemphigus
i. Rosacea
j. Stevens–Johnson syndrome

Instruction: For each scenario described below, choose the SINGLE most likely option from the above list. Each option may be used once, more than once or not at all.

1 A 28-year-old woman presents to A&E with a one-week history of shortness of breath, green sputum and cough. On further questioning she admits to having had rigors, and in the last few hours has developed a blistering rash affecting her whole body, including her lips. ☐

2 An 80-year-old man presents with a blistering rash over the right side of his head. It is sharply cut off at the tragus and includes the eye and tip of the nose. It does not cross the midline. ☐

3 A known diabetic gentleman presents to his GP with a thick, waxy rash over the anterior aspect of his shin. ☐

4 A 76-year-old lady presents to her GP with blisters over her trunk. On examination they are tense and range in size from 2 mm to 2 cm. They are not present anywhere else. ☐

5 A 19-year-old student presents with diarrhoea and weight loss over the preceding 6 weeks. He admits to having a blistering rash over his knees and elbows. Anti-endomysial antibodies are strongly positive. ☐

29. Theme: Acute abdomen

a. Acute pancreatitis
b. Acute pyelonephritis
c. Biliary peritonitis
d. Mesenteric adenitis
e. Perforated appendix
f. Perforated peptic ulcer
g. Perforated sigmoid diverticulus
h. Ruptured abdominal aortic aneurysm
i. Small bowel obstruction
j. Superior mesenteric artery occlusion

Instruction: For each scenario described below, choose the SINGLE most likely option from the above list. Each option may be used once, more than once or not at all.

1 A 25-year-old man is admitted with severe upper abdominal pain radiating to the back. He has nausea and vomiting and is clearly dehydrated. He returned from holiday in Ayia Napa earlier today. ☐

2 A 33-year-old woman is admitted with a 24-hour history of nausea, vomiting and absolute constipation. Her abdomen is tympanic with increased high-pitched bowel sounds. ☐

3 A 57-year-old man is admitted with onset of abdominal pain initially in the left upper quadrant. He has been self-medicating with ibuprofen after spraining his left ankle playing squash. On examination the abdomen is rigid with no bowel sounds. ☐

4 A 24-year-old woman is admitted with fever, rigors and severe back pain radiating to her left groin. ☐

5 A 10-year-old boy is admitted with right iliac fossa pain of several hours' duration. Initially he has guarding and rebound tenderness. Pressing in the right iliac fossa also elicits the same pain. He rapidly progresses to having a tense abdomen. ☐

30. Theme: Surgical incisions

a. Clamshell
b. Corkscrew thoracoabdominal incision
c. Gridiron
d. Kocher's
e. Lanz
f. Lower midline
g. Median sternotomy
h. Pfannenstiel
i. Rooftop
j. Rutherford–Morrison

Instruction: For each scenario described below, choose the SINGLE most likely option from the above list. Each option may be used once, more than once or not at all.

1 Incision for cadaveric liver transplantation. ☐

2 Incision used for open cholecystectomy. ☐

3 Incision for open appendectomy. ☐

4 Incision used for coronary artery bypass surgery. ☐

5 Incision used for access to the sigmoid colon. ☐

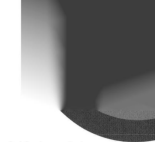

1.
a. True — The spiral arteries of the placenta are always maximally dilated so perfusion is dependent on maternal blood pressure.

b. False — In the fetal circulation, blood flows from the right to left atrium via the foramen ovale.

c. False — About 5% of the in-utero cardiac output goes through the lungs.

d. True — The oxygenated blood of the fetus arises from the placenta rather than the lungs.

e. False — Five per cent of the cardiac outflow goes round the lungs in utero and the first breath decreases the pulmonary arterial pressure.

2.
a. True — Since Eisenmenger's is a right-to-left shunt with pulmonary hypertension, any decrease in left-sided pressure (delivery) leads to an increased shunt, which often precipitates death.

b. False — Focal migraine is a contraindication to the combined oral contraceptive due to the increased risk of cerebrovascular accident.

c. False — A condom is less efficacious than the diaphragm.

d. True — Progesterone-only contraceptives lead to delayed return to fertility.

e. True — One is an intra-abdominal procedure requiring general anaesthetic, the other can be done under local anaesthetic.

3.
a. True — All aortic murmurs can radiate to the neck.

b. True — The increased work of forcing blood through a stenosed valve leads to hypertrophy of the left ventricle.

c. False — You would expect left axis deviation with aortic stenosis.

d. True — As the pressure gradient across the valve becomes greater, patients get syncopal symptoms.

e. False — A collapsing pulse is associated with aortic regurgitation.

4.
a. False — A myocardial infarction (MI) can be diagnosed if two of the following three features are present: chest pain, ST changes and enzyme rises.

b. True — Those would be criteria for thrombolysis.

c. False — An inferior MI is associated with ST elevation in II, III and aVF.

d. False — Left bundle branch block precludes useful interpretation of the ST segments.

e. True — The origins of CPR/arrest team protocols in the 1960s stem from the recognition that MI patients get arrhythmias (these are particularly common in inferior MIs).

5.
a. True — Haemoglobin carries oxygen; not enough haemoglobin, not enough oxygen, so you feel short of breath.

b. True — Again, a significant pulmonary embolism will impair oxygen exchange so not enough oxygen gets into the blood.

c. True — People with raised pulmonary arterial pressure become short of breath.

d. True — Both hypo- and hyperthyroid patients can become short of breath.

e. False — Constrictive pericarditis does not cause exertional dyspnoea.

6.
a. False — A collapsing pulse is not associated with hypertrophic cardiomyopathy.

b. True — Aortic reflux is a recognized cause of a collapsing pulse.

c. True — Depending on the site of the fistula, it may be associated with a collapsing radial pulse.

d. True — Persistent ductus arteriosus can cause a collapsing radial pulse.

e. False — A collapsing pulse is not associated with hypovolaemia; the classic signs would be a weak and thready pulse.

7.
a. True — Trauma and immobility are risk factors for a pulmonary embolus.

b. True — Pregnancy at any stage is a risk factor for a pulmonary embolus.

c. True — Malignancy leads to altered coagulation states which may precipitate deep venous thrombosis, hence pulmonary embolus.

d. True — People in atrial fibrillation get warfarinized to prevent pulmonary embolus/cerebrovascular accident.

e. False — Thrombocytopenia is not a risk factor for pulmonary embolus.

8.
a. True — If you don't put your finger in it, you'll put your foot in it!

b. True — That's also why you check a patient's haemoglobin postoperatively.

c. False — A blue venflon is small and these patients should have two large-bore cannulae.

d. False   Tachycardia is a serious sign of shock and in itself worrying.

e. False   People with varices have a worse prognosis than those who don't.

9. a. True   There are many causes of clubbing – learn them.

b. True   See above.

c. False   Rectal cancer is not a cause of clubbing.

d. False   Irritable bowel syndrome is not a cause of clubbing.

e. False   Chronic pancreatitis is not a cause of clubbing.

10. a. False   Start palpation away from any pain to gain the patient's trust.

b. False   Start in the RIF for palpation of the liver.

c. True   Start in the RIF and work up to the left upper quadrant.

d. False   Spider naevi follow the distribution of the superior vena cava and are, therefore, only seen on the upper trunk.

e. True   Fluid thrills and shifting dullness may both be used to detect ascites (an ultrasound scan is good too!).

11. a. True   It's commoner in Japan so they screen for it.

b. False   Oesophageal cancer has a poor five year survival (less than 30%).

c. True   Difficulty swallowing solids tends to be caused by an obstructive pathology whereas difficulty swallowing liquids tends to be a neurological problem.

d. False   Fifty per cent of patients have nodal involvement at presentation.

e. True   Smoking is a risk factor for most diseases.

12. a. True   If you've got a partial area of weakness, you're more likely to breach it.

b. True   If the area of weakness erodes a blood vessel, you bleed.

c. False   Surgery for a peptic ulcer will not remove so much of the bowel that 'short bowel syndrome' occurs.

d. False   Peptic ulcers are not associated with pancreatitis.

e. False   Peptic ulcers are not associated with peripheral vascular disease.

13. a. True   What people do day-to-day helps you judge not only their physical fitness but, to some extent, their quality of life.

b. True   Home oxygen should be used for 16 hours a day to prevent the development of cor pulmonale.

c. False   Home oxygen should not be used primarily to relieve breathlessness (it should be used to prevent the development of cor pulmonale), although it may do this through its effect on oxygen levels.

d. False   An FEV_1 of <1 L denotes very poor lung function and such a patient would struggle to get off a ventilator.

e. False   An arterial blood gas sample should be done two to three months post discharge.

14. a. True   Hypercapnia leads to vasodilation so you see warm peripheries and large veins.

b. True   You see a tremor with hypercapnia (all tremors are easier to see if you ask patients to close their eyes).

c. False   Hypercapnia can cause a bounding pulse.

d. True   Progressive hypercapnia leads to confusion and drowsiness. These are markers of severe disease process.

e. False   Hypercapnia does not affect liver function tests.

15. a. True   The arterial/venous engorgement you see with left ventricular failure can lead to haemoptysis.

b. True   Classically, people with pulmonary TB have haemoptysis.

c. True   Any infective process in the lungs can lead to haemoptysis.

d. False   Oesophageal cancer will not cause someone to cough up blood.

e. False   Haemoptysis is not associated with thrombophilia.

16. a. True   This is basic first aid – as a doctor you will still be asked about first aid.

b. True   However, the Heimlich manoeuvre remains in the advanced life support and advanced paediatric life support guidelines.

c. True   If the foreign body gets lodged and isn't removed, then an abscess may well develop.

d. False   Foreign bodies typically go down the right main bronchus.

e. True   People who feel they have something stuck very often do so – they should be referred for an ENT opinion as they may well warrant laryngoscopy.

17. a. True   Serial peak flows should be recorded on a peak flow chart – they are an extremely useful guide to a person's asthma and how well controlled it is.

b. True   It's true that volumatics are as good as nebulizers – hospitals like nebulizers because they take less nursing time to administer!

c. False $PaCO_2$ should drop due to hyperventilation. If it is normal it may suggest that the patient is getting tired.

d. True   If someone is so short of breath that they can't talk, they are at risk of a respiratory arrest.

e. False   The incidence of asthma is rising in the developed world.

18. a. True   Traumatic means tend to be fatal, e.g. jumping, hanging, shooting – people with borderline personalities tend not to use such methods.

b. False   Being young and female does not increase your risk of completing suicide.

c. True   Schizophrenia carries a 1 in 10 lifetime risk of suicide.

d. True   Being an alcoholic does increase your chance of completing suicide.

e. False   Having children is a protective factor.

19. a. False   A persistently low mood of two weeks' duration is suggestive of depression.

b. True   People have less energy and don't enjoy things that they used to when they are depressed.

c. False   Flight of ideas is a feature of mania.

d. False   Increased libido is associated with mania.

e. True   People with depression have a disturbance in their sleep cycle and often wake up early in the morning (NB: make sure the patient isn't supposed to get up at 4 a.m., e.g. a postman).

20. a. False   Tourette's syndrome affects people's speech.

b. False   Epilepsy doesn't affect your gait.

c. True   The shuffling gait is a classic feature of Parkinson's disease.

d. True   If you have a proximal weakness, it will affect how you walk.

e. True   If you can't coordinate your movement, walking will be difficult.

21. a. True   Temporomandibular joint dysfunction does lead to headaches – check the origin and radiation of the patient's pain.

b. True   Where do you think hangovers originate?

c. False   While an increased intracerebral pressure is associated with headache, it is not common.

d. True   You surely know someone who suffers from migraine and complains of a headache.

e. True   Middle ear problems (including infection) can cause headache.

22. a. True   Have a CT report available so that you know the patient won't cone on the end of your needle.

b. True   Causing a spinal haematoma is a 'no-no'.

c. True   You should measure the CSF opening pressure to confirm there is no intracerebral pressure.

d. False   Oligoclonal bands are associated with multiple sclerosis.

e. False   Cellulitis over the area is a contraindication as you do not want to cause a subarachnoid infection.

23. a. True   Medical management is the first-line management of virtually everything!

b. True   Exercising within the claudication distance will help slow the progression of the disease.

c. True   Do the ABPI pre and post angioplasty to document any improvement. It makes assessment of the patient in clinic in six months' time easier if they complain of their symptoms recurring.

d. False   As a house officer, you cannot consent patients for procedures.

e. True   Diabetic patients develop peripheral vascular disease.

24. a. True   That's the definition.

b. False   A fistula is an abnormal communication between two epithelial surfaces.

c. True   That's the definition.

d. True   That's the definition.

e. False   An aneurysm is an arterial pathology.

25. a. True   A patient will have a shortened and externally rotated leg.

b. False   See above.

c. False   A family history does not increase or decrease the likelihood of this pathology.

d. True   All post-op or immobile patients are at an increased risk of DVT.

e. True   A simple trip is a common cause of a fractured neck of femur. It is often associated with osteoporosis.

26. a. False   Diffuse alopecia is not associated with pernicious anaemia.

b. True   It can be caused by iron deficiency.

c. False   Zinc deficiency is associated with the disturbance of nail growth.

d. True   Intravenous heparin can disturb the growth of hair.

e. True   It is genetically determined in women.

27. a. True — Streptococcus is a bacterium, so an elevated titre suggests infection.

b. True — If your tonsils have been removed then they cannot become infected (although tonsillectomy will not prevent an acute red throat).

c. False — The most common glands affected are the submandibular glands.

d. True — With infection, you get an increase in the number of white blood cells.

e. True — Tonsillitis not only causes a sore throat, but the pain may be referred to the ear.

28. a. True — If you have an elevated BMI, you are likely to be less mobile and hence at increased risk.

b. False — Most DVTs do not break off and cause PE.

c. False — They are very difficult to diagnose clinically, hence all the ultrasound scanning that gets done.

d. False — The use of SC heparin decreases the risk but does not remove it, and they are a lot more common than you may think.

e. True — It is alarmingly common.

29. a. False — Urinary tract infections are more common in girls than boys at six years of age.

b. True — Boys are more likely to get UTIs in the newborn period compared to when they are older.

c. True — Any infection in children may present as a febrile convulsion.

d. True — Recurrent UTIs can cause failure to thrive and it is the failure to thrive that can be picked up first.

e. True — Look at the urine down the microscope – if you can see bacteria then there is a UTI.

30. a. False — Small-for-dates babies do not suffer from significant weight loss after they are born.

b. True — Small-for-dates babies are at risk of fetal distress which increases their likelihood of meconium aspiration.

c. True — See above.

d. True — Small-for-dates babies are not likely to have laid down enough glycogen stores as they have been expending all their energy on growing.

e. False — Hyaline membrane disease is a feature of prematurity.

31. a. True — Breath-holding attacks can be precipitated by frustration in the same way tantrums are.

b. False — Breath holding is not associated with epilepsy.

c. False — A one-year-old baby is too immature to attempt to breath hold.

d. True — Children can hold their breath to the point that they lose consciousness.

e. False — Phenobarbital won't help to control them. Parents need to be counselled how to cope.

32. a. False — The broad ligament does not involve the ovary.

b. False — The broad ligament does not involve the ureter.

c. True — The broad ligament does involve the uterine artery.

d. True — The broad ligament does involve the fallopian tube.

e. True — The broad ligament does involve the ovarian artery.

33. a. False — Hydrocortisone isn't effective in bronchiolitis.

b. False — As viruses cause bronchiolitis, antibiotics are of little use.

c. False — An IV aminophylline infusion does not help in infancy.

d. True — Humidified oxygen is always better tolerated as it does not lead to the drying of the upper airways.

e. True — Feeding is a lot of work for infants (you can't eat and breathe at the same time) and tube feeding decreases the amount of work.

34. a. False — Urodynamics should be done prior to surgery and cystoscopy will add nothing in stress incontinence.

b. True — With all the stretching of the pelvic floor during childbirth, it can take some time for things to return to normal.

c. True — The more stretching, the more problems.

d. True — A cystogram can distinguish urge from stress incontinence.

e. True — The two may coexist, producing a management challenge.

35. a. True — All patients who have had a CVA should have their swallowing assessed to prevent aspiration pneumonia.

b. False — Recurrent laryngeal nerve trauma will affect the vocal cords, not the oesophagus.

c. True — Carcinoma of the stomach can lead to swallowing difficulties – initially solids and then later liquids.

d. False Dementia does not impede a patient's swallow.

e. True Motor neuron disease can affect the swallow reflex.

36. a. True Syphilis can lead to nerve injury.

b. True Diabetes mellitus can affect the reflex pathways, e.g. diabetic autonomic neuropathy.

c. False Gonorrhoea will not cause a patient to lose their ankle reflexes.

d. False Parkinson's disease will not cause a patient to lose their ankle reflexes.

e. True Motor neuron disease can alter your reflexes.

37. a. False As the liver is used to metabolizing large quantities of alcohol (a sedative), it is very good at clearing other sedative drugs through enzyme induction.

b. True The liver manufactures clotting factors – if the liver is damaged, it can't make the clotting factors so you have an increased PT.

c. True See above.

d. True The degree of alcohol excess determines the onset of liver disease.

e. False While alcoholics do have many dietary insufficiencies, the alcohol excess in itself damages the liver.

38. a. True There is a release of antidiuretic hormone to try and conserve body fluid as a response to blood loss.

b. True The body loses potassium after trauma.

c. True This is because the body requires nitrogen for protein synthesis.

d. False The body needs sugar to deal with trauma so there is an increase in gluconeogenesis.

e. True The body will use lean body mass in the initial response to trauma.

39. a. True Colles' fractures are impacted.

b. False Colles' fracture is not associated with median nerve damage.

c. True Colles' fractures are deviated to the radial side.

d. False Colles' fracture will have a posterior and radial displacement; the angulation of the distal fragment will be dorsal. There may be impaction leading to the shortening of the radius compared to the ulna. Good alignment is required to prevent the development of carpal tunnel syndrome.

e. False The fragment is not torn off the intra-articular triangular disc.

40. a. False Multiple pregnancies occur in older women.

b. True Twin pregnancies are shorter (there's less space).

c. True Multiple pregnancies are at a higher risk of most complications.

d. False They are usually dizygotic following IVF.

e. False Twins occur 1 : 105 and triplets 1 : 10 000.

41. a. True This does indeed confirm ovulation.

b. False Oestrogen does not help in confirming ovulation.

c. False There is a temperature rise with ovulation at mid cycle.

d. False Spinbarkeit cervical mucus does not confirm ovulation.

e. True Histological examination of an endometrial biopsy can confirm ovulation.

42. a. False Abortion only requires two medical professionals to agree.

b. True A medical termination involves giving the woman intravaginal prostaglandins to soften the cervix and stimulate an early labour.

c. True Surprising but true.

d. True You can't have a suction termination of pregnancy beyond 14 weeks.

e. False Mifepristone can be used up to 9 weeks.

43. a. False Median nerve injury will cause weak flexion at the wrist and loss of movement at the interphalangeal joint of the thumb.

b. False The motor deficit leads to loss of opposition of the thumb.

c. True The median nerve supplies sensation to the forearm.

d. False Upper arm pain is not associated with median nerve injury.

e. False Median nerve lesions lead to loss of sensation over the palmar aspect of the thumb, index and middle fingers.

Learn your milestones!

44. a. True A 10-month-old will demonstrate opposition and will pull themselves to a sitting position.

b. True A 12-month-old will release objects on request and can walk with one hand held.

c. False A 3-month-old child will have a slight head lag when pulled to sit and will hold their hands loosely open.

277

d. True   An 18-month-old can put three words together and can build a tower of three to four blocks.

e. False   A 4-month-old child will hold their hands together, pull clothes over their head and laugh.

45. a. False   Young men have an increased chance of completing suicide.

b. False   The rates of suicide vary across the world.

c. False   A significant proportion of people who complete suicide will have seen a doctor in the week before.

d. True   People who are unemployed are more likely to complete suicide.

e. True   People who have a definable mental illness are at increased risk – many psychiatrists feel that depression is almost always present in people who commit suicide.

46. a. True   Having schizophrenia puts you at a high risk of suicide.

b. False   Schizophrenia is not generally associated with violence to others.

c. False   Patients may recover from it. Even if recovery is not complete, there may be remission for long periods.

d. True   There is evidence that high expressed emotion correlates with a poor prognosis.

e. True   Chronic schizophrenics often have a burned out appearance with cognitive impairment.

47. a. False   Ectopic thyroid tissue would not be at the angle of the mandible.

b. False   A thyroglossal cyst is a midline structure.

c. True   There are lymph nodes around the neck which can become inflamed.

d. False   A pharyngeal pouch is a midline structure.

e. True   A carotid body tumour arises in the neck.

48. a. True   Hyperthyroidism is associated with weight loss.

b. True   Atrial fibrillation can be precipitated by hyperthyroid states.

c. True   You get lid lag with hyperthyroidism.

d. True   You get onycholysis with hyperthyroidism.

e. False   Feeling tired all the time is more commonly associated with hypothyroidism.

49. a. False   Dementia does not necessarily progress.

b. False   Pre-senile means young or middle aged, usually defined as onset at less than 65 years of age.

c. False   The sleep–wake cycle is often an early sign of dementia.

d. True   A head injury may precipitate dementia.

e. False   Mini-mental test scores are more useful at monitoring progression.

50. a. False   Urinary incontinence is not an early symptom of Parkinson's.

b. True   A shuffling gate is classic of Parkinson's disease.

c. False   The onset of symptoms is usually after 50.

d. True   Cog-wheel rigidity is associated with Parkinson's disease.

e. False   Patients classically have a resting tremor; so-called 'pill rolling.' It may disappear altogether on movement.

51. a. False   Nocturnal cough is a common symptom of asthma.

b. False   No, young patients can look very well, compensate, compensate and then decompensate very quickly and can die.

c. True   The peak expiratory flow allows assessment of expiration, and acts as a surrogate marker of airway size.

d. True   Atopy is associated with asthma.

e. False   Type 4 hypersensitivity is a T-cell-mediated reaction.

52. a. False   It is most likely after 7–10 days post-op.

b. True   D-dimers are fibrin degradation products and will be present as long as there is clot breakdown, up to 6 weeks post-op.

c. True   Wells' score is an evidence-based scoring system for lower limb deep vein thrombosis (DVT).

d. True   Warfarin is the mainstay of DVT treatment but takes 48+ hours to take effect. Low molecular weight heparin takes effect immediately and anticoagulates the patient during this period.

e. False   Cancer and pelvic masses both predispose to clot formation.

53. a. False   Most gallstones are asymptomatic and are often an incidental finding.

b. True   This is true.

c. False   Ninety per cent are radiolucent while 90% of renal stones are radio-opaque.

d. False   Cholecystectomy removes the gallbladder, not the components of gallstones; they can form in the biliary system.

e. True   Any obstruction of any viscus predisposes to infection.

54. a. True   Hypercalcaemia is an important and treatable cause of confusion.

b. True   True, always check your hospital policy on empirical therapy as Microbiology will be aware which strains of bacteria are most prevalent locally.

c. False   Confused patients should be nursed in well-lit conditions to try to reduce illusions and reinforce their orientation.

d. False   Confusion is common in elderly patients with respiratory tract infections.

e. False   Sedation should only be used if the patient becomes a danger to themselves or others; it may make confusion worse.

55. a. True   There is an increase in haemoglobin but a greater increase in blood volume, leading to a decreased haematocrit.

b. False   A number of things may cause increased white cell count e.g. steroids and haematological malignancies.

c. True   Platelets are raised in inflammation.

d. True   True, this demonstrates the spread of red cell widths.

e. True   Asthma induces IL 5, leading to eosinophil recruitment.

56. a. True   This is known as prerenal renal failure.

b. True   Poor perfusion due to heart failure is a potential cause of organ dysfunction.

c. False   Rarely requires dialysis; many causes of renal failure can be reversed.

d. False   Renal failure, particularly prerenal, is common in hospitalized patients.

e. False   No, but raised potassium due to renal failure is potentially life threatening.

57. a. False   A serum calcium below 2 mmol/L is a poor prognostic marker.

b. False   Although true, it's unlikely to be significant, analgesia is much more important and should not be delayed.

c. True   Due to the inflammatory process these patients can have massive third space losses.

d. False   Classically seen in haemorrhagic pancreatitis, this is now rare but can be seen in retroperitoneal haemorrhage.

e. True   This allows the overlying contents of the abdomen to hang free of the pancreas.

58. a. False   Systemic lupus erythematosus is associated with anti-nuclear antibodies (more strongly anti-double-stranded DNA antibodies).

b. False   Anti-Ro and anti-La antibodies are associated with Sjögrens syndrome.

c. True   The M2 antigen is the epitope.

d. False   LKM antibodies are associated with autoimmune hepatitis.

e. True   Not all patients have antibodies though when present limited scleroderma is associated with anti-centromere and diffuse scleroderma with anti-SCL antibodies.

59. a. False   Hepatitis A is usually a self-limiting condition.

b. True   Hepatitis e antigen is produced by infected hepatocytes; when cleared the antigen is no longer produced.

c. True   Hepatitis s antigen is utilized in hepatitis immunization.

d. False   This is less common than other forms of spread.

e. False   Approximately 85% of hepatitis C patients become chronically infectious.

60. a. True   To a point this is true, statins also have pleiotropic effects and lead to plaque stabilization.

b. True   This is true, and why exercise tests are often supervised by doctors.

c. True   But is not a modifiable risk factor, so optimization of other risk factors is important.

d. False   Although not an indication for transplant, the sequelae of ischaemic heart disease (e.g. heart failure) are.

e. False   Beta blockers are an important treatment modality in many forms of cardiac disease.

61. a. True   This is the main diagnostic criterion for respiratory failure.

b. False   The $PaCO_2$ must be above 6.7 KPa.

c. True   Depending upon the circumstances, always check your local guidelines.

d. True   These may cause respiratory depression.

e. False   Long-term oxygen therapy reduces mortality in chronic obstructive pulmonary disease if used for >15 hours per day.

62. a. False   Type I diabetes predisposes to DKA (diabetic ketoacidosis), type II diabetes can lead to HONK (hyperosmolar non-ketotic coma).

b. False   Impaired glucose tolerance is part of the metabolic syndrome.

c. True   The exocrine functions of the pancreas are not used post transplantation.

d. True   Gastric paresis is due to autonomic neuropathy secondary to diabetes.

e. True   Diabetes also predisposes to vascular disease; always have a low threshold for performing an ECG in diabetics.

63. a. True   Abnormal red blood cells are haemolysed in the spleen leading to anaemia, removal of the spleen prevents haemolysis.

b. True   Red blood cells usually live for 120 days.

c. False   This is a potential cause but is uncommon.

d. False   Plasma haptoglobins are low in haemolytic anaemia.

e. True   Haem breakdown produces bile, excess bile can lead to bile gallstones.

64. a. False   Airborne agents tend to travel to the best aerated areas of lung, the apices.

b. False   Lung cavitation, if present, may allow Aspergilloma to form.

c. True   TB infection of the adrenal glands is the commonest cause worldwide of Addison's disease.

d. True   Named after Anton Ghon, a Czech pathologist.

e. False   It refers to the presence of acid-fast bacilli in the sputum not the absence.

65. a. True   This is the definition of septic shock; organ dysfunction must be due to poor organ perfusion.

b. True   In the absence of intrinsic renal disease, urine output is an important marker of perfusion in septic shock.

c. False   Confusion is most likely due to peripheral vasodilatation, not cardiac dysfunction.

d. False   Systemic vascular resistance is likely to be decreased due to inflammation.

e. True   This is well documented; line care and aseptic technique are vitally important.

66. a. False   Paget's disease increases serum calcium by increased bone turnover.

b. True   Boney metastases can cause hypercalcaemia.

c. True   Phosphate binds calcium and can lead to hypocalcaemia.

d. False   The described sign is Trousseau's sign; Chvostek's sign is facial spasms on tapping the facial nerve.

e. True   Vitamin D promotes calcium uptake in the intestine and resorption in the kidney.

67. a. False   The commonest cause of polycystic kidney disease (PCKD) is autosomal dominant; there is a rare form of recessive PCKD.

b. True   There is an association with liver cysts.

c. True   This is an important correlation to remember in patients with PCKD.

d. False   Valve abnormalities are associated but much less common than 40%.

e. True   Keeping blood pressure low can slow the progression of PCKD.

68. a. False   LDL cholesterol is raised in nephrotic syndrome.

b. True   This is a diagnostic feature.

c. False   Haematuria is found in nephrotic syndrome.

d. False   This is not a feature of the syndrome.

e. False   Renal loss of albumin leading to oedema is part of the syndrome.

69. a. True   Due to loss of fluid and salt during surgery as insensible losses and third space losses.

b. False   Serum osmolarity would be less than 270 mOsmol/kg.

c. False   This is likely to make the hyponatraemia worse and may be the cause.

d. True   These are features of symptomatic hyponatraemia.

e. True   But other causes of hyponatraemia should be ruled out.

70. a. True   This is a common presentation for right-sided colon cancers.

b. True   Tenesmus may be caused by a low lying tumour; the feeling of incomplete defaecation caused by the presence of the tumour.

c. False   Tumour markers are neither sensitive nor specific enough to be used as a diagnostic test.

d. True   The presence of chronic inflammation leads to a fertile soil for metaplasia and subsequent neoplasia.

e. False   The reverse is true.

71. a. True   Histologically this is the most common type.

b. False   Pancreatic cancer typically presents late and carries a poor prognosis.

c. True   The commonest site.

d. True   This is a complicated procedure with a significant mortality.

e. False   Pancreatic cancer may cause obstructive jaundice by compression of the biliary system.

72. a. False   The inflammation in Crohn's disease is transmural.

b. True   Fistula formation is not uncommon in Crohn's disease.

c. True   There may be no examination findings; history is therefore extremely important.

d. False   Surgery should be reserved for medically resistant or severe cases as the disease recurs and resection can lead to short bowel syndrome.

e. True   This is one of the extraintestinal manifestations of Crohn's disease.

73. a. True   The causes of iron deficiency are more common in the elderly.

b. True   Spooning of the nails is seen in iron deficiency anaemia.

c. True   The causes of gastrointestinal blood loss should be investigated in iron deficiency anaemia.

d. False   It would lead to a microcellular anaemia.

e. False   Intravenous iron can lead to anaphylaxis and should be used only in severe cases under the supervison of a specialist.

74. a. False   There would be coarse crackles.

b. True   Clubbing of the fingers is associated with bronchiectasis.

c. True   Persistent inflammation can lead to bronchiectasis.

d. False   Plethora may be seen in conditions such as chronic obstructive pulmonary disease, but not bronchiectasis.

e. True   Bronchiectasis is a chronic disease and normocytic, normochromic, anaemia may be a feature.

75. a. False   Venous ulcers present most commonly in the gaiter area.

b. True   The ball of the foot is an area where pressure is transmitted; this can cause ischaemia in the presence of poor perfusion.

c. True   These may be known as Marjolin's ulcers.

d. False   Eczematous change may be seen in venous ulcers.

e. False   They commonly exist, hence the need for adequate assessment.

1. Features in the history include the triad of:
   - Syncope.
   - Breathlessness on exertion.
   - Angina pectoris.

   Eventually orthopnoea or paroxysmal nocturnal dyspnoea may develop.

   Features of severe aortic stenosis on examination include:
   - Narrow pulse pressure when recording the blood pressure (e.g. 90/70 mmHg).
   - A slow-rising pulse – small in volume, rises slowly to its peak and takes a long time to pass the finger.
   - A sustained apex beat suggestive of left ventricular hypertrophy.

2. Features of heart failure include:
   - A displaced apex beat (due to left ventricular volume overload).
   - A third heart sound.
   - An elevated jugular venous pressure.
   - Peripheral oedema.

3. Features of a severe attack of asthma include:
   - Tachycardia (heart rate greater than 110/min).
   - Tachypnoea (respiratory rate greater than 30/min).
   - Pulsus paradoxus greater than 15 mmHg.
   - Peak expiratory flow rate less than 50% predicted maximum for height and weight.
   - Exhaustion.
   - Cyanosis.
   - Silent chest.
   - Bradycardia.
   - Difficulty speaking.

4. Median nerve palsy at the wrist is due to carpal tunnel syndrome. The more common causes include:
   - Pregnancy.
   - Oral contraceptive pill.
   - Hypothyroidism.
   - Rheumatoid arthritis.

   The physical signs include:
   - Wasting of the thenar eminence (if long-standing).
   - Sensory loss over the palmar aspects of the radial three-and-a-half digits.
   - Weakness of abduction, flexion and opposition of the thumb.
   - Tinel's sign – tingling sensation produced in the distribution of the median nerve by percussion over the carpal tunnel.

   - Phalen's sign – on flexing the wrists for 60 seconds, there is an exacerbation of the paraesthesiae, which is rapidly relieved when the wrist is extended.

5. The main features of Parkinson's disease are:
   - Tremor (resting, pill-rolling tremor).
   - Bradykinesia.
   - Rigidity ('cog wheel').

   Other features include an expressionless unblinking face, low-volume monotonous speech, drooling from the mouth and micrographia. Patients have a characteristic stooping shuffling gait, holding their arms by their sides.

6. The signs of an acute disc prolapse may include features of compression of the sciatic nerve roots. Straight leg raising will be impaired on one or both sides – ask the patient to lie flat and lift the leg by the ankle with the knee extended. This will reproduce pain going down the back of the leg.

   There may be:
   - Sensory loss over the back of the calf and sole of the foot.
   - Weakness of dorsiflexion of the foot.
   - A reduced ankle reflex.

7. Causes of massive splenomegaly include:
   - Chronic myeloid leukaemia.
   - Myelofibrosis.
   - Chronic malaria.

   Features of a palpable spleen on physical examination are:
   - Location in the left upper quadrant of the abdomen.
   - Movement towards the right lower quadrant on inspiration.
   - Dullness to percussion.
   - There may be a notch on the anterior surface.
   - It is not possible to get above the mass.
   - Firm consistency.
   - It cannot be felt bimanually.

8. Any of the following would be acceptable:
   - Symmetrical distribution of joint disease with redness, swelling, warmth and tenderness.
   - Predominant inflammation in the proximal interphalangeal and metacarpophalangeal joints with relative sparing of the distal interphalangeal joints.
   - Signs of carpal tunnel syndrome.
   - Rheumatoid nodules behind the elbows.
   - Wasting of the small muscles of the hand.
   - Ulnar deviation of the fingers.

- Deformity of the fingers (e.g. swan neck deformity, Boutonnière deformity).

9. The three most common causes of papilloedema are:
   - Intracranial space-occupying lesion.
   - Malignant hypertension.
   - Benign intracranial hypertension.
   Features of these causes include a focal neurological deficit, very high blood pressure and obesity.

10. Figure 1 illustrates X-linked recessive inheritance. Examples include:
    - Colour blindness.
    - Haemophilia A.
    Figure 2 illustrates autosomal dominant inheritance. Examples include:
    - Adult polycystic kidney disease.
    - Dystrophia myotonica.
    - Hereditary spherocytosis.
    - Neurofibromatosis.

11. Disturbing features on examination can be divided into local and systemic.
    Local features include:
    - Mass fixed to deeper tissues.
    - Mass tethered to the skin.
    - Hard and craggy mass.
    - Peau d'orange.
    - Inverted nipple.
    - Associated axillary lymphadenopathy.
    Systemic features include:
    - Hepatomegaly.
    - Pleural effusion.
    - Local bony tenderness.
    - Ascites.

12. a. Cytomegalovirus.
    b. *Pneumocystis carinii*, *Mycobacterium avium intracellulare*, cytomegalovirus.
    c. *Toxoplasma gondii*, *Cryptococcus neoformans*.
    d. Cryptosporidium.
    Groups of patients who are at risk of opportunistic infection include transplant recipients who have received immunosuppression and patients receiving chemotherapy or cytotoxic drugs.

13. Iron is found in most meat. This is why vegans are susceptible to iron-deficiency anaemia. Deficiency may also result from chronic blood loss (e.g. menorrhagia, hookworm) or malabsorption (e.g. gastrectomy, coeliac disease).
    Folate is found in most foodstuffs, especially liver, green vegetables and yeast. Deficiency may also arise from malabsorption (e.g. in coeliac disease, Crohn's disease) or excess use (e.g. due to pregnancy, psoriasis).
    Vitamin B$_{12}$ is found in foods of animal origin only such as liver, fish and dairy produce. Deficiency may

also result from pernicious anaemia or malabsorption in the ileum (e.g. in Crohn's disease).

14. Hypercalcaemia may present non-specifically or through its effect on end-organ damage. The principal sites of presentation include:
    - Kidney – renal stones can cause renal colic; hypercalcuria can cause polyuria and consequent polydipsia.
    - Nervous system – depression, anorexia, nausea and vomiting are common.
    - Gastrointestinal tract – abdominal pain and constipation usually result from hypercalcaemia.
    - Bones – aches and pains.
    Four causes of hypercalcaemia include:
    - Hyperparathyroidism (primary or tertiary).
    - Malignant disease of the bone.
    - Vitamin D excess (e.g. sarcoidosis, iatrogenic).
    - Milk-alkali syndrome.

15. The most likely causes are:
    - Impaired venous drainage of the legs due to the enlarged uterus.
    - Pre-eclampsia.
    - Deep vein thrombosis (increased incidence in pregnancy).
    Discriminatory features on bed-side examination would include blood pressure check, signs of periorbital oedema and urinalysis for pre-eclampsia to exclude proteinuria. Features of inflammation such as redness, warmth and tenderness would make a deep vein thrombosis more likely.
    Pre-eclampsia may be suggested by the results of a full blood count (thrombocytopenia) and urate, urea and electrolyte estimate (evidence of renal dysfunction). A Doppler ultrasound scan of the leg veins can be performed to look for a deep vein thrombosis. Venography should be avoided if possible due to the radiation exposure to the fetus.

16. The list is almost endless, but some of the more characteristic signs include:
    - Acanthosis nigricans (especially with gastrointestinal tract malignancy).
    - Dermatomyositis (heliotropic rash).
    - Herpes zoster.
    - Erythema gyratum repens (carcinoma of the breast).
    - Acquired ichthyosis (especially lymphoma).
    - Tylosis (thickening of the palms or soles associated with upper gastrointestinal malignancy).
    - Thrombophlebitis migrans (especially carcinoma of the pancreas).

17. Features in the history include:
    - A defined precipitating cause for diabetic ketoacidosis (e.g. infection, missed an insulin injection) or for hypoglycaemia (e.g. known overdosage of hypoglycaemic treatment or missing a meal).

- Time course – ketoacidosis usually develops over hours or days; hypoglycaemia is usually associated with a shorter history of illness.
  Features on examination include:
- Dehydration in patients with ketosis. Patients with hypoglycaemia are usually euvolaemic.
- Kussmaul's breathing (deep sighing breathing due to acidosis) in ketosis.
- Sweating (often profound) in a patient with hypoglycaemia.

18. Some questions to ask a patient with suspected hypothyroidism include:
- 'Do you feel more comfortable in warm or cool weather?' to reveal cold intolerance.
- 'Has your weight changed over the past six months?' to reveal weight gain.
- 'Do you have more or less energy now than you used to?' to reveal tiredness, malaise.
- 'Has anyone noticed a change in your facial appearance?'
  The most discriminatory signs are:
- Bradycardia.
- Slow-relaxing reflexes – easily demonstrated with the ankle reflex.
- Facial appearance (e.g. dry, coarse hair, periorbital swelling, thinning of eyebrows).

19. Drugs causing hyperkalaemia include:
- Potassium supplements given in excess – due to increased gastrointestinal absorption of potassium.
- Potassium-sparing diuretics (e.g. spironolactone, amiloride) – these drugs antagonize the effect of aldosterone on the distal tubule sodium–potassium exchange pump, resulting in decreased excretion of potassium into the urine.
- Angiotensin-converting enzyme (ACE) inhibitors – decreased angiotensin results in decreased aldosterone and therefore decreased urinary loss of potassium.
- Nephrotoxic drugs – for example non-steroidal anti-inflammatory drugs (NSAIDs) and aminoglycosides – may cause hyperkalaemia through impaired renal excretion of potassium.

20. Any of the following would be acceptable:
- Haemophilia.
- Blood transfusion in Africa since 1977.
- Sexual partner of a prostitute.
- Intravenous drug user.
- Anal intercourse.
- Multiple sexual partners and unprotected sexual intercourse.
- Sexual partner of an individual in any one of the above groups.

1. **Theme: Limp (musculoskeletal)**

   1. d   Fractured neck of femur. This is a classic presentation and must always be considered in an elderly patient who has pain or mobility problems after a fall, however trivial.

   2. e   Gout. Thiazide diuretics can also cause hyperglycaemia and hypokalaemia.

   3. i   Slipped femoral epiphysis. Avoid the pitfall of labelling such presentation as 'growing pains'.

   4. b   Compartment syndrome. Also known as 'shin splints', this is due to increased pressure in the anterior tibial compartment as a result of swelling or bleeding from a muscle tear.

   5. f   Osteoarthritis of hip. Another classic presentation with referred pain. The patient's knee would be normal on examination (unless he also had osteoarthritis in that joint!).

2. **Theme: Gastrointestinal bleeding**

   1. b   Colonic carcinoma. The prognosis is good if detected early, so prompt referral is necessary.

   2. e   Duodenal ulcer. Altered blood almost always comes from haemorrhage in the upper GI tract.

   3. g   Haemorrhoids. Haemorrhoids commonly follow, or are exacerbated by, childbirth but often resolve spontaneously.

   4. i   Mallory–Weiss tear. Another classic presentation due to tearing at the gastro-oesophageal junction as a result of excess vomiting.

   5. j   Oesophageal varices. Oesophageal varices are associated with portal hypertension caused, in turn, by hepatic cirrhosis.

3. **Theme: Liver problems**

   1. a   Alcoholic liver disease. Classic signs of hepatic failure.

   2. g   Metastatic liver disease. The patient needs an urgent ultrasound examination and possible liver biopsy; liver function tests would not be particularly helpful at this stage.

   3. b   Biliary atresia. Physiological jaundice.

   4. d   Hepatitis A. The disease is endemic in many parts of the world, and transmission is via the oro-faecal route, usually from consumption of contaminated water or food.

   5. c   Gallstones. In the classic presentation, the patient might also have fat intolerance and flatulence!

4. **Theme: Cardiac murmurs and added sounds**

   1. a   Aortic regurgitation. The early diastolic murmur is best heard at the left sternal edge, with the patient leaning forward and the breath held in expiration.

   2. e   Mitral stenosis. Acquired mitral stenosis is usually due to rheumatic fever, now a very rare disease in the West, but still common in the developing world.

   3. b   Aortic stenosis. Patients may also present with angina, even in the absence of severe coronary arterial disease, due to a combination of increased oxygen requirements and fixed flow obstruction.

   4. d   Mitral regurgitation. This often occurs because of papillary muscle or chordea rupture.

   5. i   Ventricular septal defect. A more common presentation is an asymptomatic ejection systolic murmur detected on routine examination.

5. **Theme: Finger clubbing**

   1. g   Pulmonary abscess. This is most usually a form of secondary pneumonia with aspiration a key aetiological factor.

   2. b   Crohn's disease. The terminal ileum and right side of the colon are the sites most frequently affected by the disease, and malabsorption is a key feature.

3. e   Fibrosing alveolitis. This is an insidious, progressive disease, for good reasons labelled 'cryptogenic'.

4. f   Infective endocarditis. An increasingly common presentation with a wide range of potential causative agents.

5. d   Familial clubbing. There is no direct association between asthma and clubbing.

6. Theme: The painful joint

1. a   Ankylosing spondylitis. The majority of affected persons carry the HLA-B27 antigen, with a male:female ratio of 4:1.

2. j   Systemic lupus erythematosus. This is the most common multi-system connective tissue disease and is characterized by a wide variety of clinical features and a diverse spectrum of autoantibodies.

3. g   Reiter's syndrome. There is a strong association with the HLA-B27 antigen, as there is for ankylosing spondylitis.

4. d   Osteoarthritis. Osteoarthritis is by far the most common form of arthritis, shows a strong association with ageing and is a major cause of pain and disability in the elderly.

5. c   Gout. Gout is a crystal arthropathy with a strong (over 10:1) male predominance.

7. Theme: Trauma

1. k   Tension pneumothorax. This develops if the communication between pleura and lung is small and acts as a one-way valve, allowing air to enter the pleural space during inspiration, but preventing it from escaping.

2. f   Open pneumothorax. This occurs when the communication between lung and pleural space does not seal and allows air to transfer freely between the two.

3. c   Flail chest. This is a clinical syndrome resulting from major trauma to the chest walls, sufficient to cause fracture of several ribs in at least two places, resulting in paradoxical movement of part of the chest on respiration.

4. a   Airway obstruction. An inhaled foreign body is more likely to enter the right main bronchus.

5. l   Traumatic aortic rupture. Similar radiological findings in association with severe chest pain radiating to the back are found in aortic dissection.

8. Theme: Heavy periods (or menstrual problems)

1. d   Fibroids. Fibroids are the commonest uterine tumour, rarely become malignant and are associated with nulliparity.

2. e   Hypothyroidism. In contrast, hyperthyroidism may be associated with amenorrhoea or oligomenorrhoea.

3. h   Physiological bleeding. There is no direct association with the oral contraceptive.

4. g   Pelvic inflammatory disease. Primary infection with chlamydia may be asymptomatic in about 80% of patients.

5. k   von Willebrand's disease. This is a common, inherited, but usually mild, bleeding disorder, characterized by a reduced level of von Willebrand factor.

9. Theme: The diabetic eye

1. b   Background retinopathy. There is no immediate threat to vision.

2. l   Vitreous haemorrhage. Vitreous haemorrhage frequently resolves, but may be recurrent.

3. i   Proliferative retinopathy. This requires urgent review and treatment by laser photocoagulation.

4. c   Central retinal artery occlusion. May have prodromal episodes of transient visual loss (so-called 'amaurosis fugax').

5. d   Central retinal vein occlusion. May lead to rubeosis of the iris with neovascular glaucoma.

10. Theme: Delirium

1. b   Diabetic ketoacidosis. This is a classic first presentation of diabetes, with children more likely to complain of abdominal pain.

2. e   Hypercalcaemia. This is a medical emergency, the mainstay of treatment being rehydration and bisphosphonates.

3. f   Hypercapnia. Respiration in this situation is driven by hypoxia, highlighting the potential dangers of oxygen therapy.

4. k   Subdural haematoma. A classic presentation which is often insidious in onset, and easily missed.

5. c   Encephalitis. Herpes simplex is the commonest cause of viral encephalitis in the UK, but insect-borne causes are important in other parts of the world.

11. Theme: Respiratory symptoms

1. d   Bronchopneumonia. This occurs more frequently in COPD patients and may present with few changes on chest X-ray.

2. e   Hayfever. Sneezing in summer with a previous history of eczema points to hayfever.

3. h   Pneumothorax. Spontaneous pneumothoraces are more common in tall individuals.

4. b   Aspiration pneumonia. This lady has motor neuron disease affecting her bulbar nerves; she has aspirated and the aspirate has caused a right basal pneumonia.

5. g   Pleural effusion. This young lady has a parapneumonic pleural effusion. This can become an empyema and should be monitored.

12. Theme: Pancreatitis

1. e   Hyperlipidaemia. This is an uncommon cause of pancreatitis and is frequently seen in conjunction with alcohol excess. High triglycerides in the blood causes pseudohyponatraemia as only the sodium in the serum, rather than lipid, is measured.

2. d   Grey-Turner's sign. Grey-Turner described his sign in acute haemorrhagic pancreatitis. This is an uncommon presentation now but the sign may still be seen in trauma where there is retroperitoneal bleeding, in this case due to trauma of the pancreas as it comes into contact with the vertebral bodies.

3. f   Pancreatic divisum. Pancreatic divisum is relatively common and most patients are asymptomatic. Drainage of the ducts of Santorini and Wirsung by a minor papilla can lead to increased intra-duct pressure leading to pancreatitis.

4. a   Alcohol. This is the commonest cause of pancreatitis. Presentation may be delayed after a significant bout of drinking, as in this case.

5. c   Gallstones. This is the other common cause of pancreatitis (alcohol and gallstones making up around 95% of pancreatitis). The patient gives a past history of biliary colic as her gallstone impacts in Hartmann's pouch.

13. Theme: Symptoms of heart failure

1. g   Pulmonary fibrosis. This patient has been treated with diuretics and is hypovolaemic. He may have had pulmonary oedema; however, this has resolved. Pulmonary fibrosis may give very similar auscultatory findings to pulmonary oedema.

2. e   Paroxysmal nocturnal dyspnoea. These are episodes of acute breathlessness that usually occur at night and may result in the patient getting out of bed and pulling the window open. It happens in left heart failure due to fluid redistribution, but there are other causes.

3. d   Orthopnoea. Breathlessness occuring on lying flat; it is often described in terms of how many pillows a patient requires to get to sleep, e.g. '4 pillow orthopnoea'. It is seen in left heart failure.

4. b   Cough. ACE inhibitors may cause a chronic cough that can persist for up to one month after cessation of treatment; this is likely due to decreased bradykinin breakdown.

5. c   Hepatomegaly. Hepatomegaly can be seen in right heart failure due to venous congestion. This can lead to a macroscopic 'nutmeg' appearance.

14. Theme: Knee injuries

1. b   Bucket handle tear of meniscus. Meniscal injuries are common in sportsmen and women; the history of locking suggests that there is a loose body in the joint.

2. c   Dislocation of the knee. This is exceedingly rare but does happen; in this case, the pulses and ligaments should all be checked as the patient may require emergency surgery to repair the vascular supply.

3. j   Tibial plateau fracture. These are common and occur with trauma to the knee in which the femoral condyles impact upon the tibia. Scatzker refers to a grading system for these types of fracture.

4. a   Anterior cruciate ligament rupture. This is demonstrated by the anterior draw test; other ligaments may be injured at the same time.

5. g   Patella fracture. Falling onto the knee may fracture the patella; this is supported by the inability to straight leg raise.

15. Theme: Hernias

1. d   Incarcerated hernia. This gentleman has an irreducible hernia that has no signs of strangulation.

2. e   Incisional hernia. These are not uncommon after abdominal surgery (≈2%).

3. g   Obturator hernia. These are uncommon but do occur in thin women and can present with features of bowel obstruction.

4. f   Inguinal hernia. This is commoner in those with a family history. Differentiating between direct and indirect hernias clinically can be attempted, but is not evidence based.

5. k   Umbilical hernia.These hernias usually resolve within two years of birth but leave a weak spot; they can reappear in the elderly and in postpartum females.

16. Theme: Nerves of the upper limb

1. c   Lower brachial plexus. This boy has arrested his fall with an outstretched arm and there has been stretching of the lower brachial plexus. This may be a neuropraxia which will hopefully resolve.

2. e   Median nerve. This lady describes a classical carpal tunnel syndrome. Symptoms may be worse at night due to fluid redistribution (especially in pregnancy) and flexion of the wrist.

3. h   Radial nerve. This injury is associated with fractures of the proximal humerus. This gentleman has both motor and sensory loss.

4. a   Axillary nerve. This injury is also associated with fractures of the proximal third of the humerus. The anaesthesia over the regimental patch can vary considerably.

5. b   Long thoracic nerve of Bell. The long thoracic nerve passes through the axilla and innervates serratus anterior which, when injured, causes winging of the scapula.

17. Theme: Headache

1. b   Benign intracranial hypertension. This is often described as affecting young obese women; the underlying cause is unknown.

2. f   Giant cell arteritis. This is a vasculitic disorder affecting the temporal artery and classically causes thick pulseless temporal arteries. Diagnosis may be by biopsy, though sampling error is common as the vasculitis doesn't affect the whole vessel. Treatment is with high-dose steroids, tapering down over months to weeks.

3. e   Cluster headache. These are migrainous headaches that occur in clusters, which may be separated by months or years. The patient describes fortification spectra pre attack.

4. j   Tension headache. This gentleman is suffering from tension headaches, or muscle contraction headaches as they are also known. This seems to be brought on by stress in his current post, although other causes should be excluded.

5. a   Bacterial meningitis. This is more common among those living in crowded conditions (such as students). Kernig's sign, headache and photophobia should always be treated as meningitis until proven otherwise. The cerebrospinal fluid findings here are consistent with bacterial infection.

18. Theme: Diabetes

1. b   Diabetes insipidus. This can be confused with diabetes mellitus due to the polydipsia and polyuria. The normal blood glucose and reduced serum osmolarity should point towards the diagnosis.

2. f   Hyperosmolar non-ketotic coma. This gentleman has presented in hyperosmolar non-ketotic coma brought on by his choice of sugary fluid to rehydrate himself. He requires rehydration and cautious correction of his blood sugar.

3. c   Diabetic ketoacidosis. This boy has presented in diabetic ketoacidosis which may have been precipitated by his chest infection. He requires fluid resuscitation, insulin and replacement of potassium.

4. h   Random serum glucose >7.1 mmol/L but below 12 mmol/L. These are the levels for impaired glucose tolerance. Sugars should be monitored as the patient may go on to develop full-blown diabetes.

5. i   Type I diabetes mellitus. This most commonly presents in the young and is characterized by absolute insulin deficiency. Type II usually presents in the older population (middle aged onwards) and is more characterized by relative insulin lack or resistance.

19. Theme: Cancer

1. f   Hypercalcaemia. This is associated with cancers due to bony metastases or excretion of parathyroid hormone related peptide (PTHrP). It should be treated with fluid resuscitation initially, then possibly a bisphosphonate infusion.

2. h   Oesophageal carcinoma. This is suggested by the dysphagia, early satiety and weight loss. The patient may also have tyelosis (hyperkeratosis of the soles of the feet, associated with certain cancers). Hoarseness suggests recurrent laryngeal nerve involvement.

3. a   Carcinoid syndrome. This is a classical presentation of carcinoid syndrome. Symptoms only become apparent once hepatic metastases have developed as 5-HT undergoes almost 100% first-pass metabolism in the liver. The survival from carcinoid syndrome is very good despite most presentations having distant metastases.

4. c   Ectopic adrenocorticotropic hormone syndrome. This lady has developed Cushing's syndrome. The most likely cause given her diagnosis is ectopic adrenocorticotropic hormone syndrome.

5. g   Mesothelioma. This is not the most common cancer in asbestos-exposed workers; the incidence is much higher, but lung cancers are still more common in this group.

20. Theme: Diarrhoea

1. g   Irritable bowel syndrome. This is the likely diagnosis in this patient; however, he should be monitored as he is the right age for a primary presentation of inflammatory bowel disease.

2. h   Metformin. This is the likely culprit. Many drugs are associated with diarrhoea; however, it is common with metformin, especially at higher doses.

3. e   Diverticular disease. Colonoscopy is indicated to rule out a malignancy due to the gentleman's age, change in bowel habit and haemochezia.

4. i   Ulcerative colitis. In a patient with the same symptoms but aged 60+, the consideration of malignancy would be much higher. This young lady has inflammatory bowel disease. Left iliac fossa tenderness would suggest ulcerative colitis rather than Crohn's disease, which more often affects the ileum.

5. j   Zollinger–Ellison syndrome. This is characterized by a duodenal or pancreatic gastrin-secreting tumour leading to profuse ulceration.

21. Theme: Electrocardiograph findings

1. g   Posterior myocardial infarction. Due to lead placement, posterior myocardial infarction changes are the reverse of those elsewhere: dominant R waves equivalent to Q waves and ST depression instead of elevation.

2. f   Pericarditis. These changes may be less marked as T wave inversion. The widespread nature and saddle shape allows distinction from myocardial infarction.

3. e   Inferior myocardial infarction.

4. d   Digitalis effect. Complete/3rd degree heart block, the ventricular escape rate of 45 bpm, broad complex QRS and presence of p waves (which will be at a rate of 70 bpm) confirm this diagnosis.

5. c   Atrial flutter with 2:1 block. Any rhythm at 150 bpm should prompt a search for p waves to exclude flutter.

22. Theme: Blood test results

1. b   High white cell and C-reactive protein. These are consistent with an infection.

2. h   Raised haemoglobin and ferritin levels. This gentleman has a tanned appearance and diabetes; he probably has hereditary haemochromatosis.

3. d   Low haemoglobin and raised C-reactive protein. This lady has lost a lot of blood in theatre. She will have a raised C-reactive protein due to the trauma of breaking her hip and the surgery, both of which stimulate an immune response.

4. i   Serum potassium of 6.6 mmol/L. A serum potassium above 6.5 mmol/L is a medical emergency and is usually treated acutely with an insulin and dextrose infusion. The calcium gluconate is cardioprotective and reduces the chance of arrhythmias.

5. f   Normal bilirubin with raised alkaline phosphatase and gamma glutamyltransferase. An obstructive biliary picture.

23. Theme: Liver and biliary tree anatomy

1. j   Rouviere's sulcus. This anatomical landmark is used in laparoscopic cholecystectomy to locate a safe plane for dissection.

2. g   Hepatoduodenal ligament. Within which runs the hepatic artery, hepatic portal vein and bile duct.

3. c   Caudate lobe. Vascular segments of the liver are used to define safe planes for dissection during liver surgery.

4. a   Bare area of the liver.

5. b   Calot's triangle. This is the safe area for dissection of the cystic artery during cholecystectomy.

24. Theme: Psychiatry

1. f   Lewy body dementia. This is suggested by the hallucinations, impairment of memory and parkinsonian features. It is possible to have Parkinson's disease dementia where the Parkinson's disease precedes the dementia.

2. c   Depression. This gentleman displays many of the classical features of depression.

3. d   Drug-induced psychosis. This is suggested by the cannabis use and paranoid symptoms. It is important to rule out underlying schizophrenia once he has stopped using drugs.

4. i   Post-traumatic stress disorder. This is suggested by the anxiety symptoms (early insomnia) accompanied by flashbacks to the precipitating incident.

5. e   Generalized anxiety disorder. This lady has many symptoms attributable to anxiety without an obvious precipitating factor (which would constitute a phobia).

25. Theme: Heart disease

1. h   Infective endocarditis. Known valve disease plus a minor procedure leading to fever should be considered infective endocarditis until proven otherwise.

2. b   Aortic stenosis. This gentleman has all three of the classical symptoms of aortic stenosis. He should be assessed for valve surgery, as the presence of these symptoms increases the risk of sudden cardiac death in these patients.

3. g   Fibrinous pericarditis. This is suggested by the speed of onset post myocardial infarction, relief by NSAIDs and classical ST changes.

4. f   Dilated cardiomyopathy. This is associated with excess alcohol consumption. The clinical findings are classic with the mitral murmur caused by dilatation of the annulus.

5. e   Cor pulmonale. This is suggested by the ankle swelling (sign of right heart failure) on top of chronic lung disease.

26. Theme: Strokes

1. g   TACS. This man has the required three features to make up a total anterior circulation syndrome.

2. a   Lacunar infarct. Isolated sensory or motor losses are described as lacunar infarcts.

3. f   PICA occlusion. Also known as the lateral medullary syndrome of Wallenberg, a very specific constellation of symptoms makes up this syndrome.

4. h   TIA. This lady has an episode of focal neurology, of presumed vascular origin, that lasted less than 24 hours. She requires a full work up and treatment of her vascular risk factors.

5. b   PACI. This man has two of the three factors for diagnosis of a total anterior syndrome, and computed tomography demonstrating an infarct makes this a PACI rather than a PACS.

27. Theme: Breast lumps

1. e   Fibroadenoma. Fibroadenoma or 'breast mouse' is common in young women as a discreet highly mobile mass.

2. i   Mastitis. This may precede breast abscess and is not uncommon in breastfeeding mothers.

3. d   Fat necrosis. This is common after trauma, especially in the breasts. This may be seen after road traffic accidents as the seatbelt passes over the breast.

4. c   Duct ectasia. This is common around the time of the menopause, and is due to a dilated, fluid-filled milk duct or ducts. These can become infected.

5. b   Cyst. This is unlikely to be serious but should be aspirated if it becomes symptomatic or there are worrying features on ultrasound.

28. Theme: Rashes

1. j   Stevens–Johnson syndrome. This woman has a serious condition secondary to her chest infection.

2. e   Herpes zoster. This man has herpes zoster in a dermatome distribution; that of the ophthalmis branch of the trigeminal nerve.

3. f   Necrobiosis lipoidica. The classical description of this condition.

4. g   Pemphigoid. This is the dermal blistering rash as opposed to pemphigus, which is the superficial (epidermal) blistering condition.

5. c   Dermatitis herpetiformis. This patient has coeliac disease. Treatment of the coeliac disease will also treat the rash.

29. Theme: Acute abdomen

1. a   Acute pancreatitis. This is consistent with the abdominal pain experienced by the patient, as well as dehydration, nausea, vomiting and the implied excess alcohol consumption that often accompanies holidays.

2. i   Small bowel obstruction. This lady has absolute constipation, nausea, vomiting and high-pitched bowel sounds, all consistent with small bowel obstruction. This is a surgical emergency.

3. f   Perforated peptic ulcer. This is suggested by the left upper quadrant pain and recent NSAID use. He is also peritonitic.

4. b   Acute pyelonephritis. This is suggested by the flank pain radiating to the groin and associated rigors.

5. e   Perforated appendix. This young man has features of peritonism related to a right iliac fossa tenderness. Other causes of right iliac fossa pain and peritonism would be unusual in a boy of this age.

30. Theme: Surgical incisions

1. i   Rooftop. Also known as a Mercedes Benz.

2. d   Kocher's. Named after Emil Kocher, an incision parallel to the costal margin on the right.

3. c   Gridiron. A muscle-splitting incision over McBurney's point.

4. g   Median sternotomy. Used for access to the heart and mediastinum.

5. f   Lower midline. Allows access to the sigmoid colon and can be extended if a left hemicolectomy is to be performed.

Index

Note: Page numbers in *italic* refer to tables or figures; (q) or (a) after page numbers indicate questions or answers to self-assessment questions.

A

ABC principles, 183–184
abdominal aortic aneurysm (AAA), 135
  acute abdominal pain, 144, 147
abdominal aortic palpation, *134*, 135, *136*
abdominal masses, 146
abdominal pain, acute, 33–36, 133, 144–147, 239, 271(q), 291(a)
abdominal radiography, 239
abdominal system
  examination, 71, 72, 131–150
    see also groin swellings; gynaecological examination; obstetric examination; scrotal swellings
  extended matching questions, 257–258, 262, 266, 268, 271
    answers, 285, 287, 289–290, 291
  history taking, *17*, 33–54
  investigations, 238–240
  multiple-choice questions, 247–248, 249, 251, 252, 253
    answers, 273–274, 277, 278, 279
  paediatric patients, 71, 72
  short-answer questions, 255, 281(a), 282(a)
abducens nerve (cranial nerve VI), 165, 166
abscesses, *147*, 158, 201
accessory nerve (cranial nerve XI), 170
activities of daily living, elderly people, 64–66
acute (surgical) abdomen, *35*, 271(q), 291(a)
adnexae, 156
adolescents, 67
  immunization schedule, *68*
  slipped upper femoral epiphysis, 87
  vital signs, *70*
aggressive patients, 6, 75
airway
  ABC principles, 183, 184
  ATOM FC, 185
alcohol consumption
  acute gastrointestinal bleeding, 43
  hangover, 58, *59*
  history taking, 14
  jaundice, 37, *39*, 40
allergies, drug-related, 12–13
amyloidosis, 148, 150
anaemia, 40–42, 140–143
  chronic renal failure, 49

  prehepatic jaundice, 37
  rheumatoid arthritis, 86
  self-assessment questions, 252, 253, 256, 279(a), 282(a)
angina, 19–22, *35*, 147
ankle:brachial pressure index (ABPI), 102, 160
ankle
  cardiovascular pathologies, 98
  motor system assessment, *171*, *174*, *175*
  respiratory pathologies, 118
ankylosing spondylitis, 204, 213, *214*
anosmia, 163
anterior circulation syndromes, 181
aorta, abdominal
  acute pain, 144, 147
  palpation, *134*, 135, *136*
aortic aneurysms, 135, 144, 147, 235
aortic coarctation, 100, 101, 102
aortic regurgitation, *102*, 106, 107–108, *109*
aortic stenosis, *102*, 105, 106–107, *109*
  self-assessment questions, 247, 255, 273(a), 281(a)
apex beat, 104, *109*
appendectomy scars, *133*
appendicitis, 144, 146, 147
arrhythmias
  cardiovascular examination, *101*, 110, 114, 115
  cardiovascular investigations, 234, 235–236
  palpitations, 22–24, 91
  stroke, 181
  thyroid status assessment, 190, 191
arterial blood gases (ABG), 128, 237
arterial pulses, 100–102, *103*, *105*, *108*
  acute abdominal pain, 145
  peripheral vascular disease, 160
arterial tree, lower limb, *160*
ascites, 135–136, *137*
asthma
  history taking, 27–29
  investigations, 237
  respiratory examination, *122*, 127–129
  self-assessment questions, 248, 251, 255, 273–274(a), 278(a), 281(a)
ataxia, 175, 205
ATOM FC, 185
atrial fibrillation, *23*, 91, *101*, 110, 181
  thyroid status assessment, 190, 234
auditory function, 168, 169

auditory hallucinations, 79
auscultation
    bowel sounds, 136, *142*
    breath sounds, 120–121, *122*, *128*
    cardiac sounds, 105–106, 107, *109*, 110–111, *112*,
        182
    reticuloendothelial examination, 193
axillary examination, 200–201
azathioprine, *86*

B
back pain
    examination, 211–214
    history taking, *8*, 83–84
    self-assessment questions, 255, 281(a)
ballottement, abdominal, 135
barium studies, 239
Barlow manoeuvre, 72
Beck Suicide Intent Scale, 80, *81*
behaviour, mental state examination, 78
biliary pain, 147
bilirubin, production and clearance, *38*
blackouts, 56–58, 60, *182*
    see also epileptic seizures
blood gases, 128, 237
blood glucose
    estimates, 89, 188
    unconscious patients, 185
blood pressure
    acute abdominal pain, 145
    ankle:brachial pressure index, 102, 160
    chronic renal failure, 148
    hypertensive retinopathy, *223*
    measurement, 70, 102
    paediatric patients, 70
    shock, 144, *146*
    stroke, 181
blood tests
    abdominal presentations, 238
    cardiac pathologies, 234
    respiratory pathologies, 237
    self-assessment questions, 251, 267, 278(a), 289(a)
BM stick analysis, 185
boutonnière deformity, *209*
bowel habit changes, 43–45
bowel sounds, 71, 136, *142*
brachial pulse, *100*, 101, *108*
bradycardia, *23*
Branhamella catarrhalis, *31*
breast examination, 197–202, 256(q), 270(q),
    282(a), 290(a)
breath sounds, 120–121, *122*, *128*
breathing, ABC principles, 183, 184
bronchial breathing, *120*, *128*
bronchitis, chronic, 29

bronchoscopy, 238
Buerger's angle, 160

C
caecum, carcinoma, 146, *147*
Calgary–Cambridge guide to interviewing, 4–5
cancer
    breast, 199, 200, 201, 202, 256(q), 282(a)
    colorectal, 42, 44, 137, 145, 146, 252(q), 279(a)
    hepatomegaly, *140*
    lung, 129, 237
    pancreatic, 37, 252(q), 279(a)
    prostatic, 137
    renal, *149*, 150
    self-assessment questions, 252, 256, 266, 279(a),
        282(a), 288–289(a)
    Troisier's sign, 132
capillary refill, 99, 160
carbon monoxide, transfer coefficient (K_{CO}), 238
cardiac arrhythmias *see* arrhythmias
cardiac catheterization, 236
    scars, 98
cardiac chest pain, 19–22
cardiac cycle, *105*
cardiac enzymes, 234
cardiac failure, 24–25, *26*, 113–115
    hepatomegaly, *140*
    investigations, 235
    self-assessment questions, 255, 263, 281(a), 287(a)
cardiac memo, 236
cardiac murmurs, 105, 106, 107, *109*, 110–111, *112*,
    182, 235
    self-assessment questions, 258, 285(a)
cardiac sounds, 105–106, 107, *109*, 110–111, *112*,
    182
    self-assessment questions, 258, 285(a)
cardiac tamponade, 185
cardiopulmonary resuscitation (CPR), 183–184
cardiovascular system
    abdominal pain, *35*
    extended matching questions, 258, 263, 267, 269,
        285(a), 287(a), 289(a), 290(a)
    history taking, *16*, 19–26, 91
    multiple-choice questions, 247, 249, 251, 273(a),
        276(a), 278(a), 279(a)
    paediatric patients, 70, 72
    physical examination, 70, 72, 97–115
    short-answer questions, 255, 281(a)
    thyroid status assessment, 190, 191, 234
carotid bruit, 181
carotid pulse, *100*, 101, *102*, *103*, *105*, *108*
carpal tunnel, *172*, 180
cataracts, 222
cerebellar pathology, 180
cerebral arteries, *63*

cervical examination, 155, 156
cervical smear test, 53, 155
chest examination
    abdominal pathologies, 132, *142*
    cardiovascular pathologies, 98
    respiratory pathologies, 117, 118, 119–121, *122*,
        129
chest infection
    abdominal pain, *35*, 147, 239
    consolidation, 126
    history taking, 30–32
    investigations, 237, 238
chest pain
    cardiac, 19–22
    non-cardiac, 19, *20*, 22, 26, 121
chest radiography, 234–235, 237, 239
children *see* paediatric patients
Chlamydia psittaci, 31
chlamydia screening swabs, 155
cholestatic (posthepatic) jaundice, 37, 39
chronic ambulatory peritoneal dialysis (CAPD), *49*,
        50
chronic obstructive pulmonary disease (COPD)
    history taking, 29–30, *31*
    investigations, 237
    respiratory examination, *122*
    self-assessment questions, 248, 274(a)
cigarette smoking, 14, 29, 30, 100
circulation, ABC principles, 183–184
cirrhosis, hepatomegaly, *140*
closed questions, 4, 5
clubbing, 98, *99*, 112, 118
    self-assessment questions, 247, 259, 274(a),
        285–286(a)
coarctation of the aorta, 100, 101, 102
cognitive assessment, 79
collapsed lung, *122, 124, 125, 128*
collateral ligaments of knee, 218
colonic polyps, 44
colorectal carcinoma, 44
    abdominal mass, 146
    anaemia, 42
    conjunctival pallor, 145
    rectal examination, 137
    self-assessment questions, 252, 279(a)
comatose patient *see* unconscious patient
common peroneal nerve, 180
communication processes, 4–5, 67, 151
compliance *see* concordance
computed tomography (CT) scans, 237, 239,
        240
concordance
    diabetes mellitus, 89
    exploration of, 12
    inhaler techniques, 28

confusional state, acute, 80–81, 261(q), 286(a)
congestive heart failure *see* heart failure
conjunctival pallor, 145
consciousness, Glasgow Coma Score, 182, *183*,
        184
    see also unconscious patients
consolidation, pulmonary, *122*, 124–126, *128*
contraception, 52, 247(q), 273(a)
cor pulmonale, 29, 30, 235
corneal reflex, 167
cortical lesions, 178, *179*
Coxiella burnetii, 31
crackles (crepitations), 121, *128*
cranial nerves, 163–170, 185
C-reactive protein (CRP), 194, 234, 237
crepitations (crackles), 121, *128*
Crohn's disease, *147*
cruciate ligaments of knee, 218–219
CT scans, 237, 239, 240
cystic hyperplasia, 201
cytomegalovirus, *31*

D

deafness, 169
deep vein thrombosis (DVT), 25, 26, 249(q), 251(q),
        276(a), 278(a)
delirium *see* confusional state
depression, 80, *82*, 248(q), 275(a)
dermatological terms, 195, *196*
dermatomes, 177, *178*, 211
developmental milestones, *69*, 250(q), 277–278(a)
diabetes mellitus
    abdominal pain, *35*, 147
    examination, 185, 187–189, 223, 224
    history taking, 89–91
    renal failure, 148
    self-assessment questions, 252, 256, 261, 265
        answers, 279, 282, 286, 288
    stroke, 181, 187
    unconscious patients, 185, 188
    urinalysis, 239
diabetic ketoacidosis (DKA), *35*, 147, 188
diabetic retinopathy, 90, 187, 223, *224*, 261(q),
        286(a)
dialysis, 49, 50, 148
diarrhoeal illness, 36–37, 266(q), 289(a)
digital rectal examination (DRE, PR), 71, 136–138
disease–illness model, 5, *6*
diverticular disease, 44
diverticulitis, 144, 146
dorsalis pedis pulse, *100*, 160
drug allergies, 12–13
drug history taking, 11–12, 14
    see also drug-related disorders; *specific
        complaints*

drug-related disorders, 12
  anaemia, 42
  bowel habit changes, 45
  haematuria, 51
  jaundice, *39*, 40
  renal dysfunction, 48
drug treatment
  diabetes mellitus, 90
  rheumatoid arthritis, 86
dysphagia, 45–46, 249(q), 276–277(a)
dysphonia, 170

E
ear examination, 71, 168–169
echocardiography, 235, *236*
ectopic beat, *23*, *101*
ectopic pregnancy, 144, 156
elbow examination, *173*, *175*, 210–211, *212*
elderly people
  acute confusional state, 80–81, 261(q), 286(a)
  fractured neck of femur, 87–88
  locomotor examination, 204, 205
  'off legs', 64–66
  social history taking, 13, 65–66
electrocardiography (ECG), 235–236, 267(q), 289(a)
electromyography (EMG), 240
electroencephalography (EEG), 240
emboli, stroke aetiology, 181, 182
  see also thromboembolism
emphysema, 29
encephalopathy, hepatic, 139, *141*
endocarditis, *99*, 100, 112–113, 234, 235
endocrine system
  examination, 187–191, 223, 224
  history taking, 89–92
  neonatal screening, 72
  self-assessment questions, 250, 252, 256, 261, 265
    answers, 278, 279, 282, 286, 288
endoscopic retrograde cholangipancreatography
    (ERCP), 240
endoscopy, 239–240
epididymis, 158, 159
epigastric pain, 147
epileptic seizures, 60–62, 182–183, 240
erythrocyte sedimentation rate (ESR), 194, 234
evoked potentials, 240
exam techniques, 95, 97, 120, 245–246
examination of patients
  abdomen, 71, 72, 131–150
  breast, 197–202
  cardiovascular system, 70, 72, 97–115
  endocrine system, 187–191
  gynaecological, 153–156
  locomotor system, 203–219
  medical note writing, 225–226, 228, 229–231

mental state, 77–79, 139, 178
  neurological, 163–186
  obstetric, 151–153
  ophthalmic, 221–224
  order of, 96
  paediatric, 68–73
  presentation of findings after, 227–228
  principles, 95–96
  psychiatric, 77–79
  respiratory system, 70, 117–129
  reticuloendothelial, 193–194, *195*
  setting, 95
  skin, 195
  structuring thoughts about, 226–227
  surgical, 157–162
exercise ECG, 236
extended matching questions (EMQs), 257–271(q),
    285–291(a)
extrapyramidal syndromes, 180
eye examination, 221–224
  abdominal pathologies, 131, 145
  cardiovascular pathologies, 98, *108*, 113
  diabetes mellitus, 90, 91, 187, 223, 224, 261(q),
    286(a)
  Graves' disease, 190
  multiple sclerosis, 185
  neurological pathologies, 163–165, 167, *168*, 181,
    184, 185
  respiratory pathologies, 118, 129
  rheumatoid arthritis, 85
eyelid
  Graves' disease, 190
  ptosis, 165

F
face examination
  abdominal pathologies, 132, 141, *142*
  cardiovascular pathologies, 98, *108*
  neurological pathologies, 166–167, 181
  respiratory pathologies, 117, 118, *122*, 129
facial nerve (cranial nerve VII), 167, *168*
fainting *see* blackouts
fallopian tube, 156
family history, 15, 255(q), 282(a)
fat necrosis, breast, 201
femoral aneurysm, 158
femoral pulse, *100*, 101, 160
femoral stretch test, 214, *215*
femur
  fractured neck of, 87–88, *241*, 249(q), 275(a)
  Perthes' disease, 87
  slipped upper femoral epiphysis, 87
fetuses, 152–153, 247(q), 273(a)
fibroadenomas, 199, 201
fibroadenosis, 201

fibrocystic disease, 201
fibrosis, pulmonary, *122*, 126, *127*, *128*
fine needle aspiration (FNA), breast lesions, 202
finger
    clubbing, 98, *99*, 112, 118
        self-assessment questions, 247, 259, 274(a),
            285–286(a)
    locomotor examination, 207, 208, *209*, *210*
    motor system assessment, *173*, 174, *175*
    thyroid acropachy, 190
finger nails
    anaemia, *142*
    cardiovascular pathologies, 98–99, 100, 112
    rheumatoid arthritis, *209*
finger–nose test, 174
fits *see* epileptic seizures
flailed chest, 185
fluid balance, renal failure, 49, 148
fluid challenge, shock, 183–184
fluid status, acute abdominal pain, 145
foot
    diabetic, 90, 91, 187, 188
    motor system assessment, *174*, 175
    peripheral vascular disease, 159–160
forced expiratory volume (FEV), 237–238
forced vital capacity (FVC), 237–238
foreign travel, 14, 37
fractured neck of femur, 87–88, *241*, 249(q), 275(a)
fundal height, 152
fundoscopy, 221–224

G

gag reflex, 169
gait assessment, 175–177, 204, 205–206
gallstones, *39*, 40, 145
Garden's classification, fractured neck of femur, *87*,
    88
gastrointestinal bleeding
    acute, 42–43, 143–144
    anaemia, 141, 143
    self-assessment questions, 247, 257, 273–274(a),
        285(a)
gastrointestinal chest pain, 19, *20*
gastrointestinal system
    examination, 134, 136–139, 141, *142*, 143–144,
        145, 146–147
    extended matching questions, 257, 258, 262, 266,
        268
        answers, 277, 279, 285, 287, 289–290
    history taking, 17, 33–39, 42–46
    investigations, 238, 239–240
    multiple-choice questions, 247–248, 249, 251, 252
        answers, 273–274, 276–277, 279
genu valgum, *218*
genu varum, *218*

Glasgow Coma Score, 182, *183*, 184
glaucoma, 223
glenohumeral joint, 210
glossopharyngeal nerve (cranial nerve IX), 169
glycaemic control, diabetes mellitus, 89, 188
gold, side effects, *86*
golfer's elbow, 211, *212*
gout, *208*
Gram-negative organisms, *31*
Graves' disease, 91, 190
Grey Turner's sign, 133
groin swellings, 157–158, 264(q), 288(a)
    see also lymph node examination
Guthrie test, 72
gynaecological examination, 153–156
gynaecological history taking, 17, 51–53
gynaecological self-assessment questions, 247, 249,
        250, 260, 273(a), 276(a), 277(a), 286(a)
gynaecomastia, 201–202

H

haematuria, 47, 50–51, 113
haemodialysis (HD), 49, 50, 148
haemolytic jaundice, 37
Haemophilus influenzae, *31*
haemothorax, 185
Hallpike's manoeuvre, 169
hallucinations, 79
hallux (thumb), *173*, *175*
hand examination, 207–208, *209–211*
    abdominal pathologies, 131, *142*
    cardiovascular pathologies, 98–100, *108*, 112–113
    neurological assessment, *173*, 207–208, *210–211*
    respiratory pathologies, 118, *122*, 129
    thyroid acropachy, 190
hangovers, 58, *59*
head examination
    abdominal pathologies, 132, 141
    cardiovascular pathologies, 98, *108*, 113
    respiratory pathologies, 117, 118, *122*, 129
headaches, 58–60, 248(q), 265(q), 275(a), 288(a)
hearing, 168, 169
heart arrhythmias *see* arrhythmias
heart block, *23*, *101*
heart failure, 24–25, *26*, 113–115
    hepatomegaly, *140*
    investigations, 235
    self-assessment questions, 255, 263, 281(a),
        287(a)
heart murmurs, 105, 106, 107, *109*, 110–111, *112*,
        182, 235
    self-assessment questions, 258, 285(a)
heart rate
    fetal, 153
    shock, 144, *146*

heart sounds, 105–106, 107, *109*, 110–111, *112*, 182
   self-assessment questions, 258, 285(a)
heaves, 104, *109*
heel–shin test, 174
hepatic encephalopathy, 139, *141*
hepatitis
   hepatomegaly, *140*
   jaundice, *39*, 40, 145
   risk factors, 14
   self-assessment questions, 251, 279(a)
hepatocellular jaundice, 37, 39
hepatomegaly, 138–139, *140*
hernias, 158
   repair scars, *133*
   self-assessment questions, 264, 288(a)
high-stepping gait, 206
hip, 214–217
   congenital dislocation, 72–73
   fractured neck of femur, 87–88, *241*, 249(q), 275(a)
   motor system assessment, *174, 175*
   Perthes' disease, 87
   slipped upper femoral epiphysis, 87
history taking
   abdominal problems, *17*, 33–54
   basic principles, 3
   Calgary–Cambridge guide, 4–5
   cardiovascular pathologies, *16*, 19–26
   drug allergies, 12–13
   drug history, 11–12
   endocrine system, 89–92
   family history, 15
   information gathering, 4–5
   introductory statement, 7
   locomotor system, 83–88
   medical note writing, 225–226
   neurological conditions, *17*, 55–66
   paediatric patients, 67–68
   past medical history, 11
   presentation of findings after, 227–228
   presenting complaint, 7–11
   psychiatric patients, 75–77
   respiratory pathologies, *16*, 27–32
   social history, 13–15
   structuring thoughts about, 226–227
   systems review, 15–18
HIV risk factors, 14, 256(q), 283(a)
hobbies, 14
Holmes–Adie pupil, 166
Holter monitor, 235–236
Horner's syndrome, 118, 129, *165*
hostile patients, 6, 75
human immunodeficiency virus (HIV), 14, 256(q), 283(a)

hydrocele, 159
hydronephrosis, *149*, 150
hydroxychloroquine, *86*
hyperosmolar non-ketotic coma, 189
hypertension, stroke, 181
hypertensive retinopathy, *223*
hyperthyroidism, 91, 92, 189–190, 234, 250(q), 278(a)
hypoglossal nerve (cranial nerve XII), 170
hypoglycaemia, 185, 188
hypoglycaemic agents, 90
hypothyroidism, 72, 91–92, 190–191, 234, 256(q), 283(a)
hysteria, *182*

I

Ideas, Concerns, Expectations (ICE), 6
illness framework, 5, *6*
imaging
   abdominal presentations, 239
   breast lesions, 202
   cardiovascular pathologies, 234–235
   neurological pathologies, 240
   respiratory pathologies, 237
immune-mediated haemolysis, 37
immunization schedule, *68*
infants, 67
   developmental milestones, *69*, 250(q), 277–278(a)
   examination, 72–73
   history taking, 71–72
   immunization schedule, *68*
   nervous system examination, 71
   vital signs, *70*
infective endocarditis, *99*, 100, 112–113, 234, 235
inflammatory bowel disease, 44
inguinal swellings, 157–158, 264(q), 288(a)
   see also lymph node examination
inherited disorders, family history, 15
insight, assessment, 79
insulin, 89, 90
internal pelvic examination, 153–156
interviews, history taking, 4–5
intracranial pressure, 58, *59*
investigations, 227, 233–241
iron deficiency anaemia, *41*, 42, 141, *142*, 253(q), 279(a)
irritable bowel disease, 43, *44*
ischaemic heart disease
   diabetes mellitus, 90, 91, 187
   investigations, 234, 235, 236
   risk factors, 21
   self-assessment questions, 251, 279(a)
   see also angina; myocardial infarction
ischaemic stroke, *64*

J

Janeway lesions, 113
jaundice, 37–40, 145
joint position sense, 178
jugular venous pressure (JVP), 103–104, *108*, 111, 145

K

kidney
    acute abdominal pain, 145
    acute failure, 46–48
    bilirubin production and clearance, *38*
    chronic failure, 48–50, 148
    diabetic nephropathy, 187–188
    palpable, 148–150
    palpation, *134, 135, 136*
    renal colic, 144, 147
    self-assessment questions, 251, 252, 279(a)
    transplantation, 49, 148
knee
    examination, 218–219
    motor system assessment, *174, 175*
    osteoarthritic changes, 240, *241*
    self-assessment questions, 263, 287(a)
knee jerk, *177*
Kocher's incision, *133*

L

lacunar infarct syndrome (LACS), *181*
language barriers, 6
laparotomy scars, *133*
lateral epicondylitis, 211
lateral ligament of knee, 218
left heart failure, 25, 113, 114–115, *235*
leg *see* lower limb
Legionella pneumophila, *31*
ligaments of knee, 218–219
likelihood ratios (LRs), 233–234
lipomas, 158
liver
    acute abdominal pain, 145
    bilirubin, *38*
    hepatomegaly, 138–139, *140*
    jaundice, 37–40, 145
    palpation, 134, 138
    self-assessment questions, 249, 251, 258, 268, 277(a), 279(a), 285(a), 289–290(a)
locomotor system
    congenital dislocation of the hip, 72–73
    examination, 72–73, 203–219
    history taking, *8, 19, 20,* 83–88
    investigations, 240, *241*
    self-assessment questions, 249, 250, 257, 259, 263, 264
        answers, 275, 277, 285, 286, 287, 288

loin pain, 147
lower limb
    locomotor examination, 206, 212, 214–219
    motor system assessment, 174, *175, 176,* 212, 214, *215*
    peripheral vascular disease, 159–160
    pretibial myxoedema, 190
    swelling, 25, *26*
    varicose veins, 161
    see also ankle; foot; hip
lower motor neuron (LMN) lesions
    facial nerve, 167, *168*
    motor system, 171, *172,* 180
lumbar puncture, 240, 248(q), 275(a)
lung disorders *see* respiratory system
lung function tests, 237–238
lymph node examination, 193–194
    axillary, 200–201
lymphadenopathy, 194, *195*
    axillary, 200–201
    supraclavicular, 132, 145
    see also lymphoproliferative disorders
lymphoproliferative disorders, *195*
    groin swellings, 158
    splenomegaly, *140, 141*

M

macular disease, 224
magnetic resonance imaging (MRI), 240
mammography, 202
medial epicondylitis (golfer's elbow), 211, *212*
medial ligament of knee, 218
median nerve lesions, 180
median sternectomy scars, *98*
mediastinal position, 119
medical history taking *see* history taking
medical note writing, 225–231
megaloblastic anaemia, *41,* 141, *142*
melaena, 42, 43, 137, 143–144
meningitis, 58, *59,* 240
menstrual histories, 51–52, 260(q), 286(a)
mental state examination, 77–79, 178
    hepatic encephalopathy, 139, *141*
methotrexate, *86*
migraine, 58, *59*
mini-mental test, 78, 178
mitral regurgitation, 105, *106,* 110–111
mitral stenosis, 105, 106, 110, 182
mitral valvectomy scars, *98*
mood, assessment, 78
motor neuron disease, *172*
motor system assessment, 170–177
    hand, *173,* 207–208
    lower limb, 174, *175, 176,* 212, 214, *215*
mouth, examination, 98, 118, 132, *142*

multiple-choice questions (MCQs), 247–253(q), 273–280(a)
multiple sclerosis, 185–186, 240
muscle bulk, 205
muscle strength, 171–172, *173–174*
muscle tone, 171
muscle wasting, myopathy, 178–179
musculoskeletal system *see* locomotor system
myasthenia gravis, *172*
Mycobacterium avium intracellulare, *31*
Mycobacterium tuberculosis, *31*, 252(q), 279(a)
Mycoplasma pneumoniae, *31*
myocardial infarction, 114–115
   abdominal pain, *35*, 147
   chest pain, 19–22
   investigations, 234
   self-assessment questions, 247, 273(a)
   stroke, 181
myopathy, 178–179

N

nailfold infarcts, 99, 112, *209*
nails
   anaemia, *142*
   cardiovascular examination, 98–99, 100, 112
narcolepsy, *182*
neck
   carotid pulse, 102, *103*, *108*
   jugular venous waveform, *103*
   respiratory examination, 118, 129
   stroke, 181
neonates, 67, *69*, 72–73, 249(q), 276(a)
nephrectomy scars, *133*
nerve root compression, *172*
nerve root entrapment, 214, *215*
neurological system
   examination, 163–186
      motor system, 170–177, 207–208, *210–211*, 212, 214, *215*
      paediatric patients, 71
      thyroid status assessment, 190, *191*
   extended matching questions, 264, 265, 269, 288(a), 290(a)
   history taking, *17*, 55–66
   investigations, 240
   multiple-choice questions, 248, 250, 275(a), 277(a), 278(a)
   rheumatoid arthritis and, 85
   short-answer questions, 255, 281(a)
neuromuscular system *see* motor system
nipple examination, 198, 199
Nocardia asteroides, *31*
note writing, 225–231
nuclear imaging, 235
nutritional status, 14

nystagmus, 166
   positional, 168–169

O

Objective Structured Clinical Examination (OSCE), 95
obstetric examination, 151–153
obstetric histories, 51, 53–54
obstetric self-assessment questions, 247, 250, 256, 273(a), 277(a), 282(a)
occupational history, 13–14, 30
oculomotor nerve (cranial nerve III), 165–166
oesophagus
   dysphagia, 45–46
   self-assessment questions, 247, 274(a)
'off legs', 64–66
olfactory nerve (cranial nerve I), 163
open questions, 4, 5
ophthalmic examination, 221–224
optic atrophy, 185, 222
optic disc, 185, 222–223
optic nerve (cranial nerve II), 163–164, 185
oral cavity examination, 98, 118, 132, *142*
oral temperature, abdominal pain, 145
orchitis, 159
Ortolani manoeuvre, *72*, 73
Osler's nodes, 100, 113
osteoarthritis
   history taking, 87
   investigations, 240, *241*
   locomotor examination, 204, *205*
      hand, 207, *208*
      hip, 214, 215, *216*
osteogenic gait, 206
ovary, examination, 156
oxygen therapy, COPD, 30

P

paediatric patients, 67–73
   Perthes' disease, 87
   self-assessment questions, 249, 250, 276(a), 277–278(a)
   shock, 183–184
   slipped upper femoral epiphysis, 87
Paget's disease of the nipple, 198, 199
pain, 7–10
   acute abdominal, 33–36, 133, 144–147, 239, 271(q), 291(a)
   back, *8*, 83–84, 211–214
   breast, 201
   chest, 19–22, 26, 121
   dysphagia, 45–46
   hip, 87
   osteoarthritis, 87
   sensory system assessment, 177

painful arc syndrome, 210
palpation
    abdomen, 133–136, 137, 138, 139, *142*, 146
    breast, 198–199
    cardiovascular examination, 104–105
    gynaecological examination, 156
    locomotor system, 206, 207, 213
    obstetric examination, 152–153
    respiratory examination, 119–120, *122*
    reticuloendothelial examination, 193
    varicose veins, 161
palpitations, 22–24, 91
Pancoast's tumour, 129
pancreatic cancer, 37, 252(q), 279(a)
pancreatitis, 145, 147, 251(q), 262(q), 279(a),
        287(a)
papilloedema, 222, *223*, 255(q), 282(a)
paralysis, after seizure, 182
parkinsonian gait, 177, 206
partial anterior circulation syndrome (PACS), *181*
patient-centred interviewing, 5
patient–doctor relationship, 5–6
    multiple sclerosis, 186
    rheumatoid arthritis, 86
peak expiratory flow rate, 237
peak flow meters, 121
pelvic examination, internal, 153–156
penicillamine, *86*
peptic ulceration, 145, 147, 248(q), 274(a)
perceptual abnormalities, 79
percussion
    abdominal examination, 134, 135–136, *137*, 138,
        142
    respiratory examination, 120, *122*, *128*
percussion test, varicose veins, 161
peripheral nerve lesions, 180
    diabetes mellitus, 187
    electromyography, 240
    self-assessment questions, 250, 255, 264, 277(a),
        281(a), 288(a)
    sensory function tests, hand, 208, *211*
peripheral vascular disease, 159–161
    diabetes mellitus, 90, 187, 223, 224
    self-assessment questions, 248, 249, 251, 253,
        275(a), 276(a), 278(a), 280(a)
peritoneal dialysis (PD), 49, 50, 148
peritoneal pain, 33
peritonitis, 144, 145, 146
persistent ductus arteriosus, *102*
Perthes' disease, 87
phenylketonuria, 72
phrenic nerve crush, *118*
physical examination *see* examination of patients
pleural effusion, 121, *122*, *123*, 126, *128*
pleural rub, 121, *122*

pleuritic chest pain, 19, *20*
    pneumothorax, 121
    pulmonary embolism, 26
Pneumocystis carinii, *31*
pneumonia
    abdominal pain, *35*, 147, 239
    consolidation, 126
    history taking, 30–32
    investigations, 237
pneumothorax, 121–124, *128*, 185
polio, *172*
polycystic kidneys, 149–150, 252, 279(a)
popliteal pulse, *100*, 160
position sense, 178, *179*
positional nystagmus, 168–169
posterior circulation syndrome (POCS), 181
posterior tibial pulse, *100*, 160
posthepatic jaundice, 37, 39
praecordium, 98, 104–106
pregnancy
    ectopic, 144, 156
    history taking in infants, 71–72
    obstetric examination, 151–153
    obstetric histories, 51, 53–54
    self-assessment questions, 247, 250, 256, 273(a),
        277(a), 282(a)
prehepatic jaundice, 37
presentation of findings, 227–228
pretibial myxoedema, 190
prostate, palpation, 137
psoas abscess, *147*, 158
psoriatic arthritis, 207, 213
psychiatric patients, 75–82
    self-assessment questions, 248, 250, 268, 275(a),
        278(a), 290(a)
psychogenic back pain, 83, 84
ptosis, 165
pulmonary disorders *see* respiratory system
pulmonary embolism (PE), 25–26
    self-assessment questions, 247, 273(a)
pulmonary oedema, 24, 25, 126
pulses
    arterial, 100–102, *103*, *105*, *108*
        acute abdominal pain, 145
        peripheral vascular disease, 160
    venous, 102–104, *108*, 111, 145
pulsus alternans, 101
pulsus paradoxus, 101, *128*
pupillary reflexes, 165–166
pupils, unconscious patients, 184
pyelonephritis, *35*, 147

Q

questioning techniques, history taking,
    4–5

R

radial nerve lesions, 180
radial pulse, 100, *108*
raised intracranial pressure, 58, *59*
RAMP, 65
rashes, 195, 204, 256(q), 270(q), 282(a),
    290–291(a)
record writing, 225–226, 228, *229–231*
recreational drug use, 14
rectal bleeding, 44, 144
rectal carcinoma *see* colorectal carcinoma
rectal examination, 136–138
    paediatric patients, 71
referred pain, 33
renal system
    acute abdominal pain, 145
    acute failure, 46–48
    bilirubin production and clearance, *38*
    chronic failure, 48–50, 148
    diabetic nephropathy, 187–188
    effects of rheumatoid arthritis, 86
    kidney transplantation, 49, 148
    osteodystrophy, 49
    palpation of kidney, *134*, 135, *136*, 148–150
    renal colic, 144, 147
    self-assessment questions, 251, 252, 279(a)
respiratory rate, shock, 144
respiratory system
    abdominal pain, *35*, 147, 239
    effects of rheumatoid arthritis, 85
    examination, 70, 117–129
    extended matching questions, 262, 287(a)
    history taking, *16*, 27–32
    integration of signs for diagnosis, 127, *128*
    investigations, 237–238
    multiple-choice questions, 248, 249, 251,
        253
      answers, 274–275, 276, 278, 279, 280
    paediatric patients, 70
    short-answer questions, 255, 281(a)
resuscitation, 183–184
reticuloendothelial examination, 193–194,
    195
retina, 90, 187, 223–224, 261(q), 286(a)
revision techniques, 245–246
rheumatoid arthritis
    history taking, 84–86
    locomotor examination, *205*, 207, *208*, *209*
    self-assessment questions, 255, 281(a)
rhonchi (wheezes), 121
right heart failure, 25, 113, 114, 115
    see also cor pulmonale
Rinne's test, 168, *169*
Romberg's test, 178, *179*
rotator cuff tears, 210

S

sacroiliac joints, 213
safety, 6, 75
saphena varix, 158, 161
scars
    abdominal examination, *133*, 145
    cardiovascular examination, 98
    respiratory examination, 117, *118*
schizophrenia, 79, 250(q), 278(a)
sclerosing cholangitis, *39*, 40
scrotal swellings, 158–159
sedation, confused patients, 80–81
seizures *see* epileptic seizures
self-assessment questions, 247–271(q), 273–291(a)
self-harm, 80
sensory ataxia, 205
sensory system assessment, 177–178
    hand function tests, 208, *208*
sexual histories, *17*, 53
shock
    fluid challenge, 183–184
    self-assessment questions, 252, 279(a)
    signs of, 144
    stages, *146*
short-answer questions (SAQs), 255–256(q),
    281–283(a)
shoulder examination, 208–210, *212*
    motor system assessment, *173*, *175*
sinus bradycardia, *23*
sinus tachycardia, *23*
skin
    dermatological terms, 195, *196*
    examination, 195, 204
    self-assessment questions, 256, 270, 282(a), 290–
        291(a)
slipped upper femoral epiphysis, 87
smear tests, 53, 155
smell, sense of, 163
smoking
    asthma, 29
    COPD, 30
    history taking, 14
    nicotine staining, 100
Snellen charts, 221, *222*
social history taking, 13–15
    elderly people, 13, 65–66
    see also specific complaints
somatic pain, 33
spastic gait, 175, 205
speculum examination, 154–156
speech
    mental state examination, 78
    neurological examination, 170
spider naevi, 132
spinal accessory nerve (cranial nerve XI), 170

spinal trauma, 212
spirometry, 237
spleen, palpation, 134–135, 139
splenomegaly, 139–140, *141*, 143, 255(q), 281(a)
splinter haemorrhages, 98, *99*, 112
sputum assessment, 237
squint, 221
Staphylococcus aureus, *31*, 113
sternomastoid muscle, 170
stools
    assessment, 240
    bowel habit changes, 43–45
    melaena, 42, 43, 137, 143–144
strabismus (squint), 221
straight leg raising, 214, *215*
Streptococcus pneumoniae, *31*
Streptococcus viridans, 113
stress, sources of, 14
stroke, *8*, 62–64, 181–182, 269, 290(a)
study groups, 245
subarachnoid haemorrhage, 58, *59*, 240
suicidal ideation, 79, 80
suicidal intent, 80, *81*
suicide, 80, 248(q), 250(q), 275(a), 278(a)
sulfasalazine, *86*
supraclavicular lymphadenopathy, 132, 145
supraventricular tachycardia, *23*
surgical (acute) abdomen, *35*, 271(q), 291(a)
surgical examination, 157–162
surgical incisions, 271(q), 291(a)
swabs, gynaecological, 155
swallowing difficulties, 45–46, 249(q), 276–277(a)
swan neck deformity, *209*
syncope, *182*
    see also blackouts
systemic lupus erythematosus (SLE), 204, *208*
systems review, 15–18
    locomotor examination, 206, 207, 208, 214
    paediatric patients, 68
    unconscious patients, 185

T
tachycardia, *23*, 190
tactile vocal fremitus (TVF), 119, 120, *128*
temporal arteritis, 58, *59*
tendon reflexes, 172, *176*, 180
tendon xanthomas, 100
tennis elbow, 211
tension pneumothorax, 123, *124*, 185
testes, 158, 159
Thomas' test, 215
thoracotomy scars, *118*
thoughts, assessment, 78–79
thrills, 105, *109*
throat examination, 71

thromboembolism, 25–26, 182
thumb, motor system assessment, *173*, *175*, *210*
thyroid acropachy, 190
thyroid disease, 72, 91–92, 189–191
    cardiovascular investigations, 190, 234
    self-assessment questions, 250, 256, 278(a),
        283(a)
toddlers, 67–71
Todd's paralysis, 182
tongue examination, 118, 132, 170
total anterior circulation syndrome (TACS), *181*
touch sense, 177
tracheal position, 118, 119, *128*
transfer coefficient of carbon monoxide (K_{CO}), 238
transplantation, renal, 49, 148
trapezius muscle, 170
travel, 14, 37
Trendelenburg test, 217
tricuspid regurgitation, *106*, 111–112
trigeminal nerve (cranial nerve V), 166–167
trochlear nerve (cranial nerve IV), 165, 166
Troisier's sign, 132
troponin T (troponin I), 234
tuberculosis, *147*, 252(q), 279(a)
tuning fork tests, 168, 169

U
ulcers, 161, *162*
    diabetes mellitus, 187, 188
    self-assessment questions, 253, 280(a)
ulnar nerve lesions, 180
ultrasound
    abdominal, 239
    breast lesions, 202
    echocardiography, 235, *236*
unconscious patients, 55–56, 183–185
    diabetes mellitus, 185, 188, 189
    Glasgow Coma Score, 182, *183*, 184
    see also blackouts; epileptic seizures
upper limb
    locomotor examination, 206, 207–211
    motor system assessment, *173*, 174, *175*, *176*
    radial nerve damage, 180
    self-assessment questions, 264, 288(a)
    wrist swelling, 118
    see also hand
upper motor neuron (UMN) lesions
    facial nerve, 167, *168*
    motor system, 171, 180
urinalysis, 188, 234, 238–239
urinary abnormalities
    continence, 53, 249(q), 276(a)
    haematuria, 47, 50–51, 113
    screening questions, *17*
uterus, examination, 152, 153, 156

V

vaginal discharge, 53
vaginal examination, 155, 156
vagus nerve (cranial nerve X), 170
varicose veins, 161
venous guttering, 160
venous incompetence, 161
venous pulse, 102–104, *108*, 111, 145
venous thrombosis, 25, 26, 249(q), 251(q), 276(a), 278(a)
ventricular tachycardia, *23*
vertigo, 169
vesicular breathing, *120*
vestibular function, 168–169
vestibulocochlear nerve (cranial nerve VIII), 168–169
vibration sense, 178
viral illness
    hepatomegaly, *140*
    HIV risk factors, 14, 256(q), 283(a)
    jaundice with, *39*, 40
Virchow's node, 132
visceral (deep) pain, 33, *34*
visual acuity, 164, 221, *222*
visual evoked potentials, 240
visual fields, 164, 224
visual hallucinations, 79
visual pathway, *164*

visual problems, diabetes, 90, 91, 223, 224, 261(q), 286(a)
vital signs
    acute abdominal pain, 145
    paediatric, *70*
    shock, 144, *146*
    see also blood pressure, measurement; pulses
vocal fremitus, 119, 120, *128*
vocal resonance, 119–120

W

waddling gait, 177, 206
ward round notes, 228, *229–231*
Weber's test, 168, *169*
wheezes (rhonchi), 121
whispering pectoriloquy, 120
wrist
    motor system assessment, *173*, *175*
    radial nerve damage, 180
    swelling, 118
writing up medical notes, 225–231

X

X-rays
    abdominal presentations, 239
    cardiovascular pathologies, 234–235
    locomotor pathologies, 240, *241*
    respiratory pathologies, 237
xanthomas, 100